D1405835

The Hutchinson Dictionary
of
Plant Names:
Common & Botanical

Compiled by
HAROLD BAGUST

Copyright © Helicon Publishing Ltd 2001

All rights reserved
Helicon Publishing Ltd
42 Hythe Bridge Street
Oxford OX1 2EP
e-mail: admin@helicon.co.uk
Web site: http://www.helicon.co.uk

Typeset by Florence Production Ltd, Stoodleigh, Devon

Printed and bound in Italy by
Officine Grafiche De Agostini, Novara

ISBN 1-85986-356-6

British Cataloguing in Publication Data
A catalogue record is available from The British Library

Papers used by Helicon Publishing Ltd are natural recyclable
products made from wood grown in sustainable forests.
The manufacturing processes of both raw materials and
paper conform to the environmental regulation of the
country of origin.

Editorial Director Hilary McGlynn	**Picture research** Sophie Evans
Managing Editor Elena Softley	**Page design** Helen Weller
Project Editor Heather Slade	**Production Manager** John Normansell
Design Manager Lenn Darroux	**Production Assistant** Stacey Penny
Illustrations Julie Williams	

Cover photograph: *Sambucus racemosa* Red berried elder © Jo Brewer
Picture credits: All photographs © Corbis except p.145 © Photodisk

CONTENTS

PREFACE

This publication covers over 30,000 plants grown in the English-speaking areas of the world together with much of Europe. It is intended as a working tool for both amateur and professional gardeners as well as for plant lovers wishing to find the botanical name when only the common name is known, or vice versa.

For easy reference the book is divided into fourteen different plant sections, with lists of common names in the first half of the book and the botanical equivalents in the second half. In order to keep the number of pages within manageable proportions, the categories and individual plants included are those most frequently purchased from nurseries and garden centres in the UK and North America. Vegetables have been excluded because very few customers use botanical terms when buying them or their seeds. A certain amount of duplication is inevitable if an unacceptable amount of cross-referencing is to be avoided. Some herbs, for instance, may appear also under 'Wild Flowers' or 'Popular Garden Plants', and 'House Plants' includes species from several categories.

Since the use of common or vernacular plant names is often restricted to a limited region or to special groups, botanical names have become more widely employed as a method of positively identifying plants. Continual revision has however, resulted in the frequent occurence of synonyms of the botanical name. These have been included in many instances to reduce confusion caused by consulting old reference books still to be found on the bookshelves of many gardeners of my generation and earlier.

The standard binomial system has been used throughout – the genus, or family name followed by the specific epithet or species name, identifying the characteristics or growing areas of that plant; for sub-species a third name can be applied and also a varietal name, for example *Pinus sibirica pumila glauca* (Dwarf blue Siberian pine). In some publications the varietal name appears in italics, in a different typeface, within quotation marks, or preceded by the abbreviation *var*. Here, where varieties are included, we have used the system widespread in the USA and elsewhere of adding

PREFACE

the varietal name after the epithet using the same typeface and without quotation marks.

Some symbols used in the book are:

× before or within a name indicates a sexually pollinated hybrid.
+ indicates a graft hybrid.

The Scottish and Irish prefixes "Mac", "Mc", and "M" are united with the rest of the name thus, *macintyre*, *macaulay*, *maginty*.

The Irish prefix "O" is united with the rest of the name thus, *oleary*.

To avoid confusion, the abbreviation St is listed alphabetically as Saint.

ACKNOWLEDGEMENTS

This volume is the product of many years in the horticultural business both as amateur and professional, during which time I have been fortunate in being befriended by many eminent gardeners and botanists from whom I have learned and am still learning. To list them all would cover a dozen pages or more, but the following call for special mention for they probably influenced my life more than they ever realized.

Percy Thrower, with whom I swapped rare plants for several years whilst he was Parks Superintendent at Shrewsbury. A knowledgable yet modest man, always willing to share his knowledge with other gardeners.

Bill Sowerbutts, a very down-to-earth character with a vast knowledge of general horticulture and, towards the end of his life, a popular radio personality.

Anthony Huxley, from whom I learned a great deal during the time I was a regular contributor to *Amateur Gardening* during his editorship.

Gervas Huxley, a lovely member of that great family who spent many hours at my nursery in the Cotswolds discussing his favourite plants and explaining their epithets. He was a cousin of Sir Julian Huxley and Aldous Huxley.

Christina Foyle, who persuaded me that I could write and actually published my first two books under her John Gifford imprint.

Joe Elliott, whose friendship and expertise I admired during the 23 years his nursery was situated near to mine and whose knowledge of botanical Latin and Greek was of immeasurable assistance.

Pierre Barandou of Agen, France, who, together with Professor Max Henry, has helped enormously in extending my knowledge of European varieties.

And finally to Professor William T Stearn, one of the leading horticulturists in the world today, who has kindly allowed me to quote from his books *Botanical Latin,* 4th ed. (David & Charles), and *Stearn's Dictionary of Plant Names* (Cassell).

I also gratefully acknowledge the help I have received from the staff of Cannington College, Bridgwater; Oxford University (Forestry), Southampton and Exeter Universities, and in the USA from Penn State and Cornell Universities.

The support of my family and friends has been invaluable during the long years of research and I thank them all for their tolerance.

Harold Bagurst
Southampton, 2000

EPITHETS

A selection of epithets frequently encountered

abruptus, -a, -um
ending suddenly

abscissus, -a, -um cut off

acaulis stemless or nearly so

acinaceus, -a, -um
sword or scimitar-shaped

acu- in compound words
signifying prickly or
sharply pointed

aduncus, -a, -um hooked

aesculifolius, -a, -um
with leaves like
horse-chestnut

aestivus, -a, -um
developing or
maturing in summer

africanus, -a, -um African

aggregatus, -a, -um clustered in
a dense mass

alatus, -a, -um winged;
having wings

alb, albi, albo in compound
words signifying white

albus, -a, -um white

alpinus, -a, -um alpine

alternans alternating

alternus, -a, -um alternate,
not opposite

altus, -a, -um tall

amarus, -a, -um bitter

americanus, -a, -um American,
north or south

ammophilus, -a, -um
sand-loving

amplexicaulis, -is, -e
stem-clasping

amplexifolius, -a, -um
leaf-clasping

anacantus, -a, -um
without thorns or prickles

androgynus, -a, -um
hermaphrodite; having
male and female flowers
separate but on the
same inflorescence

anglicus, -a, -um English

anguinus, -a, -um serpentine;
wavy

angustifolius, -a, -um
having narrow leaves

annulatus, -a, -um having rings

annuus, -a, -um annual

apiatus, -a, -um spotted

apiculatus, -a, -um terminating
in a point or spike

apricus, -a, -um sun-loving

areneosus, -a, -um cobwebby

arcticus, -a, -um from arctic
regions

arcuatus, -a, -um arched; curved;
bent like a bow

argentatus, -a, -um silver;
silvered

argutus, -a, -um sharply notched
or toothed

aridus, -a, -um growing in dry,
arid places

aristatus, -a, -um bearded

armatus, -a, -um having thorns
or spines

articulatus, -a, -um jointed

ascendens sloping upwards

asper, -a, -um rough

attenuatus, -a, -um attenuated;
pointed

aurantiacus, -a, -um
orange-coloured

aureus, -a, -um golden

auriculatus, -a, -um ear-shaped

autumnalis, -is, -e
pertaining to autumn or fall

azureus, a, -um sky blue

bathyphyllus, -a, -um
thickly leaved

betaceus, -a, -um beetlelike

bi- two

biennis, -is, -e biennial

bifidus, -a, -um
divided into two parts,
not necessarily equal

biflorus, -a, -um twin-flowered

bifolius, -a, -um	twin-leaved
bifurcatus, -a, -um	bifurcate; divided into two almost equal parts
bisectus, -a, -um	divided into two equal parts
bombycinus, -a, -um	silky
brachiatus, -a, -um	branched at right angles
bracteatus, -a, -um	having bracts
brunneus, -a, -um	brown
bryoides	resembling moss
bufonius, -a, -um	found in damp conditions
bulbosus, -a, -um	bulbous; having a swollen underground stem
buxifolius, -a, -um	box-leaved
caeruleus, -a, -um	dark blue
caesius, -a, -um	light blue
calcareus, -a, -um	Lime-loving; pertaining to lime
californicus, -a, -um	Of California
cambricus, -a, -um	Welsh; of Wales
campaniflorus, -a, -um	having bell-shaped flowers
canadensis, -is, -e	Canadian; of Canada
canariensis, -is, -e	of the Canary islands
candicans	shining white
canescens	with off-white hairs
capensis, -is, -e	of the Cape of Good Hope
caperatus, -a, um	wrinkled
capilliformis, -is, -e	hair-like
capsularis, -is, -e	having a capsule or capsules
cardinalis, -is, -e	cardinal red; scarlet
carinatus, -a, -um	keeled; having a keel
carneus, -a, -um	flesh pink
carnosus, -a, -um	fleshy
castus, -a, -um	clean; pure; chaste
cataria	pertaining to cats
caulescens	having a stem
cavus, -a, -um	hollow
centifolius, -a, -um	multi-leaved or -petalled
ceraceus, -a, -um	waxy
cereus, -a, -um	waxy
cernuus, -a, -um	drooping; nodding
chamae-	prefix indicating low growth or dwarf habit
chinensis, -is, -e	Chinese
chrysanthus, -a, -um	having golden flowers
chryseus, -a, -um	golden yellow
ciliaris, -is, -e	fringed with hairs
citrinus, -a, -um	lemon-yellow
citriodorus, -a, -um	lemon-scented
clavatus, -a, -um	club-shaped
cochlearis, -is, e	spoon-shaped
cochleatus, -a, -um	spiral
collinus, -a, -um	pertaining to hills
columnaris, -is, -e	column-shaped
comatus, -a, -um	tufted; having a tuft
communis, -is, -e	common; general
compactus, -a, -um	dense
concavus, -a, -um	concave
conifer	cone-bearing
consolidus, -a, -um	stable; firm; solid
contortus, -a, -um	twisted
cordatus, -a, -um	heart-shaped
corneus, -a, -um	horny
cornutus, -a, -um	horn-shaped or having horns
corticosus, -a, -um	having thick bark
costatus, -a, -um	ribbed or veined
crassicaulis, -is, -e	having thick stems
crassifolius, -a, -um	having thick leaves
crassus, -a, -um	thick; fleshy
crenatus, -a, -um	crenate; scalloped
crinitus, -a, -um	having long thin hairs
crispatus, -a, -um	wavy; curled
cristatus, -a, -um	crested
cruciatus, -a, -um	in the form of a cross
crustatus, -a, -um	encrusted
ctenoides	comb-like
cucculatus, -a, -um	resembling a hood; hooded
cuneatus, -a, -um	wedge-shaped
curtus, -a, -um	shortened
curvatus, -a, -um	curved

cuspidatus, -a, -um
cuspidate; having a sharp point

cyaneus, -a, -um blue

cymbiformis, -is, -e boat-shaped

dactyloides fingerlike

dealbatus, -a, -um
coated with white dust or powder

debilis, -is, -e weak; frail

declinatus, -a, -um
bent or curved downward

deformis, -is, -e
deformed; distorted

deltoides, -a, -um triangular

dentatus, -a, -um toothed

diaphanus, -a, -um transparent

diffusus, -a, -um spreading

dipterus, -a, -um two-winged

discolor of two colours

diurnus, -a, -um day-flowering

drupacius, -a, -um
producing drupes or fleshy fruits

dulcis, -is, -e sweet

durus, -a, -um hard

echinatus, -a, -um
covered in spines or prickles

edulis, -is, -e edible

effusus, -a, -um spreading

elatior taller

elatus, -a, -um tall

elegans; elegantulus, -a, -um
elegant

elongatus, -a, -um
elongated; stretched

ensiformis, -is, -e
sword-shaped with a sharp point

erosus, -a, -um irregular; jagged

esculentus, -a, -um edible

falcatus, -a, -um
falcate; sickle-shaped

fallax false

farinosus, -a, -um
mealy; powdery; floury

fasciatus, -a, -um
linked or bound together

fenestralis, -is, -e
pierced or perforated

ferox
vicious, having sharp thorns

flaccidus, -a, -um weak; limp

flavens yellow

flexilis, -is, -e
pliant; flexible; easily bent

flexuosus, -a, -um twisted; zigzag

floccosus, -a, -um woolly

floribundus, -a, -um
free-flowering

fluvialis, -is, -e
growing in running water

foetidus, -a, -um
foul-smelling; stinking

fragilis, -is, -e
brittle; fragile; delicate

fragrans fragrant

frigidus, -a, -um
growing in cold regions

fumosus, -a, -um smoky

galactinus, -a, -um
growing in cold regions

galanthus, -a, -um
having milky-white flowers

galeatus, -a, -um helmet-shaped

gallicus, -a, -um
French; appertaining to France

geniculatus, -a, -um
bent abruptly like an elbow

germanicus, -a, -um
German; appertaining to Germany

gibbosus, -a, -um
swollen on one side

gladiatus, -a, -um swordlike

glaucus, -a, -um
covered in bloom, the fine white or grey coating seen on grapes, plums, etc.

globosus, -a, -um spherical

glutinosus, -a, -um gluey; sticky

gracilis, -is, -e graceful; slender

grammopetalus, -a, -um
having striped petals

grandis, -is, -e showy; large

graveolens strongly scented

griseus, -a, -um grey

guttatus, -a, -um
speckled; spotted

halophilus, -a, -um salt-loving

hamatus, -a, -um hooked

hastatus, -a, -um spear-shaped

helix spiral or twisted

helveticus, -a, -um
Swiss; of Switzerland

hians open; gaping

hirsutus, -a, -um
hairy; covered in hairs

hispanicus, -a, -um
Spanish; of Spain

hispidus, -a, -um bristly

horizontalis, -is, -e
prostrate; horizontal; flat

horridus, -a, -um
very thorny or prickly

hortensis, -is, -e
pertaining to gardens

humilis, -is, -e   very dwarf

hybridus, -a, -um
of mixed parentage; hybrid

hypnoides   mosslike

hystrix   bristly or having
many spines or prickles

imbricatus, -a, -um
overlapping in a
regular pattern

immaculatus, -a, -um
spotless; immaculate

immersus, -a, -um
growing under water

implexus, -a, -um   tangled

incurvatus, -a, -um
incurved; bent inward

inermis, -is, -e
without thorns or prickles

inquinans
marked; stained; flecked

integrifolius, -a, -um
having entire or
uncut foliage

italicus, -a, -um   italian

jubatus, -a, -um   crested

junceus, -a, -um   rushlike

kewensis, -is, -e
of the Royal Botanic
Gardens, Kew

labiatus, -a, -um   lipped

laciniatus, -a, -um
ripped or slashed into strips

lacrimans   weeping

lanatus, -a, -um   woolly

lanceolatus, -a, -um
spear-shaped

lateritius, -a, -um   brick-red

laxus, -a, -um
relaxed; loose; open

lenticularis, -is, -e   lens-shaped

lentiginosus, -a, -um   freckled

lepidus, -a, -um
elegant; graceful; slender

lignosus, -a, -um   woody

limbatus, -a, -um
bordered; edged

lineatus, -a, -um
having stripes

lingua   tongue or tongue-like

lithophilus, -a, -um
rock-loving; growing
among or on rocks

longus, -a, -um   long

lunatus, -a, -um   shaped like a
crescent moon

luteolus, -a, -um   yellowish

luteus, -a, -um   yellow

maculatus, -a, -um   spotted

magnificus, -a, -um
magnificent; splended

magnus, -a, -um   large

majalis, -is, -e   May-flowering

maliformis, -is, -e   apple-shaped

malvinus, -a, -um   mauve

marginalis, -is, -e
margined; bordered

maritimus, -a, -um
coastal; pertaining to the
sea or seashore

marmoratus, -a, -um
mottled; marbled

maxillaris, -is, -e   of the jaws

maximus, -a, -um   largest

medicus, -a, -um   medicinal

medullaris, -is, -e   pithy

melancholicus, -a, -um
limp; sad-looking; wilted

meleagris, -is, -e   spotted

meridianus, -a, -um
blooming at noontime

militaris, -is, -e
upright; rigid; stiff

mirabilis, -is, -e
miraculous; wonderful

mollis, -is, -e
soft; having soft hairs;
velvety

moniliformis, -is, -e
like a necklace

monstrosus, -a, -um
abnormal; distorted

montanus, -a, -um
pertaining to mountains

monticola   mountain-lover,
or a plant indigenous
to mountain areas

moschatus, -a, -um   musky

mucosus, -a, -um   slimy

mucronatus, -a, -um   pointed

mucronulatus, -a, -um
terminating in a sharp point

mundulus, -a, -um   neat; tidy

muralis, -is, -e   growing on walls

muscarius, -a, -um
pertaining to flies
or flying insects

muscivorus, -a, -um   fly-eating

muscosus, -a, -um
mossy; moss-like

nanus, -a, -um   dwarf

natans   floating; reclining
on the water surface

navicularis, -is, -e   boat-shaped

nervosus, -a, -um
having conspicuous
veins or ribs

nidus   a nest

niger; nigra, -um   black

nipponicus, -a, -um   Japanese

nitidifolius, -a, -um
    having glossy leaves

nocturnus, -a, -um
    night-flowering

non-scriptus   unmarked

notatus, -a, -um   spotted

nucifera, -um   nut-bearing

nutans   nodding

nyctagineus, -a, -um
    night-blooming

obconicus, -a, -um
    formed like an inverted cone

obesus, -a, -um
    succulent; bloated

obscurus, -a, -um
    uncertain; indistinct

obtusus, -a, -um   blunt

occidentalis, -is, -e   western

odoratus, -a, -um
    fragrant; scented

officinalis, -is, -e
    used in medicine

oleraceus, -a, -um
    vegetables, potherbs, etc.

oporinus, -a, -um
    pertaining to the autumn or fall

orientalis, -is, -e   eastern; oriental

osmanthus, -a, -um
    having fragrant flowers

oxyphilus, -a, -um
    acid-loving (soil)

palliatus, -a, -um
    cloaked; wrapped

palmatus, -a, -um
    palmate, shaped like an open hand

palustris, -is, -e
    marsh-loving, found in boggy areas

pannosus, -a, -um
    torn; tattered; scruffy

papyraceus, -a, -um   papery

pendulus, -a, -um
    hanging; drooping

peregrinus, -a, -um
    wandering; extending laterally

perennial
    living more than two years

perfoliatus, -a, -um
    having leaves which enclose or wrap round the stem

persicus, -a, -um   Persian

pes   foot

petiole   leaf-stalk

picturatus, -a, -um   variegated

pileatus, -a, -um   having a cap

pilosus, -a, -um
    having a covering of long soft hairs

pinnatus, -a, -um   feathery

plebeius, -a, -um
    common, not rare

plicatus, -a, -um   pleated

plumatus, -a, -um   plumed

pogonanthus, -a, -um
    having bearded flowers

polifolius, -a, -um
    having grey leaves

praecox   very early

pratensis, -is, -e
    of the fields or meadows

procumbens   prostrate

procurrens   spreading

profusus, -a, -um
    plentiful; abundant

pruriens
    causing itching or irritation

psittacinus, -a, -um
    parrot-like, with contrasting colours

psycodes   butterfly-like

pubescens   downy

pulvinatus, -a, -um   cushion-like

punctatus, -a, -um   spotted

pungens   sharp-pointed

pusillus, -a, -um   very small

pyriformis, -is, -e   pear-shaped

pyxidatus, -a, -um
    having a lid or pyxis

racemosus, -a, -um
    having flowers in racemes

radians   radiating outward

radicans
    with roots growing from stems

radula   file; rasp

ramosus, -a, -um   branched

reclinatus, -a, -um
    curved or bent backward

rectus, -a, -um   erect; upright

reniformis, -is, -e   kidney-shaped

repens; reptans   creeping

reticulatus, -a, -um
    covered with net-like markings

rivalis, -is, -e
    growing in or near rivers or streams

rosaceus, -a, -um   like a rose

rosea   rose-like

rubens, ruber, rubra, rubrum   red

rufus, -a, -um   red

rugosus, -a, -um   wrinkled

EPITHETS

rupestris, -is, -e
rock-loving; indigenous to rocky areas

saggitatus, -a, -um arrow-shaped

salinus, -a, -um
growing in salty areas

sanguineus, -a, -um blood-red

saponaceus, -a, um soapy

sarmentosus, -a, -um
equipped with runners

sativus, -a, -um
cultivated, not wild or natural

saxatilis, -is, -e
a rock plant; one found among rocks

scaber, scabra. scabrum
rough; coarse

scandens climbing

scoparius, -a, -um broom-like

scutatus, -a, -um shield-shaped

semperflorens everblooming

sempervirens evergreen

sericeus, -a, -um silky

serpens creeping

serratus, -a, -um saw-toothed

sessilis, -is, -e
sessile; without a stalk

setaceus, -a, -um
bristly; having bristles

siliceus, -a, -um
sand-loving; growing in sand

silvaticus, -a, -um;
silvestris, -is, -e
growing in woods or wooded areas; growing wild (not cultivated)

sinensis, -is, -e chinese

solidus, -a, -um dense

spectabilis, -is, -e
showy; spectacular

spinosus, -a, -um spiny

spiralis, -is, -e spiral

stragulus, -a, -um mat-forming

striatus, -a, -um striped

strictus, -a, -um erect

sulcatus, -a, -um furrowed

sulfureus, -a, -um;
sulphureus, -a, -um
sulfur (sulphur)-yellow

superbus, -a, -um superb

supinus, -a, -um prostrate

sylvaticus, -a, -um;
sylvestris, -is, -e
growing on woods and forests; wild, not cultivated

tenuis, -is, -e slender

ternatus, -a, -um
in groups of three

tinctus, -a, -um coloured

tomentosus, -a, -um
very woolly or furry

tortilis, -is, -e; tortus, -a, -um twisted; contorted

toxicarius, -a, -um;
toxifera -um poisonous

tremulus, -a, -um
trembling; gently shaking

trivialis, -is, -e
ordinary; common; trivial

tropicus, -a, -um tropical

tuberosus, -a, -um tuberous

tumidus, -a, -um
swollen; distended

uliginosus, -a, -um
of swamps and boggy places

umbrosus, -a, -um
shade-loving

urens stinging

utilis, -is, -e useful

vacillans variable; not constant

vagans wandering

variegatus, -a, -um variegated

velaris, -is, -e veiled

velox fast-growing

velutinus, -a, -um velvety

veris spring-flowering

vernalis, -is, -e spring-flowering

vespertinus, -a, -um
evening-blooming

villosus, -a, -um
covered with soft hairs

virens green

virgatus, -a, -um
twiggy; multi-branched

viridis, -is, -e green

viscidus, -a, -um;
viscosus, -a, -um
sticky; gummy

volubilis, -is, -e
twisting; twining

vulgaris, -is, -e;
vulgatus, -a -um
common; not rare

xanthinus, -a, -um yellow

zebrinus, -a, -um zebra-striped

zibethinus, -a, -um evil-smelling

zonalis, -is, -e; zonatus, -a, -um
zoned

PERSONAL EPITHETS

When the name of a person ends in a vowel, the letter i is added, except when the name ends in a, when e is added.

When the name ends in a consonant, the letters ii are added, except when the name ends in er, when i is added.

COMMON
NAMES

COMMON
NAMES

ALPINES AND ROCKERY PLANTS

Alpines are plants whose natural habitat is the mountainous area above the tree line, but in gardening circles the term includes rock-gardening plants. Very little true alpine gardening is attempted in the English-speaking parts of the world because alpine conditions are rarely encountered naturally and it is very difficult to reproduce them artificially.

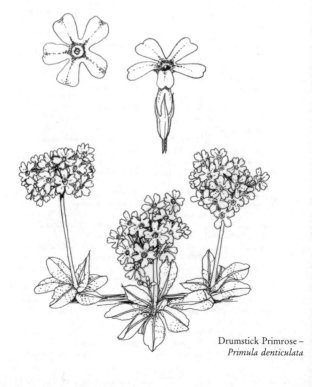

Drumstick Primrose –
Primula denticulata

Acanthus-leaved carline
thistle

Acanthus-leaved carline thistle
Carlina acanthifolia
Alaskan phlox *Phlox borealis*
Alaska violet *Viola langsdorfii*
Alpenrose
Rhododendron ferrugineum
Alpine aster *Aster alpinus*
Alpine avens *Geum montanum*
Alpine barrenwort
Epimedium alpinum
Alpine brook saxifrage
Saxifraga rivularis
Alpine buttercup
Ranunculus alpestris
Alpine calamint *Acinos alpinus*
Calamintha alpina
Alpine carline thistle
Carlina acaulis simplex
Alpine catchfly *Lychnis alpina*
Silene alpestris
Alpine cat's foot
Antennaria alpina
Alpine clematis *Clematis alpina*
Alpine columbine
Aquilegia alpina
Alpine forget-me-not
Myosotis alpestris
Alpine gentian *Gentiana alpina*
Gentiana newberryi
Alpine gypsophila
Gypsophila repens
Alpine lady's mantle
Alchemilla alpina
Alchemilla conjuncta
Alpine lychnis *Lychnis alpina*
Alpine marsh violet
Viola palustris
Alpine mouse-eared chickweed
Cerastium alpinum
Alpine pennycress
Thlaspi alpestre
Alpine phlox *Phlox douglasii*
Alpine pink *Dianthus alpinus*
Alpine poppy *Papaver alpinum*
Papaver burseri
Alpine rock cress *Arabis alpina*
Alpine rockrose
Helianthemum oelandicum
Alpine rose
Rhododendron ferrugineum
Alpine sandwort
Arenaria montana
Alpine sainfoin
Hedysarum obscurum
Alpine scullcap *Scutellaria alpina*
Alpine snowbell *Soldanella alpina*
Alpine speedwell *Veronica alpina*
Alpine spurge
Euphorbia capitulata
Alpine strawberry *Fragaria vesca*
Alpine thistle *Carlina acaulis*

Alpine veronica
Veronica alpina
Alpine violet
Cyclamen purpurascens
Alpine wallflower
Erysimum alpinum
American brooklime
Veronica americana
American dog violet
Viola conspersa
American dwarf iris *Iris lacustris*
Anemone *Anemone apennina*
Angel's eye *Veronica chamaedrys*
Apennine sandwort
Minuartia graminifolia
Appleblossom alpine anemone
Anemone narcissiflora
Arctic sandwort
Arenaria norvegica
Arrow-leaved violet
Viola sagittata
Asarabacca *Asarum europaeum*
Aubrietia *Aubrieta deltoidea*
Auricula *Primula auricula*
Primula × pubescens
Australian violet *Viola hederacea*
Autumn crocus
Colchicum bornmuelleri
Autumn-flowering gentian
Gentiana sino-ornata
Awl-leaved pearlwort
Sagina subulata
Baby joshua tree
Sedum multiceps
Baby primrose *Primula forbesii*
Primula malacoides
Baby's breath
Gypsophila repens
Bacon and eggs
Lotus corniculatus pleniforus
Baldmoney
Meum anthamanticum
Balloon flower
Platycodon grandiflorus
Barberry *Berberis candidula*
Barren strawberry
Waldsteinia geoides
Barrenwort *Epimedium alpinum*
Bastard balm
Melittis melissophyllum
Bastard jasmine
Androsace chamaejasme
Bearberry
Arctostaphylos uva-ursi
Bear's ear *Primula auricula*
Bear's foot *Alchemilla vulgaris*
Beaver's tail
Sedum morganianum
Beefsteak geranium
Saxifraga stolonifera

Bell flower
 Campanula cochleariifolia
Bethlehem sage
 Pulmonaria picta
 Pulmonaria saccharata
Bidgee-widgee
 Acaena anserinifolia
Bidi-bidi *Acaena anserinifolia*
Bididy-bid *Acaena anserinifolia*
Bird's eye *Veronica chamaedrys*
Bird's eye primrose
 Primula farinosa
 Primula laurentiana
 Primula mistassinica
Bird's foot trefoil
 Lotus corniculatus pleniforus
Bird's foot violet *Viola pedata*
Bistort *Polygonum bistorta superbum*
Biting stonecrop *Sedum acre*
Bitter cress *Cardamine pratensis*
Bitter root *Lewisia rediviva*
Bitterwort *Lewisia rediviva*
Black false helleborine
 Veratrum nigrum
Black snakeroot
 Cimicifuga racemosa
Bladder gentian
 Gentiana utriculosa
Bleeding heart
 Dicentra eximia
Blind gentian
 Gentiana clausa
Bloody cranesbill
 Geranium sanguineum
Blue alpine daisy
 Aster alpinus
Blue Chinese juniper
 Juniperus chinensis blaauw
Blue cowslip
 Pulmonaria angustifolia
Blue creeping juniper
 Juniperus horizontalis glauca
Blue dwarf spruce
 Picea abies pumila glauca
Blue-eyed grass
 Sisyrinchium angustifolium
Blue-eyed mary
 Omphalodes verna
Blue flaky juniper
 Juniperus squamata glauca
Blue monkshood
 Aconitum napellus
Blue Mountain bidi-bidi
 Acaena inermis
 Acaena microphylla
Blue phlox *Phlox divaricata*
Blue saxifrage *Saxifraga caesia*
Blue snakeroot *Liatris spicata*
Blue spruce *Picea glauca*
Blue spurge
 Euphorbia myrsinites

Bottle gentian
 Gentiana andrewsii
 Gentiana clausa
Bouncing bet *Saponaria officinalis*
Brass buttons
 Cotula coronopifolia
Breckland thyme
 Thymus serpyllum
Bridal wreath *Francoa sonchifolia*
Bristle-cone pine *Pinus aristata*
Broom *Cytisus decumbens*
 Genista lydia
Bugle *Ajuga reptans*
Bulbous buttercup
 Ranunculus bulbosus
Bulbous crowfoot
 Ranunculus bulbosus
Burmese dwarf rhododendron
 Rhododendron aperantum
Burnet saxifrage
 Pimpinella saxifraga
Burro's tail *Sedum morganianum*
Buttercup primrose
 Primula floribunda
Buttercup winter hazel
 Corylopsis pauciflora
Butter daisy *Ranunculus repens*
Butterwort *Pinguicula vulgaris*
Calathian violet
 Gentiana pneumonanthe
California golden violet
 Viola pedunculata
Californian fuchsia
 Zauschneria californica
Californian wild pansy
 Viola pedunculata
Canadian violet
 Viola canadensis
Candy mustard
 Acthionema saxatile
Candytuft
 Aethionema rotundifolium
 Iberis amara
 Iberis saxatilis
 Iberis sempervirens
Caraway thyme
 Thymus herba-barona
Carthusian pink
 Dianthus carthusianorum
Catesby's gentian
 Gentiana catesbaei
Catmint *Nepeta × faassenii*
Cat's foot *Antennaria dioica*
Chalk plant *Gypsophila repens*
Chamois cream
 Hutchinsia alpina
Cheddar pink *Dianthus caesius*
 Dianthus gratianopolitanus
Chinese primrose
 Primula sinensis
Christmas cheer
 Sedum × rubrotinctum

COMMON NAMES

Christmas rose

Christmas rose	*Helleborus niger*
Closed gentian	
	Gentiana andrewsii
	Gentiana clausa
	Gentiana linearis
Coast violet	*Viola brittoniana*
Cobweb houseleek	
	Sempervivum arachnoideum
Common broom	
	Cytisus scoparius
	Sarothamnus scoparius
Common butterwort	
	Pinguicula vulgaris
Common foxglove	
	Digitalis purpurea
Common gromwell	
	Lithospermum officinale
Common houseleek	
	Sempervivum tectorum
Common monkshood	
	Aconitum napellus
Common pasque flower	
	Anemone pulsatilla
	Pulsatilla vulgaris
Common pearlwort	
	Sagina procumbens
Common periwinkle	*Vinca minor*
Common pink	
	Dianthus plumarius
Common polypody	
	Polypodium vulgare
Common primrose	
	Primula vulgaris
Common rockrose	
	Helianthemum chamaecistus
	Helianthemum nummularium
Common speedwell	
	Veronica officinalis
Common storksbill	
	Erodium cicutarium
Common yew	*Taxus baccata*
Confederate violet	*Viola priceana*
Coral bells	
	Heuchera sanguinea alba
Corsican sandwort	
	Arenaria balearica
Cowslip	*Primula officinalis*
	Primula veris
Cream violet	*Viola striata*
Creeping buttercup	
	Ranunculus repens
Creeping common juniper	
	Juniperus communis
	hornibrookii
Creeping crowfoot	
	Ranunculus repens
Creeping evening primrose	
	Oenothera missouriensis
Creeping gypsophila	
	Gypsophila repens
Creeping jenny	
	Lysimachia nummularia

Creeping sailor	
	Saxifraga stolonifera
Creeping thyme	
	Thymus serpyllum
Crested gentian	
	Gentiana septemfida
Cross gentian	
	Gentiana cruciata
Crowfoot violet	*Viola pedata*
Cuckoo flower	
	Cardamine pratensis
Cushion pink	*Silene acaulis*
Dead men's bells	
	Digitalis purpurea
Dog's tooth violet	
	Erythronium dens-canis
Dog violet	*Viola canina*
	Viola riviniana
Dogwood	*Cornus canadensis*
Donkey's tail	
	Sedum morganianum
Dovedale moss	
	Saxifraga hypnoides
Downy yellow violet	
	Viola pubescens
Dragon's mouth	
	Horminum pyrenaicum
Dropwort	*Filipendula vulgaris*
Drumstick primrose	
	Primula denticulata
Dusty miller	*Primula auricula*
Dwarf acanthus	
	Acanthus dioscoridis
Dwarf balsam fir	
	Abies balsamea nana
Dwarf birch	*Betula nana*
Dwarf blue lawson's cypress	
	Chamaecyparis lawsoni
minima glauca	
Dwarf blue Siberian pine	
	Pinus sibirica pumila glauca
Dwarf Chinese juniper	
	Juniperus chinensis blaauw
Dwarf common juniper	
	Juniperus communis compressa
Dwarf eastern hemlock	
	Tsuga canadensis jeddeloh
Dwarf golden sawara cypress	
	Chamaecyparis pisifera filifera
	aurea nana
Dwarf Grecian fir	
	Abies cephalonica nana
Dwarf hemlock	
	Tsuga canadensis nana
Dwarf hinoki cypress	
	Chamaecyparis obtusa nana
Dwarf iris	*Iris pumila*
Dwarf Japanese yew	
	Taxus cuspidata minima
Dwarf joshua tree	
	Sedum multiceps

Gibralter candytuft

Dwarf lawson's cypress
　　　Chamaecyparis lawsoniana
　　　　　　minima
Dwarf Norway spruce
　　　Picea abies compressa
Dwarf pencil cedar
　　　Juniperus virginiana nana
　　　　　　compacta
Dwarf pine　　*Pinus mugo pumila*
Dwarf pink thrift
　　　Armeria juniperifolia
Dwarf Russian almond
　　　Prunus tenella
Dwarf sawara cypress
　　　Chamaecyparis pisifera
　　　　　filifera nana
Dwarf Scots pine
　　Pinus sylvestris beuvronensis
　　　Pinus sylvestris nana
Dwarf Siberian pine
　　　Pinus sibirica pumila
Dwarf solomon's seal
　　　Polygonatum hookeri
Dwarf Spanish columbine
　　　Aquilegia discolor
Dwarf spruce
　　　Picea abies echiniformis
Dwarf Weymouth pine
　　Pinus strobus umbraculifera
Dwarf white pine
　　　Pinus strobus nana
Dwarf willow
　　　Salix hastata wehrhahnii
Dwarf yew
　　　Taxus baccata compacta
　　　Taxus baccata nana
Early blue violet
　　　　　Viola palmata
Early yellow violet
　　　Viola rotundifolia
Eastern water violet
　　　Viola lanceolata
Edelweiss　*Leontopodium alpinum*
Edging candytuft
　　　Iberis sempervirens
Einsel's columbine
　　　Aquilegia einseleana
English primrose
　　　Primula vulgaris
English stonecrop
　　　Sedum anglicum
English violet　　*Viola odorata*
European brooklime
　　　Veronica beccabunga
European spring adonis
　　　Adonis vernalis
European wild pansy
　　　Viola tricolor
Evergreen candytuft
　　　Iberis sempervirens
Evergreen violet
　　　Viola sempervirens

Fair maids of France
　　　Saxifraga granulata
Fair maids of Kent
　　　Ranunculus aconitifolius
Fairy cups　　　*Primula veris*
Fairy primrose　*Primula malacoides*
Fairy's thimble
　　Campanula cochleariifolia
　　　Campanula pusilla
Fairy thimbles　*Digitalis purpurea*
Fanweed　　　*Thlaspi arvense*
Feather grass　　*Stipa capillata*
Field anemone
　　Pulsatilla pratensis nigricans
Field forget-me-not
　　　Myosotis arvensis
Field pansy　　*Viola rafinesquii*
　　　　　Viola tricolor
Field pennycress
　　　Thlaspi arvense
Fine-leaved sandwort
　　　Minuartia hybrida
Fingered fumitory
　　　Corydalis solida
Fleabane　　*Erigeron uniflorus*
Florist's violet　　*Viola odorata*
Flowering onion
　　　Allium oreophilum
Foam flower　*Tiarella cordifolia*
French weed　　*Thlaspi arvense*
Fringed sandwort
　　　Arenaria ciliata
Fumitory
　　Corydalis cheilanthifolia
Garden arabis　*Arabis caucasica*
Garden London pride
　　　Saxifraga × urbium
Garden pansy
　　Viola × wittrockiana
Garden thyme　*Thymus vulgaris*
Garden violet　　*Viola odorata*
Gargano bellflower
　　　Campanula garganica
Garland flower　*Daphne cneorum*
Garlic pennycress
　　　Thlaspi alliaceum
German catchfly　*Lychnis viscaria*
Germander speedwell
　　　Veronica chamaedrys
German primrose
　　　Primula obconica
Giant bellflower
　Campanula latifolia macrantha
Giant knapweed
　　Centaurea rhapontica
　　　Leuzea rhapontica
Giant solomon's seal
　　Polygonatum commutatum
Gibraltar candytuft
　　　Iberis gibraltarica

Glabrous rupturewort

Glabrous rupturewort
    Herniaria glabra
Glacier crowfoot
    Ranunculus glacialis
Globe candytuft   *Iberis umbellata*
Globe daisy
    Globularia trichosantha
Globe flower   *Trollius pumilus*
Glory-of-the-snow
    Chionodoxa luciliae
Gold alpine poppy
    Papaver kerneri
Gold dust   *Alyssum saxatile*
Golden alyssum   *Alyssum saxatile*
Golden aster   *Aster linosyris*
Golden carpet   *Sedum acre*
Golden cinquefoil
    Potentilla aurea
Golden-eye saxifrage
    Saxifraga tennesseensis
Golden garlic   *Allium moly*
Golden Japanese maple
    Acer japonicum aureum
Golden moss   *Sedum acre*
Golden rod   *Solidago virgaurea*
Golden sedum   *Sedum adolphi*
Golden star
    Chrysogonum virginianum
Golden tuft   *Alyssum saxatile*
Gold plantain lily
    Hosta fortunei aurea
Golden flax   *Linum flavum*
Goose grass   *Potentilla anserina*
Goose tansy   *Potentilla anserina*
Grape hyacinth
    Muscari botryoides
Grass-leaved buttercup
    Ranunculus gramineus
Grass-leaved day lily
    Hemerocallis minor
Grass-leaved iris   *Iris graminea*
Great basin violet
    Viola beckwithii
Great butterwort
    Pinguicula grandiflora
Great meadow rue
    Thalictrum aquilegifolium
Great spurred violet
    Viola selkirkii
Grey cinquefoil
    Potentilla arenaria
    Potentilla cinerea
Grey fescue   *Festuca cinerea*
Gromwell
    Lithospermum officinale
Gypsyweed   *Veronica officinalis*
Hacquetia   *Hacquetia epipactis*
Hairy alpine rose
    Rhododendron hirsutum
Hairy bitter cress
    Cardamine hirsuta

Hairy rock cress   *Arabis hirsuta*
Hairy stonecrop   *Sedum villosum*
Hairy thyme   *Thymus praecox*
    pseudolangiunosus
Halberd-leaved violet
    Viola hastata
Hard fern   *Blechnum spicant*
Hart's tongue fern
    Asplenium scolopendrium
    Phyllitis scolopendrium
Heartsease   *Viola × wittrockiana*
Heath pearlwort   *Sagina subulata*
Hen-and-chickens houseleek
    Jovibarba sobolifera
    Sempervivum soboliferum
    Sempervivium tectorum
Herb peter   *Primula veris*
Herringbone cotoneaster
    Cotoneaster horizontalis
Himalayan may apple
    Podophyllum hexandrum
Hinoki cypress
    Chamaecyparis obtusa
Hoary alison   *Berteroa incana*
Hoary cinquefoil
    Potentilla argentea
    Potentilla argentea calabra
Hollow fumitory   *Corydalis cava*
Hookspur violet   *Viola adunca*
Hoop-petticoat daffodil
    Narcissus bulbocodium
Horned pansy   *Viola cornuta*
Horned rampion
    Phyteuma scheuchzeri
Horned violet   *Viola cornuta*
Horse's tail
    Sedum morganianum
Host's saxifrage
    Saxifraga altissima
    Saxifraga hostii
Houseleek
    Sempervivum tectorum
Hudson balsam fir
    Abies balsamea hudsoniana
Humming-bird's trumpet
    Zauschneria californica
Iceland poppy
    Papaver nudicaule
Ice plant   *Sedum spectabile*
Indian paint
    Lithospermum canescens
Indian physic   *Gillenia trifoliata*
Indian warpaint
    Lithospermum canescens
Irish saxifrage   *Saxifraga rosacea*
Italian alyssum
    Alyssum argenteum
Italian starwort   *Aster amellus*
Ivyleaf cyclamen
    Cyclamen hederifolium
    Cyclamen neapolitanum

Ivy-leaved violet
  Viola hederacea
Japanese dwarf wormwood
  Artemisia schmidtiana nana
Japanese gentian
  Gentiana nipponica
  Gentiana scabrae
Japanese maple
  Acer japonicum aconitifolium
  Acer palmatum
Japanese toadlily
  Tricyrtis hirta
Jellybean plant
  Sedum pachyphyllum
Jellybeans *Sedum pachyphyllum*
Jersey thrift *Armeria alliacea*
Johnny-jump-up
  Viola pendunculata
  Viola tricolor
Joshua tree *Sedum multiceps*
Karst gentian *Gentiana tergestina*
Kenilworth ivy
  Cymbalaria muralis
Keyflower *Primula veris*
Kidney-leaved violet
  Viola renifolia
Kidney saxifrage
  Saxifraga hirsuta
King cup *Caltha palustris*
King's spear
  Asphodeline lutea
Knotted pearlwort
  Sagina nodosa
Labrador violet
  Viola labradorica
Ladies' delight
  Viola × wittrockiana
Lady's gloves *Digitalis purpurea*
Lady's mantle
  Alchemilla vulgaris
Lady's slipper orchid
  Cypripedium calceolus
Lady's smock
  Cardamine pratensis
Lamb's ears *Stachys byzantina*
Lamb's tail
  Chiastophyllum oppositifolium
  Sedum morganianum
Lampshade poppy
  Meconopsis integrifolia
Lance-leaved violet
  Viola lanceolata
Large bellflower
  Campanula tridentata
Large-flowered butterwort
  Pinguicula grandiflora
Large-flowered sandwort
  Arenaria grandiflora
Large self-heal
  Prunella grandiflora
Large speedwell
  Veronica austriaca teucrium

Large white plantain lily
  Hosta plantaginea
Large yellow foxglove
  Digitalis grandiflora
Large yellow loosestrife
  Lysimachia punctata
Larkspur violet *Viola pedatifida*
Lather root *Saponaria officinalis*
Lavender cotton
  Santolina chamaecyparissus
Lebanon candytuft
  Aethionema warleyense
Lemon thyme
  Thymus citriodorus
  Thymus serpyllum
Lesser celandine
  Ranunculus ficaria
Lesser periwinkle *Vinca minor*
Lily-of-the-valley
  Convallaria majalis
Limestone houseleek
  Sempervivum calcareum
  Sempervivum tectorum calcareum
Lion's foot *Alchemilla vulgaris*
  Leontopodium alpinum
Little joshua tree
  Sedum multiceps
Livelong *Sedum telephium*
Live-long saxifrage
  Saxifraga aizoon major
  Saxifraga paniculata major
Liverleaf *Hepatica nobilis*
London pride *Saxifraga umbrosa*
Long-leaved speedwell
  Veronica longifolia
Long-spurred violet
  Viola rostrata
Lungwort
  Pulmonaria angustifolia
  Pulmonaria rubra
Macedonian white pink
  Dianthus pinifolius
Maidenhair spleenwort
  Asplenium tricomanes
Maiden pink *Dianthus deltoides*
Maiden's wreath
  Francoa ramosa
Many fingers
  Sedum pachyphyllum
Marsh cinquefoil
  Potentilla palustris
Marsh five finger
  Potentilla palustris
Marsh gentian
  Gentiana pneumonanthe
Marsh marigold *Caltha palustris*
Marsh spurge *Euphorbia palustris*
Marsh violet *Viola palustris*
Matted globularia
  Globularia cordifolia
Mayflower *Cardamine pratensis*

May lily   *Maianthemum bifolium*

Meadow anemone
          Pulsatilla vulgaris

Meadow buttercup
          Ranunculus acris multiplex

Meadow cress
          Cardamine pratensis

Meadow saxifrage
          Saxifraga granulata

Mendocino gentian
          Gentiana setigera

Mexican butterwort
          Pinguicula caudata

Mexican puccoon
          Lithospermum distichum

Mezereon   *Daphne mezereum*

Michaelmas daisy   *Aster amellus*

Midsummer men
          Sedum telephium

Milfoil   *Achillea serbica*

Milkwhite rock jasmine
          Androsace lactea

Milkwort
          Euphorbia myrsinites

Miniature joshua tree
          Sedum multiceps

Miniature pansy   *Viola tricolor*

Missouri violet
          Viola missouriensis

Mithridate mustard
          Thlaspi arvense

Monkshood   *Aconitum napellus*

Moss campion   *Silene acaulis*

Moss phlox   *Phlox subulata*

Moss pink   *Phlox subulata*

Mossy rockfoil
          Saxifraga hypnoides

Mossy sandwort
          Moehringia muscosa

Mossy saxifrage
          Saxifraga × arendsii
          Saxifraga hypnoides

Mother-of-thousands
          Saxifraga stolonifera

Mountain alyson
          Alyssum montanum

Mountain alyssum
          Alyssum montanum

Mountain avens   *Dryas octopetala*

Mountain buttercup
          Ranunculus alpestris
          Ranunculus montanus

Mountain butterwort
          Pinguicula montana

Mountain cat's ear
          Antennaria dioica

Mountian cornflower
          Centaurea montana

Mountain everlasting
          Antennaria dioica

Mountain houseleek
          Sempervivum montanum

Mountain kidney vetch
          Anthyllis montana

Mountain phlox   *Phlox subulata*

Mountain pine   *Pinus montana*
          Pinus mugo

Mountain rocket
          Bellendena montana

Mountain sandwort
          Arenaria montana

Mountain sedge   *Carex montana*

Mountain soldanella
          Soldanella montana

Mountain tassel
          Soldanella montana

Mountain valerian
          Valeriana montana

Mountain willow   *Salix alpina*

Mount Atlas daisy
          Anacyclus depressus
          Anacyclus pyrethrum depressus

Mouse-eared chickweed
          Cerastium tomentosum

Mouse-ear hawkweed
          Hieracium pilosella

Musk mallow   *Malva moschata*

Musk saxifrage
          Saxifraga moschata

Musky saxifrage
          Saxifraga muscoides

Myrtle   *Vinca minor*

Nancy pretty   *Saxifraga umbrosa*

Narcissus-flowered anemone
          Anemone narcissiflora

Narrow-leaved lungwort
          Pulmonaria angustifolia
          azurea

Narrow-leaved wormwood
          Artemisia nitida

New Zealand bur
          Acaena buchananii

New Zealand burr
          Acaena microphylla

Nodding catchfly   *Silene pendula*

None-so-pretty
          Saxifraga × urbium

Northern blue violet
          Viola septentrionalis

Northern bog violet
          Viola nephrophylla

Northern downy violet
          Viola fimbriatula

Northern maidenhair fern
          Adiantum pedatum

Northern white violet
          Viola renifolia

October daphne   *Sedum sieboldii*

October plant   *Sedum sieboldii*

Oldfield cinquefoil
          Potentilla simplex

Redwood violet

Old-man-and-woman
Sempervivum tectorum

Olympian violet   *Viola gracilis*

Olympic violet   *Viola flettii*

One-flowered cushion saxifrage
Saxifraga burserana

Orange hawkweed
Hieracium aurantiacum

Oregon sunshine
Eriophyllum lanatum

Orpine   *Sedum telephium*

Oxlip   *Primula elatior*

Pale violet   *Viola striata*

Palm lily   *Yucca gloriosa*

Palsywort   *Primula veris*

Pansy   *Viola × wittrockiana*

Pansy violet   *Viola pedata*

Pasque flower
Anemone pulsatilla
Pulsatilla vulgaris

Patagonian slipper flower
Calceolaria polyrrhiza

Peach-leaved bellflower
Campanula persicifolia

Pearlwort   *Sagina pilifera*

Pellitory   *Anacyclus pyrethrum*

Pennycress   *Thlaspi arvense*

Perennial flax   *Linum perenne*

Perfoliate pennycress
Thlaspi perfoliatum

Persian buttercup
Ranunculus asiaticus

Persian ranunculus
Ranunculus asiaticus

Persian stonecress
Aethionema grandiflorum

Pilewort   *Ranunculus ficaria*

Pine-barren gentian
Gentiana autumnalis

Pink lily-of-the-valley
Convallaria majalis rosea

Pink rock jasmine
Androsace carnea

Pink sandwort
Arenaria purpurascens

Plains violet   *Viola viarum*

Poison primrose   *Primula obconica*

Polyanthus   *Primula × polyantha*

Pork and beans
Sedum × rubrotinctum

Prairie tea   *Potentilla rupestris*

Prickly conesticks
Petrophila sessilis

Prickly-pear cactus
Opuntia engelmannii

Prickly thrift
Acantholimon glumaceum

Primrose   *Primula vulgaris*

Primrose-leaved violet
Viola primulifolia

Prince of Wales' feathers
Tanacetum densum amani

Prophet flower   *Arnebia echioides*
Arnebia pulchra
Echioides longiflora

Prostrate juniper
Juniperus communis depressa

Prostrate rhododendron
Rhododendron chrysanthum

Puccoon
Lithospermum canescens
Lithospermum officinale

Purple bugle
Ajuga reptans atropurpurea

Purple columbine   *Aquilegia atrata*

Purple loosestrife
Lythrum salicaria

Purple mountain saxifrage
Saxifraga oppositifolia

Purple prairie violet
Viola pedatifida

Purple saxifrage
Saxifraga oppositifolia

Pussy's toes   *Antennaria dioica*

Pygmy lewisia   *Lewisia pygmaea*

Pygmy spruce
Picea abies pygmaea

Pyramidal saxifrage
Saxifraga cotyledon

Pyrenean cinquefoil
Potentilla pyrenaica

Pyrenean dragonmouth
Horminum pyrenaicum

Pyrenean eryngo
Eryngium bourgatii

Pyrenean primrose
Ramonda myconi
Ramonda pyrenaica

Pyrenean ramonda
Ramonda myconi

Pyrenean saxifrage
Saxifraga longifolia

Pyrenean woodruff
Asperula hirta

Quaking grass   *Briza media*

Queen's slipper orchid
Cypripedium reginae

Red baneberry   *Actaea rubra*

Red bidi-bidi
Acaena novae-zealandiae

Red burning bush
Dictamnus rubra

Reddish stonecrop
Sedum anacampseros

Red-hot poker primrose
Primula viallii

Red maids   *Calandrinia menziesii*

Red pasque flower
Pulsatilla vulgaris rubra

Redwood violet
Viola sempervirens

Reflexed stonecrop
Sedum reflexum

Reticulate willow   Salix reticulata

Rhaetian poppy
Papaver rhaeticum

Rock beauty   Petrocallis pyrenaica

Rock bells   Aquilegia canadensis

Rock buckthorn
Rhamnus saxitilis

Rock cinquefoil
Potentilla rupestris

Rock cranesbill
Geranium macrorrhizum

Rock cress   Arabis albida
Arabis ferdinandi-coburgii

Rockery daisy
Chrysanthemum arcticum

Rocket candytuft   Iberis amara

Rock jasmine
Androsace occidentalis

Rock jessamine
Androsace occidentalis

Rock purslane
Calandrinia menziesii

Rock sandwort   Arenaria stricta
Minuartia stricta

Rock soapwort
Saponaria ocymoides

Rock rose
Helianthemum nummularium

Rock stonecrop
Sedum forsteranum
Sedum reflexum

Rock violet   Viola flettii

Rock windflower
Anemone rupicola

Rocky mountain columbine
Aquilegia caerulea

Roof iris   Iris tectorum

Roseroot sedum   Sedum roseum

Round-headed rampion
Phyteuma orbiculare

Round-leaved yellow violet
Viola rotundifolia

Royal fern   Osmunda regalis

Running myrtle   Vinca minor

Sagebrush violet   Viola trinervata

Saint Bruno's lily
Paradisea liliastrum

Saint Patrick's cabbage
Saxifraga × urbium

Saint Peter's wort   Primula veris

Sampson's snakeroot
Gentiana catesbaei
Gentiana villosa

Sand phlox   Phlox bifida

Sandwort   Arenaria tetraquetra

Sargent's cedar
Cedrus libani sargentii

Satin flower
Sisyrinchium angustifolium

Scarlet bidi-bidi
Acaena microphylla

Sea alyssum   Lobularia maritima

Sea campion   Silene maritima
Silene uniflora

Sea heath   Frankenia laevis

Sea pink   Armeria maritima

Sea storksbill
Erodium maritimum

Sedge   Carex firma

Sedum live-forever
Sedum telephium

Shaggy hawkweed
Hieracium villosum

Shooting star
Dodecatheon meadia

Short-leaved gentian
Gentiana brachyphylla

Showy lady's slipper orchid
Cypripedium reginae

Shrubby speedwell
Veronica fruticulosa

Shrubby-stalked speedwell
Veronica fruticulosa

Shrubby white flax
Linum suffruticosum

Siberian bugloss
Brunnera macrophylla

Siberian iris   Iris sibirica

Siberian phlox   Phlox sibirica

Siberian whitlow grass
Draba sibirica

Siebold's stonecrop
Sedum sieboldii

Sierra gentian
Gentianopsis holopetala

Silver sage   Salvia argentea

Silver speedwell
Veronica spicata incana

Silver-spiked speedwell
Veronica spicata incana

Silver stonecrop   Sedum treleasii

Silverweed   Potentilla anserina

Silvery cinquefoil
Potentilla argentea

Silvery milfoil   Achillea clavennae

Slender loosestrife
Lythrum virgatum

Slender sandwort
Arenaria leptoclados

Slipper flower   Calceolaria biflora
Calceolaria plantaginea

Small celandine
Ranunculus ficaria

Snakeroot   Liatris elegans
Polygonum bistorta superbum

Snake's head fritillary
Fritillaria meleagris

Snow-in-summer
Cerastium tomentosum

White stonecrop

Snow-on-the-mountain *Arabis albida*

Solitary harebell *Campanula pulla*

Solomon's seal *Polygonatum × hybridum*

Southern coast violet *Viola septemloba*

Sowbread *Cyclamen purpurascens*

Spanish bluebell *Scilla hispanica*

Spanish dagger *Yucca gloriosa*

Spanish pellitory *Anacyclus pyrethrum*

Spanish thrift *Armeria welwitschii*

Speedwell *Veronica filiformis*

Spider's web houseleek *Sempervivum arachnoideum*

Spignel *Meum athamanticum*

Spotted gentian *Gentiana punctata*

Spring adonis *Adonis vernalis*

Spring anemone *Pulsatilla vernalis*

Spring cinquefoil *Potentilla verna*

Spring gentian *Gentiana verna*

Spring plantain lily *Hosta fortunei aurea*

Spring sandwort *Minuartia verna*

Spring vetchling *Lathyrus vernus*

Spurge *Euphorbia myrsinites*

Starry bellflower *Campanula elatines*

Starry saxifrage *Saxifraga stellata*

Stemless gentian *Gentiana acaulis* *Gentiana dinarica*

Stemless thistle *Cirsium acaule*

Stepmother's flower *Viola × wittrockiana*

Sticky sandwort *Minuartia viscosa*

Stinking hellebore *Helleborus foetidus*

Stinkweed *Thlaspi arvense*

Store cress *Aethionema warleyense*

Strapwort *Corrigiola litoralis*

Strawberry begonia *Saxifraga stolonifera*

Strawberry geranium *Saxifraga stolonifera*

Stream violet *Viola glabella*

Striped violet *Viola striata*

Sulphur cinquefoil *Potentilla recta warrenii*

Summer gentian *Gentiana septemfida*

Superb pink *Dianthus superbus*

Sweet alyssum *Lobularia maritima*

Sweet coltsfoot *Petasites fragrans*

Sweet violet *Viola odorata*

Sweet white violet *Viola blanda*

Sweet woodruff *Galium odoratum*

Swiss mountain pine *Pinus mugo mugo*

Swiss rock jasmine *Androsace helvetica*

Tall white violet *Viola canadensis*

Thick-leaved stonecrop *Sedum dasyphyllum suendermannii*

Thread agave *Yucca filamentosa*

Three-toothed cinquefoil *Potentilla tridentata*

Three-veined pink *Dianthus pavonius*

Three-veined sandwort *Moehringia muscosa*

Thrift *Armeria caespitosa* *Armeria maritima*

Thyme-leaved sandwort *Arenaria serpyllifolia*

Thyme-leaved speedwell *Veronica serpyllifolia*

Tibetan cowslip *Primula florindae*

Tibetan rhododendron *Rhododendron leucaspis*

Trailing phlox *Phlox nivalis*

Trailing violet *Viola hederacea*

Triangle-leaved violet *Viola × emarginata*

Triglav gentian *Gentiana terglouensis*

Trumpet gentian *Gentiana acaulis*

Tufted saxifrage *Saxifraga cespitosa*

Tufted soapwort *Saponaria caespitosa*

Tumbling ted *Saponaria ocymoides*

Tunic flower *Petrorhagia saxifraga*

Turtle head *Chelone obliqua*

Tussock bellflower *Campanula carpatica*

Two-eyed violet *Viola ocellata*

Two-leaved scilla *Scilla bifolia*

Valerian *Valeriana supina*

Vernal gentian *Gentiana verna*

Vernal sandwort *Minuartia verna*

Virginian cowslip *Mertensia virginica*

Variegated bugle *Ajuga reptans variegata*

Vegetable sheep *Raoulia eximia*

Wall pellitory *Parietaria diffusa* *Parietaria judaica*

Wallpepper *Sedum acre* *Sedum alba*

Wall spleenwort *Asplenium ruta-muraria*

White stonecrop *Sedum alba*

Water avens	*Geum rivale*
Western cranesbill	*Geranium endressii*
Western dog violet	*Viola adunca*
Western round-leaved violet	*Viola orbiculata*
Western sweet white violet	*Viola macloskeyi*
White ball primrose	*Primula denticulata alba*
White baneberry	*Actea pachypoda*
White bells	*Platycodon grandiflorus*
White bloody cranesbill	*Geranium sanguineum album*
White burning bush	*Dictamnus albus*
White buttercup	*Ranunculus aconitifolius*
Whitecrop	*Sedum alba*
White false helleborine	*Veratrum album*
White musk mallow	*Malva moschata alba*
White pasque flower	*Pulsatilla alba*
	Pulsatilla vulgaris alba
White rock	*Arabis caucasica*
White sea lavender	*Armeria maritima alba*
White sea pink	*Armeria maritima alba*
White stonecrop	*Sedum album*
White storksbill	*Erodium chamaedryoides*
	Erodium reichardii
Wild candytuft	*Iberis amara*
Wild marjoram	*Origanum vulgare compactum*
Wild okra	*Viola palmata*
Wild pink	*Dianthus plumarius*
	Silene caroliniana
Wild strawberry	*Fragaria vesca*
Wild sweet william	*Phlox canadensis*
	Phlox divaricata
	Phlox maculata
Wild thyme	*Thymus articus coccineus*
	Thymus coccineus
	Thymus serpyllum
Willow gentian	*Gentiana asclepiadea*
Winter aconite	*Eranthis hyemalis*
Winter heliotrope	*Petasites fragrans*
Winter jasmine	*Jasminum nudiflorum*
Winter savory	*Satureja montana alba*
Witch's gloves	*Digitalis purpurea*
Wood anemone	*Anemone nemorosa*
Wood forget-me-not	*Myosotis sylvatica*
Woodruff	*Asperula nitida*
Wood sorrel	*Oxalis acetosella*
Wood violet	*Viola riviniana*
Woolly blue violet	*Viola sororia*
Yarrow	*Achillea serbica*
Yellow adonis	*Adonis vernalis*
Yellow asphodel	*Asphodeline lutea*
Yellow flag	*Iris pseudacorus*
Yellow flax	*Linum flavum compactum*
Yellow forget-me-not	*Myosotis discolor*
Yellow fumitory	*Corydalis lutea*
Yellow gentian	*Gentiana lutea*
Yellow gowan	*Ranunculus repens*
Yellow iris	*Iris orientalis*
Yellow milfoil	*Achillea tomentosa*
Yellow mountain cornflower	*Centaurea montana sulphurea*
Yellow mountain saxifrage	*Saxifraga aizoides*
Yellow prairie violet	*Viola nuttallii*
Yellow scullcap	*Scutellaria orientalis pinnatifida*
Yellow soapwort	*Saponaria lutea*
Yellow stonecrop	*Sedum acre*
Yellow whitlow grass	*Draba aizoides*
Yellow wood violet	*Viola biflora*
	Viola lobata
Yew	*Taxus baccata*

AQUATICS

This section applies to plants living usually in fresh water, either rooted in soil or free-floating, also to plants living in bogs, swamps, and around the edges of ponds and lakes.

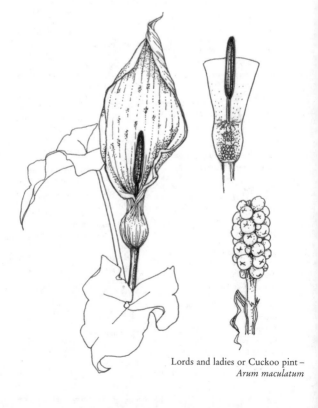

Lords and ladies or Cuckoo pint –
Arum maculatum

Amazon sword plant

Amazon sword plant
Echinodorus paniculatus
Echinodorus tenellus
Amazon water lily
Victoria amazonica
Victoria regia
Amazon water platter
Victoria amazonica
Victoria regia
American hornwort
Ceratophyllum demersum
American lotus   Nelumbo lutea
Nelumbo pentapetala
American pondweed
Potamogeton epihydrus
American spatterdock
Nuphar advena
American swamp lily
Saururus cernuus
American water willow
Justicia americana
Australian water lily
Nymphaea gigantea
Autumnal water starwort
Callitriche hermaphroditum
Awl-leaf arrowhead
Sagittaria subulata
Awlwort   Subularia aquatica
Baby tears   Bacopa monnieri
Beaked tassel pondweed
Ruppia maritima
Biblical bulrush   Cyperus papyrus
Blue Egyptian lotus
Nymphaea caerulea
Blue lotus
Nymphaea caerulea
Nymphaea stellata
Blue water lily
Nymphaea capensis
Blue water speedwell
Veronica anagallis-aquatica
Blunt-fruited water starwort
Callitriche obtusangula
Bog arum   Calla palustris
Lysichiton americanum
Bog bean   Menyanthes trifoliata
Bogmat   Wolffiella floridana
Bog pondweed
Potamogeton polygonifolius
Branched bur-reed
Sparganium erectum
Brandy bottle   Nuphar luteum
Brazilian waterweed
Anacharis densa
Elodea densa
Brook bean   Menyanthes trifoliata
Bulrush   Cyperus papyrus
Bulrush reedmace   Typha latifolia
Bur-reed   Sparganium ramosum
Canadian pondweed
Elodea canadensis

Canadian wild rice
Zizania aquatica
Cape asparagus
Aponogeton distachyus
Cape blue water lily
Nymphaea capensis
Cape pondweed
Aponogeton distachyus
Chinese water chestnut
Eleocharis dulcis
Common arrowhead
Sagittaria sagittifolia
Common boneset
Eupatorium perfoliatum
Common bur-reed
Sparganium ramosum
Common duckweed   Lemna minor
Common eel grass
Zostera marina
Common grass wrack
Zostera marina
Common spatterdock
Nuphar advena
Common water crowfoot
Ranunculus aquatalis
Comman water dropwort
Oenanthe fistulosa
Common wolffia
Wolffia columbiana
Corkscrew rush
Juncus effusus spiralis
Creeping charlie
Lysimachia nummularia
Creeping jenny
Lysimachia nummularia
Crystalwort   Riccia fluitans
Curled pondweed
Potamogeton crispus
Curly water thyme
Lagerosiphon major
Ditch moss
Anarcharis canadensis
Elodea canadensis
Duck potato   Sagittaria latifolia
Duckweed   Lemna minor
Spirodela polyrhiza
Dwarf amazon sword plant
Echinodorus magdalenensis
Dwarf eel grass   Zostera noltii
Dwarf grass wrack   Zostera noltii
Dwarf papyrus   Cyperus isocladus
East Indian lotus
Nelumbo nucifera
Nelumbo speciosa
Eel grass   Vallisneria spiralis
Zostera angustifolia
Egyptian lotus
Nymphaea caerulea
Nymphaea lotus
Egyptian water lily
Nymphaea lotus

Marsh trefoil

Engelmann's quillwort
Isoetes engelmannii

European water clover
Marsilea quadrifolia

European white water lily
Nymphaea alba
Nymphaea venusta

Fairy moss   *Azolla caroliniana*

Fanwort   *Cabomba caroliniana*

Fat duckweed   *Lemna gibba*

Fennel pondweed
Potamogeton pectinatus

Fen pondweed
Potamogeton coloratus

Fish grass   *Cabomba caroliniana*

Flag iris   *Iris pseudacorus*

Flat-stalked pondweed
Potamogeton friesii

Flexible naiad   *Najas flexilis*

Floating bur-reed
Sparganium angustifolium

Floating fern
Ceratopteris pteridoides
Salvinia auriculata

Floating marsh-wort
Apium inundatum

Floating moss
Salvinia rotundifolia

Floating pondweed
Potamogeton natans

Floating water plantain
Luronium natans

Fool's watercress
Apium nodiflorum

Fountain moss
Fontinalis antipyretica

Fragrant water lily
Castalia odorata
Nymphaea odorata

Frog-bit
Hydrocharis morsus ranae

Frog's lettuce
Potamogeton densus

Garden lysimachia
Lysimachia punctata
Lysimachia vulgaris

Giant arrowhead
Sagittaria montevidensis

Golden water dock
Rumex maritimus

Gooseneck lysimachia
Lysimachia clethroides

Gorgon   *Euryale ferox*

Grass-wrack pondweed
Potamogeton compressus

Grassy pondweed
Potamogeton obtusifolius

Great(er) duckweed
Spirodela polyrhiza

Greater naiad   *Najas marina*

Greater water thyme
Elodea callitrichoides

Great water dock
Rumex hydrolapathan

Hair grass   *Eleocharis acicularis*

Hair-like pondweed
Potamogeton trichoides

Horned pondweed
Zannichellia palustris

Hornwort
Ceratophyllum demersum

Indian blue lotus
Nymphaea stellata

Indian red water lily
Nymphaea rubra

Italian-type eel grass
Vallisneria spiralis

Ivy duckweed   *Lemna trisulca*

Japanese arrowhead
Sagittaria sagittifolia
leucopetala

Japanese mat rush   *Juncus effusus*

Jesuit's nut   *Trapa natans*

Joe-pye weed
Eupatorium maculatum

Kidney mud plantain
Heteranthera reniformis

Kingcup   *Caltha palustris*

Lace leaf
Aponogeton fenestralis
Aponogeton madagascariensis

Latticeleaf   *Aponogeton fenestralis*
Aponogeton madagascariensis

Least bulrush   *Typhe minima*

Least spike rush
Eleocharis acicularis

Lesser bulrush
Typha angustifolia

Lesser duckweed   *Lemna miniscula*
Lemna minor

Lesser water plantain
Baldellia ranunculoides

Loddon pondweed
Potamogeton nodosus

Lotus   *Nymphaea lotus*

Madagascar lace plant
Aponogeton fenestralis
Aponogeton madagascariensis

Mad-dog weed
Alisma plantago-aquatica

Magnolia water lily
Nymphaea tuberosa

Malayan sword
Aglaonema simplex

Marsh clover
Menyanthes trifoliata

Marsh dock   *Rumex palustris*

Marsh forget-me-not
Myosotis secunda

Marsh marigold   *Caltha palustris*

Marsh St John's wort
Hypericum elodes

Marsh trefoil
Menyanthes trifoliata

Ma-tai	*Eleocharis dulcis*
Miniature papyrus	
	Cyperus isocladus
Molly blobs	*Caltha palustris*
Moneywort	
	Lysimachia nummularia
Mosquito fern	*Azolla caroliniana*
Mosquito plant	*Azolla caroliniana*
Mud-midget	*Wolffiella floridana*
Myriad leaf	
	Myriophyllum verticillatum
Narrow-leaved eel grass	
	Zostera angustifolia
Narrow-leaved grass wrack	
	Zostera angustifolia
Nile blue lotus	
	Nymphaea stellata
Nodding avens	*Geum rivale*
Nodding bur-marigold	
	Bidens cernua
Nut grass	*Cyperus esculentus*
Nut sedge	*Cyperus esculentus*
Nuttall's water thyme	
	Elodea nuttallii
Old world arrowhead	
	Sagittaria sagittifolia
Opposite-leaved pondweed	
	Groenlandia densa
Paper plant	*Cyperus papyrus*
Papyrus	*Cyperus papyrus*
Parrot's feather	
	Myriophyllum aquaticum
Parsley water dropwort	
	Oenanthe lachenalis
Peacock hyacinth	
	Eichhornia azurea
Perfoliate pondweed	
	Potamogeton perfoliatus
Pickerel weed	*Pontederia cordata*
Pipewort	*Eriocaulon aquaticum*
Platterdock	*Nymphaea alba*
	Nymphaea venusta
Pond lily	*Castalia odorata*
	Nymphaea odorata
Pond nuts	*Nelumbo lutea*
	Nelumbo pentapetala
Prickly water lily	*Euryale ferox*
Purple loosestrife	
	Lythrum salicaria
Pygmy chainsword plant	
	Echinodorus intermedius
	Echinodorus martii
Pygmy water lily	
	Nymphaea pygmaea
	Nymphaea tetragona
Queen Victoria water lily	
	Victoria amazonica
	Victoria regia
Rattlebox	
	Ludwigia alternifolia
Red pondweed	
	Potamogeton alpinus
Red water milfoil	
	Myriophyllum hippuroides
Rigid hornwort	
	Ceratophyllum demersum
River crowfoot	
	Ranunculus fluitans
River water crowfoot	
	Ranunculus fluitans
River water dropwort	
	Oenanthe fluviatilis
Royal water lily	
	Victoria amazonica
	Victoria regia
Rushy pondweed	
	Potamogeton alpinus
Sacred lotus	*Nelumbo nucifera*
	Nelumbo speciosa
Salt rush	*Juncus lesuerii*
Santa Cruz water lily	
	Victoria cruziana
	Victoria trickeri
Santa Cruz water platter	
	Victoria cruziana
	Victoria trickeri
Seedbox	*Ludwigia alternifolia*
Shellflower	*Pistia stratiotes*
Shining pondweed	
	Potamogeton lucens
Skunk cabbage	
	Lysichiton americanum
Slender-leaved pondweed	
	Potamogeton filiformis
Slender spike rush	
	Eleocharis acicularis
Small bur-reed	
	Sparganium minimum
Small pondweed	
	Potamogeton pusillus
Smartweed	
	Polygonium hydropiper
Smokeweed	
	Eupatorium maculatum
Soft hornwort	
	Ceratophyllum submersum
Soft rush	*Juncus effusus*
Spatterdock	*Nuphar advena*
Spider tassel pondweed	
	Ruppia cirrhosa
Spiked lythrum	*Lythrum salicaria*
Spiked water milfoil	
	Myriophyllum spicatum
Spiny-spored quillwort	
	Isoetes echinospora
Spiral rush	*Juncus effusus spiralis*
Spring moss	
	Fontinalis antipyretica
Star duckweed	*Lemna trisulca*
Swamp lily	*Saururus cernuus*
Swamp potato	
	Sagittaria sagittifolia

Water yam

Swan potato	*Sagittaria sagittifolia*
Sweet sedge	*Acorus calamus*
Sweet water lily	*Nymphaea odorata*
Tape grass	*Vallisneria spiralis*
	Zostera angustifolia
Texas mud-baby	*Echinodorus cordifolius*
	Echinodorus radicans
Thoroughwort	*Eupatorium perfoliatum*
Triangular water fern	*Ceratopteris richardii*
Tuberous water lily	*Nymphaea tuberosa*
Tufted lysimachia	*Lysimachia thyrsiflora*
Umbrella palm	*Cyperus alternifolius*
Umbrella plant	*Cyperus alternifolius*
Umbrella sedge	*Cyperus alternifolius*
Unbranched bur reed	*Sparganium emerson*
Various-leaved pondweed	*Potamogeton gramineus*
Victoria water lily	*Victoria amazonica*
	Victoria regia
Wapato	*Sagittaria cuneata*
	Sagittaria latifolia
Washington grass	*Cabomba caroliniana*
Water aloe	*Stratiotes aloides*
Water archer	*Sagittaria sagittifolia*
Water arum	*Calla palustris*
Water avens	*Geum rivale*
Water balsam	*Hydrocera angustifolia*
Water blinks	*Montia fontana*
Water blobs	*Caltha palustris*
Water buttercup	*Ranunculus aquatalis*
Water celery	*Vallisneria americana*
Water chestnut	*Trapa natans*
Water chickweed	*Myosotis aquaticum*
Water chinquapin	*Nelumbo lutea*
	Nelumbo pentapetala
Water convolvulus	*Ipomoea aquatica*
Water cowslip	*Caltha palustris*
Water crowfoot	*Ranunculus aquatalis*
Water dock	*Rumex hydrolapathan*
Water dragon	*Saururus cernuus*

Water dropwort	*Oenanthe fistulosa*
Water feather	*Myriophyllum aquaticum*
Water fern	*Azolla caroliniana*
	Ceratopteris thalictroides
Water flag	*Iris pseudacorus*
Water flaxseed	*Spirodela polyrhiza*
Water forget-me-not	*Myosotis scorpioides*
Water hawthorn	*Aponogeton distachyus*
Water hemlock	*Cicuta virosa*
Water hyacinth	*Eichhornia crassipes*
Water lettuce	*Pistia stratiotes*
Water lobelia	*Lobelia dortmanna*
Water lovage	*Oenanthe fistulosa*
Water maize	*Victoria amazonica*
	Victoria regia
Water mint	*Mentha aquatica*
Water moss	*Fontinalis antipyretica*
Water oats	*Zizania aquatica*
Water pepper	*Polygonium hydropiper*
Water plantain	*Alisma plantago-aquatica*
Water poppy	*Hydrocleys nymphoides*
Water purslane	*Isnardia palustris*
	Ludwigia palustris
Water shamrock	*Menyanthes trifoliata*
Water shield	*Brasenia schreberi*
Water soldier	*Stratiodes aloides*
Water speedwell	*Veronica anagallis*
	Veronica anagallis-aquatica
Water sprite	*Ceratopteris thalictroides*
Water star grass	*Heteranthera dubia*
Water starwort	*Callitriche stagnalis*
Water thyme	*Anarchais canadensis*
	Elodea canadensis
Water trumpet	*Cryptocoryne affinis*
Water violet	*Hottonia palustris*
Waterweed	*Anarcharis canadensis*
	Elodea canadensis
Water willow	*Justicia americana*
Water wistaria	*Hygrophila difformis*
Waterwort	*Elatine hexandra*
Water yam	*Aponogeton fenestralis*
	Aponogeton madagascariensis

Western milfoil

Western milfoil
Myriophyllum hippuroides
White Egyptian lotus
Nymphaea lotus
White water lily   *Castalia odorata*
Nymphaea alba
Nymphaea odorata
Whorled water milfoil
Myriophyllum verticillatum
Wild celery
Vallisneria americana
Wild jonquil   *Narcissus jonquilla*
Wild rice   *Zizania aquatica*
Willow moss
Fontinalis antipyretica
Wolffia   *Wolffia columbiana*

Wonkapin   *Nelumbo lutea*
Nelumbo pentapetala
Yanquapin   *Nelumbo lutea*
Nelumbo pentapetala
Yellow flag iris   *Iris pseudacorus*
Yellow nelumbo   *Nelumbo lutea*
Nelumbo pentapetala
Yellow nut grass
Cyperus esculentus
Yellow nut sedge
Cyperus esculentus
Yellow water lily   *Castalia flava*
Nuphar luteum
Nymphaea flava
Nymphaea mexicana

BULBS

Horticulturally the term 'bulb' includes bulbs, corms, tubers, and rhizomes, but here a few stoloniferous subjects have been included (some lilies, for instance) where confusion exists in the minds of some amateur gardeners. Orchids are listed separately elsewhere in the book.

Sieber's crocus –
Crocus sieberi

Achira *Canna edulis*
Acidanthera *Gladiolus callianthus*
Adam-and-Eve *Arum maculatum*
Adder's tongue
 Erythronium americanum
Adobe lily *Fritillaria pluriflora*
African blood lily
 Haemanthus multiflorus
African corn lily *Ixia maculata*
 Ixia viridiflora
African lily *Agapanthus africanus*
 Agapanthus umbellatus
African wonder flower
 Ornithogalum thyrsoides
Alligator lily
 Hymenocallis palmeri
Alpine lily *Lilium parvum*
Alpine squill *Scilla bifolia*
Alpine violet
 Cyclamen hederifolium
Amazon lily *Eucharis amazonica*
 Eucharis grandiflora
Amberbell
 Erythronium americanum
American ornamental onion
 Allium cernuum
American swamp lily
 Saururus cernuus
American trout lily
 Erythronium revolutum
American turk's cap lily
 Lilium superbum
Angel's fishing-rod
 Dierama pendulum
 Sparaxis pendula
Angel's tears *Narcissus triandrus*
Arum lily
 Zantedeschia aethiopica
 Zantedeschia africana
Asiatic poison bulb
 Crinum asiaticum
Atamasco lily
 Zephyranthes atamasco
Australian giant lily
 Doryanthes excelsa
Autumn crocus
 Colchicum autumnale
Autumn daffodil
 Sternbergia lutea
Autumn snowdrop
 Galanthus reginae-olgae
Autumn squill *Scilla autumnalis*
Avalanche lily
 Erythronium giganteum
 Erythronium grandiflorum
 Erythronium montanum
 Erythronium obtusatum
Aztec lily *Sprekelia formosissima*
Baboon flower
 Babiana rubrocyanea
Baby cyclamen
 Cyclamen hederifolium

Bachelor's buttons
 Ranunculus acris
Barbados lily *Hippeastrum edule*
 Hippeastrum equestre
Basket flower
 Hymenocallis × festalis
 Hymenocallis narcissiflora
Bath asparagus
 Ornithogalum pyrenaicum
Bear's garlic *Allium ursinum*
Beavertail grass
 Calochortus coeruleus
Bedding dahlia *Dahlia merckii*
Beefsteak begonia
 Begonia × erythrophylla
Belladonna lily
 Amaryllis belladonna
Bell-flowered squill
 Endymion hispanicus
Bell lily *Lilium grayi*
Bell tree dahlia *Dahlia imperialis*
Berg lily *Galtonia candicans*
Bermuda buttercup *Oxalis cernua*
 Oxalis pes-caprae
Bermuda lily *Lilium longiflorum*
Black arum *Dracunculus vulgaris*
Blackberry lily
 Belamcanda chinensis
Black calla *Arum palaestinum*
 Arum pictum
Black fritillary
 Fritillaria camschatensis
 Fritillaria camtschatensis
Black lily
 Fritillaria camschatensis
 Fritillaria camtschatensis
Black sarana
 Fritillaria camschatcensis
 Fritillaria camtschatcensis
Black-throated calla lily
 Zantedeschia albomaculata
 Zantedeschia melanoleuca
Blazing star *Liatris aspera*
Blonde lilian
 Erythronium albidum
Blood lily
 Haemanthus multiflorus
 Haemanthus natalensis
 Scadoxus multiflorus
Bloody butcher
 Trillium recurvatum
Blotched panther lily
 Nomocharis pardanthina
Blue agapanthus
 Agapanthus africanus
 Agapanthus umbellatus
Blue-and-red baboon root
 Babiana rubrocyanea
Blue anemone *Anemone appenina*
Blue fall iris *Iris graeberiana*
Blue flag *Iris versicolor*
 Iris virginica

Common hyacinth

Blue lily   *Agapanthus africanus*
Blue onion   *Allium cyaneum*
Blue star   *Chamaescilla corymbosa*
Brazilian edelweiss
           Sinningia leucotricha
Brazilian gloxinia
           Sinningia speciosa
Brown beth   *Trillium erectum*
           Trillium flavum
Buckrams   *Allium ursinum*
Bulbous buttercup
           Ranunculus bulbosus
Bulbous crowfoot
           Ranunculus bulbosus
Bunch-flowered narcissus
           Narcissus tazetta
Bush lily   *Clivia mineata*
Butterfly ginger lily
        Hedychium coronarium
Butterfly iris   *Iris spuria*
           Iris ochroleuca
Butterfly lily
        Hedychium coronarium
Californian firecracker
           Brimeura ida-maia
        Dichelostemma ida-maia
Calla lily   *Zantedeschia aethiopica*
           Zantedeschia africana
Camass   *Camassia quamash*
Cambridge grape hyacinth
        Muscari tubergenianum
Camosh   *Camassia quamash*
Campernelle jonquil
           Narcissus calathinus
           Narcissus × odorus
Canada lily   *Lilium canadense*
Candelabra dahlia
           Dahlia imperialis
Candelabra flower
           Brunsvigia josephinae
Candlestick lily
           Lilium × hollandicum
           Lilium × umbellatum
Canterbury bells
           Gloxinia perennis
Cape belladonna
           Amaryllis belladonna
Cape cowslip   *Lachenalia aloides*
           Lachenalia bulbifera
           Lachenalia contaminata
           Lachenalia glaucina
           Lachenalia mutabilis
           Lachenalia ribida
           Lachenalia tricolor
Cape hyacinth
           Amphisiphon stylosa
Cape lily   *Crinum × powellii*
Cape pondweed
           Aponogeton distachyos
Cardinal flower
           Sinningia cardinalis
Carolina lily   *Lilium michauxii*

Catherine-wheel
           Haemanthus katharinae
Cat's ear   *Calochortus coeruleus*
Caucasian lily
           Lilium monadelphum
Celandine crocus
           Crocus korolkowii
Chamise lily   *Lilium rubescens*
Chaparral lily   *Lilium rubescens*
Checkered daffodil
           Fritillaria meleagris
Checkered lily
           Fritillaria meleagris
Checker lily   *Fritillaria lanceolata*
Chilean crocus
           Tecophilaea cyanocrocus
Chincherinchee
           Ornithogalum thyrsoides
Chinese chives   *Allium tuberosum*
Chinese lantern
        Narcissus poeticus physaloides
Chinese lantern lily
           Sandersonia aurantiaca
Chinese ornamental onion
           Allium amabile
Chinese sacred lily
           Narcissus canaliculatus
           Narcissus tazetta
Chinese squill   *Scilla chinensis*
           Scilla scilloides
Chinese white lily
           Lilium leucanthum
           Lilium formosum
Chive   *Allium schoenoprasum*
Christmas bells
           Sandersonia aurantiaca
Ciboule   *Allium fistulosum*
Cinderella slippers
           Sinningia regina
Cinnamon jasmine
        Hedychium coronarium
Cipollino   *Muscari comosum*
Cive   *Allium schoenoprasum*
Clanwilliam bluebell
           Ixia incarnata
Climbing lily   *Littonia modesta*
Cloth-of-gold   *Crocus angustifolius*
           Crocus susianus
Coastal lily   *Lilium maritimum*
Coast trillium   *Trillium ovatum*
Cobra lily   *Arisaema speciosum*
Columbia lily
           Lilium columbianum
Commom camass
           Camassia quamash
Common garden canna
           Canna × generalis
Common grape hyacinth
           Muscari botryoides
Common hyacinth
           Hyacinthus orientalis

Common snowdrop

Common snowdrop	*Galanthus nivalis*
Copper iris	*Iris fulva*
Coral lily	*Lilium pumilum*
	Lilium tenuifolium
Corn flag	*Gladiolus segetum*
Corn lily	*Ixia maculata*
	Ixia viridiflora
Crested iris	*Iris cristata*
Crimson flag	*Schizostylis coccinea*
Crowfoot	*Ranunculus asiaticus*
Crow garlic	*Allium vineale*
Crown imperial fritillary	
	Fritillaria imperialis
Crusaders' spears	
	Urginea maritima
Cuban lily	*Scilla peruviana*
Cuckoo pint	*Arum maculatum*
Culverkeys	
	Hyacinthoides non-scriptus
Daffodil	
	Narcissus pseudonarcissus
Daffodil garlic	*Allium neapolitanum*
Dasheen	*Colocasia esculenta*
Davis begonia	*Begonia davisii*
De caen anemone	
	Anemone coronaria
Desert candle	*Eremurus robustus*
Desert mariposa	
	Calochortus kennedyi
Dog's tooth violet	
	Erythronium dens-canis
Dove's dung	
	Ornithogalum umbellatum
Dragon arum	
	Dracunculus vulgaris
Dragonroot	
	Arisaema dracontium
	Arisaema triphyllum
Drakensberg star	
	Rhodohypoxis baurii
Drooping star-of-Bethlehem	
	Ornithogalum nutans
Dutch crocus	*Crocus vernus*
Dutch hyacinth	
	Hyacinthus orientalis
Dutch yellow crocus	*Crocus flavus*
Dwarf crested iris	*Iris cristata*
Dwarf iris	*Iris verna*
Dwarf jonquil	*Narcissus assoanus*
	Narcissus requienii
Dwarf squill	*Scilla monophyllos*
Dwarf white trillium	
	Trillium nivale
Easter lily	
	Cardiocrinum giganteum
	Lilium longiflorum
	Zephyranthes atamasco
Eastern camass	
	Camassia scilloides

Edible canna	*Canna edulis*
Elephant lily	*Crinum × powellii*
Elephant's ear begonia	
	Begonia albo-coccinea
English bluebell	
	Endymion non-scriptus
	Endymion nutans
	Hyacinthoides non-scriptus
	Hyacinthoides nutans
English iris	*Iris anglica*
	Iris latifolia
	Iris xiphioides
Eucharist lily	
	Eucharis grandiflora
Eureka lily	
	Lilium occidentale
European wood anemone	
	Anemone nemorosa
European wood sorrel	
	Oxalis acetosella
Ever-ready onion	*Allium cepa*
Eyelash begonia	*Begonia boweri*
Fair maids of February	
	Galanthus nivalis
Fairy-carpet begonia	
	Begonia versicolor
Fairy lantern	*Calochortus albus*
	Calochortus amoenus
Fairy lily	*Zephyranthes candida*
Falklands scurvy grass	
	Oxalis enneaphylla
Fall crocus	*Colchicum autumnale*
Falling stars	
	Crocosmia × crocosmiiflora
False sea onion	
	Ornithogalum caudatum
Fancy-leaved caladium	
	Caladium × hortulanum
Fawn lily	
	Erythronium californicum
Feather hyacinth	
	Muscari comosum
Few-flowered leek	
	Allium paradoxum
Field garlic	*Allium oleraceum*
	Allium vineale
Fireball lily	*Scadoxus multiflorus*
Fire-king begonia	
	Begonia goegoensis
Fire lily	*Clivia mineata*
	Cyrtanthus mackennii
	Lilium bulbiferum
Flag	*Iris × germanica*
Flame lily	*Gloriosa rothschildiana*
	Gloriosa superba
Flames	
	Homoglossum merianella
Fleur-de-lis	*Iris × germanica*
Florist's allium	
	Allium neapolitanum

Gray's lily

Florist's calla lily
    Zantedeschia aethiopica
    Zantedeschia africana
Florist's cyclamen
    Cyclamen persicum
Florist's gloxinia
    Sinningia speciosa
Flowering onion
    Allium neapolitanum
Flower of the western wind
    Zephyranthes candida
Flower of the wind
    Zephyranthes candida
Forest lily   *Veltheimia brachteata*
    Veltheimia viridiflora
Fox's grape   *Fritillaria uva-vulpis*
Foxtail lily   *Eremurus robustus*
Fragrant-flowered garlic
    Allium ramosum
French asparagus
    Ornithogalum pyrenaicum
Fumewort   *Corydalis bulbosa*
Galtonia   *Galtonia candicans*
Garden calla lily
    Zantedeschia aethiopica
    Zantedeschia africana
Garden gladiolus
    Gladiolus × hortulanus
Garden hyacinth
    Hyacinthus orientalis
Garland flower
    Hedychium coronarium
Garlic   *Allium sativum*
Garlic chive   *Allium tuberosum*
Gayfeather   *Liatris aspera*
George lily   *Cyrtanthus purpureus*
    Vallota speciosa
German garlic   *Allium senescens*
German onion
    Ornithogalum caudatum
Giant bellflower
    Ostrowskia magnifica
Giant bluebell
    Hyacinthoides hispanica
Giant chincherinchee
    Ornithogalum saundersiae
Giant garlic
    Allium scorodoprasum
Giant Himalayan lily
    Cardiocrinum giganteum
Giant lily
    Cardiocrinum giganteum
Giant ornamental onion
    Allium giganteum
Giant pineapple flower
    Eucomis pallidiflora
Giant snowdrop
    Galanthus elwesii
Giant snowflake
    Leucojum aestivum
Giant spider plant   *Cleome spinosa*

Giant stove brush
    Haemanthus magnificus
Giant summer hyacinth
    Galtonia candicans
Ginger lily
    Hedychium coronarium
Gingerwort
    Hedychium gardneranum
Gipsy onion   *Allium ursinum*
Gladwin iris   *Iris foetidissima*
Globe lily   *Calochortus albus*
Gloriosa lily
    Gloriosa rothschildiana
    Gloriosa superba
Glory lily   *Gloriosa rothschildiana*
    Gloriosa superba
Glory-of-the-snow
    Chionodoxa gigantea
    Leucocoryne ixioides
Glory-of-the-sun
    Leucocoryne ixioides
    Leucocoryne odorata
Gloxinia   *Sinningia speciosa*
Goddess mariposa
    Calochortus vestae
Gold-banded lily   *Lilium auratum*
Golden African lily
    Amaryllis aurea
    Lycoris africana
Golden-banded lily
    Lilium auratum
Golden-bowl mariposa
    Calochortus concolor
Golden calla lily
    Zantedeschia elliottiana
Golden fairy lantern
    Calochortus amabilis
Golden globe tulip
    Calochortus amabilis
Golden hurricane lily
    Amaryllis aurea
    Lycoris africana
Golden garlic   *Allium moly*
Golden lily   *Lycoris aurea*
Golden-rayed lily   *Lilium auratum*
Golden spider lily   *Amaryllis aurea*
    Lycoris africana
    Lycoris aurea
Good-luck leaf   *Oxalis deppei*
Good-luck plant   *Oxalis deppei*
Grape hyacinth
    Muscari moschatum
    Muscari neglectum
    Muscari racemosum
Grape-leaf begonia   *Begonia dregei*
    Begonia parvifolia
    Begonia × speculata
Grapevine begonia
    Begonia weltoniensis
Grassy bell   *Dierama pendulum*
    Sparaxis pendula
Gray's lily   *Lilium grayi*

Greek windflower
Anemone blanda

Green-banded mariposa
Calochortus macrocarpus

Green dragon
Arisaema dracontium

Green ixia   *Ixia viridiflora*

Guernsey lily   *Nerine bowdenii*
Nerine sarniensis

Guinea-hen tulip
Fritillaria meleagris

Hardy begonia   *Begonia discolor*
Begonia grandis

Harebell   *Endymion non-scriptus*

Harlequin flower   *Sparaxis elegans*
Sparaxis grandiflora
Sparaxis tricolor
Streptanthera elegans

Hartshorn plant
Anemone nuttalliana

Harvest brodiaea
Brodiaea coronaria
Brodiaea elegans

Heart-of-Jesus   *Caladium bicolor*

Helmet flower
Sinningia cardinalis

Herb lily
Alstroemeria haemantha

Hog's garlic   *Allium ursinum*

Hollyhock begonia
Begonia gracilis
Begonia martiana

Hoop-petticoat daffodil
Narcissus bulbocodium

Horned tulip   *Tulipa acuminata*

Hot-water plant
Achimenes longiflora

Humbold lily   *Lilium humboldtii*

Hurricane lily   *Rhodophiala bifida*

Hyacinth   *Hyacinthus orientalis*

Hyacinth-of-Peru   *Scilla peruviana*

Hyacinth squill
Scilla hyacinthoides

Hybrid tuberous begonia
Begonia × tuberhybrida

Ifafa lily   *Cyrtanthus mackennii*

Indian elephant flower
Crinum × powellii

Indian pink   *Spigelia marilandica*

Indian shot   *Canna indica*

Indian turnip   *Arisaema triphyllum*

Indigo squill   *Camassia scilloides*

Iris-flowered crocus
Crocus byzantinus
Crocus iridiflorus

Irish shamrock   *Oxalis acetosella*

Iron-cross begonia
Begonia masoniana

Italian arum   *Arum italicum*

Italian squill   *Endymion italicus*

Jack-in-the-pulpit
Arisaema triphyllum

Jacobean lily
Sprekelia formosissima

Japanese anemone
Anemone hupehensis

Japanese bunching onion
Allium fistulosum

Japanese iris   *Iris kaempferi*

Japanese jacinth   *Scilla chinensis*
Scilla scilloides

Japanese lily
Cardiocrinum cordatum
Lilium japonicum
Lilium krameri
Lilium lancifolium
Lilium makinoi
Lilium speciosum

Japanese turk's cap lily
Lilium hansonii

Jersey lily   *Amaryllis belladonna*

Jonquil   *Narcissus jonquilla*

Josephine's lily
Brunsvigia josephinae

Kaffir lily   *Schizostylis coccinea*

Kahili ginger lily
Hedychium gardneranum

Kahli ginger
Hedychium gardneranum

Kamchatka lily
Fritillaria camschatcensis
Fritillaria camtschatcensis

Keeled garlic   *Allium carinatum*

Kerry lily   *Simethis planifolia*

Kidney begonia
Begonia × erythrophylla

King begonia   *Begonia rex*

Kynassa lily   *Cyrtanthus obliquus*

Lacework lily
Amaryllis reticulata
Hippeastrum reticulatum

Lady tulip   *Tulipa clusiana*

Lady's leek   *Allium cernuum*

Lamance iris   *Iris brevicaulis*

Lavender globe lily
Allium tanguticum

Lebanon squill
Puschkinia libanotica
Puschkinia scilloides

Lemon lily   *Lilium parryi*

Lent lily
Narcissus pseudonarcissus

Leopard lily   *Lilium carolinianum*
Lilium catesbaei
Lilium pardalinum

Lesser celandine
Ranunculus ficaria

Lesser turk's cap lily
Lilium pomponium

Lettuce-leaf begonia
Begonia × crestabruchii

Lilac mariposa
  Calochortus splendens
Lily leek Allium moly
Lily-of-the-Amazon
  Eucharis grandiflora
Lily-of-the-field Sternbergia lutea
Lily-of-the-Incas
  Alstroemeria aurantiaca
Lily-of-the-Nile
  Agapanthus africanus
  Agapanthus umbellatus
Lily-of-the-palace Amaryllis aulica
  Hippeastrum aulicum
  Hippeastrum robustum
Lily-of-the-valley
  Convallaria majalis
Lily-pad begonia
  Begonia nelumbiifolia
Lily-royal Lilium superbum
Lion's beard Anemone nuttalliana
Lion's paw
  Alstroemeria leontochir ovallei
Little turk's cap lily
  Lilium pomponium
Lizard flower
  Sauromatum guttatum
Loddon lily Leucojum aestivum
Long-headed anemone
  Anemone cylindrica
Lords and ladies
  Arum maculatum
Lucky clover Oxalis deppei
Madonna lily Eucharis grandiflora
  Lilium candidum
Magic lily Amaryllis hallii
  Lycoris squamigera
Maid of the mist
  Gladiolus primulinus
Maltese cross
  Sprekelia formosissima
Manipur lily Lilium mackliniae
Maple-leaf begonia Begonia dregei
  Begonia parvifolia
  Begonia weltoniensis
Marble martagon lily
  Lilium duchartrei
March lily Amaryllis belladonna
Mariposa lily
  Calochortus nuttallii
Martagon lily Lilium martagon
Maryland pink root
  Spigelia marilandica
Meadow hyacinth
  Camassia scilloides
Meadow leek Allium canadense
Meadow lily Lilium canadense
Meadow saffron
  Colchicum × agrippinum
  Colchicum autumnale
Mediterranean lily
  Pancratium maritimum

Meshed lily Amaryllis reticulata
  Hippeastrum reticulatum
Mexican lily Hippeastrum reginae
Mexican shell flower
  Tigridia pavonia
Mexican star Milla biflora
Michigan lily Lilium michiganense
Miniature begonia Begonia boweri
Miniature pond-lily begonia
  Begonia hydrocotylifolia
Minor turk's cap lily
  Lilium pomponium
Mission bells Fritillaria biflora
Montbretia
  Crocosmia × crocosmiiflora
Morocco iris Iris tingitana
Mountain lily Lilium auratum
Mountain spiderwort
  Lloydia serotina
Mourning iris Iris basaltica
  Iris susiana
Mouse garlic Allium angulosum
Mouse plant
  Arisarum proboscideum
Multiplier onion Allium cepa
Musk hyacinth Muscari moschatum
  Muscari muscarini
  Muscari racemosum
Mysteria Colchicum autumnale
Naked ladies
  Colchicum autumnale
Nap-at-noon
  Ornithogalum umbellatum
Nankeen lily Lilium excelsum
  Lilium × testaceum
Narrow-leaved fritillary
  Fritillaria lanceolata
Nasturtium-leaf begonia
  Begonia francisii
Natal iris Curtonus paniculatus
Natal paintbrush
  Haemanthus natalensis
Netted iris Iris reticulata
Nodding nerine Nerine undulata
Nodding onion Allium cernuum
Nodding squill Scilla bifolia
Nodding star-of-Bethlehem
  Ornithogalum nutans
Nodding trillium
  Trillium cernuum
North African narcissus
  Narcissus watieri
Nutmeg hyacinth
  Muscari moschatum
  Muscari racemosum
Orange-bell lily Lilium grayi
Orange-cup lily
  Lilium philadelphicum
Orange lily Lilium bulbiferum
Orange pearl
  Polianthes geminiflora

Orange tuberose

Orange tuberose	
	Polianthes geminiflora
Orange turk's cap lily	
	Lilium davidii
Orchid amaryllis	
	Sprekelia formosissima
Orchid-flowered canna	
	Canna × orchiodes
Orchid iris	*Iris orchidoides*
Oregon lily	*Lilium columbianum*
Oriental garlic	*Allium tuberosum*
Orris	*Iris odoratissima*
	Iris pallida
Outdoor freesia	*Freesia hybrida*
Oxford grape hyacinth	
	Muscari tubergenianum
Paint brush	
	Haemanthus multiflorus
	Haemanthus natalensis
Painted-leaf begonia	*Begonia rex*
Painted trillium	
	Trillium undulatum
Palestine iris	*Iris basaltica*
	Iris susiana
Panther lily	*Lilium pardalinum*
Paperwhite narcissus	
	Narcissus papyraceus
Paradise lily	*Paradisea liliastrum*
Pasque flower	
	Anemone nuttalliana
	Anemone patens
	Anemone pulsatilla
	Pulsatilla patens
Peacock tiger flower	
	Tigridia pavonia
Pendant ornamental onion	
	Allium sikkimense
Pendulous begonia	
	Begonia × tuberhybrida
Pennywort begonia	
	Begonia hydrocotylifolia
Perfumed fairy lily	
	Chlidanthus fragrans
Persian buttercup	
	Ranunculus asiaticus
Persian iris	*Iris persica*
Persian ranunculus	
	Ranunculus asiaticus
Persian sun's eye	
	Tulipa occulus-solis
Persian violet	*Cyclamen persicum*
Peruvian daffodil	
	Chlidanthus fragrans
	Hymenocallis amancaes
	Hymenocallis narcissiflora
Peruvian jacinth	*Scilla peruviana*
Peruvian lily	
	Alstroemeria aurantiaca
Peruvian mountain daffodil	
	Pyrolirion tubiflorum
Peruvian redbird	
	Gloxinia sylvatica

Petticoat daffodil	
	Narcissus bulbocodium
Pheasant's eye narcissus	
	Narcissus poeticus
Pig lily	*Zantedeschia aethiopica*
	Zantedeschia africana
Pilewort	*Ranunculus ficaria*
Pineapple flower	
	Eucomis autumnalis
	Eucomis bicolor
	Eucomis comosa
	Eucomis pallidiflora
	Eucomis undulata
	Eucomis zambesiaca
Pine lily	*Lilium carolinianum*
	Lilium catesbaei
Pink agapanthus	
	Tulbaghia simmleri
Pink arum lily	
	Zantedeschia rehmannii
Pink calla lily	*Richardia rehmannii*
	Zantedeschia rehmannii
Pink candelabra	
	Brunsvigia radulosa
Pink fritillary	*Fritillaria pluriflora*
Pink star tulip	
	Calochortus uniflorus
Poetaz narcissus	
	Narcissus × medioluteus
Poet's narcissus	*Narcissus poeticus*
Poison flag	*Iris versicolor*
Polyanthus narcissus	
	Narcissus canaliculatus
	Narcissus tazetta
Pond-lily begonia	
	Begonia nelumbiifolia
Poppy anemone	
	Anemone coronaria
Portuguese iris	*Iris xiphium*
Portuguese paradise lily	
	Paradisea lusitanica
Potato onion	*Allium cepa*
Pot-of-gold lily	*Lilium iridollae*
Prairie onion	*Allium stellatum*
Prairie smoke	
	Anemone nuttalliana
Prickly blazing star	*Liatris aspera*
Primrose peerless narcissus	
	Narcissus × medioluteus
Prussian asparagus	
	Ornithogalum pyrenaicum
Purple globe tulip	
	Calochortus amoenus
Purple onion	*Allium cyaneum*
Purple toadshade	
	Trillium recurvatum
Purple trillium	*Trillium erectum*
	Trillium flavum
	Trillium recurvatum
Purple wake-robin	
	Trillium recurvatum

BULBS

Slender blue flag

Pussy ears *Calochortus maweanus*
Calochortus tolmiei

Pyrenean fritillary
Fritillaria pyrenaica

Quamash *Camassia quamash*

Queen lily
Amaryllis phaedranassa

Queensland arrowroot
Canna edulis

Rain lily *Zephyranthes candida*

Rakkyo *Allium bakeri*
Allium chinense

Ramp *Allium tricoccum*

Ramsons *Allium ursinum*

Ramsons wood garlic
Allium ursinum

Rattlesnake plantain
Goodyera pubescens

Red calla lily *Richardia rehmannii*
Zantedeschia rehmannii

Red ginger lily
Hedychium coccineum

Red hot poker
Kniphofia triangularis

Red iris *Iris fulva*

Red-skinned onion
Allium haematochiton

Red spider lily *Amaryllis radiata*
Lycoris radiata

Red squill *Urginea maritima*

Redwood lily *Lilium rubescens*

Redwood sorrel *Oxalis oregana*

Regal lily *Lilium regale*

Resurrection lily *Amaryllis hallii*
Lycoris squamigera

Rhodesian gladiolus
Gladiolus dalenii

Ring-of-bells
Hyacinthoides non-scriptus

River lily *Schizostylis coccinea*

Roan lily *Lilium grayi*

Rocambole *Allium sativum*
Allium scordoprasum

Rock harlequin
Corydalis sempervirens

Roman hyacinth
Bellevalia romana
Hyacinthus orientalis
Hyacinthus romanus

Roman wormwood
Corydalis sempervirens

Roof iris *Iris tectorum*

Rose leek *Allium canadense*

Rosey garlic *Allium roseum*

Round-headed garlic
Allium sphaerocephalum

Round-headed leek
Allium sphaerocephalum

Royal lily *Lilium regale*

Royal paintbrush
Haemanthus puniceus
Scadoxus puniceus

Saffron crocus *Crocus sativus*

Saint Brigid anemone
Anemone coronaria

Saint Bruno's lily
Paradisea liliastrum

Saint James's lily
Sprekelia formosissima

Saint Joseph's lily
Amaryllis × johnsonii
Hippeastrum × johnsonii

Sand crocus *Romulea columnae*

Sand leek *Allium scorodoprasum*

Scarborough lily
Cyrtanthus purpureus
Vallota speciosa

Scarlet anemone *Anemone fulgens*

Scarlet fritillary *Fritillaria recurva*

Scarlet ginger lily
Hedychium coccineum

Scarlet martagon lily
Lilium chalcedonicum

Scarlet-seeded iris *Iris foetidissima*

Scarlet turk's cap lily
Lilium chalcedonicum

Scarlet windflower
Anemone fulgens

Schnittlauch
Allium schoenoprasum

Scotch crocus *Crocus biflorus*

Scurvy grass *Oxalis enneaphylla*

Sea daffodil
Pancratium maritimum

Sea lily *Pancratium maritimum*

Sea onion
Ornithogalum caudatum
Scilla verna
Urginea maritima

Sea squill *Urginea maritima*

Sego lily *Calochortus nuttallii*

Serpent garlic *Allium sativum*

Serpent's tongue
Erythronium americanum

Shallot *Allium cepa*

Showy Japanese lily
Lilium lancifolium
Lilium speciosum

Showy lily *Lilium lancifolium*
Lilium speciosum

Shrimp begonia
Begonia limmingheiana

Siberian iris *Iris sibirica*

Siberian squill *Scilla sibirica*

Sierra iris *Iris hartwegii*

Sierra lily *Lilium parvum*

Sierra star tulip
Calochortus nudus

Slender blue flag *Iris prismatica*

Small celandine

Small celandine	*Ranunculus ficaria*
Small grape hyacinth	*Muscari botryoides*
Small tiger lily	*Lilium parvum*
Snake's head fritillary	*Fritillaria meleagris*
Snowdrop windflower	*Anemone sylvestris*
Snow trillium	*Trillium nivale*
Solomon's lily	*Arum palaestinum*
South African squill	*Scilla natalensis*
Southern blue flag	*Iris virginica*
Southern red lily	*Lilium carolinianum*
	Lilium catesbaei
Southern swamp crinum	*Crinum americanum*
Spanish bluebell	*Endymion campanulata*
	Endymion hispanica
	Hyacinthoides campanulata
	Hyacinthoides hispanicus
Spanish daffodil	*Narcissus hispanicus*
Spanish garlic	*Allium scorodoprasum*
Spanish iris	*Iris xiphium*
Spanish jacinth	*Endymion hispanicus*
Spanish onion	*Allium fistulosum*
Spider lily	*Amaryllis radiata*
	Hymenocallis × festalis
	Hymenocallis narcissiflora
	Ismene calathina
	Lycoris africana
	Lycoris radiata
Spiky gayfeather	*Liatris spicata*
Spire lily	*Galtonia candicans*
Spotted calla lily	*Zantedeschia albomaculata*
	Zantedeschia melanoleuca
Spotted Chinese lily	*Cardiocrinum cathayanum*
Spring fumitory	*Corydalis aurea*
Spring snowflake	*Leucojum vernum*
Spring squill	*Scilla verna*
Spring starflower	*Brodiaea uniflora*
	Ipheion uniflorum
	Leucocoryne uniflora
	Milla uniflora
Spuria iris	*Iris spuria*
Squawroot	*Trillium erectum*
	Trillium flavum
Stag's garlic	*Allium vineale*
Star begonia	*Begonia heracleifolia*
Star hyacinth	*Scilla amoena*
Star-leaf begonia	*Begonia heracleifolia*
Star lily	*Lilium concolor*
Star-of-Bethlehem	*Ornithogalum arabicum*
	Ornithogalum narbonense
	Ornithogalum pyrenaicum
	Ornithogalum umbellatum
Starry hyacinth	*Scilla autumnalis*
Stars-of-Persia	*Allium christophii*
Stinking benjamin	*Trillium erectum*
	Trillium flavum
Stinking gladwin iris	*Iris foetidissima*
Stinking iris	*Iris foetidissima*
Storm lily	*Zephyranthes candida*
Striped garlic	*Allium cuthbertii*
Striped squill	*Puschkinia scilloides*
Summer hyacinth	*Galtonia candicans*
	Hyacinthus candicans
Summer snowflake	*Leucojum aestivum*
	Ornithogalum umbellatum
Sunset lily	*Lilium pardalinum*
Swamp lily	*Crinum americanum*
	Lilium superbum
	Saururus cernuus
Swamp onion	*Allium validum*
Sweet beth	*Trillium vaseyi*
Sweet garlic	*Tulbaghia simmleri*
Sword-leaved iris	*Iris ensata*
Sword lily	*Gladiolus × hortulanus*
Taro	*Colocasia esculenta*
Tassel hyacinth	*Muscari comosum*
Tenby daffodil	*Narcissus obvallaris*
The pearl	*Polianthes tuberosa*
Thimble lily	*Lilium bolanderi*
Thimbleweed	*Anemone cylindrica*
	Anemone riparia
	Anemone virginiana
Tiger flower	*Tigridia pavona*
Tiger lily	*Lilium lancifolium*
	Lilium tigrinum
	Tigrida pavonia
Toadshade	*Trillium sessile*
Torch lily	*Kniphofia triangularis*
Tous-les-mois	*Canna edulis*
Tree dahlia	*Dahlia imperialis*
Triangular-stemmed garlic	*Allium triquetrum*
Triquetros leek	*Allium triquetrum*
Trout lily	*Erythronium americanum*
Trumpet lily	*Lilium longiflorum*
	Zantedeschia aethiopica
	Zantedeschia africana
Trumpet narcissus	*Narcissus pseudonarcissus*

Yellow ginger lily
Hedychium flavescens

Yellow grape hyacinth
Muscari macrocarpum

Yellow iris *Iris pseudacorus*

Yellow lily *Lilium canadense*

Yellow mariposa
Calochortus luteus

Yellow marsh afrikander
Gladiolus tristis

Yellow onion *Allium flavum*

Yellow snowdrop
Erythronium americanum

Yellow star flower
Sternbergia lutea

Yellow star ornamental
onion *Allium moly*

Yellow turk's cap lily
Lilium pyrenaicum

Zephyr lily *Zephyranthes candida*

CACTI AND SUCCULENTS

The majority of cacti are succulent, arid- or desert-area plants with thickened stems that serve the plant both for water-storage and as photosynthetic organs, replacing the leaves which are usually miniscule or totally absent. Most are armed with vicious spines and should be handled with care. They should be kept well away from children and domestic pets.

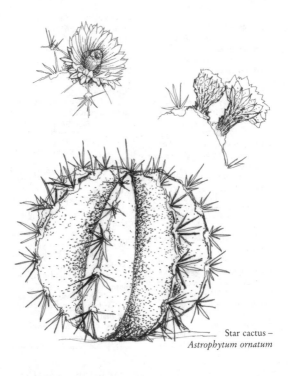

Star cactus –
Astrophytum ornatum

Aaron's beard

Aaron's beard *Opuntia leucotricha*

African living rock
Pleiospilos bolusii

Agave cactus
Leuchtenbergia principis

Agave salmiana *Pulque agave*

American aloe *Agave americana*

Annual mesembryanthemum
Dorotheanthus bellidiformis

Apple cactus *Cereus peruvianus*

Aristocrat plant
Haworthia chalwinii

Arizona giant *Carnegiea gigantea*

Artichoke cactus
Obregonia denegrii

Bald old man
Cephalocereus palmeri

Barbados aloe *Aloe barbadensis*

Barbados gooseberry
Pereskia aculeata

Barbary fig *Opuntia vulgaris*

Barrel cactus
Echinocactus grusonii
Echinopsis multiplex

Bead vine *Crassula rupestris*

Beavertail cactus *Opuntia basilaris*

Beehive cactus
*Coryphantha vivipara
arizonica*

Big nipple cactus
Coryphantha runyonii

Bird's nest cactus
Mammillaria camptotricha

Bird's nest sansevieria
Sansevieria hahnii

Bishop's cap cactus
Astrophytum myriostigma
Astrophytum ornatum

Bishop's hood cactus
Astrophytum myriostigma

Black echeveria *Echeveria affinis*

Black fingers *Opuntia clavarioides*

Blind pear
Opuntia microdasys rufida

Blue barrel cactus
Echinocactus ingens
Echinocereus horizonthalonius
Ferocactus glaucescens

Blue blade
Opuntia violacea santa-rita

Blue candle
Myrtillocactus geometrizans

Blue century plant *Agave palmeri*

Blue echeveria *Echeveria glauca*
Echeveria secunda glauca

Blue flame
Myrtillocactus geometrizans

Blue myrtle cactus
Myrtillocactus geometrizans

Bottle plant *Hatiora salicornioides*

Bowstring hemp
Sansevieria trifasciata laurentii

Boxing glove
Opuntia fulgida mamillata

Brain cactus
*Echinofossulocactus
zacatecasensis*

Bunny ears *Opuntia microdasys*

Burbank's spineless cactus
Opuntia ficus-indica

Button cactus *Epithelantha bokei*
Epithelantha micromeris

Cactus spurge
Euphorbia pseudocactus

Candelabra cactus
Lemaireocereus weberi

Candelilla
Euphorbia antisyphilitica

Candle plant *Senecio articulatus*

Cane cactus *Opuntia cylindrica*

Cantala *Agave cantala*

Cape aloe *Aloe ferox*

Caricature plant
Graptophyllum pictum

Cat claw cactus
Hamatocactus uncinatus

Cat's whiskers
*Schlumbergera gaertneri
makoyana*

Century plant *Agave americana*

Chain cactus *Rhipsalis paradoxa*

Chainlink cactus
Opuntia imbricata
Rhipsalis paradoxa

Chenille plant
Echeveria leucotricha

Chin cactus
Gymnocalycium gibbosum

Christmas cactus
Epiphyllum truncatum
Schlumbergera bridgesii
Schlumbergera truncata
Zygocactus truncatus

Cinnamon cactus
Opuntia microdasys rufida
Opuntia rufida

Claw cactus
Schlumbergera truncata

Cleftstone *Pleiospilos nelii*

Cloud grass
Aichryson × domesticum

Club cactus
Opuntia fulgida mamillata

Cob cactus *Escobaria tuberculosa*
Lobivia hertrichiana

Cobweb houseleek
Sempervivum arachnoides
Sempervivum arachnoideum

Cochineal plant
Nopalea cochenillifera

Colombian ball cactus
Wigginsia vorwerkiana

Gingham golf ball

Column-of-pearls
Haworthia chalwinii

Comb cactus *Pachycereus pecten-aboriginum*

Copper roses
Echeveria multicaulis

Coral cactus
Mammillaria heyderi
Rhipsalis cereuscula

Corncob cactus
Euphorbia mammillaris

Cotton-ball cactus *Espostoa lanata*

Cotton-pole cactus
Opuntia vestita

Crab cactus
Schlumbergera truncata

Crab's claw cactus
Epiphyllum truncatum
Schlumbergera bridgesii
Zygocactus truncatus

Cream cactus
Mammillaria heyderi

Creeping devil cactus
Lemaireocereus eruca
Machaerocereus eruca

Crested opuntia
Opuntia clavarioides

Crocodile jaws *Aloe humilis*

Crown of thorns *Euphorbia milii*
Euphorbia milii splendens

Cuban hemp *Furcraea hexapetala*

Cushion cactus *Opuntia floccosa*

Dagger cactus
Lemaireocereus gummosus

Dahlia cactus *Wilcoxia poselgeri*

Dancing bones
Hariota salicornioides
Hatiora salicornioides
Rhipsalis salicornioides

Desert candle
Dasylirion leiophyllum

Desert christmas cactus
Opuntia leptocaulis

Desert rose *Adenium obesum*
Echeveria rosea
Trichodiadema densum

Devil cactus *Opuntia schottii*

Devil's backbone
Pedilanthus tithymaloides

Devil's root cactus
Lophophora williamsii
Lothophora williamsii

Devil's tongue *Ferocactus corniger*
Ferocactus latispinus

Dog cholla *Opuntia schottii*

Dollar cactus
Opuntia violacea santa-rita

Dominoes *Opuntia erectoclada*

Dragon tree *Dracaena draco*

Dragon's head *Euphorbia gorgonis*

Drunkard's dream
Hariota salicornioides
Hatiora salicornioides
Rhipsalis salicornioides

Dry whisky
Lophophora williamsii
Lothophora williamsii

Dumpling cactus
Coryphantha runyonii
Lophophora williamsii
Lothophora williamsii

Eagle-claws cactus
Echinocactus horizonthalonius

Easter cactus
Rhipsalidopsis gaertneri

Easter-lily cactus
Echinopsis multiplex

Electrode cactus *Ferocactus histrix*

Emerald-idol *Opuntia cylindrica*

Empress of Germany
Nopalxochia phyllanthoides

English stonecrop *Sedum anglicum*

Eve's pin cactus *Opuntia subulata*

Fairy agave *Hechtia scariosa*

Fairy castles *Opuntia clavarioides*

Fairy-elephant's feet
Fritha pulchra

Fairy needles *Opuntia soehrensii*

Fairy washboard
Haworthia limifolia

Falcon feather *Aloe variegata*

Feather cactus
Mammillaria plumosa

Finger tree *Euphorbia tirucalli*

Firecracker cactus
Cleistocactus smaragdifolius

Firecracker plant *Echeveria setosa*

Fire crown cactus *Rebutia senilis*

Fishbone cactus
Epiphyllum anguliger

Fish-hook cactus
Ancistrocactus scheeri
Ferocactus latispinus
Ferocactus wislizenii
Mammillaria bocasana

Fish-hook pincushion cactus
Mammillaria wildii

Flapjack cactus *Opuntia chlorotica*

Fleshy-stalked pelargonium
Pelargonium carnosum

Frangipani tree
Plumiera acuminata

Ghost plant
Graptopetalum paraguayense

Giant barrel cactus
Echinocactus ingens

Giant cactus *Carnegiea gigantea*

Giant Mexican cereus
Pachycereus pringlei

Giant saguaro *Carnegiea gigantea*

Gingham golf ball *Euphorbia obesa*

Globe spear-lily

Globe spear-lily	*Doryanthes excelsa*
Glory of Texas	*Thelocactus bicolor*
Gnome's throne	*Opuntia clavarioides*
Goat's horn cactus	*Astrophytum capricorne*
Golden ball	*Echinocactus grusonii*
	Notocactus leninghausii
Golden barrel cactus	*Echinocactus grusonii*
Golden bird's nest cactus	*Mammillaria camptotricha*
Golden column	*Trichocereus spachianus*
Golden lily cactus	*Echinopsis aurea*
	Lobivia aurea
Golden old man	*Cephalocereus chrysacanthus*
Golden opuntia	*Opuntia microdasys*
Golden spines	*Cephalocereus chrysacanthus*
Golden star	*Mammillaria elongata*
Golden tom thumb	*Parodia aureispina*
Gold lace	*Mammillaria elongata*
Goldplush	*Opuntia microdasys*
Gorgon's head	*Euphorbia gorgonis*
Gouty pelargonium	*Pelargonium gibbosum*
Green aloe	*Furcraea foetida*
Green-flowered pitaya	*Echinocereus viridiflorus*
Green-flowered torch cactus	*Echinocereus viridiflorus*
Grizzly bear cactus	*Opuntia erinacea ursina*
Hairbrush cactus	*Pachycereus pecten-aboriginum*
Hatpin cactus	*Ferocactus rectispinus*
Hedge cactus	*Cereus peruvianus*
Hedgehog aloe	*Aloe humilis*
Hedgehog cactus	*Echinocereus pectinatus*
Hedgehog-cory cactus	*Coryphantha echinus*
Hen-and-chickens	*Sempervivum tectorum*
Henequen	*Agave fourcroydes*
Hollyhock-leaved pelargonium	*Pelargonium cotyledonis*
Honey-bunny	*Opuntia microdasys albispina*
Honolulu queen	*Hylocereus undatus*

Horned-leaf pelargonium	*Pelargonium ceratophyllum*
Hyacinth-scented rochea	*Rochea coccinea*
Ice plant	*Sedum spectabile*
Indian comb	*Pachycereus pecten aboriginum*
Indian fig	*Opuntia ficus-indica*
Irish mittens	*Opuntia vulgaris*
Jade plant	*Crassula arborescens*
	Crassula argentea
	Crassula portulacea
Jade tree	*Crassula ovata*
Japanese poinsettia	*Pedilanthus tithymaloides*
Jellybean plant	*Sedum pachyphyllum*
Jewbush	*Pedilanthus tithymaloides*
Jewel plant	*Titanopsis calcarea*
Joseph's coat cactus	*Opuntia vulgaris*
Joshua tree	*Yucca brevifolia*
Jumping cactus	*Opuntia fulgida*
Jumping cholla	*Opuntia prolifera*
Kanniedood	*Aloe variegata*
Lace cactus	*Echinocereus reichenbachii*
	Mammillaria elongata
Lace haworthia	*Haworthia setata*
Lady finger	*Mammillaria elongata*
Lamb's tail	*Chiastophyllum oppositifolium*
	Cotyledon simplicifolia
Lamb'stail cactus	*Wilcoxia schmollii*
Large barrel cactus	*Echinocactus ingens*
Leafy cactus	*Pereskia aculeata*
Lechuguilla	*Agave lophantha*
Lemon-ball cactus	*Mammillaria pringlei*
	Notocactus submammulosus
Lemon vine	*Pereskia aculeata*
Lesser old man cactus	*Echinocereus delaetii*
Link plant	*Rhipsalis paradoxa*
Lion's tongue	*Opuntia schickendantzii*
Little candles	*Mammillaria prolifera*
Little tree opuntia	*Opuntia vilis*
Living rock cactus	*Ariocarpus fissuratus*
	Pleiospilos bolusii
Living stones	*Lithops julii*
Madagascan palm	*Pachypodium lamerei*
Mauritius hemp	*Furcraea foetida*
Medicinal aloe	*Aloe barbadensis*

CACTI

Plush plant

Medusa's head spurge
    Euphorbia caput-medusae
Melon cactus
    Melocactus communis
Mescal button
    Lophophora williamsii
    Lothophora williamsii
Mexican dwarf tree cactus
    Opuntia vilis
Mexican giant
    Cephalocereus fulviceps
Mexican giant barrel cactus
    Echinocactus ingens
Mexican giant cactus
    Pachycereus pringlei
Mexican sunball   *Rebutia miniscula*
Milk bush   *Euphorbia tirucalli*
Mimicry plant   *Pleiospilos bolusii*
    Pleiospilos nelii
Missouri pincushion cactus
    Coryphantha missouriensis
Mistletoe cactus
    Rhipsalis baccifera
    Rhipsalis cassutha
Money tree   *Crassula arborescens*
Monkshood cactus
    Astrophytum myriostigma
Moon cactus   *Harrisia jusbertii*
    Harrisia martinii
Moonstones
    Pachyphytum oviferum
Mother-in-law's armchair
    Echinocactus grusonii
Mother-in-law's tongue
    Sansevieria trifasciata laurentii
Mother-of-pearl plant
    Graptopetalum paraguayense
Mountain cereus
    Borzicactus fossulatus
Mule-crippler cactus
    Echinocactus horizonthalonius
Mule's ears
    Opuntia schickendantzii
Native's comb
    Pachycereus pecten-aboriginum
Night-blooming cereus
    Epiphyllum oxypetalum
    Hylocereus undatus
    Nyctocereus serpentinus
    Peniocereus greggii
    Selenicereus grandiflorus
Nipple cactus   *Neobessya similis*
Nopal   *Opuntia megacantha*
Old-father-live-forever
    Pelargonium cotyledonis
Old lady cactus
    Mammillaria hahniana
    Mammillaria lanata
Old lady of Mexico
    Mammillaria hahniana
Old man cactus
    Cephalocereus senilis

Old man of the Andes
    Borzicactus trollii
    Oreocereus trollii
Old man of the mountains
    Borzicactus celsianus
    Pilocereus celsianus
Old man opuntia   *Opuntia vestita*
Old man's head
    Cephalocereus senilis
Old woman cactus
    Mammillaria hahniana
Orange tuna   *Opuntia elata*
Orchid cactus
    Epiphyllum ackermannii
    Epiphyllum × hybridum
    Nopalxochia ackermannii
Organ pipe cactus
    Lemaireocereus marginatus
    Lemaireocereus thurberi
    Stenocereus thurberi
Ornamental monkshood
    Astrophytum ornatum
Painted lady
    Echeveria derenbergii
Palmer spear-lily
    Doryanthes palmeri
Panda plant   *Kalanchoe tomentosa*
Panda-bear plant
    Kalanchoe tomentosa
Paper cactus   *Opuntia articulata*
Partridge-breasted aloe
    Aloe variegata
Peanut cactus
    Chamaecereus silvestrii
Pearl plant
    Haworthia margaritifera
    Haworthia pumila
Pearly dots   *Haworthia papillosa*
Pebble cactus   *Lithops lesliei*
Pencil cactus   *Opuntia ramosissima*
Pencil cholla   *Opuntia arbuscula*
Peruvian apple   *Cereus peruvianus*
Peruvian apple cactus
    Cereus peruvianus
Peyote cactus
    Lophophora williamsii
    Lothophora williamsii
Pheasant's wing   *Aloe variegata*
Pickle plant
    Trichodiadema barbatum
Pincushion cactus
    Coryphantha vivipara
Pink easter-lily cactus
    Echinopsis multiplex
Pitahaya   *Carnegiea gigantea*
Plaid cactus
    Gymnocalycium mihanovichii
Plain cactus
    Gymnocalycium mihanovichii
Plush plant   *Echeveria pulvinata*
    Kalanchoe tomentosa

Polka dots

Polka dots
Opuntia microdasys albispina
Pond-lily cactus
Nopalxochia phyllanthoides
Popcorn cactus
Rhipsalis cereuscula
Rhipsalis warmingiana
Porcupine pelargonium
Pelargonium hystrix
Powder-blue cereus
Lemaireocereus pruinosus
Powder puff
Mammillaria bocasana
Mammillaria gracilis
Prickly pear   *Opuntia ficus-indica*
Opuntia microdasys albispina
Opuntia vulgaris
Princess of the night
Selenicereus pteranthus
Prism cactus
Leuchtenbergia principis
Purple hedgehog cereus
Echinocereus sarissophorus
Purple pitaya   *Echinocereus dubius*
Purple prickly pear
Opuntia santa-rita
Pussy ears   *Kalanchoe tomentosa*
Queen of the night
Epiphyllum oxypetalum
Hylocereus undatus
Nyctocereus serpentinus
Selenicereus grandiflorus
Selenicereus macdonaldiae
Queen Victoria's aloe
Aloe victoriae reginae
Rabbit ears   *Opuntia microdasys*
Rainbow cactus
Echinocereus pectinatus rigidissimus
Rat's tail cactus
Aporocactus flagelliformis
Rat-tail plant
Crassula lycopodioides
Rattlesnake tail   *Crassula barklyi*
Red aloe   *Aloe ferox*
Redbird cactus
Pedilanthus tithymaloides
Redbird flower
Pedilanthus tithymaloides
Red bunny ears
Opuntia microdasys rufida
Opuntia rufida
Red crown cactus
Rebutia kupperana
Rebutia minuscula
Red orchid cactus
Nopalxochia ackermannii
Red spike
Cephalophyllum alstonii
Reina-de-la-noche
Peniocereus greggii
Ribbon cactus
Pedilanthus tithymaloides

Rice cactus   *Rhipsalis cereuscula*
Roof houseleek
Sempervivum tectorum
Rosary vine   *Crassula rupestris*
Rose cactus   *Pereskia grandifolia*
Rose pincushion
Mammillaria zeilmanniana
Rose-plaid cactus
Gymnocalycium quehlianum
Rose tuna   *Opuntia basilaris*
Royal-cross cactus
Mammillaria karwinskiana
Rubber spurge   *Euphorbia tirucalli*
Ruby dumpling
Mammillaria tetracantha
Saguaro   *Carnegiea gigantea*
Sahuaro   *Carnegiea gigantea*
Samphire-leaved pelargonium
Pelargonium crithmifolium
Sand dollar cactus
Astrophytum asterias
Scarlet ball cactus
Notocactus haselbergii
Scarlet bugler   *Cereus baumannii*
Cleistocactus baumannii
Scarlet crown cactus
Rebutia grandiflora
Scented cactus   *Wilcoxia poselgeri*
Sea coral   *Opuntia clavarioides*
Sea urchin cactus
Astrophytum asterias
Echinopsis eyriesii
Senita   *Lophocereus schottii*
Serpent cactus
Nyctocereus serpentinus
Seven stars   *Ariocarpus retusus*
Silver ball cactus   *Notocactus scopa*
Silver beads   *Crassula deltoides*
Crassula rhomboidea
Silver cluster cactus
Mammillaria prolifera
Silver dollar cactus
Astrophytum asterias
Silver torch   *Cleistocactus straussii*
Slipper flower
Pedilanthus tithymaloides
Small barrel cactus
Ferocactus viridescens
Snake cactus
Nyctocereus serpentinus
Snowball cactus   *Espostoa lanata*
Mammillaria bocasana
Pediocactus simpsonii
Snowball pincushion
Mammillaria candida
Snowdrop cactus
Rhipsalis houlletiana
Socotrine aloe   *Aloe perryi*
South American golden-barrel
Lobivia aurea

Zygocactus

South American old man	*Borzicactus celsianus*
Spice cactus	*Hariota salicornioides*
	Hatiora salicornioides
	Rhipsalis salicornioides
Spider aloe	*Aloe humilis*
Spider cactus	*Gymnocalycium denudatum*
Spider houseleek	*Sempervivum arachnoides*
	Sempervivum arachnoideum
Spineless cactus	*Opuntia ficus-indica*
Spiny aloe	*Aloe africana*
Split rock	*Pleiospilos nelii*
Spoon plant	*Dasylirion leiophyllum*
Sprawling cactus	*Morangaya pensilis*
Square-stemmed pelargonium	*Pelargonium tetragonum*
Staghorn cholla	*Opuntia versicolor*
Star cactus	*Ariocarpus fissuratus*
	Astrophytum ornatum
Star window plant	*Haworthia tessellata*
Stick cactus	*Euphorbia tirucalli*
Sticky moonstones	*Pachyphytum glutinicaule*
Stonecrop	*Sedum anglicum*
Stone plant	*Lithops dorotheae*
	Lithops julii
Strawberry cactus	*Echinocereus enneacanthus*
	Echinocereus salm-dyckianus
	Ferocactus setispinus
	Thelocactus setispinus
String of beads	*Senecio rowleyanus*
String of hearts	*Caralluma woodii*
String of pearls	*Senecio rowleyanus*
Sugar-almond plant	*Pachyphytum oviferum*
Sun cactus	*Heliocereus speciosus*
Sun plant	*Portulaca grandiflora*
Tasajillo	*Opuntia leptocaulis*
Teddy-bear cactus	*Opuntia bigelovii*
Teddy-bear cholla	*Opuntia bigelovii*
Texas pride	*Thelocactus bicolor*
Texas rainbow cactus	*Echinocereus dasyacanthus*
Thanksgiving cactus	*Schlumbergera truncata*
Thimble cactus	*Mammillaria fragilis*
Thimble mammillaria	*Mammillaria fragilis*
Thimble tuna	*Opuntia sphaerica*
Tiger aloe	*Aloe variegata*
Tiger jaws	*Faucaria tigrina*
Tom thumb cactus	*Parodia aureispina*
Toothpick cactus	*Stetsonia coryne*
Tortoise cactus	*Deamia testudo*
Totem-pole cactus	*Lophocereus schottii*
Turk's cap cactus	*Melocactus communis*
	Melocactus intortus
Turk's head	*Ferocactus hamatacanthus*
Turk's head cactus	*Melocactus communis*
	Melocactus maxonii
Unguentine cactus	*Aloe barbadensis*
Velvet opuntia	*Opuntia velutina*
Wallflower crown	*Rebutia pseudodeminuta*
Wandering cactus	*Morangaya pensilis*
Wax plant	*Euphorbia antisyphilitica*
Wax rose	*Pereskia bleo*
Whisker cactus	*Lophocereus schottii*
White chin cactus	*Gymnocalycium schickendantzii*
White jewel	*Titanopsis schwantesii*
White torch cactus	*Trichocereus spachianus*
Whitsun cactus	*Schlumbergera gaertneri*
Window aloe	*Haworthia cymbiformis*
Window cushion	*Haworthia cymbiformis*
Window plant	*Fenestraria rhopalophylla*
	Haworthia cymbiformis
Woolly sheep	*Opuntia floccosa*
Woolly torch cactus	*Cephalocereus palmeri*
Yellow bunny ears	*Opuntia microdasys*
Yellow old man	*Cephalocereus palmeri*
Yellow rabbit ears	*Opuntia microdasys*
Youth-and-old-age	*Aichryson × domesticum*
Zanzibar aloe	*Aloe perryi*
Zebra haworthia	*Haworthia fasciata*
Zygocactus	*Schlumbergera truncata*

CARNIVOROUS PLANTS

Otherwise known as insectivorous plants, these are plants that have developed special mechanisms for trapping and digesting mainly, but not exclusively, small insects. There are several types of carnivorous plants including the pitcher plants, the sticky-leaved sundews and butterworts, and the spring-trap leaves of Venus's flytrap. The strange nature of these plants makes them popular subjects for exhibiting at horticultural shows and school study groups. They are also extensively grown indoors and in greenhouses as a nature control for flying insects.

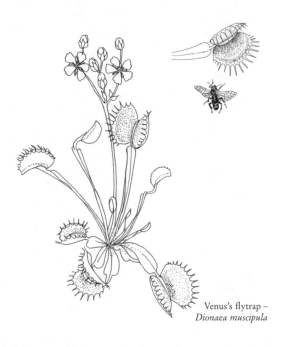

Venus's flytrap –
Dionaea muscipula

COMMON NAMES

Alabama canebrake
pitcher plant

Alabama canebrake pitcher plant
Sarracenia alabamensis

Alpine butterwort
Pinguicula alpina

Amethyst bladderwort
Utricularia amethystina

Australian pitcher plant
Cephalotus follicularis

Blue butterwort
Pinguicula caerulea

Branched butterwort
Pinguicula ramosa

Bristly flytrap   *Genlisea hispidula*

Butterwort   *Pinguicula vulgaris*

California pitcher plant
Chrysamphora californica
Darlingtonia californica

Cobra lily
Chrysamphora californica
Darlingtonia californica

Cobra orchid
Chrysamphora californica
Darlingtonia californica

Cobra plant
Chrysamphora californica
Darlingtonia californica

Common butterwort
Pinguicula vulgaris

Common pitcher plant
Sarracenia purpurea

Common sundew
Drosera rotundifolia

Creeping genlisea   *Genlisea repens*

Deep red pitcher plant
Nepenthes sanguinea

Dew thread   *Drosera filiformis*

Downy bladderwort
Utricularia pubescens

Dwarf butterwort
Pinguicula pumila

Dwarf genlisea   *Genlisea pygmaea*

Dwarf sundew   *Drosera brevifolia*

Eared sundew   *Drosera auriculata*

Erect Australian sundew
Byblis gigantea

Floated bladderwort
Utricularia inflata

Floating pondtrap
Aldrovanda vesiculosa

Flytrap sensitive
Dionaea muscipula

Funnelform pitcher plant
Nepenthes hookerana

Giant marsh pitcher plant
Heliamphora tatei

Giant sun pitcher plant
Heliamphora tatei

Golden flytrap   *Genlisea aurea*

Great bladderwort
Utricularia macrorhiza

Greater butterwort
Pinguicula grandiflora

Great sundew   *Drosera anglica*

Green pitcher plant
Sarracenia oreophila

Guayan sun pitcher plant
Heliamphora nutans

Hooded pitcher plant
Sarracenia minor

Horned bladderwort
Utricularia cornuta

Humped bladderwort
Utricularia gibba

Huntsman's cup
Sarracenia purpurea

Huntsman's horn   *Sarracenia flava*

Indian cup   *Sarracenia purpurea*

Inverted bladderwort
Utricularia resupinata

Lesser bladderwort
Utricularia minor

Marsh pitcher plant
Heliamphora nutans

Mottled purple pitcher plant
Nepenthes × atrosanguinea

Monkey pitcher
Nepenthes merrilliana

Monkey's larder
Nepenthes merrilliana

Monkey's rice pot
Nepenthes merrilliana

Mouse pitcher
Nepenthes merrilliana

New Zealand bladderwort
Utricularia nova-zealandiae

Northern pitcher plant
Sarracenia purpurea

Pale green pitcher plant
Nepenthes ventricosa

Pale pitcher   *Sarracenia alata*
Sarracenia sledgei

Pale violet butterwort
Pinguicula lilacina

Parrot pitcher plant
Sarracenia psittacina

Pink fans   *Polypompholyx tenella*

Pink petticoat
Polypompholyx multifida

Pink rainbow   *Drosera menziesii*

Pink sundew   *Drosera capillaris*

Pitcher plant
Chrysamphora californica
Darlingtonia californica
Sarracenia purpurea

Portuguese butterwort
Pinguicula lusitanica

Portuguese sundew
Drosophyllum lusitanicum

Purple pitcher plant
Sarracenia purpurea

Rainhat trumpet   *Sarracenia minor*

Round-leaved sundew
Drosera rotundifolia

CARNIVOROUS PLANTS

Yellow trumpets

Sensitive flytrap
Dionaea muscipula

Side-saddle flower
Sarracenia purpurea

Small marsh pitcher
Heliamphora minor

Small sun pitcher
Heliamphora minor

Southern pitcher plant
Sarracenia purpurea

Spikey pondtrap
Aldrovanda vesiculosa

Spiral-leaved flytrap
Genlisea filiformis

Sprawling Australian sundew
Byblis liniflora

Sundew   *Drosera capensis*

Swamp pitcher plant
Cephalotus follicularis

Sweet pitcher plant
Sarracenia purpurea
Sarracenia rubra

Sweet trumpet   *Sarracenia rubra*

Thread-leaved sundew
Drosera filiformis

Tipitiwitchet   *Dionaea muscipula*

Toothed butterwort
Pinguicula crenatiloba

Trumpet leaf   *Sarracenia flava*

Trumpets   *Sarracenia flava*

Twining bladderwort
Utricularia spiralis

Umbrella trumpets
Sarracenia flava

Variegated butterwort
Pinguicula variegata

Venus's flytrap   *Dionaea muscipula*

Violet butterwort
Pinguicula ionantha
Pinguicula lilacina

Watches   *Sarracenia flava*

Waterbug trap
Aldrovanda vesiculosa

Waterwheel plant
Aldrovanda vesiculosa

White trumpet pitcher plant
Sarracenia drummondii
Sarracenia leucophylla

Winged trumpets   *Sarracenia alata*

Yellow butterwort   *Pinguicula lutea*

Yellow pitcher plant
Sarracenia flava

Yellow trumpets   *Sarracenia alata*

CARNIVOROUS

FERNS AND FERN ALLIES

These are flowerless plants bearing leaves (fronds) and reproducing by spores on the lower surface of the mature foliage. Ferns and fern allies share a similar life-cycle and are treated horticulturally in a similar manner. For example, the so-called asparagus fern *Asparagus setaceus* is not a fern but one of the *liliaceae*.

Toothed Davallia –
Davallia denticulata

Adder's fern

Adder's fern   Polypodium vulgare
Adder's tongue
          Ophioglossum vulgatum
Alpine bladder fern
          Cystopteris regia
Alpine polypody
          Polypodium alpestre
Alpine woodsia   Woodsia alpina
Alternate spleenwort
          Asplenium germanicum
American maidenhair
          Adiantum pedatum
American oak fern
          Onoclea sensibilis
American tree fern   Ctenitis sloanei
American wall fern
          Polypodium virginianum
Annual maidenhair
          Gymnogramma leptophylla
Ash-leaf polypody
          Polypodium fraxinifolium
Australian bracken
          Pteridium esculentum
Australian brake   Pteris tremula
Australian cliff brake
          Pellaea falcata
Australian lady fern
          Athyrium australe
Australian maidenhair
          Adiantum formosum
Australian slender brake
          Pteris ensiformis
Australian tree fern
          Cyathea cooperi
Australian water fern
          Blechnum cartilagineum
Autumn fern
          Dryopteris erythrosora
Ball fern   Davallia bullata
          Davallia mariesii
Basket selaginella   Selaginella apoda
          Selaginella densa
Bat's wing fern   Histiopteris incisa
Bead fern   Onoclea sensibilis
Bear's foot fern
          Humata tyermannii
Beautiful hard-shield fern
          Polystichum aculeatum
          pulchrum
Beech fern   Phegopteris connectilis
          Polypodium phegopteris
          Thelypteris phegopteris
Berry bladder fern
          Cystopteris bulbifera
Bird's nest fern   Asplenium nidus
Black maidenhair spleenwort
          Asplenium adiantum-nigrum
Black spleenwort
          Asplenium adiantum-nigrum
Blackstem maidenhair
          Adiantum formosum
Black tree fern   Cyathea medullaris

Blue selaginella
          Selaginella uncinata
Blunt-lobed woodsia
          Woodsia obtusa
Boston fern   Nephrolepis exaltata
Boulder fern
          Dennstaedia punctilobula
          Dicksonia punctilobula
Bracken   Pteridium aquilinum
          Pteris aquilinum
Bramble fern   Hyolepis punctata
Branch-crested hartstongue
          Scolopendrium vulgare
          ramo-cristatum
Branch-crested shield fern
          Polystichum angulare
          ramulosum
Branched black maidenhair
  spleenwort
          Asplenium adiantum-
          nigrum ramosum
Branched hard fern
          Blechnum spicant ramosum
Branched maidenhair fern
          Asplenium trichomanes
          ramosum
Branched maidenhair spleenwort
          Asplenium trichomanes
          ramosum
Branched male fern
          Lastrea filix-mas ramosa
Branched sea spleenwort
          Asplenium marinum ramosum
Braun's holly fern
          Polystichum braunii
Bristle fern
          Blechnum cartilagineum
Bristly shield fern
          Lastreopsis hispida
Bristly tree fern
          Dicksonia youngiae
Brittle bladder fern
          Cystopteris fragilis
Brittle maidenhair
          Adiantum tenerum
Broad beech fern
          Thelypteris hexagonoptera
          Thelypteris phegopteris
Broad buckler fern
          Dryopteris austriaca
          Dryopteris dilatata
          Lastrea dilatata
Broad curly hartstongue
          Scolopendrium vulgare
          crispum-latum
Buckhorn   Osmunda cinnamomea
Bulbil bladder fern
          Cystopteris bulbifera
Button fern   Pellaea rotundifolia
          Tectaria cicutaria
          Tectaria gemmifera
Californian adder's tongue
          Ophioglossum californicum

FERNS

Double-fronded hartstongue

California polypody
   Polypodium californicum

Chinese brake   *Pteris longifolia*
                    Pteris vittata

Chinese fern   *Pteris multifida*
                    Pteris serrulata

Christmas fern
   Polystichum acrostichoides

Cinnamon fern
   Osmunda cinnamomea

Cleft maidenhair
   Adiantum capillus-
   veneris incisum

Cleft moonwort
   Botrychium lunaria incisum

Cleft true maidenhair
   Adiantum capillus-
   veneris incisum

Climbing bird's nest fern
   Polypodium integrifolium
   Polypodium irioides
   Polypodium punctatum

Climbing fern
   Lygodium palmatum

Climbing swamp fern
   Stenochlaena palustris

Cloven green spleenwort
   Asplenium viride multifidum

Cloven lady fern   *Athyrium filix-*
                    femina multifidum

Cloven maidenhair fern
   Asplenium trichomanes
   multifidum

Cloven maidenhair
spleenwort
   Asplenium trichomanes
   multifidum

Cloven mountain polypody
   Polypodium phegopteris
   multifidum

Cloven rock spleenwort
   Asplenium fontanum
   multifidum

Common bracken
   Pteridium aquilinum
   Pteris aquilinum

Common buckler fern
   Dryopteris filix-mas
   Lastrea filix-mas

Common fishbone fern
   Nephrolepis cordifolia

Common ground fern
   Culcita dubia

Common horsetail
   Equisetum hyemale

Common maidenhair
   Adiantum aethiopicum

Common oak fern
   Gymnocarpium dryopteris
   Polypodium dryopteris

Common polypody
   Polypodium vulgare

Common rasp fern   *Doodia media*

Common scouring brush
   Equisetum hymale

Common woodsia
   Woodsia obtusa

Confluent maidenhair fern
   Asplenium trichomanes
   confluens

Confluent maidenhair spleenwort
   Asplenium trichomanes
   confluens

Creeping shield fern
   Lastreopsis microsora

Crested bracken
   Pteris aquilina cristata

Crested broad buckler fern
   Lastrea dilatata cristata

Crested buckler fern
   Dryopteris cristata
   Lastrea cristata

Crested climbing bird's nest fern
   Polypodium integrifolium
   cristatum

Crested hartstongue
   Scolopendrium vulgare
   cristatum

Crested male fern
   Lastrea filix-mas cristata

Crested mountain buckler fern
   Lastrea montana cristata

Crested polypody
   Polypodium vulgare cristatum

Crested royal fern
   Osmunda regalis cristata

Crested soft prickly shield fern
   Polystichum angulare cristatum

Crested wood fern
   Dryopteris cristata

Cretan brake   *Pteris cretica*

Cretan fern   *Pteris cretica*

Crown fern   *Blechnum discolor*

Crow's nest fern   *Asplenium nidus*

Curly-grass fern   *Schizaea pusilla*

Dagger fern
   Polystichum acrostichoides

Deer fern   *Blechnum spicant*

Deer's foot fern
   Davallia canariensis

Delta maidenhair
   Adiantum cuneatum
   Adiantum decorum
   Adiantum raddianum

Dense holly fern
   Polystichum lonchitis
   confertum

Diamond maidenhair
   Adiantum trapeziforme

Divided soft prickly shield fern
   Polystichum angulare sem
   tripinnatum

Double-fronded hard fern
   Blechnum spicant duplex

Double-fronded hartstongue
   Scolopendrium vulgare duplex

Double-pinnuled polypody

Double-pinnuled polypody
Polypodium vulgare bifidum

Douglas's spike-moss
Selaginella douglasii

Drooping adder's-tongue
Ophioglossum pendulum

Dwarf cloven lady fern
Athyrium filix-femina multifidum nanum

Dwarf horsetail
Equisetum scirpoides

Dwarf leather fern
Polystichum tsus-simense

Dwarf lycopod
Selaginella rupestris

Dwarf scouring brush
Equisetum scirpoides

Ear-lobed polypody
Polypodium vulgare auritum

Eastern bracken
Pteridium latiusculum

Ebony spleenwort
Asplenium platyneuron

Elkhorn   *Platycerium bifurcatum*
Platycerium superbum

Engelmann's adder's tongue
Ophioglossum engelmannii

Engelmann's quillwort
Isoetes engelmannii

Erect sword fern
Nephrolepis cordifolia

European bristle fern
Trichomanes radicans

European chain fern
Woodwardia radicans

European polypody
Polypodium vulgare

European water clover
Marsilea quadrifolia

Fairy moss   *Azolla caroliniana*

False bracken   *Culcita dubia*

Fan-like hard fern
Blechnum spicant flabellata

Fiddleheads
Osmunda cinnamomea

Filmy maidenhair
Adiantum diaphanum

Fine-toothed brake   *Pteris dentata*
Pteris flabellata
Pteris flaccida

Finger fern   *Grammitis australis*
Grammitis billardieri

Fishbone fern
Nephrolepis cordifolia
Nephrolepis exaltata

Fishbone rib fern
Blechnum nudum

Fishbone water fern
Blechnum nudum

Five-fingered jack
Adiantum hispidulum

Flexible alpine polypody
Polypodium alpestre flexile

Floating fern
Ceratopteris pteridoides

Floating moss
Salvinia rotundifolia

Floating pepperwort
Marsilea crenata

Florida tree fern   *Ctenitis sloanei*

Florist's fern   *Dryopteris dilatata*

Flowering fern   *Osmunda regalis*

Forked bristle fern
Trichomanes radicans furcans

Forked brittle bladder fern
Cystopteris fragilis furcans

Forked hard fern
Blechnum spicant furcans

Forked hard shield fern
Polystichum aculeatum furcatum

Forked male fern
Lastrea filix-mas furcans

Forked mountain buckler fern
Lastrea montana furcans

Forked spleenwort
Asplenium septentrionale

Fragile bladder fern
Cystopteris fragilis

Fragrant fern
Microsorium pustulatum
Microsorium scandens
Polypodium pustulatum
Polypodium scandens

Fringed alpine polypody
Polypodium alpestre laciniatum

Giant brake   *Pteris tripartita*

Giant chain fern
Woodwardia chamissoi
Woodwardia fimbriata

Giant fern   *Angiopteris evecta*

Giant holly fern
Polystichum munitum

Giant maidenhair
Adiantum formosum

Giant wood fern
Dryopteris goldiana

Glade fern
Aglaomorpha pycnocarpon

Golden polypody
Phlebodium aureum
Polypodium aureum

Golden-scaled male fern
Dryopteris borreri

Golden tree fern
Dicksonia fibrosa

Gold fern
Pityrogramma austroamericana
Pityrogramma chrysophylla

Goldie's fern   *Dryopteris goldiana*

Grand-tasselled shield fern
Polystichum angulare grandiceps

FERNS

Maidenhair fern

Green cliff brake *Pellaea viridis*

Green spleenwort
Asplenium viride

Ground cedar
Lycopodium complanatum
Lycopodium tristachyum

Ground pine
Lycopodium clavatum
Lycopodium complanatum
Lycopodium obscurum

Hacksaw fern *Doodia aspera*
Doodia media

Hairy-lip fern *Cheilanthes lamosa*

Hammock fern
Blechnum occidentale

Hand fern *Doryopteris pedata*

Hanging spleenwort
Asplenium flaccidum
Asplenium majus
Asplenium mayi

Hapu tree fern *Cibotium glaucum*

Hard fern *Blechnum spicant*

Hard shield fern
Polystichum aculeatum

Hard water fern *Blechnum watsii*

Hare's foot fern *Davallia fejeensis*
Davallia mariesii
Phlebodium aureum
Polypodium aureum

Hartford fern
Lygodium palmatum

Hart's tongue fern
Phyllitis scolopendrium

Hawaiian tree fern
Cibotium glaucum

Hay-scented buckler fern
Dryopteris aemula
Lastrea recurva

Hay-scented fern
Dennstaedia punctilobula
Dicksonia punctilobula

Hedge fern *Polystichum setiferum*

Hen-and-chickens fern
Asplenium bulbiferum

Herringbone fern
Nephrolepis cordifolia

Holly fern *Polystichum lonchitis*

House holly fern
Cyrtomium falcatum

Huguenot fern *Pteris multifida*
Pteris serrulata

Interrupted fern
Osmunda claytoniana

Irish holly fern
Polystichum lonchitis
confertum

Irish polypody
Polypodium vulgare
semilacerum

Jagged-edge hartstongue
Scolopendrium vulgare
laceratum

Japanese buckler fern
Dryopteris erythrosora

Japanese climbing fern
Lygodium japonicum

Japanese felt fern *Pyrossia lingua*

Japanese holly fern
Cyrtomium falcatum
Polystichum falcatum

Japanese painted fern
Aglaomorpha
goeringianum pictum

Japanese shield fern
Dryopteris erythrosora

Jersey fern
Anogramma leptophylla

Jointed pine
Polypodium subauriculatum

Kangaroo fern
Microsorium diversifolium

Kidney maidenhair
Adiantum reniforme

Kidney-shaped hartstongue
Scolopendrium
vulgare reniforme

King fern *Angiopteris evecta*
Marattia fraxinea
Marattia salicina

Korau *Cyathea medullaris*

Lacy ground fern
Dennstaedia davallioides

Ladder brake *Pteris longifolia*
Pteris vittata

Ladder fern *Blechnum spicant*

Lady fern *Athyrium filix-femina*

Lanceolate spleenwort
Asplenium billottii
Asplenium lanceolatum

Leather fern *Acrostichum aureum*
Rumohra adiantiformis

Leatherwood fern
Dryopteris marginalis

Leathery moonwort
Botrychium multifidum

Leathery polypody
Polypodium scouleri

Leathery shield fern
Rumohra adiantiformis

Lemon-scented fern
Oreopteris limbosperma

Licorice fern
Polypodium glycyrrhiza

Limestone oak fern
Gymnocarpium robertianum

Limestone polypody
Polypodium calcareum

Little adder's tongue
Ophioglossum lusitanicum

Maidenhair *Adiantum capill*
veneris

Maidenhair fern
Adiantum pedatum
Asplenium trichomanes

Maidenhair spleenwort

Maidenhair spleenwort
    Asplenium trichomanes

Malayan flowering fern
    Helminthostachys zeylanica

Malay climbing fern
    Lygodium circinatum

Male fern   *Dryopteris filix-mas*
    Lastrea filix-mas

Mamaku   *Cyathea medullaris*

Many-shaped hartstongue
    Scolopendrium vulgare
    multiforme

Marginal buckler fern
    Dryopteris marginalis

Marginal shield fern
    Dryopteris marginalis

Marsh buckler fern
    Lastrea thelypteris

Marsh fern   *Dryopteris thelypteris*
    Thelypteris palustris

Mat spike moss
    Selaginella kraussiana

Mauritius spleenwort
    Asplenium daucifolium
    Asplenium viviparum

Mexican tree fern
    Cibotium schiedei

Moonwort   *Botrychium lunaria*

Moosehorn   *Platycerum superbum*

Mosquito fern   *Azolla caroliniana*

Mosquito plant   *Azolla caroliniana*

Moss fern   *Selaginella cuspidata*
    Selaginella pallescens

Mother fern
    Asplenium bulbiferum

Mother shield fern
    Polystichum proliferum

Mother spleenwort
    Asplenium bulbiferum

Mountain bladder fern
    Cystopteris montana

Mountain buckler fern
    Lastrea montana
    Thelypteris oreopteris

Mountain holly fern
    Polystichum lonchitis

Mountain male fern
    Dryopteris oreades

Mountain polypody
    Polypodium phegopteris

Multi-crested male fern
    Lastrea filix-mas multi-cristata

Multi-forked lady fern
    Athyrium filix-
    femina multifurcatum

Narrow-branched hard fern
    Blechnum spicant
    contractum-ramosum

Narrow-leaved strap fern
    Polypodium angustifolium

Narrow-lined shield fern
    Polystichum angulare lineare

Necklace fern
    Asplenium flabellifolium

New York fern
    Thelypteris noveboracensis

New Zealand cliff brake
    Pellaea rotundifolia

New Zealand tree fern
    Dicksonia squarrosa

Northern beech fern
    Phegopteris connectilis
    Thelypteris phegopteris

Northern elkhorn
    Platycerium hillii

Northern maidenhair
    Adiantum pedatum

Northern oak fern
    Gymnocarpium robertianum

Notched polypody
    Polypodium vulgare crenatum

Oak fern
    Gymnocarpium dryopteris
    Histiopteris incisa
    Polypodium dryopteris

Oak-leaf fern   *Drynaria quercifolia*
    Phymatodes quercifolia

Oblong woodsia   *Woodsia ilvensis*

One-sided filmy fern
    Hymenophyllum unilaterale

One-sided rue-leaved
spleenwort   *Asplenium ru*
    muraria unilaterale

Ostrich-feather fern
    Matteuccia struthiopteris
    Struthiopteris germanica

Pala   *Marattia douglasii*

Palm-leaf fern   *Blechnum capense*

Para fern   *Marattia fraxinea*
    Marattia salicina

Parasol fern   *Gleichnia microphylla*

Parsley fern   *Allosorus crispus*
    Cryptogramma crispa

Peacock fern
    Selaginella willdenovii

Peacock moss   *Selaginella uncinata*

Pine fern   *Aneimia adiantifolia*

Plums and custard fern
    Tricholomopsis rutilans

Ponga   *Cyathea dealbata*

Potato fern   *Marattia fraxinea*
    Marattia salicina

Pouched coral fern
    Gleichnia dicarpa

Princess pine
    Lycopodium obscurum

Prince of Wales' feather
    Leptopteris superba

Prickly buckler fern
    Lastrea spinulosa

Prickly rasp fern   *Doodia aspera*

Prickly shield fern
    Arachniodes aristata
    Polystichum vestitum

Spiny-spored quillwort

Purple cliff brake
 Pellaea atropurpurea
Queen lady fern *Athyrium filix-femina victoriae*
Rabbit's foot fern
 Davallia canariensis
 Davallia fejeensis
 Phlebodium aureum
 Polypodium aureum
Rainbow fern *Culcita dubia*
 Selaginella uncinata
Rasp fern *Doodia aspera*
Rattlesnake fern
 Botrychium virginianum
Refracted rock spleenwort
 Asplenium fontanum refractum
Resurrection fern
 Polypodium incanum
 Polypodium polypodioides
Resurrection plant
 Selaginella lepidophylla
Ribbon brake *Pteris cretica*
Ribbon fern
 Polypodium phyllitidis
 Pteris cretica
Rigid buckler fern
 Dryopteris submontana
 Lastrea rigida
Rock felt fern *Pyrossia rupestris*
Rock polypody
 Polypodium virginianum
Rock selaginella
 Selaginella rupestris
Rock spleenwort
 Asplenium fontanum
Rose-of-Jericho
 Selaginella lepidophylla
Rosy maidenhair
 Adiantum hispidulum
Rough dicksonia
 Dicksonia squarrosa
Rough maidenhair
 Adiantum hispidulum
Rough tree fern *Cyathea australis*
Royal fern *Osmunda regalis*
Rue-leaved spleenwort
 Asplenium ruta-muraria
Running pine
 Lycopodium clavatum
Rusty back fern
 Asplenium ceterach
 Ceterach officinarum
Rusty brake *Pteris longifolia*
 Pteris vittata
Rusty woodsia *Woodsia ilvensis*
Savannah fern
 Gleichnia dichotoma
 Gleichnia linearis
Scaly broad buckler fern
 Lastrea dilatata lepidota
Scaly male fern *Dryopteris affinis*

Scaly spleenwort
 Asplenium ceterach
 Ceterach officinarum
Scrambling coral fern
 Gleichnia microphylla
Sea spleenwort
 Asplenium marinum
Sensitive fern *Onoclea sensibilis*
Sharp-toothed alternate
spleenwort
 Asplenium germanicum acutidentatum
Shield hare's foot
 Rumohra adiantiformis
Shining club moss
 Lycopodium lucidulum
Shining fan fern
 Sticherus flabellatus
Short-fronded hartstongue
 Scolopendrium vulgare truncatum
Shuttlecock fern
 Matteuccia struthiopteris
 Struthiopteris germanica
Sickle fern *Pellaea falcata*
Silky fan fern *Sticherus tener*
Silver dollar fern
 Adiantum peruvianum
Silver elkhorn *Platycerium veitchii*
Silver fern
 Pityrogramma calomelanos
Silver glade fern
 Aglaomorpha thelypteroides
Silver king fern *Cyathea dealbata*
Silvery glade fern
 Athyrium thelypteroides
 Diplazium acrostichoides
Silvery spleenwort
 Athyrium thelypteroides
 Diplazium acrostichoides
Skeleton soft prickly shield
fern *Polystichum angulare depauperatum*
Slender brake *Pteris ensiformis*
Small rasp fern *Doodia caudata*
Snail fern *Tectaria cicutaria*
 Tectaria gemmifera
Snow brake *Pteris ensiformis*
Soft-prickly shield fern
 Polystichum angulare
Soft shield fern
 Polystichum angulare
 Polystichum setiferum
Soft tree fern *Dicksonia antarctica*
Southern beech fern
 Thelypteris hexagonoptera
Southern maidenhair
 Adiantum capillus-veneris
Spider fern *Pteris multifida*
 Pteris serrulata
Spiny-spored quillwort
 Isoetes echinospora

Split-fronded bracken

Split-fronded bracken
  Pteris aquilina bisulca

Squirrel's foot fern
  Davallia bullata
  Davallia mariesii
  Davallia trichomanoides

Staghorn *Platycerium bifurcatum*
  Platycerium superbum

Stemless pepperwort
  Marsilea pubescens
  Marsilea strigosa

Strap fern *Polypodium phyllitidis*

Strap rib fern *Blechnum patersonii*

Strap water fern
  Blechnum patersonii

Sweat plant *Selaginella cuspidata*
  Selaginella pallescens

Sword brake *Pteris ensiformis*

Sword fern *Nephrolepis cordifolia*
  Nephrolepis exaltata
  Polystichum munitum

Tangle fern *Gleichnia dicarpa*

Tapering polypody
  Polypodium vulgare acutum

Tasmanian tree fern
  Dicksonia antarctica

Tasselled lady fern *Athyrium filix-femina corymbiferum*

Thousand-leaved fern
  Hypolepis millefolia

Tongue fern *Pyrossia lingua*

Toothed brake *Pteris tremula*

Toothed davallia
  Davallia denticulata

Toothed wood fern
  Dryopteris austriaca spinulosa
  Dryopteris carthusiana
  Dryopteris spinulosa

Trailing selaginella
  Selaginella kraussiana
  Selaginella uncinata

Trailing spike moss
  Selaginella kraussiana

Tree fern *Cyathea arborea*

Treelet spike-moss
  Selaginella braunii

Trembling brake *Pteris tremula*

Trembling fern *Pteris tremula*

Triangular water fern
  Ceratopteris richardii

Tri-pinnate soft prickly shield
fern *Polystichum angulare tripinnatum*

Triple-branched polypody
  Polypodium dryopteris

Trisect brake *Pteris tripartita*

True maidenhair
  Adiantum capillus-veneris

Truncate shield fern
  Polystichum angulare truncatum

Tsusima holly fern
  Polystichum tsus-simense

Tunbridge filmy fern
  Hymenophyllum tunbridgense

Twin-fronded hartstongue
  Scolopendrium vulgare ramo-palmatum

Twin-fronded lanceolate
spleenwort
  Asplenium lanceolatum kalon

Umbrella fern
  Gleichnia microphylla
  Sticherus flabellatus

Variegated black maidenhair
spleenwort
  Asplenium adiantum-nigrum variegatum

Variegated black spleenwort
  Asplenium adiantum-nigrum variegatum

Variegated horsetail
  Equisetum variegatum

Variegated scouring brush
  Equisetum variegatum

Venus' hair *Adiantum capill veneris*

Virginia chain fern
  Woodwardia virginica

Virginian moonwort
  Botrychium virginianum

Walking fern
  Camptosorus rhizophyllus

Wall fern *Polypodium vulgare*

Wall polypody
  Polypodium vulgare

Wall rue *Asplenium ruta-muraria*

Wart fern
  Phymatodes scolopendrium
  Polypodium phymatodes
  Polypodium scolopendria
  Polypodium vulgare

Water fern *Azolla caroliniana*
  Ceratopteris thalictroides

Wavy hartstongue
  Scolopendrium vulgare undulato-ramosum

Weeping spleenwort
  Asplenium flaccidum
  Asplenium majus
  Asplenium mayi

Welsh polypody
  Polypodium vulgare cambricum

Western bracken
  Pteridium pubescens

Western polypody
  Polypodium hesperium
  Polypodium interjectum

West Indian tree fern
  Cyathea arborea

Wheki *Dicksonia squarrosa*

Wheki-ponga *Dicksonia fibrosa*

Whisk fern *Psilotum nudum*

Wide-fronded scaly
spleenwort
  Asplenium ceterach kalon

FERNS

Woolly tree fern

Wig tree fern   *Cyathea baileyana*
Willdenow's selaginella
          Selaginella willdenovii
Woolly tree fern
          Dicksonia antarctica
          Dicksonia fibrosa

FUNGI

A large group of simple plants lacking chlorophyll is covered by the word fungi. This section is concerned only with the larger edible, inedible, and poisonous fungi which are visible to the naked eye. Many edible fungi are matched in appearance with inedible, or even poisonous types so it is reckless to gather wild fungi unless you are experienced and familiar with the subtle differences. In the following list entries have been classified by the use of bracketed initials thus:

(E) – Edible, but some may be allergic to them.
(I) – Inedible for a variety of reasons.
(P) – Poisonous, but not usually fatal.
(F) – Can be fatal if eaten, sometimes within minutes.

Field mushroom –
Agaricus campestris

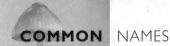

Amethyst deceiver

Amethyst deceiver
Laccaria amethystea (E)

Aniseed toadstool
Clytocybe odora (E)

Bachelor's buttons
Bulgaria inquinans (E)

Bare-tooth russala
Russala vesca (E)

Basket fungus
Clathrus cancellatus (E)
Clathrus ruber (E)

Bay boletus *Boletus badius* (E)
Xerocomus badius (E)

Beech tuft
Oudemansiella mucida (I)

Beechwood sickener
Russula mairei (I)

Beefsteak fungus
Fistulina hepatica (E)

Bigelow's blewit *Lepista irina* (E)

Birch bracket fungus
Piptoporus betulinus (E)
Polyporus betulinus (E)

Birch polypore
Piptoporus betulinus (E)

Bird's nest fungus
Crucibulum crucibuliforme (I)
Crucibulum laeve (I)

Bitter bolete *Tylopilus felleus* (I)

Bitter boletus
Gyroporus castaneus (E)

Bitter cep *Tylopilus felleus* (I)

Bitter hydnum
Hydnum scabrosum (I)

Bitter-sweet fungus
Hebeloma sacchariolens (P)

Black birch boletus
Leccinum melaneum (E)

Black bulgar
Bulgaria inquinans (E)

Blackening russula
Russula nigricans (I)

Black helvella
Helvella lacunosa (P)

Black saddle helvella
Helvella fusca (P)

Blewits *Lepista saeva* (E)
Tricholoma personatum (E)

Blue and yellow russula
Russula cyanoxantha (E)

Blushing bracket fungus
Daedaleopsis confragosa (I)

Bonnet mycena
Mycena galericulata (I)

Bootlace fungus
Armillaria mellea (E)
Clitocybe mellea (E)

Brain fungus *Sparassis crispa* (E)

Branched oyster fungus
Pleurotus cornucopiae (E)

Brick-red agaric
Hypholoma sublateritium (I)

Bronze boletus *Boletus aereus* (E)

Brown birch boletus
Boletus leucophareus (E)
Leccinum scabrum (E)

Brown roll-rim
Paxillus involutus (P)

Brown star fungus
Peziza ammophila (P)

Brown wood mushroom
Agaricus sylvaticus (E)
Psalliota sylvatica (E)

Buckler agaric
Entoloma clypeatum (I)

Butter cap *Collybia butyracea* (E)

Buttery collybia
Collybia butyracea (E)

Caesar's mushroom
Amanita caesarea (I)

Candlesnuff fungus
Xylaria hypoxylon (I)

Carpet fungus
Thelephora terrestris (I)

Cauliflower fungus
Masseola crispa (E)
Sparassis crispa (E)
Sparassis ramosa (E)

Cep *Boletus edulis* (E)

Changeable agaric
Kuehneromyces mutabilis (E)

Changeable mutabilis
Kuehneromyces mutabilis (E)

Chanterelle
Cantharellus cibarius (E)

Chestnut boletus
Gyroporus castaneus (E)

Cinnabar polypore
Pycnoporus cinnabarinus (I)

Clouded agaric
Clitocybe nebularis (E)
Lepista nebularis (E)

Clouded clitocybe
Clitocybe nebularis (E)
Lepista nebularis (E)

Clover windling
Marasmius oreades (E)

Club foot *Clitocybe clavipes* (I)

Clustered tough shank
Collybia confluens (I)

Coconut-scented milk cap
Lactarius glyciosmus (I)

Coffin filler *Amanita virosa* (F)

Common earthball
Scleroderma citrinum (P)

Common grisette
Amanita vaginata (E)

Common morel
Morchella esculenta (E)

Common puffball
Lycoperdon gemmatum (E)
Lycoperdon perlatum (E)

Common white helvella
Helvella crispa (P)

Goat's lip mushroom

Common white inocybe
Inocybe geophylla (P)

Conical inocybe
Inocybe fastigiata (P)

Conical morel
Morchella conica (E)

Conical wax cap
Hygrocybe conica (I)

Copper trumpet
Clitocybe illudens (P)
Clitocybe olearia (P)
Omphalotus olearius (P)

Coral spot fungus
Nectria cinnabarina (I)

Corn smut   *Ustilago maydis* (P)

Cow boletus   *Suillus bovinus* (E)

Cow fungus
Boletus leucophareus (E)
Leccinum scabrum (E)

Crazed boletus
Boletus chrysenteron (I)

Crested coral fungus
Clavulina coralloides (P)
Clavulina cristata (P)

Cultivated mushroom
Agaricus bisporus (E)

Cupped vellosa
Helvella vellosa (P)

Dark red russula
Russula xerampelina (E)

Dead man's fingers
Xylaria polymorpha (I)

Death cap   *Amanita phalloides* (F)

Deceiver   *Laccaria laccata* (E)

Deer mushroom
Pluteus atricapillus (E)
Pluteus cervinus (E)

Destroying angel
Amanita virosa (F)

Devil's boletus   *Boletus satanas* (P)

Devil's egg fungus
Phallus hadriani (I)

Dingy agaric
Tricholoma portentosum (E)

Distorted helvella
Helvella elastica (P)

Dog stinkhorn   *Mutinus caninus* (I)

Donkey's ear fungus
Otidea onotica (I)

Downy boletus
Boletus subtomentosus (E)
Xerocomus subtomentosus (E)

Dryad's saddle
Polyporus squamosus (E)

Ear-pick fungus
Auriscalpium vulgare (I)

Earthfan   *Thelephora terrestris* (I)

Earthstar   *Geastrum fimbriatum* (I)
Geastrum fornicatum (I)
Geastrum pectinatum (I)
Geastrum sessile (I)
Geastrum triplex (I)

Elf cup   *Sarcoscypha coccinea* (E)

Emetic russula   *Russula emetica* (P)

Ergot   *Claviceps purpurea* (P)

Eyelash cup fungus
Scutellinia scutellata (I)

Eyelash fungus
Scutellinia scutellata (I)

Faded russala
Russala decolorans (E)

Fairy bonnets
Coprinus disseminatus (P)

Fairy cake fungus
Hebeloma crustuliniforme (F)

Fairy ring mushroom
Marasmius oreades (E)

Fairy sunshade   *Lepiota procera* (E)
Macrolepiota procera (E)

False blusher
Amanita pantherina (P)

False chanterelle
Hygrophoropsis aurantiaca (I)

False death cap   *Amanita citrina* (I)
Amanita mappa (I)

False morel
Gyromitra esculenta (P)

Fawn agaric   *Pluteus atricapillus* (E)
Pluteus cervinus (E)

Field mushroom
Agaricus campestris (E)
Psalliota campestris (E)

Firwood agaric
Tricholoma equestre (E)
Tricholoma flavovirens (E)

Fleecy milk cap
Lactarius vellereus (I)

Fly agaric   *Amanita muscaria* (P)

Fool's mushroom
Amanita verna (F)

Foxy spot   *Collybia maculata* (I)

Fragile russula   *Russula fragilis* (P)

French truffle
Tuber melanosporum (E)

Gas tar fungus
Trichloma sulphureum (P)

Giant club
Clavariadelphus pistillaris (E)

Giant giromitra
Neogyromitra gigas (E)

Giant polypore
Meripilus giganteus (I)

Giant puffball   *Calvatia gigantea* (E)
Langermannia gigantea (E)
Lycoperdon maximum (E)

Gipsy mushroom
Pholiota caperata (E)
Rozites caperata (E)

Glistening inkcap
Coprinus micaceus (I)

Goat's lip mushroom
Boletus subtomentosus (E)
Xerocomus subtomentosus (E)

Goaty-smell cortinarius
Cortinarius traganus (I)
Golden russula *Russula aurata* (E)
Grass-green russula
Russula aeruginea (E)
Green agaric
Russula cyanoxantha (E)
Russula virescens (E)
Green cracking russula
Russula virescens (E)
Green earth tongue
Microglossum viride (I)
Greenstain fungus
Chlorociboria aeruginacens (I)
Chlorosplenium aeruginosum (I)
Green wood-cup
Chlorociboria aeruginacens (I)
Chlorospenium aeruginacens (I)
Grey agaric
Tricholoma terreum (E)
Grey coral fungus
Clavulina cinerea (P)
Grey inkcap
Coprinus atramentarius (E)
Hairy stereum *Sterum hirsutum* (I)
Hare's ear fungus *Otidea onotica* (I)
Hedgehog fungus
Hericium erinaceum (E)
Hydnum repandum (E)
Lycoperdon echinatum (E)
Hen-of-the-woods
Grifola frondosa (E)
Polyporus frondosus (E)
Honey fungus *Armillaria mellea* (E)
Clitocybe mellea (E)
Hoof fungus *Fomes fomentarius* (I)
Hornbeam boletus
Leccinum carpini (E)
Leccinum griseum (E)
Horn of plenty
Cantharellus cornucopioides (E)
Craterellus cornucopioides (E)
Horse agaric *Agaricus arvensis* (E)
Psalliota arvensis (E)
Horse-dung fungus
Coprinus niveus (I)
Horsehair fungus
Marasmius androsaceus (I)
Horse mushroom
Agaricus arvensis (E)
Psalliota arvensis (E)
Indian paint fungus
Echinodontium tinctorium (I)
Indigo boletus
Boletus cyanescens (E)
Gyroporus cyanescens (E)
Inkcap *Coprinus atramentarius* (E)
Coprinus comatus (E)
Ivory wax-cap
Hygrophorus eburneus (I)
Japanese sunshade
Coprinus plicatilis (P)
Jellybaby fungus *Leotia lubrica* (E)

Jellybean fungus *Leotia lubrica* (E)
Jellybrain fungus
Tremella mesenterica (I)
Jello tongue
Pseudohydnum gelatinosum (P)
Jelly tongue
Pseudohydnum gelatinosum (P)
Jersey cow boletus
Suillus bovinus (E)
Jew's ear
Hirneola auricula-judae (P)
Judas's ear
Auricularia auricula-judae (E)
King Alfred's balls
Daldinia concentrica (I)
King Alfred's cakes
Daldinia concentrica (I)
King Alfred's cramp balls
Daldinia concentrica (I)
Knotted fungus
*Gymnosporangium
clavariaeforme* (I)
Larch boletus *Suillus grevillei* (E)
Lawyer's wig *Coprinus comatus* (E)
Leathery russula
Russula alutacea (E)
Liberty cap
Psilocybe semilanceata (P)
Little Japanese umbrella
Coprinus plicatilis (P)
Little nail fungus *Mycena ribula* (P)
Livid entoloma
Entoloma lividum (P)
Entoloma sinuatum (P)
Long root mushroom
Oudemansiella radicata (E)
Lurid boletus *Boletus luridus* (E)
Magpie fungus
Coprinus picaceus (P)
Man-on-horseback
Tricholoma equestre (E)
Tricholoma flavovirens (E)
March mushroom
Hygrophorus marzuolus (E)
Maze gill *Daedalea quercina* (P)
Meadow wax cap
Hygrocybe pratensis (I)
Mica inkcap *Coprinus micaceus* (E)
Milk agaric *Lactarius deliciosus* (E)
Milk cap *Lactarius deliciosus* (E)
Milky conocybe *Conocybe lactea* (I)
Miller mushroom
Clitopilus prunulus (E)
Moor club *Clavaria argillacea* (I)
Morel *Morchella elata* (E)
Mortician cap
Amanita phalloides (F)
Multi-branched fungus
Ramaria formosa (P)

Shaggy pholiota

Multi-zoned bracket fungus
Coricolus versicolor (P)
Polyporus versicolor (P)
Trametes versicolor (P)

Multi-zoned polypore
Coricolus versicolor (P)
Polyporus versicolor (P)
Trametes versicolor (P)

Oak milk cap *Lactarius quietus* (I)

Oaktree boletus
Leccinum quercinum (E)

Oaktree collybia
Collybia dryophila (E)

Olive boletus *Boletus calopus* (I)

Olive-green russula
Russula olivacea (E)

Orange birch boletus
Leccinum testaceoscabrum (E)
Leccinum versipelle (E)

Orange cap boletus
Leccinum aurantiacum (E)

Orange-peel fungus
Aleuria aurantia (E)
Peziza aurantia (E)

Oyster mushroom
Pleurotus ostreatus (E)

Padi-straw fungus
Volvaria volvacea (E)
Volvariella volvacea (E)

Panther *Amanita pantherina* (P)

Panther cap
Amanita pantherina (P)

Parasol mushroom
Lepiota procera (E)
Macrolepiota procera (E)

Parrot fungus
Hygrocybe psittacina (E)
Hygrophorus psittacinus (E)

Parrot wax cap
Hygrocybe psittacina (E)

Pavement mushroom
Agaricus bitorquis (E)

Penny bun fungus *Boletus edulis* (E)

Perigord truffle
Tuber melanosporum (E)

Piedmont truffle
Tuber magnatum (E)

Pine fire fungus
Rhyzina undulata (P)

Pine forest mushroom
Gomphidius rutilis (E)

Pink bottom mushroom
Agaricus campestris (E)
Psalliota campestris (E)

Pixie cup *Sarcoscypha coccinea* (E)

Pixie's cap fungus *Mycena vitilis* (I)

Plum agaric *Clitopilus prunulus* (E)

Plums and custard
Tricholomopsis rutilans (E)

Poached egg fungus
Oudemansiella mucida (I)

Poison pie
Hebeloma crustuliniforme (F)

Poor man's sweetbread
Lycoperdon gemmatum (E)
Lycoperdon perlatum (E)

Purple-black russula
Russula undulata (P)

Purple boletus
Boletus rhodoxanthus (P)

Rainfall boletus *Suillus bovinus* (E)

Razor-strop fungus
Piptoporus betulinus (E)
Polyporus betulinus (E)

Red cracked boletus
Boletus chrysenteron (I)

Red-leg boletus
Boletus erythropus (E)

Red-staining inocybe
Inocybe patouillardii (F)

Ringed boletus *Suillus luteus* (E)

Roll-rim fungus
Paxillus involutus (P)

Root fomes
Heterobasidion annosum (I)

Rose-gilled grisette
Volvaria speciosa (E)
Volvariella speciosa (E)

Rose mildew
Sphaerotheca pannosa (I)

Royal amanita *Amanita regalis* (P)

Royal boletus *Boletus regius* (E)

Rusty agaric
Kuehneromyces mutabilis (E)

Rusty milk cap *Lactarius rufus* (I)

Rusty oak fungus
Fistulina hepatica (E)

Saffron milk cap
Lactarius deliciosus (E)

Saint George's mushroom
Calocybe gambosa (E)
Tricholoma gambosa (E)
Tricholoma georgii (E)

Satan's mushroom
Boletus satanas (P)

Scaly hydnum
Hydnum imbricatum (E)
Sarcodon imbricatum (E)

Scaly polypore
Polyporus squamosus (E)

Scarlet caterpillar fungus
Cordyceps militaris (P)

Scarlet cup
Sarcoscypha coccinea (E)

Scarlet elf cap
Sarcoscypha coccinea (E)

Shaggy cap *Coprinus comatus* (E)

Shaggy inkcap
Coprinus comatus (E)

Shaggy milk cap
Lactarius torminosus (I)

Shaggy parasol *Lepiota rhacodes* (E)
Macrolepiota rhacodes (E)

Shaggy pholiota
Pholiota squarrosa (I)

Sheathed agaric

Sheathed agaric   *Amanita fulva* (E)

Shiitake fungus
     Lentinula edodes (E)

Shining inkcap
     Coprinus micaceus (I)

Showy mushroom
     Boleta speciosus (E)

Sickener mushroom
     Russula emetica (P)

Silverleaf fungus
Chondrostereum purpureum (I)

Skullcap inocybe
     Inocybe napipes (P)

Slippery jack   *Suillus luteus* (E)

Smokey-gilled woodlover
     Hypholoma capnoides (I)

Snowy wax cap
     Hygrocybe virginea (E)

Spindle shank   *Collybia fusipes* (E)

Splash cup   *Cyathus striatus* (I)

Split gill fungus
     Schizophyllum commune (P)

Spotted tough-shank
     Collybia maculata (I)

Spring amanita   *Amanita verna* (F)

Spurred rye
     Claviceps purpurea (P)

Stinging-nettle fungus
     Calyptella capula (I)

Stinkhorn   *Phallus impudicus* (E)

Sulfur polypore
     Laetiporus sulphureus (E)

Sulfur tuft
     Hypholoma fasciculare (P)
     Namatoloma fasciculare (P)

Sulphur polypore
     Laetiporus sulphureus (E)

Sulphur tuft
     Hypholoma fasciculare (P)
     Namatoloma fasciculare (P)

Summer boletus
     Boletus aestivalis (E)
     Boletus reticulatus (E)

Summer truffle   *Tuber aestivum* (E)

Sweetbread mushroom
     Clitopilus prunulus (E)

Sycamore tarspot
     Rhytisma acerinum (I)

Tasselated toadstool
     Lepiota castanea (P)

Tawny grisette   *Amanita fulva* (E)

The blusher
     Amanita rubescens (E)

The deceiver   *Laccaria laccata* (E)

Tiger tricholoma
     Tricholoma pardalotum (P)
     Tricholoma pardinum (P)

Tinder fungus
     Fomes fomentarius (I)

Tripe fungus
     Auricularia mesenterica (I)

Trooping crumble cap
     Coprinus disseminatus (P)
     Pseudocoprinus disseminatus (P)

Truffle   *Tuber aestivum* (E)

Trumpet agaric
     Clitocybe geotropa (E)

Trumpet of death
     Craterellus cornucopioides (E)

Turban fungus
     Gyromitra infula (E)
     Hevella infula (E)

Ugly mushroom
     Lactarius necator (E)
     Lactarius plumbeus (I)
     Lactarius turpis (I)

Urban mushroom
     Agaricus bitorquis (E)

Variegated boletus
     Suillus variegatus (E)

Velvet cap   *Lacrymaria velutina* (I)

Velvet shank
     Collybia velutipes (E)
     Flammulina velutipes (E)

Verdigris fungus
     Stropharia aeruginosa (E)

Weeping widow
     Lacrymaria lacrymabunda (I)
     Lacrymaria velutina (I)

White birch boletus
     Leccinum holopus (E)

White poplar mushroom
     Leccinum duriusculum (E)

White truffle
     Tuber magnatum (E)

Winter mushroom
     Collybia velutipes (E)
     Flammulina velutipes (E)

Witch's broom
     Taphrina betulina (P)

Witch's butter
     Exidia glandulosa (I)

Wood agaric
     Collybia dryophila (E)

Wood blewits
     Lapista nuda (E)
     Tricholoma nudum (E)

Wood hedgehog
     Dentinum repandum (E)
     Hydnum repandum (E)

Woodland black spot
     Gomphidius glutinosus (E)

Woodland pink mushroom
     Amanita rubescens (E)

Wood mushroom
     Agaricus silvicola (E)
     Psalliota silvicola (E)

Wood woollyfoot
     Collybia peronata (I)

Woolly milk cap
     Lactarius torminosus (I)

Wrinkled club
     Clavulina rugosa (P)

Yellow agaric
> *Tricholoma sulphureum* (P)

Yellow brain fungus
> *Tremella mesenterica* (I)

Yellow russula
> *Russula ochroleuca* (I)

Yellow stainer
> *Agaricus xanthodermus* (P)

Yellow swamp russula
> *Russula claroflava* (E)
> *Russula flava* (E)

GRASSES, REEDS, SEDGES, BAMBOOS, VETCHES, ETC.

True grasses are members of the family *Gramineae* and are found in almost every corner of the planet in one form or another, but in this section we are concerned with the ornamentals, meadow grasses, timber grasses (bamboos), soil-holding or sandbinding grasses, as well as wetland subjects such as sedges and reeds. Some grasses may also be listed under 'Aquatics'.

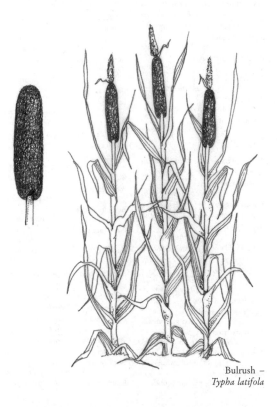

Bulrush –
Typha latifola

Abyssinian feathertop

Abyssinian feathertop
    Pennisetum villosum
Adrue   *Cyperus articulatus*
African Bermuda grass
    Cynodon transvaalensis
African fountain grass
    Pennisetum setaceum
African millet   *Eleusine coracana*
    Pennisetum americanum
Alaska wheat   *Triticum turgidum*
Aleppo grass   *Sorghum halepense*
Alfalfa   *Medicago sativa*
Alpine cat's tail   *Phleum alpinum*
    Phleum commutatum
Alpine clover   *Trifolium alpinum*
Alpine foxtail   *Alopecurus alpinus*
Alpine hair grass
    Deschampsia alpina
Alpine meadow grass   *Poa alpina*
Alpine rush
    Juncus alpinoarticulatus
Alpine sedge   *Carex norvegica*
Alpine timothy   *Phleum alpinum*
    Phleum commutatum
Alsike clover   *Trifolium hybridum*
Alta fescue   *Festuca elatior*
Altai wild rye   *Elymus angustus*
American beach grass
    Ammophila breviligulata
American cord grass
    Spartina alterniflora
American wild oats
    Chasmanthium latifolium
Amur silver grass
    Miscanthus sacchariflorus
Angleton bluestem
    Dichanthium aristatum
Angleton grass
    Dichantium aristatum
Animated oat grass   *Avena sterilis*
Animated oats   *Avena sterilis*
Annual beard grass
    Polypogon monspeliensis
Annual bluegrass   *Poa annua*
Annual meadow-grass   *Poa annua*
Annual vernal grass
    Anthroxanthum puelii
Annual wild rice   *Zizania aquatica*
Aral wild rye   *Elymus aralensis*
Arrow bamboo
    Arundinaria japonica
    Pseudosasa japonica
Arrow cane   *Gynerium sagittatum*
Australian bluestem
    Bothriochloa intermedia
Australian danthonia
    Danthonia semiannularis
Australian feather grass
    Stipa elegantissima
Australian oat grass
    Danthonia semiannularis

Australian rye grass
    Lolium multiflorum
Australian windmill grass
    Chloris ventricosa
Autumn bent   *Agrostis perennans*
Awnless brome   *Bromus inermis*
Awnless sheep's fescue
    Festuca tenuifolia
Bahia grass   *Paspalum notatum*
Balfour's meadowgrass
    Poa balfouri
Balkan bluegrass
    Sesleria heufleriana
Baltic rush   *Juncus ballicus*
Bambusa   *Bambusa beecheyana*
Barley   *Hordeum vulgare*
Barn grass   *Echinochloa crus-galli*
Barnyard grass   *Echinochloa crus-galli*
Barnyard millet   *Echinochloa crus-galli*
Barren brome   *Bromus sterilis*
Basket grass   *Oplismenus hirtellus*
Beach pea   *Lathyrus japonicus*
    Lathyrus littoralis
Beaked sedge   *Carex rostrata*
Bearded couch
    Agropyron caninum
    Elymus caninus
Bearded fescue   *Vulpia ambigua*
Bearded twitch
    Agropyron caninum
Bear grass   *Xerophyllum tenax*
Beechey bamboo
    Bambusa beecheyana
Bengal grass   *Setaria italica*
Bermuda grass   *Cynodon dactylon*
Big bluegrass   *Poa ampla*
Big bluestem grass
    Andropogon gerardii
Big galleta   *Hilaria rigida*
Birdseed grass
    Phalaris canariensis
Bird's foot sedge
    Carex ornithopoda
Bird's foot trefoil
    Lotus corniculatus
Bird vetch   *Vicia cracca*
Bitter vetch   *Vicia ervilia*
Black bamboo   *Phyllostachys nigra*
Black bent   *Agrostis gigantea*
Black bog rush   *Schoenus nigricans*
Black grama   *Bouteloua eriopoda*
Black grass
    Alopecurus myosuroides
Black medick   *Medicago lupulina*
Black sedge   *Carex atrata*
Black twitch
    Alopecurus myosuroides
Bladder sedge   *Carex vesicaria*

Carrizo

Blue bent	*Molinia caerulea*
Bluebunch wheatgrass	*Agropyron spicatum*
Blue couch grass	*Digitaria didactyla*
Blue fescue	*Festuca ovina glauca*
	Festuca glauca
Blue finger grass	*Digitaria didactyla*
Blue grama	*Bouteloua gracilis*
Blue love grass	*Eragrostis chloromelas*
Blue moor grass	*Sesleria caerulea*
	Sesleria albicans
Blue panic grass	*Panicum antidotale*
Blue stem	*Schizachyrium scoparium*
Blue wild rye	*Elymus glaucous*
Blunt-flowered rush	*Juncus subnodulosus*
Boer love grass	*Eragrostis chloromelas*
Bog hair grass	*Deschampsia setacea*
Bog sedge	*Carex limosa*
Borrer's saltmarsh grass	*Puccinellia fasciculata*
Bottle sedge	*Carex rostrata*
Bowles' golden grass	*Milium effusum aureum*
Brahman grass	*Dichanthium annulatum*
Branched bur reed	*Sparganium erectum*
Branched cup grass	*Eriochloa aristata*
Bread wheat	*Triticum aestivum*
Bristle bent	*Agrostis curtisii*
Bristle club rush	*Scirpus setaceus*
Bristle grass	*Setaria viridis*
Bristle oat	*Avena strigosa*
Bristle scirpus	*Scirpus setaceus*
Bristly bent	*Agrostis setacea*
Broad bean	*Vicia faba*
Broad-leaved cotton grass	*Eriophorum latifolium*
Broad-leaved meadow grass	*Poa chaixii*
Broad-leaved mud sedge	*Carex paupercula*
Broad red clover	*Trifolium pratense praecox*
Broom	*Schizachyrium scoparium*
Broom beard grass	*Schizachyrium scoparium*
Broomcorn	*Panicum miliaceum*
	Sorghum vulgare
Broomcorn millet	*Panicum miliaceum*
Brown beak sedge	*Rhynchospora fusca*
Brown bent	*Agrostis canina*
Brown bent grass	*Agrostis perennans*
Brown bog rush	*Schoenus ferrugineus*
Brown cyperus	*Cyperus fuscus*
Brown durra	*Sorghum vulgare durra*
Brown sedge	*Carex disticha*
Brown top	*Agrostis tenuis*
Browntop millet	*Panicum ramosum*
Buddham bamboo	*Bambusa ventricosa*
Buddha's belly bamboo	*Bambusa ventricosa*
Buffalo grass	*Buchloe dactyloides*
	Stenotaphrum secundatum
Buffel grass	*Pennisetum ciliare*
Bulbous barley	*Hordeum bulbosum*
Bulbous bluegrass	*Poa bulbosa*
Bulbous foxtail	*Alopecurus bulbosus*
Bulbous meadow grass	*Poa bulbosa*
Bulbous panic grass	*Panicum bulbosum*
Bulbous rush	*Juncus bulbosus*
Bulrush	*Typha latifolia*
Bulrush (biblical)	*Cyperus papyrus*
Bunch grass	*Schizachyrium scoparium*
Bur clover	*Medicago hispida*
Burma reed	*Neyraudia reynaudiana*
Bush grass	*Calamagrostis epigejos*
Caithness sedge	*Carex recta*
Calcutta bamboo	*Dendrocalamus strictus*
Caley pea	*Lathyrus hirsutus*
Californian brome	*Bromus carinatus*
Cana brava	*Arundo donax*
Canada pea	*Vicia cracca*
Canadian bluegrass	*Poa compressa*
Canadian wild rye	*Elymus canadensis*
Canary grass	*Phalaris canariensis*
Canebrake bamboo	*Arundinaria gigantea*
Cane reed	*Arundinaria gigantea*
Capitate rush	*Juncus capitatus*
Carib grass	*Eriochloa polystachya*
Carnation grass	*Carex panicea*
Carnation sedge	*Carex panicea*
Carpet grass	*Axonopus affinis*
Carrizo	*Arundo donax*
	Phragmites australis

Cat grass

Cat grass	*Leersia oryzoides*
Cat's tail grass	*Phleum pratense*
Caucasian bluestem	*Bothriochloa caucasica*
Centipede grass	*Eremochloa ophiuroides*
Chalk false brome	*Brachypodium pinnatum*
Chalk sedge	*Carex polyphylla*
Chamois grass	*Hutchinsia alpina*
Chee grass	*Stipa splendens*
Chestnut rush	*Juncus castaneus*
Chewing's fescue	*Festuca rubra*
China grass	*Boehmeria nivea*
Chinese fountain grass	*Pennisetum alopecuroides*
Chinese mat grass	*Cyperus tagetiformis*
Chinese pennisetum	*Pennisetum alopecuroides*
Chinese silver grass	*Miscanthus sinensis*
Chinese sugar maple	*Sorghum saccharatum*
Chinese sweet cane	*Saccarum sinense*
Chinese water chestnut	*Eleocharis dulcis*
	Eleocharis tuberosa
Chinese wild rye	*Elymus chinensis*
Citronella grass	*Cymbopogon nardus*
Chufa	*Cyperus esculentus*
Cloud bent	*Agrostis nebulosa*
Cloud grass	*Agrostis nebulosa*
Club rush	*Scirpus lacustris*
Club sedge	*Carex buxbaumii*
Club wheat	*Triticum compactum*
Cocksfoot	*Dactylis glomerata*
Cockspur grass	*Echinochloa crus-galli*
Colonial bent	*Agrostis tenuis*
Colorado grass	*Panicum texanum*
Common bamboo	*Bambusa vulgaris*
Common barley	*Hordeum vulgare*
Common bent	*Agrostis tenuis*
Common bird's foot	*Ornithopus perpusillus*
Common bulrush	*Scirpus lacustris*
Common carpet grass	*Axonopus affinis*
Common cat-tail	*Typha latifolia*
Common cord grass	*Spartina anglica*
Common cotton grass	*Eriophorum angustifolium*
Common couch	*Elymus repens*

Common feather grass	*Stipa pennata*
Common foxtail	*Alopecurus pratensis*
Common horsetail	*Equisetum hyemale*
Common meadow grass	*Poa pratensis*
Common millet	*Panicum miliaceum*
Common quaking grass	*Briza media*
Common reed	*Phragmites australis*
	Phragmites communis
Common rush	*Juncus subuliflorus*
	Juncus communis
Common rye	*Secale cereale*
Common saltmarsh grass	*Puccinellia maritima*
Common scouring brush	*Equisetum hyemale*
Common sedge	*Carex nigra*
Common spike rush	*Eleocharis palustris*
Common timothy	*Phleum pratense*
Common vetch	*Vicia angustifolia*
	Vicia sativa
Common wheat	*Triticum aestivum*
Common wild oat	*Avena fatua*
Common wood rush	*Luzula campestris*
Compact brome	*Bromus madritensis*
Cordgrass	*Spartina pectinata*
Corn	*Zea mays*
Corn silk	*Zea mays*
Cossack asparagus	*Typha latifolia*
Couch grass	*Agropyron repens*
	Elymus repens
Cow hop clover	*Trifolium procumbens*
Cow vetch	*Vicia cracca*
Crab grass	*Digitaria sanguinalis*
Creeping bent	*Agrostis stolonifera*
Creeping brown sedge	*Carex disticha*
Creeping dog's tooth grass	*Cynodon dactylon*
Creeping fescue	*Festuca rubra*
Creeping finger grass	*Cynodon dactylon*
	Digitaria serotina
Creeping foxtail	*Alopecurus arundinaceus*
Creeping signal grass	*Brachiaria subquadripara*
Creeping soft grass	*Holcus mollis*

European feather grass

Creeping twitch grass
Agropyron repens
Elymus repens
Creeping windmill grass
Chloris truncata
Crested dogstail
Cynosurus cristatus
Crested hair grass *Koeleria cristata*
Koeleria gracilis
Koeleria macrantha
Crested wheatgrass
Agropyron cristatum
Crimson clover
Trifolium incarnatum
Crinkled hair grass
Deschampsia flexuosa
Culeu *Chusquea culeou*
Curly mesquite *Hilaria belangeri*
Curved sea hard grass
Parapholis incurva
Curved sedge *Carex maritima*
Curved wood rush *Luzula arcuata*
Cuscus *Pennisetum glaucum*
Cut grass *Leersia oryzoides*
Cyperus sedge
Carex pseudocyperus
Dallis grass *Paspalum dilatatum*
Dark sedge *Carex buxbaumii*
Darnel *Lolium temulentum*
Darnel sedge *Carex loliacea*
Deer grass *Scirpus cespitosus*
Dense silky bent *Apera interrupta*
Desert wheatgrass
Agropyron sibiricum
Diaz bluestem
Dichanthium annulatum
Dioecious sedge *Carex dioica*
Distant-flowered sedge
Carex remota
Distant sedge *Carex distans*
Doddering dillies *Briza media*
Dog grass *Agropyron caninum*
Dog hair grass
Deschampsia setacea
Dogstooth grass
Dichanthium ischaemum
Don's twitch *Agropyron donianum*
Doob grass *Cynodon dactylon*
Dotted sedge *Carex punctata*
Double-grain spelt
Triticum dicoccon
Double-grain wheat
Triticum dicoccon
Double-stemmed sedge
Carex diandra
Downy-fruited sedge
Carex filiformis
Carex lasiocarpa
Downy oat grass
Helictotrichon pubescens
Drake *Avena fatua*

Drooping brome *Bromus tectorum*
Drooping love grass
Eragrostis curvula
Drooping sedge *Carex pendula*
Dune fescue *Vulpia membranacea*
Vulpia fasciculata
Dune grass *Elymus arenarius*
Durra *Sorghum vulgare durra*
Durum wheat *Triticum durum*
Dutch mice *Lathyrus tuberosus*
Dutch rush *Equisetum hyemale*
Dutch white clover
Trifolium repens
Dwarf bamboo
Arundinaria pumila
Dwarf fern leaf bamboo
Arundinaria disticha
Dwarf meadow grass *Poa annua*
Dwarf papyrus *Cyperus isocladus*
Dwarf rush *Juncus capitatus*
Dwarf sedge *Carex humilis*
Dwarf spike-rush
Eleocharis parvula
Dwarf wheat *Triticum compactum*
Dwarf white-stripe bamboo
Arundinaria variegata
Early hair grass *Aira praecox*
Early meadow grass *Poa infirma*
Early sand grass *Mibora minima*
Earth nut pea *Lathyrus tuberosus*
Earth almond *Cyperus esculentus*
Eel grass *Vallisneria spiralis*
Egyptian millet *Sorghum halepense*
Egyptian paper rush
Cyperus papyrus
Einkorn *Triticum monococcum*
Elephant grass
Pennisetum purpureum
Typha elephanta
Elk grass *Xerophyllum tenax*
Elongated sedge *Carex elongata*
Emmer *Triticum dicoccon*
English bean *Vicia faba*
English bluegrass *Festuca pratensis*
English rye grass *Lolium perenne*
English wheat *Triticum turgidum*
Esparcet *Onobrychis viciifolia*
Erect brome *Bromus erectus*
Ervil *Vicia ervilia*
Esparto grass *Stipa tenacissima*
Eulalia *Miscanthus sinensis*
European beach grass
Ammophila arenaria
European bean *Vicia faba*
European dune grass
Elymus arenarius
European feather grass
Stipa pennata

Everlasting pea

Everlasting pea
Lathyrus grandiflorus
Lathyrus latifolius
Lathyrus sylvestris

Eyelash pearl grass   Melica ciliata

Fairy crested wheatgrass
Agropyron cristatum

False brome
Brachypodium sylvaticum

False fox sedge   Carex otrubae

False oat grass
Arrhenatherum elatius

False sedge   Kobresia simpliciuscula

False wheatgrass   Elymus chinensis

Fanwort   Cabomba caroliniana

Feather bunchgrass   Stipa viridula

Feather grass   Stipa pennata

Feather love grass
Eragrostis amabilis

Feathertop   Pennisetum villosum

Fenland sedge   Cladium mariscus

Fenland wood rush
Luzula pallescens

Fern grass   Catapodium rigidum
Desmazeria rigida

Feterita
Sorghum vulgare caudatum

Fevergrass   Cymbopogon citratus

Few-flowered sedge
Carex pauciflora
Carex rariflora

Field bean   Vicia faba

Field brome   Bromus arvensis

Field wood rush   Luzula campestris

Fine bent   Agrostis tenuis

Fine-leaved sheep's fescue
Festuca tenuifolia

Fingered sedge   Carex digitata

Finger millet   Eleusine coracana

Fire lily   Xerophyllum tenax

Fish grass   Cabomba caroliniana

Fishpole bamboo
Phyllostachys aurea

Flat pea   Lathyrus sylvestris

Flat sedge   Blysmus compressus

Flat-stalked meadow grass
Poa compressa

Flaver   Avena fatua

Flawn   Zoysia matrella

Flea sedge   Carex puticaris

Floating bur reed
Sparganium angustifolium

Floating foxtail
Alopecurus geniculatus

Floating mud rush   Scirpus fluitans

Floating sweet grass
Glyceria fluitans

Flowering rush
Butomus umbellatus

Flying bent   Molinia caerulea

Forage bamboo
Phyllostachys aureosulcata

Forster's wood rush
Luzula forsteri

Fountain grass
Pennisetum setaceum

Fox sedge   Carex vulpina

Foxtail barley   Hordeum jubatum

Foxtail bristle grass   Setaria italica

Foxtail chess   Bromus rubens

Foxtail grass   Setaria italica

Foxtail millet   Setaria italica

Freshwater cordgrass
Spartina pectinata

Fringed brome   Bromus canadensis

Galingale   Cyperus longus

Galleta   Hilaria jamesii

Gardener's garters
Phalaris arundinacea picta

German wheat   Triticum dicoccon

Giant bamboo
Dendrocalamus gigantea

Giant cane   Arundinaria gigantea

Giant fescue   Festuca gigantea

Giant finger grass   Chloris berroi

Giant panic grass
Panicum antidotale

Giant reed   Arundo donax

Giant rye   Triticum polonicum

Giant thorny bamboo
Bambusa arundinacea

Giant timber bamboo
Phyllostachys bambusoides

Giant wild rye   Elymus condensatus

Glaucous bristle grass
Setaria glauca

Glaucous club rush
Scirpus tabernaemontani

Glaucous meadow grass   Poa glauca

Glaucous sedge   Carex flacca

Glaucous sweet grass
Glyceria declinata

Golden bamboo
Phyllostachys aurea

Golden feathergrass
Stipa pulcherrima

Golden foxtail
Alopecurus pratensis 'aureus'

Golden-groove bamboo
Phyllostachys aureosulcata

Golden oat grass
Trisetum flavescens

Golden oats   Stipa gigantea

Goldentop   Lamarckia aurea

Goose grass   Eleusine indica

Grass sorghum   Sorghum halepense

Grassy rush   Butomus umbellatus

Great brome   Bromus diandrus

Greater tussock sedge
Carex paniculata

Japanese millet

Great fenland sedge	*Cladium mariscus*
Great millet	*Sorghum halepense*
Great pond sedge	*Carex riparia*
Great tussock sedge	*Carex paniculata*
Great wood rush	*Luzula sylvatica*
Green bristle grass	*Setaria viridis*
Green needlegrass	*Stipa viridula*
Green-ribbed sedge	*Carex binervis*
Grey fescue	*Festuca glauca*
Grey hair grass	*Corynephorus canescens*
Grey sedge	*Carex divulsa*
Guinea grass	*Panicum maximum*
Guinea rush	*Cyperus articulatus*
Hair fescue	*Festuca tenuifolia*
Hair grass	*Eleocharis acicularis*
Hair sedge	*Carex capillaris*
Hairy brome	*Bromus ramosus*
Hairy crab grass	*Digitaria sanguinalis*
Hairy cup grass	*Eriochloa villosa*
Hairy finger grass	*Digitaria sanguinalis*
Hairy grama	*Bouteloua hirsuta*
Hairy oat grass	*Helictotrichon pubescens*
Hairy sedge	*Carex hirta*
Hairy vetch	*Vicia villosa*
Hairy wood rush	*Luzula pilosa*
Hard fescue	*Festuca longifolia*
Hard grass	*Parapholis strigosa*
Hard rush	*Juncus inflexus*
Hard wheat	*Triticum durum*
Hardy bamboo	*Pseudosasa japonica*
Hardy timber bamboo	*Phyllostachys bambusoides*
Hare's foot sedge	*Carex lachenalii*
Hare's tail	*Lagurus ovatus*
Hare's tail cotton grass	*Eriophorum vaginatum*
Hare's tail grass	*Eriophorum vaginatum*
	Lagurus ovatus
Heath false brome grass	*Brachypodium pinnatum*
Heath grass	*Danthonia decumbens*
	Sieglingia decumbens
Heath pea	*Lathyrus japonicus*
Heath rush	*Juncus squarrosus*
Heath sedge	*Carex ericetorum*
Heath wood rush	*Luzula multiflora*
Hedge bamboo	*Bambusa glaucesens*
	Bambusa multiplex
Hedgehog wheat	*Triticum compactum*
Hegari	*Sorghum vulgare caffrorum*
Henon bamboo	*Phyllostachys nigra henon*
Herd's grass	*Phleum pratense*
Himalaya fairy grass	*Miscanthus nepalensis*
Himalayan bamboo	*Arundinaria anceps*
Hog millet	*Panicum miliaceum*
Holy clover	*Onobrychis viciifolia*
Holy grass	*Hierochloe odorata*
Hop clover	*Medicago lupulina*
	Trifolium agrarium
Horse bean	*Vicia faba*
Horsetail	*Equisetum hymale*
Hunangemoho	*Chionochloa conspicua*
Hungarian brome	*Bromus inermis*
Hungarian clover	*Trifolium pannonicum*
Hungarian grass	*Setaria italica*
Hybrid fescue	*Festulolium loliaceum*
Hybrid marram grass	*Ammocalamagrostis baltica*
Hybrid sweet grass	*Glyceria pedicellata*
Indian basket grass	*Xerophyllum tenax*
Indian corn	*Zea mays*
Indian grass	*Molinia caerulea*
	Sorghastrum avenaceum
Indian millet	*Oryzopsis hymenoides*
	Pennisetum americanum
Indian rice	*Oryzopsis hymenoides*
	Zizania aquatica
Intermediate wheatgrass	*Agropyron intermedium*
Interrupted brome	*Bromus interruptus*
Irish shamrock	*Trifolium procumbens*
Italian clover	*Trifolium incarnatum*
Italian millet	*Setaria italica*
Italian rye grass	*Lolium multiflorum*
Japanese brome	*Bromus japonicus*
Japanese carpet grass	*Zoysia matrella*
Japanese chess	*Bromus japonicus*
Japanese lawn grass	*Zoysia japonica*
Japanese love grass	*Eragrostis amabilis*
Japanese mat rush	*Juncus effusus*
Japanese millet	*Echinochloa crus-galli*
	Setaria italica

Japanese timber bamboo	
Phyllostachys bambusoides	
Jersey club rush	*Scirpus americanus*
Job's tears	*Coix lacryma-jobi*
Johnson grass	*Sorghum album*
	Sorghum halepense
Jointed rush	*Juncus articulatus*
June grass	*Poa pratensis*
Kaffir corn	
Sorghum vulgare caffrorum	
Kentucky bluegrass	*Poa pratensis*
Khas-khas	*Vetiveria zizanioides*
Khus-khus	*Vetiveria zizanioides*
Kidney vetch	*Anthyllis vulneraria*
Kleberg grass	
Dichanthium annulatum	
Korakan	*Eleusine coracana*
Korean grass	*Zoysia japonica*
	Zoysia tenuifolia
Korean lawn grass	*Zoysia japonica*
	Zoysia tenuifolia
Korean velvet grass	*Zoysia japonica*
	Zoysia tenuifolia
Kuma bamboo grass	*Sasa veitchii*
Kura clover	*Trifolium ambiguum*
Kweek grass	*Cynodon dactylon*
Lace grass	*Eragrostis capillaris*
Lady's fingers	*Anthyllis vulneraria*
Large hop clover	
Trifolium campestre	
Large quaking grass	*Briza maxima*
Large Russian vetch	*Vicia villosa*
Late-flowering red clover	
Trifolium pratense serotinum	
Late-flowering sedge	
Carex serotina	
Lazy-man's grass	
Eremochloa ophiuroides	
Least spike rush	
Eleocharis acicularis	
Lehmann's love grass	
Eragrostis lehmanniana	
Lemongrass	*Cymbopogon citratus*
Lesser burnet	
Poterium sanguisorba	
Lesser cat's tail	*Phleum bertolonii*
Lesser pond sedge	
Carex acutiformis	
Lesser quaking grass	*Briza minor*
Lesser tussock sedge	
Carex appropinquata	
Carex diandra	
Little bluestem	
Schizachyrium scoparium	
Little quaking grass	*Briza minor*
Long-bracted sedge	*Carex extensa*
Loose silky bent	
Apera spica-venti	
Loose-spiked wood sedge	
Carex strigosa	
Love grass	*Eragrostis elegans*
Low clover	*Trifolium campestre*
Low sedge	*Carex demissa*
Low spear grass	*Poa annua*
Lucerne	*Medicago sativa*
Lyme grass	*Elymus arenarius*
	Lolium perenne
Mace sedge	*Carex grayi*
Madake	
Phyllostachys bambusoides	
Madrid brome	*Bromus madritensis*
Maize	*Zea mays*
Malacca cane	*Calamus scipionum*
Male bamboo	
Dendrocalamus strictus	
Malogilla	*Eriochloa polystachya*
Malojilla	*Eriochloa polystachya*
Malojillo	*Eriochloa polystachya*
Manila grass	*Zoysia matrella*
Many-flowered wood rush	
Luzula multiflora	
Many-stemmed spike rush	
Eleocharis multicaulis	
Marram grass	
Ammophila arenaria	
Marsh foxtail	
Alopecurus geniculatus	
Marsh grass	*Spartina maritima*
Marsh meadow grass	*Poa palustris*
Mascarene grass	*Zoysia tenuifolia*
Ma-tai	*Eleocharis dulcis*
Mat grass	*Nardus stricta*
Mat-grass fescue	
Vulpia unilateralis	
Mauritania vine reed	
Ampelodesmos mauritanicus	
Meadow barley	
Hordeum secalinum	
Meadow brome	
Bromus commutatus	
Meadow cat's tail	*Phleum pratense*
Meadow fescue	*Festuca pratensis*
Meadow foxtail	
Alopecurus pratensis	
Meadow grass	*Poa pratensis*
Meadow oat grass	
Arenula pratensis	
Helictotrichon pratense	
Meadow soft grass	*Holcus lanatus*
Meadow vetchling	
Lathyrus pratensis	
Means grass	*Sorghum halepense*
Mediterranean brome	
Bromus lanceolatus	
Mediterranean wheat	
Triticum turgidum	
Metake	*Arundinaria japonica*
	Pseudosasa japonica
Mexican everlasting grass	
Eriochloa aristata	

Meyer's bamboo *Phyllostachys meyeri*
Millet *panicum miliaceum*
Miniature papyrus *Cyperus isocladus*
Moa grass *Gynerium sagittatum*
Molasses grass *Melinis minutiflora*
Moor grass *Molinia caerulea*
Moor mat grass *Nardus stricta*
Moso bamboo *Phyllostachys pubescens*
Mosquito grass *Bouteloua gracilis*
Mountain brome *Bromus marginatus*
Mountain heath grass *Sieglingia decumbens*
Mountain melick *Melica nutans*
Mountain sedge *Carex montana*
Mountain timothy *Phleum pratense*
Mountain vetch *Anthyllis montana*
Mountain water sedge *Carex aquatilis*
Mud sedge *Carex limosa*
Multi-spiked cord grass *Spartina alterniflora*
Muriel bamboo *Thamnocalamus spathaceus*
Napier grass *Pennisetum purpurea*
Nail-rod *Typha latifolia*
Nard grass *Cymbopogon nardus*
Narihira bamboo *Semiarundinaria fastuosa*
Narrow-leaved cat-tail *Typha angustifolia*
Narrow-leaved meadow grass *Poa angustifolia*
Narrow-leaved reedmace *Typha angustifolia*
Narrow small-reed *Calamagrostis stricta*
Narrow-spiked sedge *Carex acuta*
Natal grass *Tricholaena rosea*
Rhynchelytrum repens
Needle-and-thread *Stipa comata*
Needle spike rush *Eleocharis acicularis*
Nepal barley *Hordeum vulgare*
Nepal silver grass *Miscanthus nepalensis*
New Zealand bent *Agrostis tenuis*
Nit grass *Gastridium ventricosum*
Nodding melick *Melica nutans*
Nonesuch *Medicago lupulina*
Nut grass *Cyperus esculentus*
Nut sedge *Cyperus esculentus*
Oat grass *Arrhenatherum elatius*
Helictotrichon sempervirens
Oats *Avena sativa*
Oldham bamboo *Bambusa oldhamii*

Old-witch grass *Panicum capillare*
One-grain wheat *Triticum monococcum*
Orange foxtail *Alopecurus aequalis*
Orchard grass *Dactylis glomerata*
Oriental hedge bamboo *Bambusa glaucescens*
Oval sedge *Carex ovalis*
Pale sedge *Carex curta*
Carex pallescens
Palm grass *Setaria palmifolia*
Pampas grass *Cortaderia argentea*
Cortaderia selloana
Pangola grass *Digitaria decumbens*
Paper plant *Cyperus papyrus*
Paper reed *Cyperus papyrus*
Papyrus *Cyperus papyrus*
Para grass *Panicum purpurascens*
Pearl barley *Hordeum distichon*
Pearl grass *Briza maxima*
Pearl millet *Pennisetum americanum*
Pennisetum glaucum
Pendulous sedge *Carex pendula*
Pentz finger grass *Digitaria pentzii*
Perennial beard grass *Agropogon littoralis*
Perennial oat grass *Avena pratensis*
Perennial pea *Lathyrus latifolius*
Lathyrus sylvestris
Perennial rye grass *Lolium perenne*
Perennial veldt grass *Ehrharta calycina*
Persian clover *Trifolium resupinatum*
Peruvian paspalum *Paspalum racemosum*
Pheasant grass *Stipa arundinacea*
Pheasant-tail grass *Stipa arundinacea*
Pigmy bamboo *Arundinaria pygmaea*
Pigmy rush *Juncus mutabilis*
Pill sedge *Carex pilulifera*
Plains bristle grass *Setaria macrostachya*
Plicate sweet grass *Glyceria plicata*
Plume grass *Erianthus ravennae*
Poiret's bristle grass *Setaria poiretiana*
Polish wheat *Triticum polonicum*
Pond sedge *Carex riparia*
Potato oat *Avena fatua*
Poulard wheat *Triticum turgidum*
Prairie beard grass *Schizachyrium scoparium*
Prairie brome *Bromus unioloides*
Prairie cordgrass *Spartina pectinata*

Prickly sedge	*Carex muricata*
Pride of California	
	Lathyrus splendens
Proso	*Panicum miliaceum*
Pubescent wheatgrass	
	Agropyron trichophorum
Punting-pole bamboo	
	Bambusa tuldoides
Purple moor grass	
	Molinia caerulea
Purple small reed	
	Calamagrostis canescens
Purple-stemmed cat's tail	
	Phleum phleoides
Purple vetch	*Vicia benghalensis*
Quack grass	*Agropyron repens*
Quaking grass	*Briza media*
Quick grass	*Agropyron repens*
Quitch grass	*Agropyron repens*
Rabbit's foot	
	Polypogon monspeliensis
Rabbit-foot grass	
	Polypogon monspeliensis
Rabbit's tail grass	*Lagurus ovatus*
Ragi	*Eleusine coracana*
Ramie fibre	*Boehmeria nivea*
Rancheria grass	
	Elymus arenarius
Rat's tail fescue	*Vulpia myuros*
Rattan	*Calamus rotang*
Rattan cane	*Calamus rotang*
Rattlesnake brome	
	Bromus briziformis
Rattlesnake chess	
	Bromus briziformis
Ravenna grass	*Erianthus ravennae*
Red blysmuss	*Blysmus rufus*
Red clover	*Trifolium pratense*
Red fescue	*Festuca rubra*
Red millet	*Digitaria ischaemum*
Red top	*Agrostis gigantea*
Reed canary grass	
	Phalaris arundinacea
Reed fescue	*Festuca altissima*
	Festuca elatior
Reed foxtail	
	Alopecurus arundinaceus
Reedmace	*Typha latifolia*
Reed sweet grass	*Glyceria maxima*
Reflexed salt-marsh grass	
	Puccinellia distans
Rescue brome	*Bromus unioloides*
	Bromus wildenowii
Rescue grass	*Bromus unioloides*
Reversed clover	
	Trifolium resupinatum
Rhode Island bent	*Agrostis tenuis*
Rhodes grass	*Chloris gayana*
Ribbed paspalum	
	Paspalum malacophyllum

Ribbon grass	
	Phalaris arundinacea picta
Rice	*Leersia oryzoides*
	Oryza sativa
Rice grass	*Spartina townsendii*
Rice wheat	*Triticum dicoccon*
Ringal	*Arundinaria anceps*
Ringed beard grass	
	Dichanthium annulatum
River wheat	*Triticum turgidum*
Rock sedge	*Carex rupestris*
Rough bluegrass	*Poa trivialis*
Rough dog's tail	
	Cynosurus echinatus
Rough meadow-grass	*Poa trivialis*
Rough pea	*Lathyrus hirsutus*
Rough-stalked bluegrass	
	Poa trivialis
Rough-stalked meadow grass	
	Poa trivialis
Round-fruited rush	
	Juncus compressus
Round-headed club rush	
	Scirpus holoschoenus
Ruby grass	*Rhynchelytrum repens*
	Tricholaena rosea
Rush-leaved fescue	
	Festuca juncifolia
Rush nut	*Cyperus esculentus*
Russet sedge	*Carex saxatilis*
Russian wild rye	*Elymus junceus*
Rye brome	*Bromus secalinus*
Rye grass	*Lolium perenne*
Sainfoin	*Onobrychis sativa*
Saint Augustine's grass	
	Stenotaphrum secundatum
Salt meadow sedge	*Carex divisa*
Salt mud rush	*Juncus gerardii*
Salt rush	*Juncus lesueurii*
Sand bent	*Mibora minima*
Sandberg's bluegrass	
	Poa sandbergii
Sand bluestem grass	
	Andropogon hallii
Sand cat's tail	*Phleum arenarium*
Sand couch	
	Agropyron junceiforme
	Elymus farctus
Sand dropseed	
	Sporobolus cryptandrus
Sand love grass	*Eragrostis trichodes*
Sand sedge	*Carex arenaria*
Sand twitch	
	Agropyron junceiforme
Sainfoin	*Onobrychis vicifolia*
Saw grass	*Eleocharis effusus*
Scented vernal grass	
	Anthoxanthum odoratum
Scilly Isles meadow grass	
	Poa infirma

Southern cane

Scottish small reed
    Calamagrostis scotica

Scouring brush
    Equisetum hyemale

Scutch grass   *Agropyron repens*

Sea arrow grass *Triglochia palustris*

Sea barley   *Hordeum marinum*

Sea club rush   *Scirpus maritimus*

Sea couch   *Agropyron pungens*
    Elymus pycnanthus

Sea hard grass *Parapholis strigosa*

Sea lyme grass *Elymus arenarius*

Sea oats   *Uniola paniculata*

Sea rush   *Juncus maritimus*

Seaside pea   *Lathyrus japonicus*

Sea twitch   *Agropyron pungens*

Sedge grass   *Carex pendula*

Sedge-like club rush
    Blysmus compressus

Shade fescue   *Festuca rubra*

Shallu
    Sorghum vulgare roxburghii

Shamrock *Trifolium procumbens*
    Trifolium repens

Sharp-flowered rush
    Juncus acutiflorus

Sharp rush   *Juncus acutus*

Sharp sea rush   *Juncus acutus*

Sheep's fescue   *Festuca ovina*

Shore grass
    Stenotaphrum secundatum

Shoreline cordgrass
    Spartina maritima

Short-awned barley
    Hordeum brevisubulatum

Short-awned foxtail
    Alopecurus aequalis

Siberian melic grass
    Melica altissima

Siberian wheatgrass
    Agropyron sibiricum

Siberian wild rye *Elymus sibiricus*

Sickle bamboo
    Chimonobambusa falcata

Sickle grass   *Parapholis incurva*

Sideoats grama
    Bouteloua curtipendula

Silk grass *Oryzopsis hymenoides*

Silky melic grass   *Melica ciliata*

Silky-spike melica   *Melica ciliata*

Silver beard grass
    Bothriochloa saccharoides

Silvery hair grass
    Aira caryophyllea

Simon bamboo
    Arundinaria simonii

Single-cut cow grass
    Trifolium pratense serotinum

Single-glumed spike rush
    Eleocharis uniglumis

Single-grain wheat
    Triticum monococcum

Singletary pea *Lathyrus hirsutus*

Sitka vetch   *Vicia gigantea*

Six-rowed barley
    Hordeum vulgare

Six-weeks grass   *Poa annua*

Slender brome *Bromus lepidus*

Slender cotton grass
    Eriophorum gracile

Slender foxtail
    Alopecurus myosuroides

Slender grama *Bouteloua repens*

Slender rush   *Juncus tenuis*

Slender-spiked sedge *Carex acuta*

Slender spike rush
    Eleocharis acicularis
    Eleocharis uniglumis

Slender wheatgrass
    Agropyron trachycaulum

Slender wild oat *Avena barbata*

Slough grass *Spartina pectinata*

Small bulrush *Typha angustifolia*

Small bur reed
    Sparganium minimum

Small cane *Arundinaria tecta*

Small cat's tail *Phleum nodosum*

Small hop clover
    Trifolium procumbens

Small oat   *Avena strigosa*

Small quaking grass *Briza minor*

Small reed *Calamagrostis nana*

Smilo grass *Oryzopsis miliacea*

Smooth brome *Bromus inermis*
    Bromus racemosus

Smooth cord grass
    Spartina alterniflora

Smooth finger grass
    Digitaria ischaemum

Smooth meadow grass
    Poa pratensis

Smooth sedge *Carex laevigata*

Smooth-stalked meadow grass
    Poa pratensis

Snowy woodrush *Luzula nivea*

Soft brome grass
    Bromus hordeaceus
    Bromus mollis

Soft chess   *Bromus mollis*

Soft flag *Typha angustifolia*

Soft grass   *Holcus mollis*

Soft rush   *Juncus effusus*

Somerset grass *Koeleria vallesiana*

Sorghum   *Holcus sorghum*
    Sorghum bicolor
    Sorghum vulgare

Sorgo
    Sorghum vulgare saccharatum

Southern cane
    Arundinaria gigantea

Spanish vetch	*Anthyllis montana*
Spear grass	*Poa pratensis*
Spike grass	*Dezmazeria sicula*
	Uniola paniculata
Spiked sedge	*Carex spicata*
Spiked wood rush	*Luzula spicata*
Spreading meadow grass	
	Poa subcaerulea
Spring sedge	*Carex caryophyllea*
Spring vetch	*Lathyrus vernus*
	Vicia sativa
Spring wild oat	*Avena fatua*
Square bamboo	
	Chimonobambusa quadrangularis
Square-stem bamboo	
	Chimonobambusa quadrangularis
Square-stemmed bamboo	
	Arundinaria quadrangularis
Squaw grass	*Xerophyllum tenax*
Squirrel tail barley	
	Hordeum jubatum
Squirrel tail fescue	
	Vulpia bromoides
Squirrel tail grass	
	Hordeum jubatum
	Hordeum marinum
Stake bamboo	
	Phyllostachys aureosulcata
Standard crested wheatgrass	
	Agropyron sibiricum
Starch wheat	*Triticum dicoccon*
Star grass	*Chloris truncata*
Star sedge	*Carex echinata*
Starved wood sedge	
	Carex depauperata
Stiff brome	*Bromus rigidus*
Stiff-hair wheatgrass	
	Agropyron trichophorum
Stiff salt-marsh grass	
	Puccinellia rupestris
Stiff sand grass	
	Catapodium marimum
Stiff sedge	*Carex bigelowii*
Strand wheat	*Lolium perenne*
Strawberry clover	
	Trifolium fragiferum
Strawberry-headed clover	
	Trifolium fragiferum
Subclover	*Trifolium subterraneum*
Subterranean clover	
	Trifolium subterraneum
Sudan grass	*Sorghum sudanese*
Sugarcane	*Saccharum officinarum*
Sugar sorghum	
	Sorghum vulgare saccharatum
Swamp grass	*Scolochloa festucacea*
Swamp meadow grass	
	Poa palustris
Sweet galingale	*Cyperus longus*
Sweetshoot bamboo	
	Phyllostachys dulcis
Sweet sorghum	
	Sorghum vulgare saccharatum
Sweet vernal grass	
	Anthoxanthum odoratum
Switch cane	*Arundinaria tecta*
Switch grass	*Panicum virgatum*
Tall fescue	*Festuca arundinacea*
	Festuca elatior
Tall festuca	*Festuca arundinacea*
Tall oat grass	
	Arrhenantherum elatius
Tall wheatgrass	
	Agropyron elongatum
Tare	*Vicia sativa*
Tartarian oat	*Avena fatua*
Tawny sedge	*Carex hostiana*
Teosinte	*Zea mexicana*
Terrell grass	*Lolium perenne*
Texas bluegrass	*Poa arachnifera*
Texas needlegrass	*Stipa leucotricha*
Texas winter grass	*Stipa leucotricha*
Thin-glumed sedge	
	Carex stenolepis
Thin-spiked wood sedge	
	Carex strigosa
Thread rush	*Juncus filiformis*
Three-flowered rush	
	Juncus triglumis
Three-leaved rush	*Juncus trifidus*
Tick bean	*Vicia flaba*
Tiger grass	
	Miscanthus sinensis zebrinus
Tiger nut	*Cyperus esculentus*
Timber bamboo	
	Phyllostachys bambusoides
Timothy grass	*Phleum pratense*
Toad rush	*Juncus bufonius*
Tobosa grass	*Hilaria mutica*
Toe-toe	*Cortaderia richardii*
Tonkin bamboo	
	Arundinaria amabilis
Tonkin cane	*Arundinaria amabilis*
Toothed bur clover	
	Medicago hispida
Tor grass	
	Brachypodium pinnatum
	Brachypodium sylvaticum
Tortoiseshell bamboo	
	Phyllostachys heterocycla
Totter grass	*Briza media*
Townsend's cord grass	
	Spartina townsendii
Transvaal dogtooth grass	
	Cynodon transvaalensis
Trefoil	*Medicago lupulina*

Winter wild oat

Triangular club rush	*Scirpus triquetrus*
Trifolium	*Trifolium incarnatum*
Tsingli cane	*Arundinaria amabilis*
Tuberous foxtail	*Alopecurus bulbosus*
Tuberous vetch	*Lathyrus tuberosus*
Tufted fescue	*Festuca amethystina*
Tufted hair grass	*Deschampsia caespitosa*
Tufted salt-marsh grass	*Puccinellia fasciculata*
Tufted sedge	*Carex elata*
Tufted soft grass	*Holcus lanatus*
Tufted vetch	*Vicia cracca*
Tunis grass	*Holcus virgatus*
Turkestan bluestem	*Bothriochloa ischaemum*
Twitch grass	*Agropyron repens*
	Elymus repens
Two-flowered pea	*Lathyrus grandifloris*
Two-flowered rush	*Juncus biglumis*
Umbrella grass	*Cyperus alternifolius*
Umbrella palm	*Cyperus alternifolius*
Umbrella plant	*Cyperus alternifolius*
Umbrella sedge	*Cyperus alternifolius*
Unbranched bur reed	*Sparganium emerson*
Upland bent	*Agrostis perennans*
Upland bent grass	*Agrostis perennans*
Upright brome	*Bromus erectus*
Uruguay finger grass	*Chloris berroi*
Uruguay pennisetum	*Pennisetum latifolium*
Uva grass	*Gynerium sagittatum*
Vanilla grass	*Hierochloe odorata*
Variegated cord grass	*Spartina pectinata aureo-marginata*
Variegated creeping soft grass	*Holcus mollis variegatus*
Variegated oat grass	*Arrhenatherum bulbosum*
	Arrhenatherum elatius
Various-leaved fescue	*Festuca heterophylla*
Vasey grass	*Paspalum urvillei*
Velvet bent	*Agrostis canina*
Velvet grass	*Holcus lanatus*
Vetiver	*Vetiveria zizanioides*
Vine mesquite	*Panicum obtusum*
Virginian wild rye	*Elymus virginicus*
Volga wild rye	*Elymus racemosus*
Wallaby grass	*Danthonia setacea*

Wall barley	*Hordeum murinum*
Washington grass	*Cabomba caroliniana*
Water bent	*Agrostis semiverticillata*
	Polypogon viridis
Water celery	*Vallisneria americana*
Water chestnut	*Eleocharis dulcis*
Water gladiolus	*Butomus umbellatus*
Water oats	*Zizania aquatica*
Water sedge	*Carex aquatilis*
Water star grass	*Heteranthera dubia*
	Heteranthera graminea
Water whorl grass	*Catabrosa aquatica*
Wavy hair grass	*Deschampsia flexuosa*
Wavy meadow grass	*Poa flexuosa*
Weeping love grass	*Eragrostis curvula*
Weeping widow	*Lacrymaria velutina*
Western wheatgrass	*Agropyron smithii*
West Indian lemongrass	*Cymbopogon citratus*
White beak sedge	*Rhynchospora alba*
White clover	*Trifolium repens*
White durra	*Sorghum vulgare cernuum*
White dutch clover	*Trifolium repens*
White sedge	*Carex curta*
Whitlow grass	*Erophila versa*
Whorl grass	*Catabrosa aquatica*
Wide-sheathed sedge	*Carex vaginata*
Wild barley	*Hordeum murinum*
Wild cane	*Gynerium sagittatum*
Wild celery	*Vallisneria americana*
Wild fescue	*Festuca altissima*
Wild oat	*Avena fatua*
Wild oats	*Chasmanthium latifolium*
Wild red clover	*Trifolium pratense spontaneum*
Wild winter pea	*Lathyrus hirsutus*
Wind grass	*Apera spica-venti*
Windsor bean	*Vicia faba*
Winter pea	*Lathyrus hirsutus*
Winter vetch	*Vicia villosa*
Winter wild oat grass	*Avena ludoviciana*
Winter wild oat	*Avena ludoviciana*
	Avena sterilis

Wire grass

Wire grass	*Eleusine indica*
	Poa compressa
	Schizachyrium scoparium
Witch grass	*Agropyron repens*
	Elymus repens
	Panicum capillare
Wood barley	
	Hordelymus europaeus
Wood bluegrass	*Poa nemoralis*
Wood club rush	*Scirpus sylvaticus*
Wood false brome	
	Brachypodium sylvaticum
Wood fescue	*Festuca altissima*
Wood grass	
	Sorghastrum avenaceum
Woodland brome	
	Bromus ramosus
Woodland meadow grass	
	Poa nemoralis
Wood meadow grass	
	Poa nemoralis
Wood melick	*Melica uniflora*
Wood millet	*Milium effusum*
Wood rush	*Luzula campestris*
Wood sedge	*Carex sylvatica*
Wood small reed	
	Calamagrostis epigejos
Woolly beard grass	
	Erianthus ravennae
Woolly-pod vetch	*Vicia dasycarpa*

Woolly vetch	*Vicia dasycarpa*
Woundwort	*Anthyllis vulneraria*
Yellow bluestem	
	Bothriochloa ischaemum
Yellow bristle grass	*Setaria glauca*
Yellow clover	*Trifolium agrarian*
	Trifolium procumbens
Yellow-groove bamboo	
	Phyllostachys aureosulcata
Yellow oat grass	
	Trisetum flavescens
Yellow nut grass	
	Cyperus esculentus
Yellow nut sedge	
	Cyperus esculentus
Yellow sedge	*Carex flava*
Yellow suckling	
	Trifolium procumbens
Yellow suckling clover	
	Trifolium dubium
	Trifolium procumbens
Yellow trefoil	*Medicago lupulina*
Yellow vetchling	*Lathyrus pratensis*
Yorkshire fog	*Holcus lanatus*
Zebra grass	
	Miscanthus sinensis zebrinus
Zigzag bamboo	
	Phyllostachys flexuosa
Zoysia grass	*Zoysia matrella*
Zulu nut	*Cyperus esculentus*

HERBS

Horticulturally, not botanically, the word 'herb' covers a range of plants used in cooking for flavouring and seasoning, as garnishes, and as domestic remedies. They are also widely used in orthodox and homeopathic medicines. The following list includes many subjects that are extremely poisonous, and drastic reactions, even occasional deaths are not unknown from the ignorant use of them. Experimentation by the general public cannot be too strongly condemned. In addition to the common culinary and medicinal herbs, a few herbal trees have been included.

English lavender –
Lavandula angustifolia

Aaron's rod

Aaron's rod	*Verbascum thapsus*
Absinthe	*Artemisia absinthium*
Aconite	*Aconitum napellus*
Adam's flannel	*Verbascum thapsus*
Adonis	*Adonis vernalis*
Agrimony	*Agrimonia eupatoria*
Alecost	*Balsamita major*
	Chrysanthemum balsamita
	Chrysanthemum vulgare
	Tanacetum balsamita
	Tanacetum vulgare
Alehoof	*Glechoma hederacea*
	Nepeta hederacea
Alder buckthorn	*Frangula alnus*
Alespice	
	Chrysanthemum balsamita
	Tanacetum balsamita
Alexanders	*Smyrnium olusatrum*
Alkanet	*Anchusa officinalis*
All-heal	*Valeriana officinalis*
Allspice	*Calycanthus floridus*
Almond	*Prunus dulcis*
Aloe	*Aloe barbadensis*
	Aloe vera
	Aloe vulgaris
Alpine strawberry	*Fragaria vesca*
Alpine wormwood	*Artemisia laxa*
Althea	*Althaea officinalis*
Alum root	*Heuchera richardsonii*
Ambrosia	*Chenopodium botrys*
American centaury	
	Sabatia angularis
Anchusa	*Anchusa sempervirens*
	Pentaglottis sempervirens
Angelica	*Angelica archangelica*
Anise	*Anisum vulgare*
	Foeniculum vulgare azoricum
	Myrrhis odorata
	Pimpinella anisum
Aniseed	*Anisum vulgare*
	Pimpinella anisum
Anise fern	*Myrrhis odorata*
Anise hyssop	
	Agastache anethiodora
	Agastache foeniculum
Annual marjoram	
	Origanum majorana
Apothecary's rose	
	Rosa gallica officinalis
Apple mint	*Mentha rotundifolia*
	Mentha suaveolens
Arnica	*Arnica montana*
Artemisia	*Artemisia abrotanum*
	Artemisia absinthium
	Artemisia dracunculus
Arugula	*Eruca vesicaria*
Asafetida	*Ferula foetida*
Ass's ear	*Symphytum officinale*
Autumn crocus	
	Colchicum autumnale

Bachelor's buttons	
	Ranunculus acris
	Tanacetum parthenium
Balm	*Melissa officinalis*
	Monarda didyma
Balm of Gilead	
	Cedronella canariensis
	Cedronella triphylla
	Commiphora opobalsamum
Balsam herb	
	Chrysanthemum balsamita
Balsamita	
	Chrysanthemum balsamita
Balsam poplar	*Populus balsamifera*
Baneberry	*Actaea spicata*
Barbados aloe	*Aloe barbadensis*
	Aloe vera
	Aloe vulgaris
Basil	*Ocimum basilicum*
Basil thyme	*Acinos arvensis*
	Calamintha acinos
Basin sagebrush	
	Artemisia tridentata
Bastard balm	
	Melittis melissophyllum
Bastard saffron	
	Carthamus tinctorus
Bay	*Laurus nobilis*
Bay leaf	*Laurus nobilis*
	Do not confuse with
	Kalmia latifolia (poisonous)
Bay laurel	*Laurus nobilis*
Beach morning glory	
	Ipomoea pes-caprae
Beach wormwood	
	Artemisia stellerana
Bean herb	
	Satureia (Satureja) hortensis
	Satureia montana
Bear's foot	*Alchemilla vulgaris*
	Alchemilla xanthochlora
Beaver poison	*Conium maculatum*
Bee balm	*Melissa officinalis*
	Monarda didyma
Bee bread	*Borago officinalis*
Beefsteak plant	
	Perilla frutescens crispa
Beggar's buttons	*Arctium lappa*
Belladonna	*Atropa belladonna*
Bengal root	*Zingiber cassumunar*
Bergamot	*Monarda didyma*
Bergamot mint	*Mentha citrata*
Betony	*Stachys officinalis*
Bhang	*Cannabis sativa*
Bible leaf	
	Chrysanthemum balsamita
Bilberry	*Vaccinium myrtillus*
Birdlime mistletoe	*Viscum alba*
Bird's foot	
	Trigonella foenum-graecum
Bishopswort	*Stachys officinalis*

Cheese rennet

Bistort	*Polygonum bistorta*
Bitter aloe	*Aloe barbadensis*
Bitter broom	*Sabatia angularis*
Bitter buttons	
	Chrysanthemum vulgare
	Tanacetum vulgare
Bitter clover	*Sabatia angularis*
Bitter cress	*Cardamine pratensis*
Bittersweet	*Solanum dulcamara*
Bittersweet nightshack	
	Solanum dulcamara
Black American willow	*Salix nigra*
Black bryony	*Tamus communis*
Black dogwood	*Frangula alnus*
Blackeye root	*Tamus communis*
Black hellebore	*Helleborus niger*
Black horehound	*Ballota nigra*
Black mulberry	*Morus nigra*
Black mustard	*Brassica nigra*
	Sinapis nigra
Black sugar	*Glycyrrhiza glabra*
Black wort	*Symphytum officinale*
Blanket leaf	*Verbascum thapsus*
Blessed thistle	*Cnicus benedictus*
Blood flower	
	Asclepias curassavica
Bloody fingers	*Digitalis purpurea*
Blowball	*Leontodon taraxacum*
	Taraxacum officinale
Blue gum	*Eucalyptus globulus*
Blue rocket	*Aconitum napellus*
Blue sage	*Salvia azurea*
Boneset	*Symphytum officinale*
Borage	*Borago officinalis*
Border catmint	*Nepeta mussini*
Bore tree	*Sambucus nigra*
Bouncing bet	*Saponaria officinalis*
Bowman	*Anthemis nobilis*
Bramble	*Rubus fruticosus*
Brandy mint	*Mentha × piperita*
Bread-and-cheese tree	
	Crataegus oxyacantha
British myrrh	*Myrrhis odorata*
Broad-leaved dock	
	Rumex obtusifolius
Broad-leaved thyme	
	Thymus pulegioides
Broom	*Cytisus scoparius*
	Sarothamnus scoparius
Brown mustard	*Brassica juncea*
	Sinapis juncea
Bruisewort	*Bellis perennis*
	Saponaria officinalis
Buckeye	*Aesculus hippocastanum*
Buckler-leaf sorrel	*Rumex scutatus*
Buckthorn	*Rhamnus catharticus*
Buffalo herb	*Medicago sativa*
Bugbane	*Actaea spicata*
Bugle	*Ajuga reptans*
Bugleweed	*Ajuga reptans*
Bugloss	*Anchusa officinalis*
Bullsfoot	*Tussilago farfara*
Burdock	*Arctium lappa*
Burnet	*Poterium sanguisorba*
	Sanguisorba minor
Burnet saxifrage	
	Pimpinella saxifraga
Burning bush	*Dictamnus albus*
	Dictamnus fraxinella
	Eonymus europaeus
Burrage	*Borago officinalis*
Bush basil	*Ocinum minimum*
Bush lawyer	*Rubus australis*
Bushy mint	*Mentha × gentilis*
Buttercup	*Ranunculus acris*
Butter dock	*Rumex obtusifolius*
Buttered haycocks	
	Linaria vulgaris
Buttons	*Tanacetum vulgare*
Calamint	*Calamintha acinos*
	Calamintha grandiflora
Calamus	*Acorus calamus*
Calendula	*Calendula officinalis*
California sagebrush	
	Artemisia californica
Call-me-to-you	*Viola tricolor*
Camphor plant	*Balsamita major*
	tomentosum
Camphor tree	
	Cinnamomum camphora
Cape ginger	*Costus speciosus*
Caraway	*Carum carvi*
Caraway thyme	
	Thymus herba-barona
Cardoon	*Cynara cardunculus*
Carob	*Ceratonia siligua*
Carolina allspice	
	Calycanthus floridus
Carpenter's square	
	Scrophularia marilandica
Cassilata	*Hyoscyamus niger*
Catmint	*Nepeta cataria*
Catnep	*Nepeta cataria*
Catnip	*Nepeta cataria*
Cat's peas	*Cytisus scoparius*
	Sarothamnus scoparius
Cat's valerian	*Valeriana officinalis*
Centaury	*Centaurium minus*
	Centaurium umbellatum
	Erythraea centaurium
Chamomile	*Anthemis nobilis*
	Chamaemelum nobile
Charity	*Polemoneum coeruleum*
Checkerberry	
	Gaultheria procumbens
Cheese rennet	*Galium verum*

Cherry pie

Cherry pie
  Heliotropium arborescens
  Heliotropium corymbosum
  Heliotropium peruvianum
Chervil   *Anthriscus cerefolium*
  Chaerophyllum temulentum
Chickweed   *Stellaria media*
Chicory   *Cichorium intybus*
Chinese parsley
  Coriandrum sativum
Chiretta   *Swertia chirata*
Chive   *Allium schoenoprasum*
Christmas rose   *Helleborus niger*
Christ's ladder   *Centaurium minus*
Church steeples
  Agrimonia eupatoria
Churnstaff   *Linaria vulgaris*
Cinnamon
  Cinnamomum zeylanicum
Clary   *Salvia sclarea*
Clary sage   *Salvia sclarea*
Clear eye   *Salvia sclarea*
Clot-bur   *Arctium lappa*
Cloves   *Eugenia aromatica*
Cobbler's bench
  Lamuim maculatum
Cockle-bur   *Arctium lappa*
Cockleburr   *Agrimonia eupatoria*
Coltsfoot   *Tussilago farfara*
Columbine   *Aquilegia vulgaris*
Comfrey   *Symphytum officinale*
Common balm   *Melissa officinalis*
Common buckthorn
  Rhamnus catharticus
Common chamomile
  Anthemis nobilis
Common comfrey
  Symphytum officinale
Common dandelion
  Leontodon taraxacum
  Taraxacum officinalis
Common heliotrope
  Heliotropium arborescens
  Heliotropium corymbosum
  Heliotropium peruvianum
Common horehound
  Marrubium vulgare
Common jasmine
  Jasminum officinale
Common mallow   *Malva sylvestris*
Common mint   *Mentha spicata*
Common oak   *Quercus robur*
Common sagebrush
  Artemisia tridentata
Common thyme   *Thymus vulgaris*
Common tormentil
  Potentilla erecta
  Potentilla tormentilla
Common wormwood
  Artemisia absinthium

Compass weed
  Rosmarinus lavandulaceus
  Rosmarinus officinalis
Cone flower
  Echinacea angustifolia
Consound   *Symphytum officinale*
Cool tankard   *Borago officinalis*
Coral bells   *Heuchera richardsonii*
Coriander   *Coriandrum sativum*
Corn salad   *Valerianella locusta*
  Valerianella olitoria
Corsican mint   *Mentha requienii*
Costmary
  Chrysanthemum balsamita
  Tanacetum balsamita
Cotton lavender
  Santolina chamaecyparissus
  Santolina incana
  Santolina tomentosa
Cotton thistle
  Onopordon acanthium
Coughwort   *Tussilago farfara*
Cowslip
  Primula officinalis
  Primula veris
Crape ginger   *Costus speciosus*
Creeping charlie   *Nepeta hederacea*
Creeping comfrey
  Symphytum grandiflorum
Creeping savory
  Satureia (Satureja) repanda
Creeping thyme
  Thymus serpyllum
Crème de menthe plant
  Mentha requienii
Crete dittany   *Amaracus dictamnus*
  Origanum dictamnus
Crimson clover
  Trifolium incarnatum
Cuckoo flower
  Cardamine pratensis
Cuckoo's meat   *Rumex acetosa*
Cuckoo sorrow   *Rumex acetosa*
Cudweed   *Artemisia ludoviciana*
Cumin   *Cuminum cyminum*
  Cuminum odorum
Culverwort   *Aquilegia vulgaris*
Curdwort   *Galium verum*
Curled dock   *Rumex crispus*
Curly mint   *Mentha crispa*
Curly parsley
  Petroselinum crispum
Curry plant
  Helichrysum angustifolium
  Helichrysum italicum
Daisy   *Bellis perennis*
Dame's violet   *Hesperis matronalis*
Dandelion   *Leontodon taraxacum*
  Taraxacum officinalis
Dark opal basil
  Ocimum basilicum purpurescens

Florentine iris

Deadly nightshade
Atropa belladonna

Dead man's bells
Digitalis purpurea

Dead nettle   *Lamium maculatum*

Devil's apple
Mandragora officinarum

Devil's bit scabious
Succisa pratensis

Devil's cherries   *Atropa belladonna*

Devil's nettle   *Achillea millefolium*

Didi   *Thymus mastichina didi*

Dill   *Anethum graveolens*
Peucedanum graveolens

Dock   *Rumex obtusifolius*

Dog poison   *Aethusa cynapium*

Dog violet   *Viola riviana*

Donkey's ears   *Stachys lanata*
Stachys olympica

Double mint   *Mentha × piperita*

Dropwort   *Filipendula hexapetala*
Filipendula vulgaris
Spiraea filipendula
Ulmaria filipendula

Durmast oak   *Quercus petraea*
Quercus sessilis

Dusty miller   *Artemisia stellerana*

Dwarf purple foxglove
Digitalis thrapsi

Dyer's weed   *Isatis tinctoria*

Eau-de-cologne mint
Mentha citrata
Mentha × piperita citrata

Echinacea   *Echinacea angustifolia*

Eggs and bacon   *Linaria vulgaris*

Egyptian mint   *Mentha rotundifolia*

Egyptian onion
Allium cepa aggregatum
Allium cepa viviparum

Elder   *Sambucus nigra*

Elecampane   *Inula helenium*

Elephant garlic
Allium ampeloprasum

English daisy   *Bellis perennis*

English lavender
Lavandula angustifolia
Lavandula officinalis
Lavandula spica
Lavandula vera

English mace   *Achillea decolorans*

English mandrake   *Bryonia dioica*

English pennyroyal
Mentha pulegium

English thyme   *Thymus vulgaris*

English yew   *Taxus baccata*

Estragon   *Artemisia dracunculus*

European crowfoot
Aquilegia vulgaris

European vervain
Verbena officinalis

European white hellebore
Veratrum album

European willow   *Salix alba*

Evening primrose
Oenothera biennis

Eyebright
Euphrasia officinalis

Everlasting onion   *Allium perutile*

Fairy thimbles   *Digitalis purpurea*

False dittany   *Dictamnus albus*
Dictamnus fraxinella

False hellebore   *Veratrum viride*

False mistletoe
Loranthus europaeus

False saffron   *Carthamus tinctorius*

False tarragon
Artemisia dracunculoides
Artemisia dranunculus

Fat hen   *Chenopodium album*

Featherfew
Chrysanthemum parthenium
Matricaria eximia
Primula parthenium

Featherfoil
Chrysanthemum parthenium
Matricaria eximia
Primula parthenium

Felon herb   *Artemisia vulgaris*

Felonwood   *Solanum dulcamara*

Fenkel   *Foeniculum officinale*
Foeniculum vulgare

Fennel   *Foeniculum officinale*
Foeniculum vulgare

Fennel hyssop
Agastache anethiodora
Agastache foeniculum

Fenugreek   *Trigonella foenu
graecum*

Feverfew
Chrysanthemum parthenium
Matricaria eximia
Primula parthenium
Tanacetum parthenium

Field balm   *Nepeta hederacea*

Field horsetail   *Equisetum arvense*

Field mint
Mentha arvensis piperascens

Fine-leaved basil
Ocimum minimum

Finocchio   *Foeniculum dulce*
Foeniculum vulgare azoricum

Flag iris   *Iris germanica florentina*

Flax   *Linum usitatissimum*

Flaxweed   *Linaria vulgaris*

Flirtwort
Chrysanthemum parthenium
Matricaria eximia
Primula parthenium

Florence fennel   *Foeniculum dulce*
Foeniculum vulgare azoricum

Florentine iris
Iris germanica florentina

Florist's violet

Florist's violet	*Viola odorata*
Fluellin	*Linaria vulgaris*
Foalfoot	*Tussilago farfara*
Fool's parsley	*Aethusa cynapium*
Forget-me-not	*Myosotis arvensis*
	Myosotis sylvatica
Foxglove	*Digitalis purpurea*
Fox's clote	*Arctium lappa*
Fraxinella	*Dictamnus albus*
	Dictamnus fraxinella
French lavender	
	Lavandula dentata
	Lavandula stoechas
	Santolina chamaecyparissus
	Santolina incana
	Santolina tomentosa
French parsley	
	Petroselinum neapolitanum
French sorrel	*Rumex scutatus*
French spinach	*Atriplex hortensis*
French tarragon	
	Artemisia dracunculus
French thyme	*Thymus vulgaris*
Friar's cap	*Aconitum napellus*
Fringed lavender	
	Lavandula dentata
Fuller's herb	*Saponaria officinalis*
Fusoria	*Eonymus europaeus*
Galbanum	*Ferula gabaniflua*
Garden angelica	
	Angelica archangelica
Garden columbine	
	Aquilegia vulgaris
Garden cress	*Lepidium sativum*
Garden heliotrope	
	Valeriana officinalis
Garden mint	*Mentha spicata*
	Mentha viridis
Garden rhubarb	
	Rheum rhabarbarum
Garden sage	*Salvia officinalis*
Garden sorrel	*Rumex acetosa*
Garden thyme	*Thymus vulgaris*
Garlic	*Allium sativum*
Garlic chive	*Allium tuberosum*
Garlic mustard	*Alliaria petiolata*
Gas plant	*Dictamnus albus*
	Dictamnus fraxinella
Gentian	*Gentiana lutea*
German chamomile	
	Chamomilla recutira
	Matricaria chamomilla
	Matricaria recutira
Germander	*Teucrium chamaedrys*
Giant chive	
	Allium schoenoprasum sibiricum
Giant fennel	*Ferula communis*
Gill-go-over-the-ground	
	Nepeta hederacea
Gillyflower	*Dianthus caryophyllus*

Ginger	*Zingiber officinale*
Ginger mint	*Mentha × gentilis*
Ginseng	*Panax ginseng*
Glabrous rupture-wort	
	Herniaria glabra
Goat's leaf	*Lonicera periclymenum*
Goat's rue	*Galega officinalis*
Golden balm	
	Melissa officinalis aurea
Golden buttons	
	Tanacetum vulgare
Golden chain	
	Laburnum anagyroides
Golden chair	*Cytisus scoparius*
	Sarothamnus scoparius
Golden-edged thyme	
	Thymus × citriodorus aureus
Golden hop	
	Humulus lupulus aureus
Golden sage	*Salvia aurea*
	Salvia officinalis icterina
Golden seal	*Hydrastis canadensis*
Golden thyme	
	Thymus × citriodorus aureus
	Thymus vulgaris aureus
Gold knots	*Ranunculus acris*
Golds	*Calendula officinalis*
Good King Henry	
	Chenopodium bonus-henricus
Goose tansy	*Potentilla anserina*
Gran's bonnet	*Aquilegia vulgaris*
Grape ginger	*Costus speciosus*
Grape hyacinth	*Muscari botryoides*
Gravelroot	
	Eupatorium purpureum
Greasewood	*Salvia apiana*
Great chervil	*Myrrhis odorata*
Greater burnet saxifrage	
	Pimpinella major
Greater periwinkle	*Vinca major*
Great-headed garlic	
	Allium ampeloprasum
Great mullein	*Verbascum thapsus*
Great spurred violet	*Viola selkirkii*
Greek basil	*Ocimum minimum*
Greek hayseed	*Trigonella foenum-graecum*
Greek valerian	
	Polemoneum coeruleum
Green ginger	
	Artemisia absinthium
Green hellebore	*Veratrum viride*
Green sauce	*Rumex acetosa*
Grey santolina	
	Santolina chamaecyparissus
	Santolina incana
	Santolina tomentosa
Ground ivy	*Glechoma hederacea*
	Nepeta hederacea
Gum tree	*Eucalyptus gunnii*

Lamb's ears

Gypsy's rhubarb	*Arctium lappa*
Hack matack	*Juniperus communis*
Hag taper	*Verbascum thapsus*
Hamburg parsley	
	Carum petroselinum tuberosum
	Petroselinum crispum fusiformis
	Petroselinum tuberosum
Hartshorn	*Rhamnus catharticus*
Hawthorn	*Crataegus oxyacantha*
Haymaids	*Nepeta hederacea*
Heal-all	*Valeriana officinalis*
Healing herb	*Symphytum officinale*
Heartsease	*Viola tricolor*
Heather	*Calluna vulgaris*
Hedge hyssop	*Gratiola officinalis*
Helmet flower	*Aconitum napellus*
	Scutellaria galericulata
	Scutellaria lateriflora
Hemlock	*Conium maculatum*
Hemp agrimony	
	Eupatorium cannabinum
Hen-and-chickens	
	Sempervivum tectorum
Henbane	*Hyoscyamus niger*
Herb christopher	*Actaea spicata*
Herb constancy	*Viola tricolor*
Herb louisa	*Aloysia triphylla*
	Lippia citriodora
	Lippia triphylla
	Verbena triphylla
Herb masticke	*Thymus mastichina*
Herb of gladness	*Borago officinalis*
Herb of grace	*Ruta graveolens*
	Verbena officinalis
Herb of the cross	
	Verbena officinalis
Highway thorn	
	Rhamnus catharticus
Himalayan parsley	
	Selinum tenuifolium
Hoarhound	*Marrubium vulgare*
Hog's bean	*Hyoscyamus niger*
Holly	*Ilex aquifolium*
Holy basil	*Ocimum sanctum*
Honeysuckle	
	Lonicera periclymenum
Hop	*Humulus lupulus*
Hop marjoram	
	Amaracus dictamnus
	Origanum dictamnus
Horehound	*Marrubium vulgare*
Horse chestnut	
	Aesculus hippocastanum
Horseheal	*Inula helenium*
Horsehoof	*Tussilago farfara*
Horse mint	*Mentha longifolia*
	Monarda punctata
Horseradish	*Armoracia rusticana*
	Cochlearia armoracia
Horse savin	*Juniperus communis*

Houseleek	*Sempervivum tectorum*
Hyssop	*Hyssopus officinalis*
Iceland moss	*Cetraria islandica*
Indian arrowroot	
	Eonymus europaeus
Indian cress	*Tropaeolum majus*
Indian elm	*Ulmus fulva*
Indian plume	*Monarda didyma*
Italian fitch	*Galega officinalis*
Italian lovage	*Levisticum officinale*
	Ligusticum paludapifolium
Italian parsley	
	Petroselinum neapolitanum
Jack-by-the-hedge	*Alliaria petiolata*
Jacob's ladder	
	Polemoneum coeruleum
Jacob's staff	*Verbascum thapsus*
Jamaica pepper	
	Calycanthus floridus
Japanese mint	
	Mentha arvensis piperascens
Jasmine	*Jasminum officinale*
Jerusalem cowslip	
	Pulmonaria officinalis
Jimson weed	*Datura stramonium*
July flower	*Dianthus caryophyllus*
Juniper	*Juniperus communis*
Kangaroo apple	
	Solanum laciniatum
Kecksies	*Conium maculatum*
Kex	*Conium maculatum*
Kingcup	*Caltha palustris*
King's cure-all	*Oenothera biennis*
Kiss-her-in-the-buttery	
	Viola tricolor
Kit runabout	*Viola tricolor*
Knight's spur	*Delphinium ajacis*
Knitbone	*Symphytum officinale*
Knotted marjoram	
	Majorana hortensis
	Origanum majorana
Korean mint	*Agastache rugosa*
Laburnum	*Laburnum anagyroides*
Ladder to heaven	
	Convallaria majalis
Ladies' meat	
	Crataegus oxyacantha
Ladies' seal	*Bryonia dioica*
Lad's love	*Artemisia abrotanum*
Lady's bedstraw	*Galium verum*
Lady's mantle	*Alchemilla vulgaris*
	Alchemilla xanthochlora
Lady's slipper	*Cytisus scoparius*
	Sarothamnus scoparius
Lady's smock	*Cardamine pratensis*
Lamb mint	*Mentha spicata*
	Mentha viridis
Lamb's ears	*Stachys lanata*
	Stachys olympica

Lamb's lettuce

Lamb's lettuce	*Valerianella locusta*
	Valerianella olitoria
Lark's heel	*Delphinium ajacis*
Larkspur	*Delphinium ajacis*
Lavender cotton	
	Santolina chamaecyparissus
	Santolina incana
	Santolina tomentosa
Leek	*Allium porrum*
Lemon basil	*Ocimum americanum*
Lemon balm	*Melissa officinalis*
	Ocimum americanum
Lemon basil	*Ocimum citriodorum*
Lemon bergamot	
	Monarda citriodora
Lemon catmint	
	Nepeta cataria citriodorum
Lemon thyme	*Thymus serphyllum*
	Thymus × citriodorus
Lemon verbena	*Aloysia citriodora*
	Aloysia triphylla
	Lippia citriodora
	Lippia triphylla
	Verbena triphylla
Leopard's bane	*Arnica montana*
Lesser hemlock	*Aethusa cynapium*
Lettuce-leaved basil	
	Ocimum basilicum lactucafolium
Licorice	*Glycyrrhiza glabra*
Lignum crucis	*Viscum alba*
Lily-of-the-valley	
	Convallaria majalis
Lion's foot	*Alchemilla vulgaris*
	Alchemilla xanthochlora
Little dragon	
	Artemisia dracunculus
Liverwort	*Agrimonia eupatoria*
Lizzy-run-up-the-hedge	
	Nepeta hederacea
London lily	*Allium ursinum*
Lovage	*Levisticum officinale*
	Ligusticum paludapifolium
Love-in-idleness	*Viola tricolor*
Love leaves	*Arctium lappa*
Love-lies-bleeding	*Viola tricolor*
Low sagebrush	
	Artemisia arbuscula
Lungwort	*Pulmonaria officinalis*
Lupin	*Lupinus polyphyllus*
Lurk-in-the-ditch	
	Mentha pulegium
Mace	*Myristica fragrans*
Mackerel mint	*Mentha spicata*
	Mentha viridis
Madder root	*Rubia tinctoria*
Madonna lily	*Lilium candidum*
Madweed	*Scutellaria galericulata*
	Scutellaria lateriflora
Maid's hair	*Galium verum*
Mandrake	
	Mandragora officinarum

Manzanilla	*Anthemis nobilis*
Marigold	*Calendula officinalis*
Marihuana	*Cannabis sativa*
Marjoram	*Origanum vulgare*
Marshmallow	*Althaea officinalis*
Marsh marigold	*Caltha palustris*
Marsh samphire	
	Arthrocnemum perenne
Marygold	*Calendula officinalis*
Mary gowles	*Calendula officinalis*
May	*Crataegus oxyacantha*
May lily	*Convalleria majalis*
Maythen	*Anthemis nobilis*
Meadowsweet	
	Filipendula hexapetala
	Filipendula ulmaria
	Filipendula vulgaris
	Spiraea filipendula
	Ulmaria filipendula
Melilot	*Melilotus officinalis*
Menthella	*Mentha requienii*
Mignonette	*Reseda odorata*
Milfoil	*Achillea millefolium*
Miner's lettuce	
	Claytonia perfoliata
Mint geranium	*Balsamita major*
	Chrysanthemum balsamita
Mistletoe	*Viscum album*
Mock orange	
	Philadelphus coronarius
Monkshood	*Aconitum napellus*
Monk's pepper	*Vitex agnus-castus*
Monster-leaved basil	
	Ocimum basilicum lactucafolium
Moonflower	*Oenothera biennis*
Moose elm	*Ulmus fulva*
Moss rose	*Rosa centifolia muscosa*
Mother of thyme	
	Calamintha acinos
Motherwort	*Leonurus cardiaca*
Mountain balm	*Calamintha acinos*
Mountain radish	
	Armoracia rusticana
	Cochlearia armoracia
Mountain tea	
	Gaultheria procumbens
Mountain tobacco	*Arnica montana*
Mugwort	*Artemisia vulgaris*
Mullein	*Verbascum thapsus*
Myrtle	*Myrtus communis*
	Myrtus communis tarentina
Nasturtium	*Tropaeolum majus*
Nettle	*Urtica diorica*
Nine hooks	*Alchemilla vulgaris*
	Alchemilla xanthochlora
Nosebleed	*Achillea millefolium*
Nutmeg	*Myristica fragrans*
Nutmeg-scented geranium	
	Pelargonium × fragrans
Oak	*Quercus rober*

Russian comfrey

Old man	*Artemisia abrotanum*
Old man's pepper	
	Achillea millefolium
Old warrior	*Artemisia pontica*
Old woman	*Artemisia absinthium*
	Artemisia stellerana
Opium poppy	*Papaver somniferum*
Orach	*Atriplex hortensis*
Orache	*Atriplex hortensis*
Orange mint	*Mentha citrata*
Orange root	*Hydrastis canadensis*
Origano	*Origanum vulgare*
Organy	*Origanum vulgare*
Oriental chive	*Allium tuberosum*
Orris	*Iris florentina*
	Iris germanica
	Iris germanica florentina
Orris root	*Iris florentina*
	Iris germanica
	Iris germanica florentina
Oswega tea	*Monarda didyma*
Our Lady's bedstraw	
	Galium verum
Our Lady's candle	
	Verbascum thapsus
Our Lady's tears	
	Convallaria majalis
Parma violet	*Viola pallida plena*
Parsley	*Petroselinum crispum*
Pattens and clogs	*Linaria vulgaris*
Peasant's clock	
	Leontodon taraxacum
	Taraxacum officinale
Peasant's mattress	*Galium verum*
Pennyroyal	*Mentha pulegium*
Pepper	*Capsicum annuum*
Pepper cress	*Lepidium sativum*
Peppermint	*Mentha × piperita*
Peppermint-scented geranium	
	Pelargonium tomentosum
Perfoliate honeysuckle	
	Lonicera caprifolium
Periwinkle	*Vinca major*
Petersylinge	*Petroselinum crispum*
Phew plant	*Valeriana officinalis*
Phu	*Valeriana officinalis*
Pigeon's grass	*Verbena officinalis*
Pimento	*Capsicum frutescens*
Pineapple mint	
	Mentha suaveolens variegata
Pineapple sage	*Salvia elegans*
	Salvia rutilans
Pineapple shrub	
	Calycanthus floridus
Pink	*Dianthus caryophyllus*
Pink-of-my-john	*Viola tricolor*
Pipe tree	*Sambucus nigra*
Pixie's slipper	*Cytisus scoparius*
	Sarothamnus scoparius

Poison hemlock	
	Conium maculatum
Poison parsley	*Conium maculatum*
Pokeweed	*Phytolacca americana*
Polar plant	
	Rosmarinus lavandulaceus
	Rosmarinus officinalis
Poor man's treacle	*Allium sativum*
Pot marigold	*Calendula officinalis*
Pot marjoram	
	Origanum heracleoticum
	Origanum onites
	Origanum vulgare
Priest's crown	
	Leontodon taraxacum
	Taraxacum officinale
Primrose	*Primula vulgaris*
Pudding grass	*Mentha pulegium*
Purple basil	
	Ocimum basilicum aurauascens
Purple perilla	
	Perilla frutescens crispa
Purple sage	*Salvia purpurescens*
Purslane	*Portulaca oleracea*
Pussy willow	*Salix nigra*
Quickthorn	*Crataegus oxycantha*
Ram's horn	*Rhamnus catharticus*
Ramsons	*Allium ursinum*
Ram's thorn	*Rhamnus catharticus*
Red centaury	*Centaurium minus*
Red clover	*Trifolium incarnatum*
Red cole	*Armoracia rusticana*
	Cochlearia armoracia
Red elm	*Ulmus fulva*
Red mint	*Mentha × gentilis*
Red orache	
	Atriplex hortensis rubra
Red sage	
	Salvia officinalis purpurescens
Restharrow	*Ononis spinosa*
Rhubarb	*Rheum palmatum*
Rocket	*Hesperis matronalis*
Rock hyssop	*Hyssopus aristatus*
Roman chamomile	
	Anthemis nobilis
Roman laurel	*Laurus nobilis*
Roquette	*Hesperis matronalis*
Rosemary	
	Rosmarinus lavandulaceus
	Rosmarinus officinalis
Round-leaved mint	
	Mentha rotundifolia
Ruddes	*Calendula officinalis*
Rue	*Ruta graveolens*
Run-by-the-ground	
	Mentha pulegium
Russian comfrey	
	Symphytum perigrinum
	Symphytum uplandicum

Russian tarragon	
	Artemisia dracunculoides
	Artemisia dranunculus
Safflower	*Carthamus tinctorius*
Saffron	*Crocus sativus*
Saffron thistle	*Carthamus tinctorius*
Sage	*Salvia officinalis*
Sagebrush	*Artemisia tridentata*
Sage of Bethlehem	*Mentha spicata*
	Mentha viridis
Sainfoin	*Onobrychis viciifolia*
Saint John's plant	
	Artemisia vulgaris
Saint John's wort	
	Hypericum perforatum
Saint Patrick's cabbage	
	Sempervivum tectorum
Salad burnet	*Poterium sanguisorba*
	Sanguisorba minor
Salad rocket	*Eruca vesicaria*
Sambal	*Ferula suaveolens*
Samphire	*Crithmum maritimum*
Sand sage	*Artemisia filifolia*
Santolina	
	Santolina chamaecyparissus
	Santolina incana
	Santolina tomentosa
Satan's apple	
	Mandragora officinarum
Savory	
	Satureia (Satureja) hortensis
	Satureia montana
Scabwort	*Inula helenium*
Scarlet monarda	*Monarda didyma*
Scented geranium	
	Pelargonium graveolens
Scented mayweed	
	Chamomilla recutira
	Matricaria chamomilla
Scotch mint	*Mentha × gentilis*
Scots lovage	*Levisticum scoticum*
Scullcap	*Scutellaria galericulata*
	Scutellaria laterifolia
Sea fennel	*Crithmum maritimum*
Sea holly	*Eryngium maritimum*
Sea purslane	*Atriplex hortensis*
Semsem	*Sesamum indicum*
	Sesamum orientale
Sesame	*Sesamum indicum*
	Sesamum orientale
Setwall	*Valeriana officinalis*
Shallot	*Allium ascalonicum*
Shepherd's club	*Verbascum thapsus*
Shepherd's needle	*Myrrhis odorata*
Silver birch	*Betula pendula*
	Betula verrucosa
Silver posy	*Thymus vulgaris*
Silver queen	*Thymus citriodorus*
Silver sage	*Salvia argentea*
Silver weed	*Potentilla anserina*

Simpler's joy	*Verbena officinalis*
Skewerwood	*Eonymus europaeus*
Skirret	*Sium sisarum*
Skullcap	*Scutellaria galericulata*
	Scutellaria lateriflora
Slippery elm	*Ulmus fulva*
Smallage	*Apium graveolens*
Small-leaved lime	*Tilia cordata*
Smearwort	*Chenopodium bonus-*
	henricus
Smooth cicely	*Myrrhis odorata*
Sneezewort	*Achillea millefolium*
Soapwort	*Saponaria officinalis*
Sorrel	*Rumex acetosa*
Sour sabs	*Rumex acetosa*
South African wood sage	
	Buddleia salviifolia
Southernwood	
	Artemisia abrotanum
Spanish juice	*Glycyrrhiza glabra*
Spanish sage	
	Salvia barrelieri
Spanish lavender	
	Lavandula stoechas
Spearmint	*Mentha crispa*
	Mentha spicata
	Mentha viridis
Speedwell	*Veronica officinalis*
Spindle tree	*Eonymus europaeus*
Spire mint	*Mentha spicata*
	Mentha viridis
Spotted alder	
	Hamamelis virginiana
Spotted dead nettle	
	Lamium maculatum
Spotted hemlock	
	Conium maculatum
Squirting cucumber	
	Echallium elaterium
Star flower	*Borago officinalis*
Starch hyacinth	*Muscari botryoides*
Staunchweed	*Achillea millefolium*
Sticklewort	*Agrimonia eupatoria*
Stinging nettle	*Urtica dioica*
Stink bomb	*Allium ursinum*
Stinking goosefoot	
	Chenopodium vulvaria
Stinking lily	*Allium ursinum*
Stinking motherwort	
	Chenopodium vulvaria
Stinking nanny	*Allium ursinum*
Stinking nightshade	
	Hyoscyamus niger
Strawberry shrub	
	Calycanthus floridus
Straw foxglove	*Digitalis lutea*
Succory	*Cichorium intybus*
Sumbul	*Ferula sumbul*
Summer coleus	
	Perilla frutescens crispa

White mustard

Summer purslane	*Portulaca oleracea*
Summer savory	*Satureia (Satureja) hortensis*
Sundew	*Drosera rotundifolia*
Sunflower	*Helianthus annuus*
Sweet balm	*Melissa officinalis*
Sweet basil	*Ocimum basilicum*
Sweet bay	*Laurus nobilis*
	Melissa officinalis
Sweet bracken	*Myrrhis odorata*
Sweetbriar	*Rosa eglanteria*
Sweet chervil	*Myrrhis odorata*
Sweet cicely	*Myrrhis odorata*
Sweet elder	*Sambucus nigra*
Sweet flag	*Acorus calamus*
Sweet marjoram	*Majorana hortensis*
	Origanum majorana
Sweet olive	*Osmanthus fragrans*
Sweet rocket	*Hesperis matronalis*
Sweet violet	*Viola odorata*
Sweet woodruff	*Asperula odorata*
	Galium odoratum
Sweet wormwood	*Artemisia annua*
Swine's snout	*Leontodon taraxacum*
	Taraxacum officinale
Tailwort	*Borago officinalis*
Tall agrimony	*Agrimonia eupatoria*
Tansy	*Chrysanthemum vulgare*
	Tanacetum vulgare
Tarentum myrtle	*Myrtus communis tarentina*
Tarragon	*Artemisia dracunculus*
Tasmanian blue gum	*Eucalyptus globulus*
Teaberry	*Gaultheria procumbens*
Tea jasmine	*Jasminum officinale*
Teasel	*Dipsacus fullonum*
	Dipsacus sylvestris
Tetterbury	*Bryonia dioica*
Thornapple	*Datura stramonium*
Thunder plant	*Sempervivum tectorum*
Thyme	*Thymus vulgaris*
Toadflax	*Linaria vulgaris*
Toadroot	*Actaea spicata*
Tonka bean	*Dipterix odorata*
Tonquin bean	*Dipterix odorata*
Toothache weed	*Achillea millefolium*
Torches	*Verbascum thapsus*
Tree onion	*Allium cepa proliferum*
Treefoil	*Trifolium incarnatum*
Tricolor sage	*Salvia officinalis tricolor*

True aloe	*Aloe barbadensis*
	Aloe vera
	Aloe vulgaris
True laurel	*Laurus nobilis*
True lavender	*Lavandula angustifolia*
	Lavandula officinalis
	Lavandula spica
	Lavandula vera
Trumpet weed	*Eupatorium purpureum*
Turmeric	*Curcuma longa*
Turnip-rooted parsley	*Petroselinum crispum fusiformis*
Valerian	*Valeriana officinalis*
Vanilla	*Vanilla fragrans*
Variegated applemint	*Mentha suaveolens variegata*
Variegated lemon balm	*Melissa officinalis variegata*
Variegated sage	*Salvia officinalis tricolor*
Velvet dock	*Inula helenium*
Vervain	*Verbena officinalis*
Vesper flower	*Hesperis matronalis*
Violet	*Viola odorata*
Violet bloom	*Solanum dulcamara*
Viper's bugloss	*Echium vulgare*
Virginian scullcap	*Scutellaria galericulata*
	Scutellaria lateriflora
Wahoo	*Eonymus europaeus*
Wall germander	*Teucrium chamaedrys*
Watercress	*Nasturtium officinale*
	Rorippa nasturtium-aquaticum
Water mint	*Mentha aquatica*
Water pepper	*Polygonum hydropiper*
Welsh onion	*Allium fistulosum*
Western mugwort	*Artemisia ludoviciana*
Western yew	*Taxus brevifolia*
White bryony	*Bryonia dioica*
White dittany	*Dictamnus albus*
	Dictamnus fraxinella
White hellebore	*Veratrum album*
	Veratrum viride
White horehound	*Marrubium incanum*
	Marrubium vulgare
White jasmine	*Jasminum officinale*
White lavender	*Lavandula alba*
White mignonette	*Reseda alba*
White mugwort	*Artemisia lactiflora*
White mulberry	*Morus alba*
White mustard	*Brassica alba*
	Brassica hirta
	Sinapis alba

White periwinkle

White periwinkle	*Vinca major alba*
White sage	*Artemisia ludoviciana*
	Salvia apiana
White thyme	*Thymus alba*
White violet	*Viola odorata alba*
White willow	*Salix alba*
Wild arrach	
	Chenopodium vulvaria
Wild artichoke	
	Cynara cardunculus
Wild bergamot	*Monarda fistulosa*
Wild celery	*Apium graveolens*
Wild chamomile	
	Matricaria chamomilla
Wild dagga	*Leonotus leonurus*
Wild garlic	*Allium ursinum*
Wild marjoram	*Origanum vulgare*
Wild mignonette	*Reseda lutea*
Wild pansy	*Viola tricolor*
Wild parsnip	*Pastinaca sativa*
Wild strawberry	*Fragaria vesca*
Wild succory	*Sabatia angularis*
Wild sweet william	
	Saponaria officinalis
Wild thyme	*Thymus serphyllum*
Wild vine	*Bryonia dioica*
Willow	*Salix nigra*
Winterbloom	
	Hamamelis virginiana
Wintergreen	
	Gaultheria procumbens
Winter marjoram	
	Origanum heracleoticum
Winter purslane	
	Claytonia perfoliata

Winter savory	*Satureia montana*
Wintersweet marjoram	
	Origanum heracleoticum
Witches' gloves	*Digitalis purpurea*
Witch-hazel	
	Hamamelis virginiana
Woad	*Isatis tinctoria*
Wood betony	*Stachys officinalis*
Woodbine	*Lonicera periclymenum*
Wood garlic	*Allium ursinum*
Woodrova	*Asperula odorata*
	Galium odoratum
Woodruff	*Asperula odorata*
	Galium odoratum
Wood sage	*Teucrium scorodonia*
Woody nightshade	
	Solanum dulcamara
Woolly betony	*Stachys lanata*
	Stachys olympica
Woolly mint	*Mentha rotundifolia*
Wormseed	*Artemisia maritima*
Wormwood	*Artemisia absinthium*
Woundwort	*Stachys officinalis*
Wuderove	*Asperula odorata*
	Galium odoratum
Yarrow	*Achillea millefolium*
Yellow bedstraw	*Galium verum*
Yellow gentian	*Gentiana lutea*
Yellow mustard	*Brassica alba*
	Sinapis alba
Yellow puccoon	
	Hydrastis canadensis
Yellow root	*Hydrastis canadensis*
Yellow starwort	*Inula helenium*
Yew	*Taxus baccata*
	Taxus brevifolia

HOUSE PLANTS

This section deals with those plants normally associated with hotel foyers, civic halls, restaurants, and public functions in addition to those found in private houses, sun rooms, and conservatories. Many subjects have been duplicated elsewhere in this book which, in their natural environment, would grow too large for indoor use but by regular pruning and root restriction can be controlled within acceptable limits. With a few exceptions bulbs and cacti are listed elsewhere under those headings.

Flaming katy –
Kalanchoe blossfeldiana

Aerial yam

Aerial yam   *Dioscorea bulbifera*

African hemp   *Sparmannia africana*

African marigold   *Tagetes erecta*

Air potato   *Dioscorea bulbifera*

Akeake   *Dodonaea viscosa*

Albany catspaw
   Anigozanthos preissii

Alexander palm
   Ptychosperma elegans
   Seaforthia elegans

Alexandra palm
   Archontophoenix alexandrae

Algaroba   *Ceratonia siliqua*

Aluminium plant   *Pilea cadierei*

Amaryllis
   Hippeastrum × *ackermannii*
   Hippeastrum × *acramanii*

Amazon lily   *Eucharis amazonica*
   Eucharis grandiflora

American aloe   *Agave americana*

American lotus   *Nelumbium lutea*
   Nelumbo lutea
   Nelumbo pentapetala

American marigold   *Tagetes erecta*

Angel's trumpet   *Datura* × *candida*

Angel's wing begonia
   Begonia coccinea

Angel's wings
   Caladium × *hortulanum*

Apple-scented geranium
   Pelargonium odoratissimum

Apple-scented pelargonium
   Pelargonium odoratissimum

Arabian coffee   *Coffea arabica*

Arabian jasmine   *Jasminum sambac*

Areca palm
   Chrysalidocarpus lutescens

Arrowhead vine
   Syngonium podophyllum

Arrowroot   *Maranta arundinacea*

Artillery plant   *Pilea microphylla*
   Pilea muscosa

Asparagus fern
   Asparagus plumosus
   Asparagus setaceus

Ass's tail   *Sedum morganianum*

Aubergine   *Solanum melongena*

Australian banyan
   Ficus macrophylla

Australian bluebell
   Sollya fusiformis
   Sollya heterophylla

Australian bluebell creeper
   Sollya fusiformis
   Sollya heterophylla

Australian fountain palm
   Livistona australis

Australian maidenhair fern
   Adiantum formosum

Australian pea
   Dolichos lablab lignosus
   Dolichos lignosus

Australian rosemary
   Westringia fruticosa
   Westringia rosmariniformis

Autograph tree   *Clusia rosea*

Autumn cattleya   *Cattleya labiata*

Autumn crocus
   Colchicum autumnale

Avocado   *Persea americana*
   Persea gratissima

Aztec lily   *Amaryllis formosissima*
   Sprekelia formosissima

Baboon flower   *Babiana stricta*

Baby blue eyes
   Nemophila menziesii

Baby rubber plant
   Peperomia clusiifolia

Baby smilax
   Asparagus asparagoides myrtifolius

Baby's tears   *Helxine soleirolii*
   Soleirolia soleirolii

Ball fern   *Davallia mariesii*

Balloon vine
   Cardiospermum halicacabum

Balm mint bush
   Prostanthera melissifolia

Balsam apple   *Clusia rosea*
   Momordica balsamina

Balsam pear   *Momordica charantia*

Bamboo palm
   Chamaedorea erumpens
   Rhapis excelsa

Banana passion fruit
   Passiflora mollissima
   Tacsonia mollissima

Banyan   *Ficus benghalensis*

Barbados cherry   *Malpighia glabra*

Barbados gooseberry
   Pereskia aculeata

Barbados pride
   Caesalpinia pulcherrima
   Poinciana pulcherrima

Barberton daisy   *Gerbera jamesonii*

Basket flower
   Hymenocallis calathina
   Hymenocallis narcissiflora

Basket grass   *Oplismenus hirtellus*

Bat flower   *Tacca chantrieri*

Bay   *Laurus nobilis*

Bead plant   *Nertera granadensis*

Bead tree   *Melia azederach*

Bead vine   *Crassula rupestris*

Beefsteak begonia
   Begonia × *erythrophylla*
   Begonia × *feastii*

Beefsteak plant   *Iresine herbstii*

Belladonna lily
   Amaryllis belladonna

Bell pepper   *Capsicum annuum*

Cactus geranium

Bermuda buttercup *Oxalis cernua*
Oxalis pes-caprae
Bilimbi *Averrhoa bilimbi*
Bird-catching tree
Pisonia umbellifera
Bird of paradise flower
Caesalpinia gilliesii
Strelitzia reginae
Bird's eyes *Gilia tricolor*
Bird's nest bromeliad
Nidularium innocentii
Bird's nest fern *Asplenium nidus*
Asplenium nidus-avis
Birth wort *Aristolochia elegans*
Black bean *Kennedia nigricans*
Blackboy *Xanthorrhoea preissii*
Black echeveria *Echeveria affinis*
Black-eyed susan *Thunbergia alata*
Black-gold philodendron
Philodendron andreanum
Philodendron melanochrysum
Black leaf panamiga *Pilea repens*
Black pepper *Piper nigrum*
Black sarana
Fritillaria camtschatcensis
Black tree fern *Cyathea medullaris*
Blackwood acacia
Acacia melanoxylon
Bleeding heart *Dicentra spectabilis*
Bleeding heart vine
Clerodendrum thomsonae
Blood flower *Asclepias curassavica*
Haemanthus katherinae
Haemanthus multiflorus
Blood leaf *Iresine lindenii*
Blue amaryllis
Hippeastrum procerum
Worsleya rayneri
Blue cape leadwort
Plumbago auriculata
Plumbago capensis
Blue cape plumbago
Plumbago auriculata
Plumbago capensis
Blue echeveria *Echeveria glauca*
Echeveria secunda glauca
Blue glory bower
Clerodendrum ugandense
Blue gum *Eucalyptus globulus*
Blue-leaved wattle
Acacia cyanophylla
Blue lotus *Nymphaea stellata*
Blue marguerite *Agathaea coelestis*
Felicia amelloides
Blue passion flower
Passiflora caerulea
Blue potato bush
Solanum rantonnetii
Blue sage
Daedalacanthus nervosum
Eranthemum nervosum
Eranthemum pulchellum

Blue screw pine *Pandanus baptisii*
Blue shamrock pea
Parochetus communis
Blue taro *Xanthosma violaceum*
Blue thimble flower *Gilia capitata*
Blue tiger lily *Cypella plumbea*
Blue trumpet vine
Thunbergia grandiflora
Blue water lily *Nymphaea capensis*
Blushing bromeliad
Guzmania picta
Nidularium fulgens
Nidularium pictum
Blushing philodendron
Philodendron erubescens
Boat lily *Rhoeo discolor*
Rhoeo spathacea
Botany bay gum
Xanthorrhoea arborea
Bo tree *Ficus religiosa*
Bottlebrush *Callistemon speciosus*
Bottle gourd *Lagenaria siceraria*
Lagenaria vulgaris
Bower plant *Pandorea jasminoides*
Brazilian coleus
Plectranthus oertendahlii
Brazilian potato tree
Solanum macranthum
Bread tree cycad
Encephalartos altensteinii
Bridal flower
Stephanotis floribunda
Bridal wreath
Francoa appendiculata
Bronze inch plant *Zebrina purpusii*
Brush cherry *Eugenia australis*
Eugenia myrtifolia
Eugenia paniculata
Syzygium paniculatum
Bullock's heart *Annona reticulata*
Burning bush *Kochia scoparia*
Burn plant *Aloe barbadensis*
Burro's tail *Sedum morganianum*
Bush violet *Browallia speciosa*
Busy lizzie *Impatiens holstii*
Impatiens sultanii
Impatiens walleriana
Butcher's broom
Ruscus hypoglossum
Butterfly ginger lily
Hedychium coronarium
Butterfly orchid *Oncidium papilio*
Butterfly vine
Stigmaphyllon ciliatum
Butterwort *Pinguicula vulgaris*
Button fern *Pellaea rotundifolia*
Cabbage palm *Cordyline australis*
Cabbage tree *Cordyline australis*
Cactus geranium
Pelargonium echinatum

Cactus pelargonium
Pelargonium echinatum

Cactus spurge
Euphorbia pseudocactus

Cajeput   *Melaleuca leucodendron*

Calabash gourd   *Lagenaria siceraria*
Lagenaria vulgaris

Calamondin   × *Citrofortunella mitis*
Citrus mitis
Citrus reticulata × fortunella

Calico flower   *Aristolochia elegans*

Calico hearts
Adromischus maculatus
Cotyledon maculata
Crassula maculata

Calico plant
Alternanthera bettzickiana

Californian pitcher plant
Darlingtonia californica

Californian poppy
Eschscholzia californica

Canary bird bush
Crotalaria agatiflora

Canary creeper
Tropaeolum canariensis
Tropaeolum peregrinum

Canary Islands ivy
Hedera canariensis

Canary Islands date palm
Phoenix canariensis

Candelabra tree
Araucaria angustifolia

Candle plant   *Kleinia articulatus*
Plectranthus coleoides
Senecio articulatus

Cape cowslip   *Lachenalia aloides*
Lachenalia tricolor

Cape grape   *Cissus capensis*
Rhoicissus capensis
Vitis capensis

Cape honeysuckle
Bignonia capensis
Tecoma capensis
Tecomaria capensis

Cape ivy   *Senecio macroglossus*

Cape jasmine   *Ervatamia coronaria*
Gardenia florida
Gardenia grandiflora
Gardenia jasminoides
Tabernaemontana coronaria
Tabernaemontana divaricata

Cape myrtle   *Lagerstroemia indica*

Cape primrose   *Streptocarpus rexii*

Cape tulip   *Homeria breyniana*
Homeria collina

Carambola tree
Averrhoa carambola

Cardinal flower   *Gesneria cardinalis*
Sinningia cardinalis

Cardinal's guard   *Justicia coccinea*
Pachystachys cardinalis
Pachystachys coccinea

Caricature plant
Graptophyllum pictum

Carnation   *Dianthus caryophyllus*

Carob   *Ceratonia siliqua*

Carolina yellow jasmine
Gelsemium sempervirens

Cassava   *Manihot esculenta*
Manihot utilissima

Cast iron plant   *Aspidistra elatior*
Aspidistra lurida

Castor oil plant   *Ricinus communis*

Cathedral bells   *Cobaea scandens*

Cat's claw vine
Doxantha unguis-cati

Cat's jaws   *Faucaria felina*

Catspaw   *Anigozanthos humilis*

Cat's whiskers   *Tacca chantrieri*

Century plant   *Agave americana*

Ceriman   *Monstera deliciosa*
Monstera pertusa
Philodendron pertusum

Chandelier plant
Kalanchoe tubiflora

Chenille plant   *Acalypha hispida*
Echeveria leucotricha

Cherimoya   *Annona cherimolia*

Cherry tomato
Lycopersicon lycopersicum cerasif orme

Chestnut dioon   *Dioon edule*

Chestnut vine
Cissus voinierianum
Tetrastigma voinierianum
Vitis voinierianum

Chilean bellflower   *Lapageria rosea*

Chilean glory flower
Eccremocarpus scaber

Chilean jasmine   *Mandevilla laxa*
Mandevilla suaveolens

Chilean potato tree
Solanum crispum

Chilean wine palm   *Jubaea chilensis*
Jubaea spectabilis

Chillies   *Capsicum annuum*

Chimney bellflower
Campanula pyramidalis

China grass   *Boehmeria nivea*

Chincherinchee
Ornithogalum thyrsoides

Chinese evergreen
Aglaonema modestum

Chinese fountain palm
Livistona chinensis

Chinese jasmine
Jasminum polyanthum

Chinese lantern plant
Ceropegia woodii

Chinese primrose   *Primula sinensis*

Chinese yam   *Dioscorea batatus*
Dioscorea opposita

Chives   *Allium schoenoprasum*

Christmas begonia
Begonia × cheimantha

Christmas fern
Polystichum acrostichoides

Christmas orchid
Cattleya labiata trianaei
Cattleya trianaei

Christmas palm *Veitchia merrillii*

Chufa nut *Cyperus esculentus*

Chusan palm
Trachycarpus fortunei

Cigar flower *Cuphea ignea*
Cuphea platycentra

Cinnamon yam *Dioscorea batatus*
Dioscorea opposita

Citron *Citrus medica*

Climbing aloe *Aloe ciliaris*

Climbing fig *Ficus pumila*
Ficus repens

Climbing onion *Bowiea volubilis*

Cluster cattleya
Cattleya bowringiana

Cobra plant
Darlingtonia californica

Cockle-shell orchid
Epidendrum cochleatum

Cockscomb
Celosia argentea cristata
Celosia cristata

Cockspur coral tree
Erythrina crista-galli

Cocktail orchid
Cattleya intermedia

Coconut palm *Cocos nucifera*

Cocoyam *Colocasia esculenta*

Coin leaf *Peperomia polybotrya*

Common banana
Musa × paradisiaca
Musa × sapientum

Common blackboy
Xanthorrhoea preissii

Common dumb cane
Dieffenbachia maculata

Common gardenia
Gardenia florida
Gardenia grandiflora
Gardenia jasminoides

Common ginger
Zingiber officinale

Common green kangaroo paw
Anigozanthos manglesii

Common heliotrope
Heliotropium arborescens
Heliotropium peruvianum

Common hop *Humulus lupulus*

Common hydrangea
Hydrangea hortensis
Hydrangea macrophylla

Common ivy *hedera helix*

Common lavender
Lavandula angustifolia

Common maidenhair fern
Adiantum capillus-veneris

Common marigold
Calendula officinalis

Common morning glory
Ipomoea purpurea
Pharbitis purpurea

Common myrtle
Myrtus communis

Common passion flower
Passiflora caerulea

Common pitcher plant
Sarracenia purpurea

Common polypody
Polypodium vulgare

Common sage *Salvia officinalis*

Common screw pine
Pandanus utilis

Common snowdrop
Galanthus nivalis

Common stag's horn fern
Platycerium alcicorne
Platycerium bifurcatum

Common sundew
Drosera rotundifolia

Common throatwort
Trachelium caeruleum

Common trigger plant
Stylidium graminifolium

Common wallflower
Cheiranthus cheiri

Common white jasmine
Jasminum officinale

Coontie *Zamia floridana*

Cootamunda wattle
Acacia baileyana

Copperleaf *Acalypha wilkesiana*

Copper roses
Echeveria multicaulis

Coquito *Jubaea chilensis*
Jubaea spectabilis

Coral aloe *Aloe striata*

Coralberry *Aechmea fulgens*
Ardisia crispa

Coral drops *Bessera elegans*

Coral gem *Lotus berthelotii*

Corallita *Antigonon leptopus*

Coral moss *Nertera depressa*

Coral plant *Berberidopsis corallina*
Jatropha multifida
Russelia equisetiformis
Russelia juncea

Coral vine *Antigonon leptopus*

Cornflower *Centaurea cyanus*

Corn plant *Dracaena fragrans*

Cow's horn
Euphorbia grandicornis

Creeping charlie
Pilea nummulariifolia

Creeping fig *Ficus pumila*
Ficus repens

Creeping peperomia

Creeping peperomia
Peperomia prostrata
Peperomia rotundifolia pilosior

Crepe ginger Costus speciosus

Cretan brake Pteris cretica

Croton
Codiaeum variegatum pictum

Crown imperial
Fritillaria imperialis

Crown of thorns Euphorbia milii

Crown stag's horn
Platycerium biforme
Platycerium coronarium

Cruel plant Araujia sericofera

Crystal anthurium
Anthurium crystallinum

Cuban royal palm Roystonea regia

Cup-and-saucer creeper
Cobaea scandens

Cupid peperomia
Peperomia scandens

Cup of gold Solandra guttata
Solandra hartwegii
Solandra maxima
Solandra nitida

Curly palm Howeia belmoreana
Kentia belmoreana

Currant tomato
Lycopersicon pimpinellifolium

Curry leaf Murraya koenigii

Curry plant
Helichrysum angustifolium
Helichrysum italicum
Helichrysum serotinum

Curuba Passiflora mollissima
Tacsonia mollissima

Custard apple Annona reticulata

Cypress vine Ipomoea quamoclit

Damper's pea Clianthus formosus
Clianthus dampieri

Dancing doll orchid
Oncidium flexuosum

Darling River pea
Swainsona greyana

Dasheen Colocasia esculenta

Date palm Phoenix dactylifera

Day flower Commelina coelestis

Desert fan palm Pritchardia filifera
Washingtonia filifera

Desert privet
Peperomia magnoliifolia

Desert rose Adenium obesum
Echeveria rosea

Devil's backbone
Kalanchoe diagremontiana

Devil's ivy Epipremnum aureum
Pothos aureus
Raphidophora aurea
Scindapsus aureus

Devil's tail
Amorphophallus bulbifer

Devil's tongue
Amorphophallus rivieri
Sansevieria zeylanica

Dewflower
Drosanthemum speciosum

Dinner-plate aralia
Polyscias balfouriana

Donkey's tail Sedum morganianum

Dove orchid
Oncidium ornithorhynchum
Peristeria elata

Dragon plant Arum dracunculus
Dracunculus vulgaris

Dragon's tree Dracaena draco

Drooping star of Bethlehem
Ornithogalum nutans

Dusky coral pea
Kennedia rubicunda

Dusty miller Centaurea cineraria
Centaurea gymnocarpa
Cineraria maritima
Senecio bicolor cineraria
Senecio cineraria
Senecio maritimus

Dutchman's breeches
Dicentra spectabilis

Dwarf fan palm
Chamaerops humilis

Dwarf ginger lily
Kaempferia roscoana

Dwarf mountain palm
Chamaedorea elegans
Neanthe elegans

Easter orchid
Cattleya labiata mossiae
Cattleya mossiae

East Indian arrowroot
Tacca leontopetaloides
Tacca pinnatifida

East Indian holly fern
Arachnoides aristata
Polystichum aristata

East Indian rosebay
Ervatamia coronaria
Tabernaemontana coronaria
Tabernaemontana divaricata

Ebony spleenwort
Asplenium platyneuron

Egg-plant Solanum melongena

Egyptian blue lotus
Nymphaea caerula

Egyptian paper reed
Cyperus papyrus

Egyptian star cluster Pentas carnea
Pentas lanceolata

Elephant foot
Beaucarnea recurvata
Nolina recurvata
Nolina tuberculata

Elephant's ear Colocasia esculenta
Philodendron domesticum
Philodendron hastatum

German ivy

Elephant's ear begonia
Begonia haageana
Begonia scharffii

Elephant's ear fern
Acrostichum crinitum
Elaphoglossum crinitum
Platycerium angolense

Elkhorn fern   *Platycerium hillii*

Emerald ripple   *Peperomia caperata*

English ivy   *Hedera helix*

English lavender
Lavandula angustifolia

Fairy lachenalia
Lachenalia mutabilis

Fairy primrose   *Primula malacoides*

False African violet
Streptocarpus saxorum

False castor-oil plant
Aralia japonica
Aralia sieboldii
Fatsia japonica

False heather   *Cuphea hyssopifolia*

False yellow jasmine
Gelsemium sempervirens

Fan begonia   *Begonia rex*

Fancy geraniums
Pelargonium × domesticum

Fancy pelargoniums
Pelargonium × domesticum

Fat pork tree   *Clusia rosea*

Fern-leaf aralia   *Polyscias filicifolia*

Fernleaf begonia   *Begonia foliosa*

Fern palm   *Cycas circinalis*

Fiddle-leaf fig   *Ficus lyrata*
Ficus pandurata

Fingernail plant
Neoregelia spectabilis

Finger tree   *Euphorbia tirucalli*

Fire cracker   *Brodiaea coccinea*
Brodiaea ida-maia

Firecracker plant
Crossandra infundibuliformis
Crossandra undulifolia
Echeveria setosa

Firecracker vine   *Manettia bicolor*
Manettia inflata
Manettia luteo-rubra

Fishpole bamboo
Phyllostachys aurea

Five fingers   *Philodendron auritum*
Syngonium auritum

Five spot   *Nemophila maculata*

Flame flower   *Bignonia venusta*
Pyrostegia ignea
Pyrostegia venusta

Flame ivy   *Hemigraphis alternata*
Hemigraphis colorata

Flame nettle   *Coleus blumei*

Flame-of-the-woods   *Ixora coccinea*

Flame vine   *Bignonia venusta*
Pyrostegia ignea
Pyrostegia venusta

Flame violet   *Episcia cupreata*

Flaming katy
Kalanchoe blossfeldiana

Flamingo flower
Anthurium scherzerianum

Flaming sword   *Vriesea splendens*

Flaming torch
Guzmania berteroana

Flaming trumpets *Bignonia venusta*
Pyrostegia ignea
Pyrostegia venusta

Floating fern
Ceratopteris pteridoides

Floating moss   *Salvinia auriculata*

Florida arrowroot
Zamia furfuracea
Zamia pumila

Florida ribbon fern *Vittaria lineata*

Florist's genista *Cytisus canariensis*
Genista canariensis
Teline canariensis

Florist's mimosa   *Acacia dealbata*

Flowering banana   *Musa coccinea*

Flowering mignonette
Peperomia fraseri

Flower of the west wind
Zephyranthes candida

Fountain bush
Russelia equisetiformis
Russelia juncea

Fountain flower
Ceropegia sandersonii

Fox brush orchid   *Aerides fieldingii*

Frangipani   *Plumeria acuminata*
Plumeria rubra

Freckle face
Hypoestes phyllostachya

French hydrangea
Hydrangea hortensis
Hydrangea macrophylla

French lavender
Lavandula stoechas

French marigold   *Tagetes patula*

Friendship plant   *Billbergia nutans*
Pilea involucrata

Fuchsia begonia
Begonia fuchsioides

Gardenia   *Gardenia florida*
Gardenia grandiflora
Gardenia jasminoides

Garden pansy   *Viola × hortensis*
Viola × wittrockiana

Garland flower
Hedychium coronarium

Gebang   *Corypha elata*
Corypha gembanga

Geraldton wax flower
Chamaelaucium uncinatum

German ivy   *Senecio mikanioides*

Ghost plant

Ghost plant
    Echeveria paraguayense
    Graptopetalum paraguayense
    Sedum weinbergii

Giant chincherinchee
    Ornithogalum saundersiae

Giant dumb cane
    Dieffenbachia amoena

Giant elephant's ear
    Alocasia macrorrhiza

Giant granadilla
    Passiflora quadrangularis

Giant holly fern
    Polystichum munitum

Giant pineapple flower
    Eucomis pole-evansii

Giant potato vine
    Solanum wendlandii

Giant reed   *Arundo donax*

Giant snowdrop *Galanthus elwesii*

Ginger   *Zingiber officinale*

Gippsland fountain palm
    Livistona australis

Globe amaranth
    Gomphrena globosa

Glory bower
    Clerodendrum thomsonae

Glory bush
    Tibouchina semidecandra
    Tibouchina urvilleana

Glory lily   *Gloriosa rothschildiana*

Glory-of-the-sun
    Leucocoryne ixioides

Glory pea   *Clianthus dampieri*
    Clianthus formosus

Gloxinia   *Gloxinia speciosa*
    Sinningia speciosa

Golden bamboo
    Phyllostachys aurea

Golden butterfly orchid
    Oncidium varicosum

Golden creeper
    Stigmaphyllon ciliatum

Golden dewdrop *Duranta plumieri*
    Duranta repens

Golden feather palm
    Chrysalidocarpus lutescens

Golden male fern
    Dryopteris borreri
    Dryopteris pseudomas

Golden pothos
    Epipremnum aureum
    Pothos aureus
    Raphidophora aurea
    Scindapsus aureus

Golden rain   *Acacia prominens*

Golden stars   *Bloomeria crocea*

Golden trumpet
    Allamanda cathartica

Golden vine
    Stigmaphyllon ciliatum

Gold dust dracaena
    Dracaena godseffiana
    Dracaena surculosa

Goldfish plant *Columnea gloriosa*
    Columnea microphylla

Goosefoot plant
    Syngonium podophyllum

Gosford wattle *Acacia prominens*

Gout plant *Jatropha podagrica*

Grapefruit   *Citrus paradisi*

Grape ivy   *Cissus rhombifolia*

Grass nut   *Brodiaea laxa*
    Triteleia laxa

Grass trigger plant
    Stylidium graminifolium

Greater butterwort
    Pinguicula grandiflora

Great sundew *Drosera anglica*

Green brake fern
    Pellaea adiantoides
    Pellaea hastata
    Pellaea viridis

Green earth star
    Cryptanthus acaulis

Green kangaroo paw
    Anigozanthos viridis

Green pepper *Capsicum annuum*

Ground ivy *Glechoma hederacea*

Groundnut *Arachis hypogaea*

Ground rattan *Rhapis excelsa*

Guernsey lily *Nerine sarniensis*
    Vallota purpurea
    Vallota speciosa

Guinea flower *Hibbertia scandens*
    Hibbertia volubilis

Hairy toad plant *Stapelia hirsuta*

Hard shield fern
    Polystichum aculeatum

Hare's foot fern
    Davallia canariensis
    Phlebodium aureum
    Polypodium aureum

Harlequin flower *Sparaxis tricolor*

Hart's tongue fern
    Asplenium scolopendrium
    Phyllitis scolopendrium

Heart leaf *Philodendron cordatum*
    Philodendron scandens

Heart pea
    Cardiospermum halicacabum

Heart seed
    Cardiospermum halicacabum

Heavenly bamboo
    Nandina domestica

Holly fern *Aspidium falcatum*
    Cyrtomium falcatum
    Polystichum falcatum

Holly-leaved begonia
    Begonia cubensis

Holy ghost plant *Peristeria elata*

Honeybush *Melianthus major*

Ladder fern

Honey flower	*Protea mellifera*
	Protea repens
Honey palm	*Jubaea chilensis*
	Jubaea spectabilis
Hong Kong kumquat	
	Fortunella hindsii
Hoop pine	
	Araucaria cunninghamii
Hop bush	*Dodonaea viscosa*
Horse-shoe geranium	
	Pelargonium zonale
Horse-shoe pelargonium	
	Pelargonium zonale
Hottentot fig	
	Carpobrotus acinaciformis
	Carpobrotus edulis
Humble plant	*Mimosa pudica*
Huntsman's horn	*Sarracenia flava*
Hyacinth bean	*Dolichos lablab*
Ifafa lily	*Cyrtanthus mackenii*
Illawarra palm	*Archontophoenix*
	cunninghamiana
Indian bean	*Dolichos lablab*
Indian day flower	
	Commelina benghalensis
Indian ginger	*Alpinia calcarata*
Indian kale	*Xanthosma lindenii*
Indian pink	*Dianthus chinensis*
Indian shot	*Canna* × *generalis*
Indoor fig	*Ficus diversifolia*
Iron cross begonia	
	Begonia masoniana
Italian bellflower	
	Campanula isophylla
Ithuriel's spear	*Brodiaea laxa*
	Triteleia laxa
Ivyleaf peperomia	
	Peperomia griseoargentea
	Peperomia hederifolia
Ivy-leaved geraniums	
	Pelargonium peltatum
Ivy-leaved pelargoniums	
	Pelargonium peltatum
Ivy-leaved toadflax	
	Cymbalaria muralis
	Linaria cymbalaria
Ivy-leaved violet	*Viola hederacea*
Jacobean lily	
	Amaryllis formosissima
	Sprekelia formosissima
Jacob's ladder	
	Pedilanthes tithymaloides smallii
Jade plant	*Crassula argentea*
	Crassula portulacea
Japanese azalea	*Azalea obtusa*
Japanese banana	*Musa basjoo*
Japanese buckler fern	
	Dryopteris erythrosora
Japanese hop	*Humulus japonicus*
Japanese lantern	
	Hibiscus schizopetalus

Japanese pittosporum	
	Pittosporum tobira
Japanese sago palm	*Cycas revoluta*
Japanese sedge	*Carex morrowii*
Japanese shield fern	
	Dryopteris erythrosora
Jasmine nightshade	
	Solanum jasminoides
Java glorybean	
	Clerodendrum speciosissimum
	Clerodendrum fallax
Jerusalem cherry	
	Solanum pseudocapsicum
Jewel plant	*Bertolonia hirsuta*
Josephine's lily	
	Brunsvigia josephinae
Joseph's coat	
	Codiaeum variegatum pictum
Joshua tree	*Yucca brevifolia*
Jungle geranium	*Ixora javanica*
Kahili ginger	
	Hedychium gardnerianum
Kaffir lily	*Clivia miniata*
	Imantophyllum miniatum
Kaka beak	*Clianthus puniceus*
Kangaroo apple	*Solanum aviculare*
Kangaroo thorn	*Acacia armata*
Kangaroo vine	*Cissus antarctica*
Karaka	*Corynocarpus laevigatus*
Karo	*Pittosporum crassifolium*
Kenilworth ivy	*Cymbalaria muralis*
	Linaria cymbalaria
Kentia palm	*Howeia forsterana*
	Kentia forsterana
Key palm	*Thrinax microcarpa*
	Thrinax morrisii
King anthurium	*Anthurium veitchii*
Kingfisher daisy	*Aster bergeriana*
	Felicia bergeriana
King of the bromeliads	
	Vriesea hieroglyphica
King plant	*Anoectochilus regalis*
King's crown	*Jacobinia carnea*
	Jacobinia velutina
	Justicia carnea
Knife acacia	*Acacia cultriformis*
Kohuhu	*Pittosporum tenuifolium*
Kris plant	*Alocasia sanderiana*
Lablab bean	*Dolichos lablab*
Lace flower	*Episcia dianthiflora*
Lace orchid	
	Odontoglossom crispum
Lacy ground fern	
	Dennstaedtia davallioides
Lacy pine fern	
	Polypodium subauriculatum
Lacy tree philodendron	
	Philodendron selloum
Ladder fern	*Nephrolepis cordifolia*
	Pteris vittata

Lady-in-the-bath
  Dicentra spectabilis
Lady of the night
  Brunfelsia americana
  Brunfelsia violacea
Lady Washington geranium
  Pelargonium × domesticum
Lady Washington pelargonium
  Pelargonium × domesticum
Lamp flower   *Ceropegia caffrorum*
Lance copperleaf
  Acalypha godseffiana
Laurel fig   *Ficus macrocarpa*
  Ficus nitida
  Ficus retusa
Laurustinus   *Viburnum tinus*
Leather fern   *Acrostichum aureum*
  Rumohra adiantiformis
Lehua   *Metrosideros collina*
Lemon bottlebrush
  Callistemon citrinus
  Callistemon lanceolatus
Lemon geranium
  Pelargonium crispum
Lemon pelargonium
  Pelargonium crispum
Lemon-scented gum
  Eucalyptus citriodora
Lemon verbena   *Aloysia citriodora*
  Lippia citriodora
  Verbena triphylla
Leopard orchid   *Ansellia africana*
Levant cotton
  Gossypium herbaceum
Lily of the Incas
  Alstroemeria pelegrina alba
Lily-of-the-valley
  Convallaria majalis
Lily-of-the-valley orchid
  Odontoglossum pulchellum
Lily-of-the-valley tree
  Clethra arborea
Lime   *Citrus aurantifolia*
  Citrus limetta
  Limonia aurantifolia
Lipstick vine
  Aeschynanthus lobbianus
Liquorice plant
  Helichrysum petiolatum
Little lady palm   *Rhapis excelsa*
Livingstone daisy
  Dorotheanthus bellidiformis
  Mesembryanthemum criniflorum
Lobster claw   *Clianthus puniceus*
Lobster claws   *Vriesea carinata*
Locust   *Ceratonia siliqua*
Lollipop plant   *Pachystachys lutea*
London pride   *Saxifraga × urbium*
Loofah   *Luffa cylindrica*
  Luffa aegyptica
Loquat   *Eriobotrya japonica*

Lorraine begonia
  Begonia × cheimantha
Love charm   *Bignonia purpurea*
  Clytostoma binatum
  Clytostoma purpureum
Love-lies-bleeding
  Amaranthus caudatus
Lucky clover   *Oxalis deppei*
Madagascar dragon tree
  Dracaena marginata
Madagascar jasmine
  Stephanotis floribunda
Madagascar periwinkle
  Catharanthus roseus
  Vinca rosea
Magic plant   *Aloe barbadensis*
Maguey   *Agave americana*
Maidenhair fern
  Adiantum capillus-veneris
Maidenhair vine
  Muehlenbeckia complexa
Maid of the mist
  Gladiolus primulinus
Malabar plum   *Eugenia jambos*
  Syzygium jambos
Malayan urn vine
  Dischidia rafflesiana
Malay ginger   *Costus speciosus*
Male fern   *Dryopteris filix-mas*
Mandarin   *Citrus nobilis*
  Citrus reticulata
Mangle's kangaroo paw
  Anigozanthos manglesii
Manioc   *Manihot esculenta*
  Manihot utilissima
Manuka   *Leptospermum scoparium*
Marble plant   *Aregelia marmorata*
  Neoregelia marmorata
Marlberry   *Ardisia crispa*
Marmalade bush
  Streptosolen jamesonii
Martha Washington geranium
  Pelargonium × domesticum
Martha Washington pelargonium
  Pelargonium × domesticum
Mask flower   *Alonsoa warscewiczii*
Mauritius hemp   *Furcraea foetida*
  Furcraea gigantea
Meadow saffron
  Colchicum autumnale
Medicine plant   *Aloe barbadensis*
Medusa's head
  Euphorbia caput-medusae
Meiwa kumquat
  Fortunella crassifolia
Metal leaf begonia
  Begonia metallica
Mexican bread fruit
  Monstera deliciosa
  Monstera pertusa
  Philodendron pertusum

Oak-leaved pelargonium

Mexican dwarf palm
Chamaedorea elegans

Mexican fern palm *Dioon edule*

Mexican flame vine
Senecio confusus

Mexican foxglove
Allophyton mexicanum
Tetranema mexicanum
Tetranema roseum

Mexican orangeblossom
Choisya ternata

Mexican violet
Allophyton mexicanum
Tetranema mexicanum
Tetranema roseum

Mickey mouse plant
Ochna serrulata

Milk bush *Euphorbia tirucalli*

Mind your own business
Helxine soleirolii
Soleirolia soleirolii

Miniature eyelash begonia
Begonia bowerii

Miniature fan palm *Rhapis excelsa*

Miniature fishtail palm
Chamaedorea metallica

Miniature holly
Malpighia coccigera

Miniature maple-leaf begonia
Begonia dregei

Miniature wax plant *Hoya bella*

Mistletoe fig *Ficus deltoidea*
Ficus diversifolia

Monarch of the east
Sauromatum guttatum

Monkey flower *Mimulus guttatus*
Mimulus luteus

Monkey musk *Mimulus guttatus*
Mimulus luteus

Monkey plant *Ruellia mackoyana*

Moonflower
Calonyction bona-nox
Ipomoea bona-nox
Ipomoea noctiflora

Moreton Bay fig *Ficus macrophylla*

Moreton Bay pine
Araucaria cunninghamii

Morning glory *Ipomoea purpurea*
Pharbitis purpurea

Moses-in-a-boat *Rhoeo discolor*
Rhoeo spathacea

Mosquito plant *Azolla caroliniana*

Mother fern
Asplenium daucifolium
Asplenium viviparum

Mother-of-pearl plant
Echeveria paraguayense
Graptopetalum paraguayense
Sedum weinbergii

Mother of thousands
Saxifraga sarmentosa
Saxifraga stolonifera

Mother spleenwort
Asplenium bulbiferum

Mottlecah *Eucalyptus macrocarpa*

Mountain flax *Phormium colensoi*
Phormium cookianum

Mountain pepper
Drimys aromatica
Drimys lanceolata
Tasmannia latifolia
Winterania latifolia

Mount Morgan wattle
Acacia podalyriifolia

Mouse plant
Arisarum proboscideum

Mulberry fig *Ficus sycamorus*

Musk *Mimulus moschatus*

Nagami kumquat
Fortunella margarita

Narrow-leaved ribbon fern
Polypodium angustifolium

Narrow-leaved strap fern
Polypodium angustifolium

Nasturtium *Tropaeolum majus*

Natal ivy *Senecio macroglossus*

Natal plum *Carissa grandiflora*

Nerve plant *Fittonia verschaffeltii*

New Zealand christmas tree
Metrosideros excelsa
Metrosideros tomentosa

New Zealand flax *Phormium tenax*

New Zealand laurel
Corynocarpus laevigatus

New Zealand tea tree
Leptospermum scoparium

Nodding pincushion
Leucospermum cordifolium
Leucospermum nutans

Norfolk Island pine
Araucaria excelsa
Araucaria heterophylla

Northern bungalow palm
Archontophoenix alexandrae

Notch-leaf statice
Limonium sinuatum

Nun's hood orchid *Phaius bicolor*
Phaius blumei
Phaius grandiflorus
Phaius gravesii
Phaius tankervillae
Phaius wallichii

Nutmeg geranium
Pelargonium × fragrans

Nutmeg pelargonium
Pelargonium × fragrans

Oak fern *Drynaria quercifolia*

Oak-leaved geranium
Pelargonium quercifolium

Oak-leaved pelargonium
Pelargonium quercifolium

Octopus tree	*Brassaia actinophylla*
	Schefflera actinophylla
Oleander	*Nerium oleander*
Olive	*Olea europaea*
Opal lachenalia	
	Lachenalia glaucina
Orange jasmine	*Murraya exotica*
	Murraya paniculata
Orange kaleidoscope flower	
	Streptanthera cuprea
Ornamental pepper	
	Piper crocatum
	Piper ornatum crocatum
Ornamental yam	
	Dioscorea discolor
Our Lord's candle	
	Hesperoyucca whipplei
	Yucca whipplei
Oval kumquat	
	Fortunella margarita
Oven's wattle	*Acacia pravissima*
Owl eyes	*Huernia zebrina*
Ox tongue	*Gasteria verrucosa*
Paco	*Athyrium esculentum*
	Diplazium esculentum
Pagoda flower	
	Clerodendrum paniculatum
Pagoda tree	*Plumeria acutifolia*
	Plumeria rubra acutifolia
Paintbrush	*Haemanthus albiflos*
Painted feathers	*Vriesea psittacina*
Painted lady	*Echeveria derenbergii*
	Gladiolus blandus
	Gladiolus carneus
Painted leaf begonia	*Begonia rex*
Painted net leaf	
	Fittonia verschaffeltii
Painted nettle	*Coleus blumei*
Painted tongue	*Salpiglossis sinuata*
Palmella	*Yucca elata*
Palmetto thatch palm	
	Thrinax parviflora
Palm grass	*Curculigo capitulata*
	Curculigo recurvata
Palm-leaf begonia	
	Begonia luxurians
Panamiga	*Pilea involucrata*
Panda plant	*Kalanchoe pilosa*
	Kalanchoe tomentosa
Pansy	*Viola × hortensis*
	Viola × wittrockiana
Papaya	*Carica papaya*
Paper flower	*Bougainvillea glabra*
Paprika	*Capsicum annuum*
Papyrus	*Cyperus papyrus*
Parachute plant	
	Ceropegia sandersonii
Paradise palm	*Howeia forsterana*
	Kentia forsterana
Parana pine	*Araucaria angustifolia*
Parapara	*Pisonia umbellifera*

Parlour palm	
	Chamaedorea elegans
	Neanthe elegans
Parrot leaf	*Alternanthera ficoidea*
Parrot's beak	*Lotus berthelotii*
Parrot's bill	*Clianthus puniceus*
Passion fruit	*Passiflora edulis*
Patient lucy	*Impatiens holstii*
	Impatiens sultanii
	Impatiens walleriana
Pawpaw	*Carica papaya*
Peace lily	*Spathiphyllum wallisii*
Peacock moss	*Selaginella uncinata*
Peacock plant	*Calathea makoyana*
	Kaempferia roscoana
Peanut	*Arachis hypogaea*
Pearl plant	
	Haworthia margaritifera
	Haworthia pumila
Peppermint-scented geranium	
	Pelargonium tomentosum
Peppermint-scented pelargonium	
	Pelargonium tomentosum
Pepper tree	*Drimys colorata*
	Pseudowintera colorata
Pepul	*Ficus religiosa*
Persian ivy	*Hedera colchica*
Persian lilac	*Melia azederach*
Persian shield	
	Strobilanthes dyerianus
Peruvian daffodil	
	Hymenocallis calathina
	Hymenocallis narcissiflora
Peruvian mastic tree	
	Schinus molle
Peruvian pepper tree	
	Schinus molle
Petticoat palm	*Pritchardia filifera*
	Washingtonia filifera
Philippines sugar plum	
	Corypha elata
	Corypha gembanga
Philippine wax flower	
	Alpinia magnifica
	Nicolaia elatior
	Phaeomeria magnifica
Physic nut	*Jatropha multifida*
Piccabeen bungalow palm	
	Archontophoenix
	cunninghamiana
Piccabeen palm	*Archontophoenix*
	cunninghamiana
Pigeon berry	*Duranta plumieri*
	Duranta repens
Piggyback plant	*Tolmiea menziesii*
Pineapple flower	*Eucomis comosa*
Pineapple guava	*Acca sellowiana*
	Feijoa sellowiana
Pineapple sage	*Salvia elegans*
	Salvia rutilans
Pine fern	*Anemia adiantifolia*

HOUSE PLANTS

Redbird cactus

Pink jasmine
   Jasminum polyanthum

Pink porcelain lily   *Alpinia nutans*
   Alpinia speciosa
   Alpinia zerumbet

Pink rainbow   *Drosera menziesii*

Pitanga   *Eugenia uniflora*

Plume albizia   *Acacia lophantha*
   Albizia distachya
   Albizia lophantha

Plush plant   *Echeveria pulvinata*

Pocket book plant
   Calceolaria × herbeohybrida

Poet's jessamine
   Jasminum officinale

Poinsettia   *Euphorbia pulcherrima*

Polka dot plant
   Hypoestes phyllostachya

Polyanthus   *Primula × tommasinii*

Pomegranate   *Punica granatum*

Ponga   *Alsophila tricolor*
   Cyathea dealbata

Pony tail   *Beaucarnea recurvata*
   Nolina recurvata
   Nolina tuberculata

Poor man's orchid
   Schizanthus pinnatus

Poppy anemone
   Anemone coronaria

Porcelain berry
   Ampelopsis brevipedunculata
   Ampelopsis heterophylla

Port Jackson fig   *Ficus rubiginosa*

Potato vine   *Solanum jasminoides*

Pot marigold   *Calendula officinalis*

Prairie gentian
   Eustoma grandiflorum
   Eustoma russellianum
   Lisianthus russellianus

Prayer plant   *Maranta leuconeura*

Prickly moses   *Acacia verticillata*

Prickly shield fern
   Polystichum aculeatum

Primrose   *Primula vulgaris*

Primrose jasmine
   Jasminum mesneyi
   Jasminum primulinum

Prince of Wales feathers
   Celosia argentea pyramidalis
   Celosia plumosa
   Celosia pyramidalis

Prince Rupert geranium
   Pelargonium crispum

Prince Rupert pelargonium
   Pelargonium crispum

Prince's feather
   Amaranthus hybridus
   Amaranthus hypochondriacus

Princess vine   *Cissus sicyoides*

Prostrate rosemary
   Rosmarinus lavandulaceum
   Rosmarinus officinalis prostratus

Pummelo   *Citrus decumanus*
   Citrus grandis
   Citrus maxima

Purple bell vine
   Rhodochiton atrosanguineum
   Rhodochiton volubile

Purple granadilla   *Passiflora edulis*

Purple ragwort   *Senecio elegans*

Purple-stemmed cliff brake
   Pellaea atropurpurea

Purple viper's bugloss
   Echium lycopsis
   Echium plantagineum

Purple wreath   *Petrea volubilis*

Pussy ears   *Cyanotis somaliensis*
   Kalanchoe pilosa
   Kalanchoe tomentosa

Pygmy date palm
   Phoenix roebelenii

Pygmy water lily
   Nymphaea pygmaea
   Nymphaea tetragona

Queen anthurium
   Anthurium warocqueanum

Queen cattleya
   Cattleya labiata dowiana

Queen of the night
   Hylocereus undatus

Queensland silver wattle
   Acacia podalyriifolia

Queensland tassel fern
   Lycopodium phlegmaria

Queensland umbrella tree
   Brassaia actinophylla
   Schefflera actinophylla

Queen's tears   *Billbergia nutans*

Queen's wreath   *Petrea volubilis*

Rabbit's foot   *Maranta leuconeura kerchoveana*

Rabbit's foot fern   *Davallia fijiensis*
   Davallia solida fijiensis

Rabbit tracks   *Maranta leuconeura kerchoveana*

Radiator plant
   Peperomia maculosa

Rainbow fern   *Selaginella uncinata*

Rainbow star
   Cryptanthus bromelioides

Ramie   *Boehmeria nivea*

Rangoon creeper   *Quisqualis indica*

Rat's tail statice
   Limonium suworowii

Rat tail plant
   Crassula lycopodioides

Rattlesnake plant   *Calathea insignis*
   Calathea lancifolia

Rattlesnake tail   *Crassula barklyi*
   Crassula teres

Red alder   *Cunonia capensis*

Redbird cactus
   Pedilanthes tithymaloides

Red-flowering gum

Red-flowering gum	
	Eucalyptus ficifolia
Red ginger	*Alpinia purpurata*
Red ginger lily	
	Hedychium coccineum
Red granadilla	*Passiflora coccinea*
Red-hot cat's tail	*Acalypha hispida*
Red ivy	*Hemigraphis alternata*
	Hemigraphis colorata
Red kangaroo paw	
	Anigozanthos rufus
Red latan palm	*Latania borbonica*
	Latania commersonii
	Latania lontaroides
Red-leaf philodendron	
	Philodendron domesticum
	× erubescens
	Philodendron × mandaianum
Red morning glory	
	Ipomoea coccinea
Red nodding bells	
	Streptocarpus dunnii
Red passion flower	
	Passiflora coccinea
	Passiflora racemosa
Red pepper	*Capsicum annuum*
Red pineapple	*Ananas bracteatus*
Reed palm	*Chamaedorea seifrizii*
Regal elkhorn fern	
	Platycerium grande
Regal geraniums	
	Pelargonium × domesticum
Regal pelargoniums	
	Pelargonium × domesticum
Resurrection plant	
	Anastatica hierochuntica
	Selaginella lepidophylla
Rex begonia vine	*Cissus discolor*
Ribbon fern	
	Campyloneurum phyllitidis
	Polypodium phyllitidis
	Pteris cretica
Ribbon plant	*Dracaena sanderiana*
Rice-paper plant	*Aralia papyrifera*
	Fatsia papyrifera
	Tetrapanax papyriferus
River tea tree	
	Melaleuca leucodendron
Rosary vine	*Ceropegia woodii*
	Crassula rupestris
Rose apple	*Eugenia jambos*
	Syzygium jambos
Rose balsam	*Impatiens balsamina*
Rose bay	*Nerium oleander*
Rose geranium	
	Pelargonium graveolens
Rose grape	*Medinilla magnifica*
Rose maidenhair fern	
	Adiantum hispidulum
	Adiantum pubescens
Rosemary	*Rosmarinus officinalis*

Rose of China	
	Hibiscus rosasinensis
Rose of Jericho	
	Anastatica hierochuntica
	Selaginella lepidophylla
Rose pelargonium	
	Pelargonium graveolens
Rose-scented geranium	
	Pelargonium capitatum
Rose-scented pelargonium	
	Pelargonium capitatum
Rouge plant	*Rivina humilis*
	Rivina laevis
Royal red bugler	
	Aeschynanthus pulcher
Rubber plant	*Ficus elastica*
Rubber spurge	*Euphorbia tirucalli*
Running postman	
	Kennedia prostrata
Russian statice	
	Limonium suworowii
Sacred bamboo	
	Nandina domestica
Sacred fig tree	*Ficus religiosa*
Sage rose	*Turnera trioniflora*
	Turnera ulmifolia
Sago fern	*Cyathea medullaris*
Sago fern palm	*Cycas circinalis*
Sago palm	*Caryota urens*
Saint Augustine grass	
	Stenotaphrum secundatum
Saint John's bread	*Ceratonia siliqua*
Salad tomato	
	Lycopersicon lycopersicum
	cerasiforme
Satin leaf	*Philodendron gloriosum*
Satsuma	*Citrus nobilis*
	Citrus reticulata
Scarborough lily	*Vallota purpurea*
	Vallota speciosa
Scarlet banana	*Musa coccinea*
Scarlet ginger lily	
	Hedychium coccineum
Scarlet leadwort	*Plumbago indica*
	Plumbago rosea
Scarlet plume	*Euphorbia fulgens*
Scarlet sage	*Salvia splendens*
Scarlet star	*Guzmania lingulata*
Scarlet trompetilla	
	Bouvardia ternifolia
	Bouvardia triphylla
Scorpion senna	*Coronilla emerus*
Scotch attorney	
	Clusia grandiflora
Sea daffodil	
	Pancratium maritimum
Sea fig	*Carpobrotus chilensis*
Sea lily	*Pancratium maritimum*
Sealing wax palm	
	Cyrtostachys lakka
Seaside grape	*Coccoloba uvifera*

Squirrel's foot fern

Seersucker plant	*Geogenanthus undatus*
Seminole bread	*Zamia floridana*
Sensitive plant	*Biophytum sensitivum*
	Mimosa pudica
Sentry palm	*Howeia belmoreana*
	Howeia forsterana
	Kentia belmoreana
	Kentia forsterana
Seville orange	*Citrus aurantium*
	Citrus bigaradia
Shaddock	*Citrus decumanus*
	Citrus grandis
	Citrus maxima
Shell ginger	*Alpinia nutans*
	Alpinia speciosa
	Alpinia zerumbet
Shingle plant	*Monstera latevaginata*
	Rhaphidophora celatocaulis
	Rhaphidophora decursiva
Shining gum	*Eucalyptus nitens*
Shiny tea tree	*Leptospermum nitidum*
Show geranium	*Pelargonium × domesticum*
Show pelargonium	*Pelargonium × domesticum*
Shrimp plant	*Beloperone guttata*
Siberian squill	*Scilla sibirica*
Siberian wallflower	*Cheiranthus × allionii*
Sicklethorn	*Asparagus falcatus*
Signet marigold	*Tagetes signata*
	Tagetes tenuifolia
Silk tree	*Albizia julibrissin*
Silky oak	*Grevillea robusta*
Silver beads	*Crassula deltoidea*
	Crassula rhomboidea
Silver heart	*Peperomia marmorata*
Silver latan palm	*Latania loddigesii*
Silverleaf peperomia	*Peperomia griseoargentea*
	Peperomia hederifolia
Silver net leaf	*Fittonia argyroneura*
Silver tree	*Leucadendron argenteum*
Silver tree fern	*Alsophila tricolor*
	Cyathea dealbata
Silver vase	*Aechmea fasciata*
Silver wattle	*Acacia dealbata*
Silvery inch plant	*Zebrina pendula*
Skyflower	*Duranta plumieri*
	Duranta repens
Skyrocket	*Gilia coronipifolia*
	Gilia rubra
	Ipomopsis rubra
Slipper cactus	*Pedilanthes tithymaloides*
Slipper flower	*Calceolaria × herbeohybrida*

Smilax	*Asparagus asparagoides*
	Asparagus medeoloides
	Smilax asparagoides
Snake gourd	*Trichosanthes anguina*
Snake lily	*Brodiaea volubilis*
	Stropholirion californicum
Snake plant	*Sansevieria trifasciata*
Snake's head fritillary	*Fritillaria meleagris*
Snake vine	*Hibbertia scandens*
	Hibbertia volubilis
Snapdragon	*Antirrhinum majus*
Snow bush	*Breynia nivosa*
	Phyllanthus nivosa
Snowdrop	*Galanthus nivalis*
Snowflake aralia	*Trevesia palmata*
Snowy mint bush	*Prostanthera nivea*
Soap tree	*Yucca elata*
Soft shield fern	*Polystichum setiferum*
Soft tree fern	*Alsophila smithii*
	Cyathea smithii
	Dicksonia antarctica
	Hemitelia smithii
Solitaire palm	*Ptychosperma elegans*
	Seaforthia elegans
South sea arrowroot	*Tacca leontopetaloides*
	Tacca pinnatifida
Spade leaf	*Philodendron domesticum*
	Philodendron hastatum
Spanish dagger	*Yucca gloriosa*
Spanish moss	*Tillandsia usneioides*
Spiceberry	*Ardisia crispa*
Spider fern	*Pteris multifida*
	Pteris serrulata
Spider plant	*Anthericum comosum*
	Anthericum elatum
	Chlorophytum capense
	Chlorophytum comosum
	Chlorophytum elatum
Spineless yucca	*Yucca elephantipes*
	Yucca gigantea
	Yucca guatemalensis
Spiral ginger	*Costus igneus*
Sponge gourd	*Luffa aegyptica*
	Luffa cylindrica
Spotted dead nettle	*Lamium maculatum*
Spring cattleya	*Cattleya labiata mossiae*
	Cattleya mossiae
Spring star flower	*Ipheion uniflorum*
Squirrel's foot fern	*Davallia dissecta*
	Davallia mariesii
	Davallia pyxidata
	Davallia trichomanoides

Squirting cucumber
Ecballium elaterium

Standing cypress *Gilia coronipifolia*
Gilia rubra
Ipomopsis rubra

Star begonia *Begonia heracleifolia*

Star cluster *Pentas carnea*
Pentas lanceolata

Star ipomoea *Ipomoea coccinea*

Star jasmine *Jasminum nitidum*

Star of Bethlehem
Ornithogalum arabicum
Ornithogalum umbellatum

Star sansevieria
Sansevieria grandicuspis

Star window plant
Haworthia tessellata

Stock *Matthiola incana*

Strap fern
Campyloneurum phyllitidis
Polypodium phyllitidis

Strap water fern
Blechnum patersonii

Strawberry geranium
Saxifraga sarmentosa
Saxifraga stolonifera

Strawberry tree *Arbutus unedo*

String of beads *Senecio rowleyanus*

String of hearts *Ceropegia woodii*

String of pearls *Senecio rowleyanus*

Sturt's desert rose
Gossypium sturtianum
Gossypium sturtii

Sugar apple *Annona squamosa*

Sugarbush *Protea mellifera*
Protea repens

Summer cypress *Kochia scoparia*

Summer hyacinth
Galtonia candicans

Summer jasmine
Jasminum officinale

Sunn hemp *Crotalaria juncea*

Sun plant *Portulaca grandiflora*

Surinam cherry *Eugenia uniflora*

Swamp lily *Crinum × powellii*

Swan orchid
Cycnoches ventricosum

Swedish ivy
Plectranthus oertendahlii

Sweet bay *Laurus nobilis*

Sweet flag *Acorus gramineus*

Sweet garlic *Tulbaghia fragrans*

Sweet granadilla *Passiflora ligularis*

Sweetheart plant
Philodendron scandens

Sweet orange
Citrus aurantium sinense
Citrus sinensis

Sweet pepper *Capsicum annuum*

Sweet potato *Ipomoea batatas*

Sweet sop *Annona squamosa*

Sweet sultan *Centaurea moschata*

Swiss cheese plant
Monstera deliciosa
Monstera pertusa
Philodendron pertusum

Sword brake *Pteris ensiformis*

Sword fern *Nephrolepis cordifolia*
Nephrolepis exaltata

Sycamore fig *Ficus sycamorus*

Sydney golden wattle
Acacia longifolia

Table fern *Pteris cretica*

Tagetes *Tagetes signata*
Tagetes tenuifolia

Tailflower *Anthurium andreanum*

Talipot palm
Corypha umbraculifera

Tall kangaroo paw
Anigozanthos flavidus

Tangerine *Citrus nobilis*
Citrus reticulata

Tapioca *Manihot esculenta*
Manihot utilissima

Taro *Colocasia esculenta*

Taro vine *Epipremnum aureum*
Pothos aureus
Raphidophora aurea
Scindapsus aureus

Tartogo *Jatropha podagrica*

Tasmanian blue gum
Eucalyptus globulus

Teddy bear vine *Cyanotis kewensis*

Temple bells
Smithiana cinnabarina

Thatchleaf palm *Howeia forsterana*
Kentia forsterana

Thread agave *Agave filifera*

Tiger jaws *Faucaria tigrina*

Tiger nut *Cyperus esculentus*

Tiger orchid *Oncidium tigrinum*

Ti tree *Cordyline terminalis*

Toadflax *Linaria maroccana*

Tobira *Pittosporum tobira*

Toddy palm *Caryota urens*

Tomato *Lycopersicon esculentum*
Lycopersicon lycopersicum

Tonkin bamboo
Arundinaria amabilis

Torch ginger *Alpinia magnifica*
Nicolaia elatior
Phaeomeria magnifica

Touch-me-not
Impatiens balsamina

Trailing water-melon begonia
Pellionia daveauana

Transvaal daisy *Gerbera jamesonii*

Traveller's tree
Ravenala madagascariensis

Tree cotton *Gossypium arboreum*

Wonga-wonga vine

Tree philodendron
  Philodendron bipinnatifidum
  Philodendron eichleri

Tree tomato
  Cyphomandra betacea

Trigger plant
  Stylidium graminifolium

Tropical crocus
  Kaempferia rotunda

Trumpet vine
  Bignonia callistegioides
  Bignonia speciosa
  Clytostoma callistegioides

Tsusina holly fern
  Polystichum tsus-simense

Tuberose   *Polianthes tuberosa*

Tulip cattleya   *Cattleya citrina*

Tulip orchid   *Anguloa clowesii*
  Anguloa uniflora

Turkish cotton
  Gossypium herbaceum

Turmeric   *Curcuma domestica*
  Curcuma longa

Umbrella grass
  Cyperus alternifolius

Umbrella plant
  Cyperus alternifolius
  Cyperus diffusus

Urn plant   *Aechmea fasciata*

Variegated ginger   *Alpinia sanderae*

Vegetable fern
  Athyrium esculentum
  Diplazium esculentum

Velour philodendron
  Philodendron andreanum
  Philodendron melanochrysum

Velvet leaf   *Kalanchoe beharensis*

Velvet plant   *Gynura aurantiaca*

Venus flytrap   *Dionaea muscipula*

Victorian box
  Pittosporum undulatum

Voodoo lily   *Sauromatum guttatum*

Wallflower   *Cheiranthus cheiri*

Wandering jew   *Callisia elegans*
  Setcreasea striata
  Tradescantia albiflora
  Tradescantia fluminensis
  Zebrina pendula

Wand flower   *Sparaxis tricolor*

Waratah   *Telopea speciosissima*

Water clover
  Marsilea drummondi
  Marsilea quadrifolia

Water feather
  Myriophyllum aquaticum
  Myriophyllum brasiliense
  Myriophyllum proserpinacoides

Water fern   *Azolla caroliniana*
  Ceratopteris cornuta
  Ceratopteris pteridoides
  Ceratopteris siliquosa
  Ceratopteris thalictroides

Water hyacinth
  Eichhornia crassipes
  Eichhornia speciosa

Water lettuce   *Pistia stratiotes*

Water-melon begonia
  Peperomia argyreia
  Peperomia sandersii

Wax begonia
  Begonia semperflorens

Wax plant   *Hoya carnosa*

Wax privet   *Peperomia glabella*

Wax torch   *Aechmea bromeliifolia*

Wax vine   *Senecio macroglossus*

Waxy jasmine
  Stephanotis floribunda

Weeping fig   *Ficus benjamina*

Western sword fern
  Polystichum munitum

West Indian tree fern
  Cyathea arborea

White kaleidoscope flower
  Streptanthera elegans

White lily turf   *Mondo jaburan*
  Ophiopogon jaburan

White rain tree
  Brunfelsia undulata

White sails   *Spathiphyllum wallisii*

Whorled peperomia
  Peperomia pulchella
  Peperomia verticillata

Wild coffee   *Gardenia citriodora*
  Mitriostigma axillare
  Polyscias guilfoylei

Wild pineapple   *Ananas bracteatus*

Wild sarsparilla
  Hardenbergia comptoniana

Windmill jasmine
  Jasminum nitidum

Windmill palm
  Trachycarpus fortunei

Wine palm   *Caryota urens*

Winged pea   *Lotus berthelotii*

Winged statice
  Limonium sinuatum

Winter cattleya
  Cattleya labiata trianaei
  Cattleya trianaei

Winter cherry
  Solanum capsicastrum

Winter heath   *Erica carnea*
  Erica herbacea

Winter's bark   *Drimys winteri*
  Wintera aromatica

Wintersweet
  Acokanthera spectabilis

Wire vine
  Muehlenbeckia complexa

Wirilda   *Acacia retinodes*

Wishbone flower   *Torenia fournieri*

Wonga-wonga vine
  Pandorea pandorana

Wood forget-me-not
Myosotis sylvatica
Woolly bear *Begonia leptotricha*
Wormwood cassia
Cassia artemisioides
Yautia *Xanthosma violaceum*
Yellow bells *Bignonia stans*
Stenolobium stans
Tecoma stans
Yellow elder *Bignonia stans*
Stenolobium stans
Tecoma stans
Yellow flax *Reinwardtia indica*
Reinwardtia trigyna
Yellow latan palm *Latania aurea*
Latania verschaffeltii
Yellow oleander
Thevetia peruviana

Yellow pagoda tree
Plumeria rubra lutea
Yellow pitcher plant
Sarracenia flava
Yellow sage *Lantana camara*
Yesterday, today and tomorrow
Brunfelsia calycina
Brunfelsia paucifolia calycina
Zebra haworthia
Haworthia fasciata
Zebra plant *Aphelandra squarrosa*
Calathea zebrina
Zonal geranium
Pelargonium × hortorum
Zonal pelargonium
Pelargonium × hortorum

ORCHIDS

Orchids are members of the very large family *Orchidaceae*. Various genera are grown by commercial specialists for the cut flower market, and by hobbyists as garden plants, in greenhouses or the home. Many artificial hybrids have been produced and are often to be found in private collections. The following list includes those most commonly offered for sale in nurseries and garden centres.

Bee orchid –
Ophrys apifera

Adam-and-Eve

Adam-and-Eve *Aplectrum hymale*
Aplectrum spicatum
Adder's mouth
Pogonia ophioglossoides
Adder's tongue-leaved pogonia
Pogonia ophioglossoides
Adder's tongue *Malaxis unifolia*
Alaskan orchid
Habenaria unalascensis
Alaska piperia
Habenaria unalascensis
Autumn cattleya *Cattleya labiata*
Autumn coralroot
Corallorhiza odontorhiza
Autumn lady's tresses
Spiranthes spiralis
Baby orchid
Epidendrum × obrienianum
Bamboo orchid
Arundina bambusifolia
Arundina graminifolia
Bastard helleborine
Epipactis helleborine
Epipactis latifolia
Bat orchid *Coryanthes speciosa*
Beardflower
Pogonia ophioglossoides
Bee orchid *Ophrys apifera*
Bee-swarm orchid
Cyrtopodium punctatum
Bird's nest orchid
Neottia nidus-avis
Black orchid *Coelogyne pandurata*
Blue butcher *Orchis mascula*
Blue orchid *Vanda coerulea*
Blunt-leaf orchid
Habenaria obusata
Bog candle *Habenaria dilatata*
Bog orchid *Arethusa bulbosa*
Calypso bulbosa
Habenaria dilatata
Bog rose orchid *Arethusa bulbosa*
Bog torch *Habenaria nivea*
Bog twayblade *Liparis loeselii*
Booth's epidendrum
Epidendrum boothianum
Epidendrum erythronioides
Bottle orchid *Physosiphon tubatus*
Broadleaved helleborine
Epipactis helleborine
Epipactis latifolia
Broad-leaved marsh orchid
Dactylorhiza majalis
Broad-leaved twayblade
Listera convallarioides
Broad-lipped twayblade
Listera convallarioides
Bucket orchid
Coryanthes macrantha
Coryanthes speciosa
Bug orchid *Orchis coriophora*

Burnt orchid *Orchis ustulata*
Burnt-tip orchid *Orchis ustulata*
Butterfly orchid
Epidendrum × obrienianum
Epidendrum tampense
Habenaria psycodes
Oncidium krameranum
Oncidium papilio
Platanthera chlorantha
Calypso *Calypso borealis*
Calypso bulbosa
Chatterbox *Amesia gigantea*
Epipactis gigantea
Checkered rattlesnake plantain
Goodyera tesselata
Chicken toes
Corallorhiza odontorhiza
Chocolate orchid
Epidendrum phoeniceum
Christmas cattleya *Cattleya labiata*
Cattleya percivaliana
Christmas orchid *Cattleya trianaei*
Cigar orchid
Cyrtopodium punctatum
Clam-shell orchid
Epidendrum cochleatum
Cockle-shell orchid
Epidendrum cochleatum
Colombia buttercup
Oncidium cheirophorum
Common lady's tresses
Spiranthes cernua
Common spotted orchid
Dactylorhiza fuchsii
Common twayblade *Listera ovata*
Cooktown orchid
Dendrobium bigibbum
Coral orchid *Rodriguezia secunda*
Coralroot orchid
Corallorhiza trifida
Cow-horn orchid
Cyrtopodium punctatum
Cradle orchid *Anguloa clowesii*
Anguloa uniflora
Cranefly orchid *Tipularia discolor*
Tipularia unifolia
Crawley root
Corallorhiza odontorhiza
Creeping lady's tresses
Goodyera repens
Crested ettercap
Pogonia ophioglossoides
Crested fringed orchid
Habenaria cristata
Crested rein orchid
Habenaria cristata
Crested yellow orchid
Habenaria cristata
Crippled cranefly *Tipularia discolor*
Tipularia unifolia

Hooker's orchid

Cypripedium
 Paphiopedilum concolor
 Paphiopedilum insigne
Cytherea *Calypso borealis*
 Calypso bulbosa
Daffodil orchid *Ipsea speciosa*
Dancing-doll orchid
 Oncidium flexuosum
Dancing-lady orchid
 Oncidium flexuosum
Dark red helleborine
 Epipactis atrorubans
Dead man's fingers *Orchis mascula*
Dense-flowered orchid
 Neotinea maculata
Dollar orchid
 Epidendrum boothianum
 Epidendrum erythronioides
Dove flower *Peristeria elata*
Dove orchid *Peristeria elata*
Downy rattlesnake orchid
 Goodyera pubescens
Downy rattlesnake plantain
 Goodyera pubescens
Dragon's claw
 Corallorhiza odontorhiza
Dragon's mouth orchid
 Arethusa bulbosa
Dune helleborine
 Epipactis dunensis
Dwarf orchid *Orchis ustulata*
Dwarf rattlesnake plantain
 Goodyera repens
Early marsh orchid
 Dactylorhiza incarnata
Early purple orchid *Orchis mascula*
Early spider orchid
 Ophrys sphegodes
Easter cattleya *Cattleya mossiae*
Elfin-spur *Tipularia discolor*
 Tipularia unifolia
Esmaralda *Arachnis clarkei*
Ettercap *Pogonia ophioglossoides*
Fairy fringe *Habenaria psycodes*
Fairy-slipper orchid
 Calypso borealis
 Calypso bulbosa
False musk orchid
 Chamorchis alpina
Fen orchid *Liparis loeselii*
Florida butterfly orchid
 Epidendrum tampense
Fly orchid *Ophrys insectifera*
Foxtail orchid *Aerides odorata*
Fragrant orchid
 Gymnadenia conopsea
Fried-egg orchid
 Dendrobium chrysotoxum
Fringed orchis *Habenaria ciliaris*
Frog orchid *Coeloglossum viride*
Frog spear *Habenaria nivea*

Frog spike *Habenaria clavellata*
Funnel-crest orchid
 Cleistes divaricata
 Pogonia divaricata
Gandergoose *Orchis morio*
Ghost orchid *Epipogium aphyllum*
Giant helleborine *Amesia gigantea*
 Epipactis gigantea
Giant lady's tresses
 Spiranthes praecox
Grass-leaved lady's tresses
 Spiranthes praecox
Giant orchid *Amesia gigantea*
 Epipactis gigantea
Giant rattlesnake plantain
 Goodyera oblongifolia
Golden chain orchid
 Dendrochilum filiforme
Golden fringed orchid
 Habenaria cristata
Gold-lace orchid
 Haemaria discolor
Grass pink *Calopogon tuberosus*
Greater butterfly orchid
 Platanthera chlorantha
Green adderling *Isotria verticillata*
 Pogonia verticillata
Green adder's mouth
 Malaxis unifolia
Green-flowered helleborine
 Epipactis phyllanthes
Green-fly orchid
 Epidendrum conopseum
Green fringed orchid
 Habenaria lacera
Green malaxis *Malaxis unifolia*
Green man orchid
 Aceras anthropothora
Green pearl-twist
 Spiranthes gracilis
Green rein orchid
 Habenaria clavellata
Green-winged orchid *Orchis morio*
Green woodland orchid
 Habenaria clavellata
Ground coco *Eulophia alta*
Hay-scented orchid
 Dendrochilum glumaceum
Heal-all *Habenaria orbiculata*
 Lysias orbiculata
Heath spotted orchid
 Dactylorhiza maculata
Heart-leaf twayblade
 Listera cordata
Helmet orchid *Galeandra lacustris*
Holy Ghost flower *Peristeria elata*
Hooded lady's tresses
 Spiranthes romanzoffiana
Hooker's orchid
 Habenaria hookeri

Hyacinth orchid
    Arpophyllum giganteum
    Arpophyllum spicatum
Indian crocus   *Pleione maculata*
Irish lady's tresses
    Spiranthes romanzoffiana
Jersey orchid   *Orchis laxifolia*
Jewel orchid *Anoectochilus setaceus*
Jumping orchid
    Catasetum macrocarpum
    Catasetum tridentatum
King-of-the-forest
    Anoectochilus setaceus
Kirtle-pink orchid
    Orchis spectabilis
Lace orchid
    Odontoglossum crispum
Lady's slipper
    Paphiopedilum concolor
    Paphiopedilum insigne
Lady's tresses   *Spiranthes cernua*
Lady-of-the-night
    Brassavola nodosa
Lady orchid   *Orchis purpurea*
Lady's slipper orchid
    Cypridium calceolus
    Phragmipedium caricinum
Large coralroot
    Corallorhiza maculata
    Corallorhiza multiflora
Large twayblade *Liparis liliifolia*
Late coralroot
    Corallorhiza odontorhiza
Late spider orchid *Ophrys fuciflora*
    Ophrys holoserica
Leafy northern green orchid
    Habenaria hyperborea
    Limnorchis hyperborea
Leafy white orchid
    Habenaria dilatata
Lesser butterfly orchid
    Habenaria bifolia
    Platanthera bifolia
Lesser purple-fringed orchid
    Habenaria psycodes
Lesser rattlesnake plantain
    Goodyera repens
Lesser twayblade *Listera cordata*
Lily-leaved pogonia
    Cleistes divaricata
    Pogonia divaricata
Lily-of-the-valley orchid
    Odontoglossum pulchellum
Link vine   *Vanilla articulata*
    Vanilla barbellata
Little bog orchid
    Hammarbya paludosa
Little club-spur orchid
    Habenaria clavellata
Little lady's tresses *Spiranthes grayi*
Little pearl-twist *Spiranthes grayi*

Lizard orchid
    Himantoglossum hircinum
Loesel's twayblade *Liparis loeselii*
Long tresses *Spiranthes gracilis*
Madeira orchid *Orchis maderensis*
Man orchid
    Aceras anthropophorum
Marsh helleborine
    Epipactis palustris
Marsh orchid
    Dactylorhiza latifolia
    Orchis latifolia
Mauve sleekwort *Liparis liliifolia*
Meadow orchid
    Dactylorhiza incarnata
Menzies' rattlesnake plantain
    Goodyera oblongifolia
Mirror-of-venus *Ophrys speculum*
Mirror orchid   *Ophrys speculum*
Moccasin flower
    Cypripedium acaule
Moccasin orchid
    Cypripedium acaule
Monkey orchid
    Coryanthes macrantha
    Orchis simia
Moon-set *Habenaria orbiculata*
    Lysias orbiculata
Moth orchid *Phalaenopsis amabilis*
Mottled cranefly *Tipularia discolor*
    Tipularia unifolia
Musk orchid
    Herminium monorchis
Musk orchis
    Herminium monorchis
Narrow-lipped orchid
    Epipactis leptochila
Nerveroot *Cypripedium acaule*
Nodding lady's tresses
    Spiranthes cernua
Northern green orchid
    Habenaria hyperborea
    Limnorchis hyperborea
Northern marsh orchid
    Dactylorhiza purpurella
Northern rattlesnake plantain
    Goodyera repens
Northern small bog orchid
    Habenaria obusata
Nun orchid   *Lycaste skinneri*
    Lycaste virginalis
Nun's hood orchid
    Phaius tankervilliae
Nun's orchid *Phaius tankervilliae*
Olive scutcheon *Liparis loeselii*
One-leaf rein orchid
    Habenaria obusata
Orange crest *Habenaria cristata*
Orange fringe *Habenaria ciliaris*
Orange plume *Habenaria ciliaris*

Palm-polly *Polyradicion lindenii*
            Polyrrhiza lindenii
Pansy orchid *Miltonia candida*
              Miltonia flavescens
              Miltonia spectabilis
              Miltonia vexillaria
Pigeon orchid
            Dendrobium crumenatum
Pine pink orchid *Bletia alta*
                  Bletia purpurea
Pink lady's slipper
            Cypripedium acaule
Pink scorpion orchid
            Arachnis × maingayi
Pink-slipper orchid
            Calypso borealis
            Calypso bulbosa
Pompona vanilla
            Vanilla grandiflora
            Vanilla pompona
Pride-of-the-peak
            Habenaria peramoena
Purple five-leaved orchid
            Isotria verticillata
            Pogonia verticillata
Purple fret-lip
            Habenaria peramoena
Purple fringeless orchid
            Habenaria peramoena
Purple-hooded orchid
            Orchis spectabilis
Purple scutcheon *Liparis liliifolia*
Purple spire orchid
            Habenaria peramoena
Puttyroot *Aplectrum hymale*
           Aplectrum spicatum
Pyramidal orchid
            Anacamptis pyramidalis
Queen cattleya *Cattleya dowiana*
                Cattleya labiata
Queen of orchids
            Grammatophyllum sanderanum
            Grammatophyllum speciosum
Ragged fringed orchid
            Habenaria lacera
Ragged orchid *Habenaria lacera*
Rainbow orchid
            Epidendrum prismatocarpum
Rattlesnake orchid
            Pholidota imbricata
Rattlesnake plantain
            Goodyera oblongifolia
Red helleborine
            Cephalanthera rubra
Rock lily *Dendrobium speciosum*
Romanzoff's lady's tresses
            Spiranthes romanzoffiana
Rosebud orchid *Cleistes divaricata*
                Pogonia divaricata
Rose crest-lip
            Pogonia ophioglossoides
Rose pogonia
            Pogonia ophioglossoides

Round-headed orchid
            Traunsteinera globosa
Round-leaved orchid
            Habenaria orbiculata
            Lysias orbiculata
Russet witch *Liparis loeselii*
Salep orchid *Orchis morio*
Savannah orchid *Habenaria nivea*
Sawfly orchid
            Ophrys tenthredinifera
Scarlet orchid
            Epidendrum × obrienianum
Scent bottle *Habenaria dilatata*
Scorpion orchid *Arachnis cathcartii*
Screw-augur *Spiranthes cernua*
Scrofula weed *Goodyera pubescens*
Showy orchid *Orchis spectabilis*
Silver chain
            Dendrochilum glumaceum
Single-leaf rein orchid
            Habenaria obusata
Slender bog orchid
            Habenaria gracilis
            Habenaria saccata
Slender lady's tresses
            Spiranthes gracilis
Slipper orchid
            Paphiopedilum concolor
            Paphiopedilum insigne
Small bog orchid
            Habenaria obusata
Small coralroot
            Corallorhiza odontorhiza
Small purple-fringed orchid
            Habenaria psycodes
Small round-leaved orchid
            Orchis rotundifolia
Small white orchid
            Leucorchis albida
Smooth rattlesnake plantain
            Goodyera tesselata
Snake's mouth orchid
            Pogonia ophioglossoides
Snowy orchid
            Habenaria blephariglottis
            Habenaria nivea
Soldier orchid *Orchis militaris*
Soldier's plume
            Habenaria psycodes
Southern lady's tresses
            Spiranthes gracilis
Southern marsh orchid
            Dactylorhiza praetermissa
Southern rein orchid
            Habenaria clavellata
Southern small white orchid
            Habenaria nivea
Spice orchid
            Epidendrum atropurpureum
            Epidendrum macrochilum

Spider orchid

Spider orchid	*Arachnis flos-aeris*
	Arachnis × maingayi
	Arachnis moschifera
Spotted coralroot	
	Corallorhiza maculata
	Corallorhiza multiflora
Spotted kirtle-pink orchid	
	Orchis rotundifolia
Spreading pogonia	
	Cleistes divaricata
	Pogonia divaricata
Spring cattleya	*Cattleya mossiae*
Star-of-Bethlehem orchid	
	Angraecum sesquipedale
Stream orchid	*Amesia gigantea*
	Epipactis gigantea
Summer cattleya	
	Cattleya gaskelliana
Summer lady's tresses	
	Spiranthes aestivalis
Swamp pink	*Calopogon tuberosus*
Swamp pink orchid	
	Arethusa bulbosa
Swan orchid	
	Cychnoches pentadactylon
Sweet-scented orchid	
	Gymnadenia conopsea
Tall white bog orchid	
	Habenaria dilatata
Tenderwort	*Malaxis unifolia*
Tiger orchid	
	Odontoglossum grande
Toothed orchid	*Orchis tridentata*
Tulip cattleya	*Cattleya citrina*
Two-leaved lady's slipper	
	Cypripedium acaule
Vanilla	*Vanilla planifolia*
Violet bird's nest orchid	
	Limodorum abortiva
Violet helleborine	
	Epipactis purpurata
Virgin Mary orchid	
	Caularthron bicornutum
	Diacrium bicornutum

Virgin orchid	
	Caularthron bicornutum
	Diacrium bicornutum
Water tresses	*Spiranthes praecox*
Western coralroot	
	Corallorhiza mertensiana
West Indian vanilla	
	Vanilla grandiflora
	Vanilla pompona
White butterfly orchid	
	Polyradicion lindenii
	Polyrrhiza lindenii
White fringed orchid	
	Habenaria bulephariglottis
White frog-arrow	*Habenaria nivea*
White helleborine	
	Cephalanthera damasonium
White nun orchid	*Lycaste skinneri*
	Lycaste virginalis
White rein orchid	
	Habenaria nivea
Whorled pogonia	
	Isotria verticillata
	Pogonia verticillata
Wide adder's mouth	
	Malaxis unifolia
Widow orchid	
	Pleurothallis macrophylla
Wild coco	*Eulophia alta*
Wild pink orchid	*Arethusa bulbosa*
Windmill orchid	
	Bulbophyllum refractum
Winter cattleya	*Cattleya trianaei*
Woodland orchid	*Orchis spectabilis*
Worm vine	*Vanilla articulata*
	Vanilla barbellata
Wormwood	*Vanilla articulata*
	Vanilla barbellata
Yellow fringed orchid	
	Habenaria ciliaris
Yellow lady's slipper	
	Cypripedium pubescens
Yellow twayblade	*Liparis loeselii*

POPULAR GARDEN PLANTS

Annuals, biennials, and perennials that are not usually listed under specific sections elsewhere. Popular flowering plants grown in many gardens as bedding plants, container plants, or climbers are covered.

Marguerite –
Argyranthemum frutescens

Aaron's rod *Verbascum thapsus*
Abcess root *Polemonium reptans*
Absinthe *Artemisia absinthium*
Aconite-leaved buttercup
 Ranunculus aconitifolius
Adam's flannel *Verbascum thapsus*
Adriatic bellflower
 Campanula elatines garganica
 Campanula garganica
African daisy *Arctotis breviscapa*
 Arctotis grandis
 Dimorphotheca osteospermum
African marigold *Tagetes erecta*
African violet *Saintpaulia ionantha*
Agrimony *Agrimonia eupatoria*
Ague root *Aletris farinosa*
Algerian statice
 Limonium bonduellii
 Statice bonduellii
Alkanet *Anchusa officinalis*
Allegheny monkey flower
 Mimulus ringens
Allegheny spurge
 Pachysandra procumbens
Alpine anemone *Pulsatilla alpina*
Alpine auricula *Primula auricula*
Alpine avens *Geum montanum*
Alpine bistort
 Polygonum viviporum
Alpine buttercup
 Ranunculus alpestris
Alpine campanula
 Campanula allionii
 Campanula alpestris
Alpine campion *Lychnis alpina*
 Silene alpestris
 Silene quadrifida
 Viscaria alpina
Alpine cinquefoil *Potentilla crantzii*
Alpine coltsfoot *Homogyne alpina*
Alpine fleabane *Erigeron borealis*
Alpine forget-me-not
 Myosotis alpestris
Alpine gypsophila
 Gypsophila repens
Alpine lady's mantle
 Alchemilla alpina
Alpine meadow rue
 Thalictrum alpinum
Alpine moon daisy
 Leucanthemopsis alpina
Alpine pennycress *Thlaspi alpestre*
Alpine penstemon
 Penstemon alpinus
Alpine phlox *Phlox douglasii*
Alpine pink *Dianthus alpinus*
Alpine poppy *Papaver alpinum*
 Papaver burseri
Alpine snowbell *Soldanella alpina*
Alpine speedwell *Veronica alpina*
Alpine toadflax *Linaria alpina*

Alpine wallflower
 Erysimum alpinum
Alpine yarrow *Achillea tomentosa*
Amaranth *Amaranthus retroflexus*
American barrenwort
 Vancouveria hexandra
American bluebell *Mertensia ciliata*
American brooklime
 Veronica americana
American bugbane
 Cimicifuga americanum
American burnet
 Poterium canadensis
 Sanguisorba canadensis
American cranesbill
 Geranium maculatum
American fumitory *Fumaria indica*
American Greek valerian
 Polemonium reptans
American marigold *Tagetes erecta*
American monkey flower
 Mimulus guttatus
American sea lavender
 Limonium carolinianum
American speedwell
 Veronica peregrina
American sundrops
 Oenothera fruticosa
 Oenothera linearis
American trumpet creeper
 Campsis radicans
American white hellebore
 Veratrum viride
American wild columbine
 Aquilegia canadensis
American willowherb
 Epilobium ciliatum
Angel's eye *Veronica chamaedrys*
Annual anchusa *Anchusa capensis*
Annual aster *Callistephus chinensis*
Annual blue flax
 Linum usitatissimum
Annual candytuft *Iberis umbellata*
Annual cape marigold
 Dimorphotheca annua
Annual clary *Salvia horminum*
Annual gaillardia
 Gaillardia pulchella
Annual gypsophila
 Gypsophila muralis
Annual mallow *Lavatera trimestris*
Annual phlox *Phlox drummondii*
Annual pink *Dianthus chinensis*
Annual poinsettia
 Euphorbia heterophylla
Annual rudbeckia *Rudbeckia hirta*
Annual sunflower
 Helianthus annuus
Annual wall rocket
 Diplotaxis muralis

POPULAR GARDEN PLANTS

Black hellebore

Antarctic forget-me-not
    Myosotidium hortensia
    Myosotidium nobile

Antwerp hollyhock
    Althaea ficifolia

Apache beads
    Anemonopsis californica

Apple of Peru *Nicandra physalodes*

Apple of sodom
    Solanum carolinense
    Solanum sosomeum

Arabian thistle
    Onopordum arabicum
    Onopordum nervosum

Arabian violet   *Exacum affine*

Archangel   *Lamium album*

Arctic poppy *Papaver nudicaule*
    Papaver radicatum

Arizona poppy
    Kalistroemia grandiflora

Armstrong   *Polygonum ariculare*

Ass's tail   *Sedum morganianum*

Asthma weed   *Euphorbia hirta*
    Lobelia inflata

Auricula   *Primula auricula*

Australian fleabane
    Erigeron karkinskianus

Australian violet *Viola hederacea*

Austrian speedwell
    Veronica austriaca

Auvergne pink
    Dianthus × arvernensis

Avens   *Geum urbanum*

Aztec marigold   *Tagetes erecta*

Baby blue-eyes *Nemophila insignis*
    Nemophila menziesii

Baby's breath   *Gypsophila elegans*
    Gypsophila paniculata

Bachelor's buttons
    Ranunculus acris
    Tanacetum parthenium

Bacon and eggs *Lotus corniculatus*

Balkan cranesbill
    Geranium macrorrhizum

Balloon flower
    Platycodon grandiflorum

Balsam   *Impatiens balsamina*

Balsam weed   *Impatiens aurea*

Banana passionfruit
    Passiflora mollissima
    Tacsonia mollissima

Band plant   *Vinca major*

Barberton daisy *Gerbera jamesonii*

Barren strawberry
    Potentilla sterilis

Basket flower
    Centaurea americana

Bastard box
    Polygala chamaebuxus

Bastard jasmine
    Androsace chamaejasme

Bats-in-the-belfry
    Campanula trachelium
    Campanula urticifolia

Beach pea   *Lathyrus littoralis*
    Lathyrus maritimus

Beach wormwood
    Artemisia stellerana

Beaked hawk's beard
    Crepis vesicaria

Bearded bellflower
    Campanula barbata

Bear's breeches *Primula auricula*

Bear's ear   *Primula auricula*

Bear's foot *Alchemilla vulgaris*

Beaver tail *Sedum morganianum*

Bee balm   *Monarda didyma*

Bee nettle   *Lamium album*

Bedding begonia
    Begonia semperflorens

Bedding dahlia *Dahlia merckii*

Bedding lobelia *Lobelia erinus*

Bellflower
    Wahlenbergia albomarginata

Bells of Ireland *Moluccella laevis*

Bellwort *Uvularia grandiflora*

Belvedere
    Kochia scoparia trichophylla

Bergamot   *Monarda didyma*

Bethlehem sage
    Pulmonaria saccharata

Betony   *Betonica grandiflora*
    Betonica macrantha
    Stachys betonica
    Stachys officinalis

Bidgee-widgee
    Acaena anserinifolia

Big marigold   *Tagetes erecta*

Bird's eye *Veronica chamaedrys*

Bird's-eye primrose
    Primula farinosa

Bird's eyes   *Gilia tricolor*

Bird's foot trefoil
    Lotus corniculatus

Bird's tongue *Polygonum aviculare*

Bishop's flower   *Ammi majus*

Bishop's wort *Betonica officinalis*
    Stachys betonica
    Stachys officinalis

Biting stonecrop   *Sedum acre*

Bitter root *Lewisia rediviva*

Bittersweet *Solanum dulcamara*

Bitter vetch *Lathyrus montanus*

Black baneberry *Actaea spicata*

Black cosmos
    Cosmos atrosanguineus

Black-eyed susan *Rudbeckia hirta*
    Thunbergia alata

Black false hellebore
    Veratrum nigrum

Black hellebore *Helleborus niger*

Black helleborine	*Veratrum nigrum*
Black knapweed	*Centaurea nigra*
Black medick	*Medicago lupulina*
Black nightshade	*Solanum nigrum*
Black rampion	*Phyteuma nigrum*
Black root	*Veronica virginica*
	Veronicastrum virginica
Black samson	*Echinacea purpurea*
Black snakeroot	*Cimicifuga racemosa*
Bladder campion	*Silene vulgaris*
Bladder cherry	*Physalis alkekengi*
Bladder fumitory	*Fumaria vesicaria*
Bladder gentian	*Gentiana utriculosa*
Blanket flower	*Gaillardia aristata*
	Gaillardia pulchella
Blazing star	*Aletris farinosa*
	Bartonia aurea
	Liatris spicata
	Mentzelia lindleyi
Bleeding heart	*Dicentra spectabilis*
Blessed thistle	*Cnicus benedictus*
	Silybum marianum
Blind nettle	*Lamium album*
Blood drop	*Stylomecon heterophylla*
Blood-red geranium	*Geranium sanguineum*
Bloodroot	*Potentilla erecta*
	Sanguinaria canadensis
Bloody cranesbill	*Geranium sanguineum*
Bluebeard	*Salvia horminum*
Bluebell	*Campanula rotundifolia*
Blue bells	*Polemonium reptans*
Blue boneset	*Eupatorium coelestinum*
Bluebottle	*Centaurea cyanus*
Blue bugle	*Ajuga genevensis*
Blue buttons	*Vinca major*
Blue cornflower	*Centaurea cyanus*
Blue cowslip	*Pulmonaria angustifolia*
Blue cupidone	*Catananche coerulea*
Blue daisy	*Agathaea coelestis*
	Felicia amelloides
Blue dawn flower	*Ipomoea acuminata*
Blue-eyed grass	*Sisyrinchium angustifolium*
Blue-eyed mary	*Collinsia grandiflora*
	Omphalodes verna
Blue fescue	*Festuca glauca*
	Festuca ovina glauca
Blue flax	*Linum perenne*
Blue fleabane	*Erigeron acre*

Blue lace flower	*Didiscus caerulea*
	Trachymene caerulea
Blue lips	*Collinsia grandiflora*
Blue lobelia	*Lobelia siphilitica*
	Lobelia urens
Blue marguerite	*Agathaea coelestis*
	Felicia amelloides
Blue mountains bidi-bidi	*Acaena inermis*
	Acaena microphylla
Blue passionflower	*Passiflora caerulea*
Blue phlox	*Phlox canadensis*
	Phlox divaricata
Blue pimpernel	*Anagallis foemina*
	Anagallis linifolia
	Anagallis monelli linifolia
Blue poppy	*Meconopsis betonicifolia*
Blue sage	*Salvia azurea*
	Salvia patens
Blue saxifrage	*Saxifraga caesia*
Blue spurge	*Euphorbia myrsinites*
Blue succory	*Catananche coerulea*
Blue thimble flower	*Gilia capitata*
Bluets	*Hedyotis caerulea*
Blue throatwort	*Trachelium caeruleum*
Blue vervain	*Verbena hastata*
Blueweed	*Echium vulgare*
Blue wings	*Torenia fournieri*
Bog pimpernel	*Anagallis tenella*
Bog sage	*Salvia uliginosa*
Bog spurge	*Euphorbia palustris*
Bokhara fleece flower	*Bilderdykia baldschuanicum*
	Polygonum baldschuanicum
Boneset	*Symphytum officinale*
Border carnation	*Dianthus caryophyllus*
Border pink	*Dianthus plumarius*
Border sea lavender	*Limonium latifolium*
	Statice latifolium
Border thrift	*Armeria plantaginea*
	Armeria pseudarmeria
Boule d'or	*Trollius europaeus*
Bouncing bess	*Centranthus ruber*
Bouncing bet	*Saponaria officinalis*
Box honeysuckle	*Lonicera nitida*
Branched larkspur	*Consolida regalis*
Brass buttons	*Cotula coronopifolia*
Breckland thyme	*Thymus serpyllum*
Bridal wreath	*Francoa sonchifolia*
Brideweed	*Linaria vulgaris*
Brilliant scabious	*Scabiosa lucida*
Broad-leaved ragwort	*Senecio fluviatilis*

POPULAR GARDEN PLANTS

Cathedral bells

Broad-leaved sea lavender
   Limonium latifolium
   Statice latifolium

Broad-leaved spurge
   Euphorbia platyphyllos

Broad-leaved willowherb
   Epilobium montanum

Brompton stock *Matthiola incana*

Brooklime *Veronica beccabunga*

Broom cypress
   Kochia scoparia trichophylla

Brown-eyed susan
   Rudbeckia triloba

Bruisewort *Symphytum officinale*

Bugle   *Ajuga reptans*

Bugleweed *Lycopus virginicus*

Bugloss *Anchusa arvensis*

Bulbous buttercup
   Ranunculus bulbosus

Bull nettle *Solanum carolinense*

Bullock's eye
   Sempervivum tectorum

Bunny rabbits *Linaria maroccana*

Bupleurum *Bupleurum fruticosum*

Bur medick *Medicago minima*

Burning bush *Dictamnus albus*
   Dictamnus fraxinella
   Kochia scoparia trichophylla
   Kochia trichophylla

Burro's tail *Sedum morganianum*

Bush monkey flower
   Diplacus glutinosus
   Mimulus aurantiacus
   Mimulus glutinosus

Bush morning glory
   Convolvulus cneorum

Busy lizzie *Impatiens sultanii*
   Impatiens wallerana

Butter and eggs *Linaria vulgaris*

Butter daisy *Verbesina encelioides*

Butterfly flag *Diplarrhena moraea*

Butterwort *Pinguicula vulgaris*

Button pink *Dianthus × latifolius*

Buttons *Tanacetum vulgare*

Button snakeroot *Liatris spicata*

Buxbaum's speedwell
   Veronica persica

Cactus dahlia *Dahlia juarezii*

Calamint *Calamintha grandiflora*
   Calamintha nepeta nepeta

Calathian violet
   Gentiana pneumonanthe

Calico plant *Aristolochia elegans*

California bluebell
   Nemophila phacelia
   Phacelia campanularia
   Phacelia whitlavia

California blue-eyed grass
   Sisyrinchium bellum

Californian fuchsia
   Zauschneria californica

Californian golden bells
   Emmenanthe penduliflora

Californian pea *Lathyrus splendens*

Californian poppy
   Eschscholzia californica

Californian whispering bells
   Emmenanthe penduliflora

Calvary clover *Medicago echinus*

Calve's snout *Linaria vulgaris*

Canadian burnet
   Poterium canadensis
   Sanguisorba canadensis

Canadian golden rod
   Solidaga canadensis

Canary-bird flower
   Tropaeolum peregrinum

Canary-bird vine
   Tropaeolum peregrinum

Canary creeper
   Tropaeolum canariensis
   Tropaeolum peregrinum

Candlewick *Verbascum thapsus*

Candytuft *Iberis amara*
   Iberis umbellata

Canterbury bell
   Campanula grandiflora
   Campanula medium

Cape daisy *Venidium fastuosum*

Cape forget-me-not
   Anchusa capensis

Cape marigold
   Dimorphotheca annua

Cape primrose
   Streptocarpus hybridus

Caper spurge *Euphorbia lathyris*

Cape stock *Heliophila longifolia*

Caraway thyme
   Thymus herba-barona

Cardinal climber
   Ipomoea cardinalis
   Ipomoea × multifida

Cardinal flower *Lobelia cardinalis*
   Lobelia fulgens

Cardinal sage *Salvia fulgens*

Cardinal salvia *Salvia fulgens*

Carnation *Dianthus caryophyllus*

Carolina lupin
   Thermopsis caroliniana

Carpathian bellflower
   Campanula carpatica

Carpathian harebell
   Campanula carpatica

Carpenter's herb *Ajuga reptans*

Carpet plant *Ionopsidium acaule*

Castor bean *Ricinus communis*

Castor-oil plant *Ricinus communis*

Catchfly *Lychnis silene*
   Lychnis viscaria
   Lychnis vulgaris
   Viscaria vulgaris

Cathedral bells *Cobaea scandens*

Catmint	*Nepeta cataria*
	Nepeta × faassenii
	Nepeta mussinii
Catnep	*Nepeta cataria*
Catnip	*Nepeta cataria*
Cat's hair	*Euphorbia pilulifera*
Cat's valerian	*Valeriana officinalis*
Cat-tail gayfeather	
	Liatris pycnostachya
Celandine poppy	
	Stylophorum diphyllum
Celery-leaved crowfoot	
	Ranunculus sceleratus
Chaff weed	*Anagallis minima*
Chalk milkwort	*Polygala calcarea*
Chalk plant	*Gypsophila paniculata*
Chatham Island lily	
	Myosotidium hortensia
	Myosotidium nobile
Cheddar pink	*Dianthus caesius*
	Dianthus gratianopolitanus
Cherry pie	
	Heliotropium arborescens
	Heliotropium corymbosum
	Heliotropium hybridum
	Heliotropium peruvianum
Chickling pea	*Lathyrus sativus*
Chickweed willowherb	
	Epilobium alsinifolium
Chilean avens	*Geum chiloense*
	Geum coccineum
	Geum quellyon
Chilean bellflower	
	Nolana acuminata
	Nolana rupicola
Chilean potato tree	
	Solanum crispum
Chimney bellflower	
	Campanula pyramidalis
China aster	*Callistephus chinensis*
China fleece flower	
	Bilderdykia aubertii
	Polygonum aubertii
Chinaman's breeches	
	Dicentra spectabilis
Chinese artichoke	*Stachys affinis*
Chinese bellflower	
	Platycodon grandiflorum
Chinese forget-me-not	
	Cynoglossom amabile
Chinese foxglove	
	Rehmannia angulata
Chinese houses	*Collinsia bicolor*
	Collinsia heterophylla
Chinese jasmine	
	Jasminum polyanthum
Chinese lantern	*Physalis alkekengi*
Chinese loosestrife	
	Lysimachia clethroides
Chinese paeony	*Paeonia albiflora*
	Paeonia lactiflora

Chinese pink	*Dianthus chinensis*
	Dianthus sinensis
Chinese trumpet creeper	
	Campsis chinensis
	Campsis grandiflora
Chinese trumpet flower	
	Incarvillea delavayi
	Incarvillea grandiflora
Chinese wax flower	
	Dregea sinensis
	Wattakaka chinensis
	Wattakaka sinensis
Chinese woodbine	
	Lonicera tragophylla
Chocolate cosmos	
	Cosmos atrosanguineus
Chorogi	*Stachys affinis*
Christmas horns	
	Delphinium nudicaule
Christmas rose	*Helleborus niger*
Christmas star	
	Euphorbia pulcherrima
Church steeples	
	Agrimonia eupatoria
Cigar flower	*Cuphea ignea*
Circle flower	*Lysimachia punctata*
Clarkia	*Clarkia unguiculata*
Clary	*Salvia sclarea*
Clear eye	*Salvia sclarea*
Climbing gazania	*Mutisia decurrens*
Clock vine	*Thunbergia grandiflora*
Clove pink	*Dianthus caryophyllus*
Cloveroot	*Geum urbanum*
Clump verbena	
	Verbena canadensis
Clustered bellflower	
	Campanula glomerata
Cobweb houseleek	
	Sempervivum arachnoideum
Cockle	*Vaccaria pyramidata*
Cockscomb	
	Celosia argentea cristata
	Celosia cristata childsii
Codlins and cream	
	Epilobium hirsutum
Colic root	*Aletris farinosa*
Comfrey	*Symphytum officinale*
Common agrimony	
	Agrimonia eupatoria
Common amaranth	
	Amaranthus retroflexus
Common aubrieta	
	Aubrieta deltoides
Common bleeding heart	
	Dicentra spectabilis
Common blue flax	
	Linum usitatissimum
Common blue passionflower	
	Passiflora caerulea
Common bugle	*Ajuga reptans*

POPULAR GARDEN PLANTS

Cow cress

Common butterwort
Pinguicula vulgaris

Common candytuft
Iberis umbellata

Common chamomile
Anthemis nobilis

Common comfrey
Symphytum officinale

Common daisy   *Bellis perennis*

Common edelweiss
Leontopodium alpinum

Common foxglove
Digitalis purpurea

Common fumitory
Fumaria officinalis

Common globe flower
Trollius europaeus

Common hawkweed
Hieracium vulgatum

Common honeysuckle
Lonicera periclymenum

Common houseleek
Sempervivum tectorum

Common immortelle
Xeranthemum anuum

Common knotgrass
Polygonum aviculare

Common ladybell
Adenophora lilifolia

Common lady's mantle
Alchemilla mollis
Alchemilla vulgaris

Common mallow   *Malva sylvestris*

Common marjoram
Origanum vulgare

Common meadow rue
Thalictrum flavum

Common meadowsweet
Filipendula ulmaria

Common millwort
Polygala vulgaris

Common monkshood
Aconitum napellus

Common morning glory
Ipomoea purpurea

Common mullein
Verbascum thapsus

Common orach   *Atriplex patula*

Common pearlwort
Sagina procumbens

Common periwinkle   *Vinca minor*

Common persicaria
Polygonum persicaria

Common pink
Dianthus plumarius

Common ramping fumitory
Fumaria muralis

Common red poppy
Papaver rhoeas

Common sage   *Salvia officinalis*

Common shooting star
Dodecatheon meadia

Common speedwell
Veronica officinalis

Common sunflower
Helianthus annuus

Common thrift   *Armeria maritima*

Common throatwort
Trachelium caeruleum

Common thyme   *Thymus vulgaris*

Common toadflax   *Linaria vulgaris*

Common tobacco
Nicotiana tabacum

Common valerian
Valeriana officinalis

Common wallflower
Cheiranthus cheiri

Common white jasmine
Jasminum officinale

Common yarrow
Achillea millefolium

Common yellow alyssum
Alyssum saxatile
Aurinia saxatilis

Common zinnia   *Zinnia elegans*

Cone flower   *Rudbeckia fulgida*

Coolwort   *Tiarella cordifolia*

Coral bells   *Heuchera sanguinea*

Coral flower   *Heuchera sanguinea*

Corn buttercup
Ranunculus arvensis

Corn campion
Agrostemma githago

Corn cockle   *Agrostemma githago*

Corn crowfoot
Ranunculus arvensis

Cornflower   *Centaurea cyanus*

Cornflower aster   *Stokesia cyanea*
Stokesia laevis

Corn lily   *Veratrum californicum*

Corn marigold
Chrysanthemum segetum

Corn mignonette
Reseda phyteuma

Corn pink   *Agrostemma githago*

Corn poppy   *Papaver rhoeas*

Corn woundwort   *Stachys arvensis*

Corsican hellebore
Helleborus argutifolius
Helleborus corsicus
Helleborus lividus

Corsican speedwell
Veronica repens

Cottage pink
Dianthus caryophyllus
Dianthus plumarius

Cotton thistle
Onopordum acanthium

Coventry bells
Campanula trachelium
Campanula urticifolia

Cowbells   *Uvularia grandiflora*

Cow cress   *Veronica beccabunga*

Cow herb	*Saponaria vaccaria*
	Vaccaria pyramidata
Cowslip	*Primula officinalis*
	Primula veris
Cream cups	
	Platystemon californicus
Creeping avens	*Geum reptans*
Creeping buttercup	
	Ranunculus repens
Creeping cinquefoil	
	Potentilla reptans
Creeping forget-me-not	
	Omphalodes verna
Creeping jacob's ladder	
	Polemonium reptans
Creeping jenny	
	Lysimachia nummularia
Creeping phlox	*Phlox reptans*
	Phlox stolonifera
Creeping Saint John's wort	
	Hypericum humifusum
Creeping thyme	*Thymus serpyllum*
Creeping vervain	*Verbena aubletia*
	Verbena canadensis
Creeping zinnia	
	Sanvitalia procumbens
Crested gentian	
	Gentiana septemfida
Crested poppy	
	Argemone platyceras
Cretan bear's tail	*Celsia arcturus*
Cretan dittany	
	Amaracus dictamnus
	Origanum dictamnus
Cretan mullein	*Celsia arcturus*
Crimson star glory	*Ipomoea lobata*
	Mina lobata
	Quamoclit lobata
Cross gentian	*Gentiana cruciata*
Crow flower	
	Geranium sylvaticum
Crown daisy	
	Chrysanthemum coronarium
Cudweed	*Artemisia gnaphalodes*
	Artemisia ludoviciana
	Artemisia purshiana
Culver's root	
	Veronicastrum virginicum
	Veronica virginica
Cup-and-saucer canterbury	
bell	*Campanula medium*
	calycanthema
Cup-and-saucer vine	
	Cobaea scandens
Cupid's dart	*Catananche coerulea*
Curled mallow	*Malva crispa*
Curly mallow	*Malva crispa*
Curry plant	
	Helichrysum angustifolium
	Helichrysum seotinum
Curuba	*Passiflora mollissima*
	Tacsonia mollissima
Cushion pink	*Silene acaulis*

Cushion spurge	
	Euphorbia epithymoides
	Euphorbia polychroma
Cut-leaved cranesbill	
	Geranium dissectum
Cut-leaved mallow	*Malva alcea*
Cyphel	*Minuartia sedoides*
Cypress spurge	
	Euphorbia cyparissias
Cypress vine	*Ipomoea quamoclit*
Dairy pink	*Saponaria vaccaria*
	Vaccaria pyramidata
Dakota vervain	
	Verbena bipinnatifida
Dalmatian pellitory	
	Tanacetum cinerariifolium
Dalmatian pyrethrum	
	Tanacetum cinerariifolium
Damask violet	*Hesperis matronalis*
Dame's rocket	*Hesperis matronalis*
Dame's violet	*Hesperis matronalis*
Dandelion	*Taraxacum officinale*
Dark mullein	*Verbascum nigrum*
Dead men's bells	
	Digitalis purpurea
Dead nettle	*Lamium maculatum*
Deer's tongue	*Liatris odoratissima*
Deer weed	*Lotus scoparius*
Deptford pink	*Dianthus armeria*
Devil's fig	*Argemone mexicana*
Devil's paintbrush	
	Hieracium aurantiacum
Diamond flower	
	Ionopsidium acaule
Digger's speedwell	
	Veronica perfoliata
Dittany	*Dictamnus albus*
	Dictamnus fraxinella
Dog daisy	*Leucanthemum vulgare*
Dog poison	*Aethusa cynapium*
Donkey plant	*Onosma pyramidale*
Donkey's tail	
	Sedum morganianum
Dotted loosestrife	
	Lysimachia punctata
Double orange daisy	
	Erigeron aurantiacus
Dovedale moss	
	Saxifraga hypnoides
Dove's foot cranesbill	
	Geranium columbinum
Downy lupin	*Lupinus pubescens*
Downy Saint John's wort	
	Hypericum lanuginosum
Downy thistle	
	Onopordum acanthium
Downy woundwort	
	Stachys germanica
Dragon mouth	
	Horminum pyranaicum

POPULAR GARDEN PLANTS

Field forget-me-not

Dropwort   *Filipendula hexapetala*
            Filipendula vulgaris

Drumstick primula
            Primula denticulata

Drunken sailor   *Centranthus ruber*

Dun daisy   *Leucanthemum vulgare*

Dusky cranesbill
            Geranium phaeum

Dusty miller   *Artemisia stellerana*
            Centaurea candidissima
            Centaurea cineraria
            Centaurea gymnocarpa
            Centaurea ragusina
            Primula auricula

Dutch agrimony
            Eupatorium cannabinum

Dutchman's breeches
            Dicentra spectabilis

Dutchman's pipe
            Aristolochia durior
            Aristolochia macrophylla
            Aristolochia sipho

Dwarf cape marigold
            Dimorphotheca barberae
            Osteospermum barberae

Dwarf globe flower   *Trollius pumilis*

Dwarf jasmine   *Jasminum parkeri*

Dwarf mallow   *Malva neglecta*

Dwarf meadow rue
            Thalictrum coreanum

Dwarf morning glory
            Convolvulus minor
            Convolvulus tricolor

Dwarf nasturtium
            Tropaeolum minus

Dwarf snowbell   *Soldenella pusilla*

Dwarf spurge   *Euphorbia exigua*

Dwarf sundrops
            Oenothera perennis
            Oenothera pumila

Dwarf valerian   *Valeriana saxatilis*

Dyer's rocket   *Reseda luteola*

Early meadow rue
            Thalictrum dioicum

Earth smoke   *Fumaria officinalis*

Eastern rocket
            Sisymbrium orientale

Edelweiss   *Leontopodium alpinum*

Edging lobelia   *Lobelia erinus*

Eggs and bacon   *Linaria vulgaris*

Elecampane   *Inula helenium*

English daisy   *Bellis perennis*

English harebell
            Campanula rotundifolia

English primrose   *Primula acaulis*
            Primula vulgaris

English stonecrop   *Sedum anglicum*

English wallflower
            Cheiranthus cheiri

European brooklime
            Veronica beccabunga

European honeysuckle
            Lonicera periclymenum

European wild columbine
            Aquilegia vulgaris

European wild paeony
            Paeonia officinalis

European wild thyme
            Thymus drucei
            Thymus serpyllum

Evening primrose
            Oenothera biennis

Evening stock   *Matthiola bicornis*

Everlasting daisy
            Helichrysum bellidiodes

Everlasting pea   *Lathyrus latifolius*

Ewe daisy   *Potentilla erecta*

Fair maids of France
            Saxifraga granulata

Fair maids of Kent
            Ranunculus aconitifolius

Fairy clock   *Taraxacum officinale*

Fairy cups   *Primula veris*

Fairy flax   *Linum catharticum*

Fairy foxglove   *Erinus alpinus*

Fairy primrose   *Primula malacoides*

Fairy's thimbles
            Campanula cochleariifolia
            Campanula pusilla

Fairy thimbles   *Digitalis purpurea*

False African violet
            Streptocarpus saxorum

False anemone
            Anemonopsis macrophylla

False bishop's weed   *Ammi majus*

False clary   *Salvia viridis*

False dragonhead
            Physostegia virginiana

False London rocket
            Sisymbrium loeselii

Fanweed   *Thlaspi arvense*

Farewell-to-spring   *Clarkia amoena*
            Godetia amoena

Feathered bronze leaf
            Rodgersia pinnata

Fen spurge   *Euphorbia palustris*

Fernleaf yarrow
            Achillea eupatorium
            Achillea filipendulina

Fern-leaved paeony
            Paeonia tenuifolia

Feverfew
            Chrysanthemum parthenium
            Matricaria eximia
            Tanacetum coccineum
            Tanacetum parthenium

Fibrous-rooted begonia
            Begonia semperflorens

Field daisy   *Leucanthemum vulgare*

Field fleawort   *Senecio integrifolius*

Field forget-me-not
            Myosotis arvensis

Field pennycress	Thlaspi arvense
Field poppy	Papaver rhoeas
Field primrose	Oenothera biennis
Field scabious	Knautia arvensis
Field speedwell	Veronica agrestis
Field woundwort	Stachys arvensis
Figleaf hollyhock	Althaea ficifolia
Fire bush	
	Kochia scoparia trichophylla
Firecracker plant	Cuphea ignea
Fire on the mountain	
	Euphorbia heterophylla
Fire plant	Euphorbia pulcherrima
Fireweed	Epilobium angustifolium
	Erechtites hieracifolia
Five-faced bishop	
	Adoxa moschatellina
Five fingers	Potentilla reptans
Five-leaved grass	Potentilla reptans
Five-spot	Nemophila maculata
Five-spot nemophila	
	Nemophila maculata
Flame creeper	
	Tropaeolum speciosum
Flame nasturtium	
	Tropaeolum speciosum
Flame violet	Episcia cupreata
Flaming poppy	
	Stylomecon heterophylla
Flanders poppy	Papaver rhoeas
Flannel flower	Actinotus helianthi
Flannel mullein	Verbascum thapsus
Flax	Linum usitatissimum
Flax-leaved Saint John's wort	
	Hypericum linarifolium
Flaxweed	Linaria vulgaris
Flore pleno	Clarkia elegans
Florist's scabious	
	Scabiosa caucasica
Floss flower	
	Ageratum houstonianum
	Ageratum mexicanum
Flowering spurge	
	Euphorbia corollata
Flowering tobacco	
	Nicotiana sylvestris
Flower of Jove	
	Agrostemma flos-jovis
	Lychnis flos-jovis
Fly honeysuckle	
	Lonicera xylosteum
Foam flower	Tiarella cordifolia
	Tiarella polyphylla
Fool's parsley	Aethusa cynapium
Forget-me-not	Myosotis alpestris
	Myosotis oblongata
	Myosotis sylvatica
Forked catchfly	Silene dichotoma
Forking larkspur	Consolida regalis
Four o'clock plant	Mirabilis jalapa

Foxbane	
	Aconitum lycotonum vulparia
Foxglove	Digitalis purpurea
Fragrant evening primrose	
	Oenothera stricta
French cranesbill	
	Geranium endressii
French hawk's beard	
	Crepis nicaeensis
French lilac	Galega officinalis
French marigold	Tagetes patula
Frenchweed	Thlaspi arvense
Friar's cap	Aconitum napellus
Fringe-bell	
	Schizocodon soldanelliodes
	Shortia soldanelliodes
Fringed bleeding heart	
	Dicentra eximia
Fringed pink	Dianthus superbus
Frosted orache	Atriplex lociniata
Fumitory	Fumaria officinalis
Garden chamomile	
	Anthemis nobilis
Garden globe flowers	
	Trollius × cultorum
	Trollius × hybridus
Garden heliotrope	
	Valeriana officinalis
Garden hollyhock	
	Althaea chimensis
	Althaea rosea
Garden loosestrife	
	Lysimachia punctata
Garden lupin	Lupinus polyphyllus
Garden nasturtium	
	Tropaeolum majus
Garden phlox	Phlox decussata
	Phlox paniculata
Garden pinks	Dianthus × allwoodii
Garden thyme	Thymus vulgaris
Garden verbenas	
	Verbena × hortensis
	Verbena × hybrida
Garlic pennycress	
	Thlaspi alliaceum
Gas plant	Dictamnus albus
	Dictamnus fraxinella
Gayfeather	Liatris spicata
Gentian sage	Salvia patens
German catchfly	Lychnis viscaria
	Lychnis vulgaris
	Viscaria vulgaris
Germander speedwell	
	Veronica chamaedrys
German primrose	
	Primula obconica
German violet	Exacum affine
Ghost flower	Monotropa uniflora
Giant bellflower	
	Campanula latifolia
Giant cowslip	Primula florindae

POPULAR GARDEN PLANTS

Green chamomile

Giant daisy
    Chrysanthemum serotinum
    Chrysanthemum uliginosum
Giant deadnettle   *Lamium orvala*
Giant forget-me-not
    Myosotidium hortensia
    Myosotidium nobile
Giant honeysuckle
    Lonicera hildebrandiana
Giant inula   *Inula magnifica*
Giant kingcup   *Caltha polypetala*
Giant knapweed
    Centaurea rhaponticum
Giant marigold *Helianthus annuus*
Giant potato vine
    Solanum wendlandii
Giant ragwort *Ligularia japonica*
Giant scabious
    Cephalaria gigantea
    Cephalaria tatarica
Giant spaniard
    Aciphylla scott-thomsonii
Giant stapelia *Stapelia gigantea*
Giant starfish *Stapelia gigantea*
Giant toad plant *Stapelia gigantea*
Gilliflower *Dianthus caryophyllus*
    Matthiola incana
Gipsyweed *Lycopus virginicus*
Gipsywort *Lycopus europaeus*
Glacier crowfoot
    Ranunculus glacialis
Glacier pink *Dianthus glacialis*
    Dianthus neglectus
    Dianthus pavonius
Globe amaranth
    Gomphrena globosa
Globe candytuft *Iberis umbellata*
Globe centaurea
    Centaurea macrocephala
Globe flower *Trollius europaeus*
Globe thistle *Echinops ritro*
Glory-of-the-marsh
    Primula helodoxa
Glory pea *Clianthus dampieri*
    Clianthus formosus
    Clianthus speciosus
Goat-leaf honeysuckle
    Lonicera caprifolium
Goat root *Ononis natrix*
Goat's beard *Aruncus dioicus*
    Aruncus sylvester
    Aruncus vulgaris
    Spiraea aruncus
Goat's rue *Galega officinalis*
Godetia *Godetia grandiflora*
Golden crown beard
    Verbesina encelioides
Gold cup *Ranunculus acris*
Gold dust *Alyssum saxatile*
    Aurinia saxatilis

Golden deadnettle
    Galeobdolon luteum
    Lamium galeobdolon
    Lamiastrum galeobdolon
Golden drop *Onosma tauricum*
Golden eardrops
    Dicentra chrysantha
Golden-eyed grass
    Sisyrinchium californicum
Golden flax *Linum flavum*
Golden groundsel *Senecio aureus*
Golden marguerite
    Anthemis tinctoria
Golden moss *Sedum acre*
Golden rod *Solidago canadensis*
    Solidago virgaurea
Golden samphire
    Inula crithmoides
Golden sedum *Sedum adolphi*
Golden spaniard *Aciphylla aurea*
Goldilocks *Aster linosyris*
    Helichrysum stoechas
    Lynosyris vulgaris
    Ranunculus auricumus
Golds *Calendula officinalis*
Good-luck plant *Oxalis deppei*
Goosewort *Achillea ptarmica*
Grand bellflower
    Adenophora lilifolia
Granny's bonnet *Aquilegia vulgaris*
Grass pea *Lathyrus sativus*
Grass pink *Dianthus plumarius*
Grass vetchling *Lathyrus nissolia*
Grass widow
    Sisyrinchium douglasii
Gravel root
    Eupatorium purpureum
Gravel weed
    Eupatorium purpureum
Greater butterwort
    Pinguicula grandiflora
Greater hawk's beard
    Crepis biennis
Greater periwinkle *Vinca major*
Great eyebright *Euphorbia arctica*
Great golden knapweed
    Centaurea macrocephala
Great knapweed
    Centaurea scabiosa
Great leopard's bane
    Doronicum pardalianches
Great lobelia *Lobelia siphililea*
Great spearwort
    Ranunculus lingua
Great willowherb
    Epilobium hirsutum
Greek valerian
    Polemonium caeruleum
Green amaranthus
    Amaranthus hybridus
Green chamomile *Anthemis nobilis*

Green hellebore
          Helleborus viridus
Grey cinquefoil   *Potentilla cinerea*
Grim collier
          Hieracium aurantiacum
Ground box
          Polygala chamaebuxus
Ground pine   *Ajuga chamaepitys*
Groundsel   *Senecio vulgaris*
Gypsyweed   *Veronica officinalis*
Hag's taper   *Verbascum thapsus*
Hairy Saint John's wort
          Hypericum hirsutum
Hairy saxifrage   *Saxifraga hirsuta*
Hairy spurge   *Euphorbia villosa*
Hairy starfish flower
          Stapelia variegata
Hairy stonecrop   *Sedum villosum*
Hairy thyme   *Thymus praecox*
Hairy toad plant   *Stapelia hirsuta*
Hairy vetchling   *Lathyrus hirsutus*
Hairy willowherb
          Epilobium parviflorum
Hardheads   *Centaurea nigra*
Hardy age   *Eupatorium rugosum*
Hardy ageratum
          Eupatorium coelestinum
Harebell   *Campanula rotundifolia*
Harebell bellflower
          Campanula rotundifolia
Harebell poppy
          Meconopsis quintuplinerva
Hawk's beard   *Crepis rubra*
Hawkweed saxifrage
          Saxifraga hieracifolia
Haybells   *Uvularia grandiflora*
Heart-leaved valerian
          Valeriana pyrenaica
Heartsease   *Viola tricolor*
Heath aster   *Aster ericoides*
Heather   *Calluna vulgaris*
          Erica vulgaris
Heath lobelia   *Lobelia urens*
Heath milkwort
          Polygala serpyllifolia
Heath pearlwort   *Sagina subulata*
Heath speedwell
          Veronica officinalis
Hedge mustard
          Sisymbrium officinale
Hedgerow cranesbill
          Geranium pyrenaicum
Hedge woundwort
          Stachys sylvatica
Helenium   *Helenium autumnale*
Heliopsis   *Heliopsis helianthoides*
          Heliopsis scabra
Hemp agrimony
          Eupatorium cannabinum

Hen-and-chickens houseleek
          Jovibarba soboliferum
          Sempervivum soboliferum
Hen-and-chickens marigold
          Calendula officinalis prolifera
Henbit   *Lamium amplexicaule*
Henbit deadnettle
          Lamium amplexicaule
Heraldic thistle
          Onopordum arabicum
          Onopordum nervosum
Herald of heaven
          Eritrichium nanum
Herb bennet   *Geum urbanum*
Herb of grace   *Ruta graveolens*
Herb peter   *Primula veris*
Herb robert
          Geranium robertianum
Himalayan balsam
          Impatiens glandulifera
Himalayan blue poppy
          Meconopsis baileyi
          Meconopsis betonicifolia
Himalayan comfrey
          Onosma pyramidale
Himalayan cowslip
          Primula florindae
          Primula sikkimensis
Himalayan elecampane
          Inula royleana
Himalayan knotweed
          Polygonum campanulatum
Himalayan mayflower
          Podophyllum emodii
Himalayan touch-me-not
          Impatiens glandulifera
          Impatiens roylei
Himalayan whorlflower
          Morina longifolia
Hoary cinquefoil
          Potentilla argentea
Hoary ragwort   *Senecio erucifolius*
Hoary mullein
          Verbascum pulverulentum
Hoary vervain   *Verbena stricta*
Hoary willowherb
          Epilobium parviflorum
Hollyhock   *Althaea chinensis*
          Althaea rosea
Hollyhock mallow   *Malva alcea*
Holy rope
          Eupatorium cannabinum
Holy thistle   *Cnicus benedictus*
          Silybum marianum
Honesty   *Lunaria annua*
          Lunaria biennis
          Lunaria rediviva
Honeysuckle
          Lonicera periclymenum
Horned poppy   *Glaucium flavum*
Horned rampion
          Phyteuma comosum

Horned violet	*Viola cornuta*
Horse daisy	*Leucanthemum vulgare*
Horsefly	*Baptisia tinctoria*
Horse nettle	*Solanum carolinense*
Horse's tail	*Sedum morganianum*
Hound's tongue	*Cynoglossum amabile*
Humming bird's trumpet	*Zauschneria californica*
Hungarian daisy	*Chrysanthemum serotinum*
	Chrysanthemum uliginosum
	Leucanthemella serotina
Hyssop	*Hyssopus aristatus*
Iberian cranesbill	*Geranium ibericum*
Iceland poppy	*Papaver nudicaule*
Ice plant	*Cryophytum crystallinum*
	Mesembryanthemum crystallinum
	Sedum maximum
	Sedum spectabile
Immortelle	*Xeranthemum anuum*
Inch plant	*Tradescantia zebrina*
Indian balsam	*Impatiens glandulifera*
Indian chocolate	*Geum rivale*
	Geum urbanum
Indian cress	*Tropaeolum majus*
Indian pink	*Dianthus chinensis*
Indian pipe	*Monotropa uniflora*
Indian poke	*Veratrum viride*
Indian tobacco	*Lobelia inflata*
Indian valerian	*Valeriana walichii*
Innocence	*Collinsia bicolor*
	Hedyotis caerulea
Inside-out flower	*Vancouveria planipetala*
Irish lace	*Tagetes filifolia*
Irish saxifrage	*Saxifraga rosacea*
Irish spurge	*Euphorbia hyberna*
Italian aster	*Aster amellus*
Italian bellflower	*Campanula isophylla*
Italian bugloss	*Anchusa azurea*
	Anchusa italica
Italian catchfly	*Silene italica*
Italian verbena	*Verbena tenera*
Itchweed	*Veratrum viride*
Ivy-leaved bellflower	*Wahlenbergia hederacea*
Ivy-leaved crowfoot	*Ranunculus hederaceus*
Ivy-leaved geraniums	*Pelargonium peltatum*
Ivy-leaved harebell	*Wahlenbergia hederacea*
Ivy-leaved pelargoniums	*Pelargonium peltatum*
Ivy-leaved speedwell	*Veronica hederifolia*
Jack-by-the-hedge	*Alliaria petiolata*
Jacobea	*Senecio jacobaea*
Jacob's ladder	*Polemonium caeruleum*
Jacob's staff	*Verbascum thapsus*
Jamaica honeysuckle	*Passiflora laurifolia*
Jamaica vervain	*Verbena jamaicensis*
Japanese artichoke	*Stachys affinis*
Japanese gentian	*Gentiana scabrae*
Japanese honeysuckle	*Lonicera japonica*
Japanese jasmine	*Jasminum mesnyi*
	Jasminum primulinum
Japanese morning glory	*Ipomoea nil*
Japanese pink	*Dianthus × heddewigii*
Japanese primrose	*Primula japonica*
Japanese spurge	*Pachysandra terminalis*
Jasmine nightshade	*Solanum jasminoides*
Jelly-bean plant	*Sedum pachyphyllum*
Jerusalem cowslip	*Pulmonaria officinalis*
Jerusalem cross	*Lychnis chalcedonica*
Jerusalem sage	*Pulmonaria officinalis*
Jewel weed	*Impatiens aurea*
	Impatiens biflora
	Impatiens capensis
Joe-pye weed	*Eupatorium maculatum*
	Eupatorium purpureum
Joseph's coat	*Amaranthus tricolor*
Jumping jack	*Impatiens glandulifera*
Jupiter's beard	*Centranthus ruber*
Jupiter's distaff	*Salvia glutinosa*
Jupiter's staff	*Verbascum thapsus*
Kangaroo apple	*Solanum laciniatum*
Kangaroo vine	*Cissus antarctica*
Kansas gayfeather	*Liatris callilepis*
	Liatris pycnostachya
Kansas niggerhead	*Echinacea angustifolia*
Kerosene bush	*Helichrysum ledifolium*
	Ozothamnus ledifolium
Keyflower	*Primula veris*
Kidney saxifrage	*Saxifraga hirsuta*
Kidneywort	*Umbilicus rupestris*
Kingcup	*Caltha palustris*

Kingfisher daisy	*Felicia bergerana*
King of the alps	
	Eritrichium nanum
King-of-the-meadow	
	Thalictrum polygamum
Knapweed	*Centaurea nigra*
Knitbone	*Symphytum officinale*
Knotgrass	*Polygonum aviculare*
Knotroot	*Stachys affinis*
Knotted pearlwort	*Sagina nodosa*
Korean chrysanthemum	
	Chrysanthemum rubellum
Lace flower	*Episcia dianthiflora*
Lad's love	*Artemisia abrotanum*
Ladybell	*Adenophora lilifolia*
Ladybird poppy	
	Papaver commutatum
Lady rue	*Thalictrum clavatum*
Lady's foxglove	*Verbascum thapsus*
Lady's gloves	*Digitalis purpurea*
Lady's locket	*Dicentra spectabilis*
Lady's mantle	*Alchemilla mollis*
	Alchemilla vulgaris
Lady Washington geraniums	
	Pelargonium × domesticum
Lady Washington pelargoniums	
	Pelargonium × domesticum
Lamb's ear	*Stachys byzantina*
	Stachys lanata
	Stachys olympica
Lamb's lugs	*Stachys byzantina*
Lamb's tail	*Sedum morganianum*
Lamb's tongue	*Stachys byzantina*
	Stachys lanata
	Stachys olympica
Lampshade poppy	
	Meconopsis integrifolia
Large blue alkanet	*Anchusa azurea*
Large-flowered butterwort	
	Pinguicula grandiflora
Large-leaved evening primrose	
	Oenothera glazioviana
Large pink	*Dianthus superbus*
Larger periwinkle	*Vinca major*
Large speedwell	*Veronica austriaca*
Large thyme	*Thymus pulegioides*
Large yellow foxglove	
	Digitalis grandiflora
Large yellow ox-eye	
	Telekia speciosa
Larkspur	*Consolida ambigua*
	Delphinium consolida
Lavender cup	
	Nierembergia caerulea
	Nierembergia hippomanica
Leafy hawkweed	
	Hieracium umbellatum
Leafy spurge	*Euphorbia esula*
Lemon-scented geranium	
	Pelargonium citriodorum
	Pelargonium crispum

Lemon-scented pelargonium	
	Pelargonium citriodorum
	Pelargonium crispum
Lemon-scented thyme	
	Thymus × citriodorus
Lemon verbena	*Aloysia triphylla*
	Lippia citriodora
Lenten rose	*Helleborus orientalis*
Leopard's bane	
	Doronicum caucasicum
Lesser bugloss	*Lycopsis arvensis*
Lesser celandine	
	Ranunculus ficaria
Lesser dandelion	
	Taraxacum erythrospermum
	Taraxacum laevicatum
Lesser herb robert	
	Geranium purpurea
Lesser knapweed	*Centaurea nigra*
Lesser meadow rue	
	Thalictrum minus
Lesser periwinkle	*Vinca minor*
Lesser plume poppy	
	Bocconia microcarpa
	Macleaya microcarpa
Lesser spearwort	
	Ranunculus flammula
Lilac pink	*Diantus superbus*
Ling	*Calluna vulgaris*
	Erica vulgaris
Linseed	*Linum usitatissimum*
Lion's foot	*Alchemilla vulgaris*
	Leontopodium alpinum
Lion's heart	*Physostegia virginiana*
Lion's teeth	*Taraxacum officinale*
Liquorice plant	
	Helichrysum petiolatum
Live forever	*Sedum telephium*
Livelong	*Sedum telephium*
Livingstone daisy	
	Cleretum bellidiforme
	Dorotheanthus bellidiflorus
	Mesembryanthemum criniflorum
Lobster claw	*Clianthus puniceus*
Lobster plant	
	Euphorbia pulcherrima
London pride	*Saxifraga umbrosa*
	Saxifraga × urbium
London rocket	*Sisymbrium irio*
Long-headed poppy	
	Papaver dubium
Lord Anson's pea	
	Lathyrus nervosus
Lousewort	*Pedicularis sylvatica*
Love-in-a-mist	*Nigella damascena*
Love-lies-bleeding	
	Amaranthus caudatus
Lyre flower	*Dicentra spectabilis*

POPULAR GARDEN PLANTS

Moon daisy

Madagascar periwinkle
Amsonia roseus
Catharanthus roseus
Vinca rosea
Maiden pink   *Dianthus deltoides*
Maiden's wreath   *Francoa ramosa*
Mallowwort   *Malope trifolia*
Maltese cross *Lychnis chalcedonica*
Marguerite
Argyranthemum frutescens
Leucanthemum vulgare
Marigold of Peru
Helianthus annuus
Marjoram   *Origanum vulgare*
Marsh cinquefoil
Potentilla palustris
Marsh fleawort   *Senecio palustris*
Marsh flower
Limnanthes douglasii
Marsh gentian
Gentiana pneumonanthe
Marsh hawk's beard
Crepis paludosa
Marsh marigold   *Caltha palustris*
Marsh pea   *Lathyrus palustris*
Marsh ragwort   *Senecio aquaticus*
Marsh speedwell
Veronica scutellata
Marsh valerian   *Valeriana dioica*
Marsh violet   *Viola cucullata*
Marsh willowherb
Epilobium palustra
Marsh woundwort
Stachys palustris
Martha Washington geraniums
Pelargonium × domesticum
Martha Washington pelargoniums
Pelargonium × domesticum
Marvel of Peru   *Mirabilis jalapa*
Mask flower *Alonsoa warscewiczii*
Matted sea lavender
Limonium bellidifolium
May apple *Podophyllum peltatum*
Meadow anemone
Pulsatilla vulgaris
Meadow buttercup
Ranunculus acris
Meadow bloom   *Ranunculus acris*
Meadow clary   *Salvia pratensis*
Meadow cranesbill
Geranium pratense
Meadow daisy   *Bellis perennis*
Meadow foam
Limnanthes douglasii
Meadow geranium
Geranium pratense
Meadow rue
Thalictrum aquilegiifolium
Meadow saxifrage
Saxifraga granulata
Meadowsweet *Filipendula ulmaria*

Mealy-cup sage   *Salvia farinacea*
Medusa's head
Euphorbia caput-medusae
Merry bells *Uvularia grandiflora*
Mexican aster *Cosmos bipinnatus*
Mexican fire plant
Euphorbia heterophylla
Mexican flame vine
Senecio confusus
Mexican fleabane
Erigeron karkinskianus
Mexican foxglove
Tetranema mexicana
Tetranema roseum
Mexican ivy   *Cobaea scandens*
Mexican poppy
Argemone mexicana
Mexican red sage   *Salvia fulgens*
Mexican sunflower
Tithonia rotundifolia
Tithonia speciosa
Mexican violet *Tetranema roseum*
Mexican zinnia   *Zinnia haageana*
Michaelmas daisy
Aster novae-angliae
Midsummer men *Sedum telephium*
Mignonette   *Reseda odorata*
Milfoil   *Achillea millefolium*
Milk purslane *Euphorbia maculata*
Milk thistle   *Silybum marianum*
Milkwhite rock jasmine
Androsace lactea
Milkwort   *Polygala calcarea*
Milky bellflower
Campanula lactiflora
Milky rock jasmine
Androsace lactea
Mistflower
Eupatorium coelestinum
Eupatorium rugosum
Mithridate mustard
Thlaspi arvense
Mock cypress
Kochia scoparia trichophylla
Mole plant   *Euphorbia lathyris*
Molucca balm   *Moluccella laevis*
Monarch of the veldt
Venidium fastuosum
Monastery bells *Cobaea scandens*
Money plant   *Lunaria annua*
Lunaria biennis
Moneywort
Lysimachia nummularia
Monkey flower   *Mimulus luteus*
Monkey musk   *Mimulus luteus*
Monkshood   *Aconitum napellus*
Moon daisy
Chrysanthemum leucanthemum
Leucanthemella serotina
Leucanthemum vulgare

Moonflower

Moonflower	*Ipomoea alba*
	Ipomoea bona-nox
	Ipomoea noctiflora
	Ipomoea roxburghii
Moonlight primula	*Primula alpicola*
	Primula microdonta alpicola
Moonvine	*Ipomoea alba*
	Ipomoea bona-nox
	Ipomoea noctiflora
	Ipomoea roxburghii
Moonwort	*Lunaria annua*
	Lunaria biennis
Morning glory	*Ipomoea acuminata*
	Ipomoea purpurea
Moschatel	*Adoxa moschatellina*
Moss campion	*Silene acaulis*
Moss phlox	*Phlox setacea*
	Phlox subulata
Moss pink	*Phlox setacea*
	Phlox subulata
Moss verbena	*Verbena tenuisecta*
Mossy cyphel	*Minuartia sedoides*
Mossy rockfoil	*Saxifraga hypnoides*
Mossy saxifrage	
	Saxifraga hypnoides
Mother of thousands	
	Saxifraga sarmentosa
	Saxifraga stolonifera
Moth mullein	*Verbascum blattaria*
Mountain bluet	
	Centaurea montana
Mountain cranesbill	
	Geranium pyrenaicum
Mountain fleece	
	Polygonum amplexicaule
Mountain foxglove	
	Ourisia macrophylla
Mountain garland	
	Clarkia unguiculata
Mountain hollyhock	
	Iliamna rivularis
Mountain houseleek	
	Sempervivum montanum
Mountain knapweed	
	Centaurea montana
Mountain pansy	*Viola lutea*
Mountain phlox	
	Linanthus grandiflorus
Mountain sandwort	
	Minuartia rubella
Mountain snow	
	Euphorbia marginata
Mountain snowbell	
	Soldanella montana
Mountain sorrel	*Oxyria dignya*
Mountain speedwell	
	Veronica montana
Mountain tassel	
	Soldanella montana
Mountain valerian	
	Valeriana montana

Mount Atlas daisy	
	Anacyclus depressus
Mournful widow	
	Scabiosa atropurpurea
Mourning widow	
	Geranium phaeum
Moutan paeony	*Paeonia arborea*
	Paeonia moutan
	Paeonia suffruticosa
Mullein pink	
	Agrostemma coronaria
	Lychnis coronaria
Musk	*Mimulus moschatus*
Musk mallow	*Malva moschata*
Muskrat weed	
	Thalictrum polygamum
Musky saxifrage	
	Saxifraga moschata
Muster-john-henry	*Tagetes minuta*
Namaqualand daisy	
	Dimorphotheca pluvialis
	Dimorphotheca sinuata
	Venidium fastuosum
Nancy pretty	*Saxifraga × urbium*
Narrow-leaved inula	*Inula ensifolia*
Narrow-leaved zinnia	
	Zinnia angustifolia
	Zinnia haageana
Navelwort	
	Omphalodes cappadocica
	Omphalodes umbilicus
	Umbilicus rupestris
Nettle-leaved bellflower	
	Campanula trachelium
	Campanula urticifolia
New England aster	
	Aster novae-angliae
New York aster	*Aster novi-belgii*
New Zealand burr	
	Acaena microphylla
New Zealand flax	
	Linum monogynum
New Zealand lilac	*Hebe hulkeana*
	Veronica hulkeana
New Zealand willowherb	
	Epilobium brunnescens
Night-flowering catchfly	
	Silene noctiflora
Night-scented stock	
	Matthiola bicornis
Nine-joints	*Polygonum aviculare*
Nodding avens	*Geum rivale*
Nodding catchfly	*Silene pendula*
None-so-pretty	
	Saxifraga × urbium
Nonsuch	*Medicago lupulina*
Northern hawk's beard	
	Crepis mollis
Northern shore wort	
	Mertensia maritima
Northern wolfsbane	
	Aconitum lycotonum lycotonum

POPULAR GARDEN PLANTS

Pincushion flower

Nosebleed   *Achillea millefolium*
Notch-leaf statice
          Limonium sinuatum
          Statice sinuatum
Nottingham catchfly   *Silene nutans*
Obedient plant
          Physostegia virginiana
Old man   *Artemisia abrotanum*
Old woman   *Artemisia stellerana*
Opium poppy   *Papaver somniferum*
Orange balsam   *Impatiens capensis*
Orange daisy   *Erigeron aurantiacus*
Orange hawkweed
          Hieracium aurantiacum
          Hieracium brunneocroceum
Orange mullein
          Verbascum phlomoides
Oregano   *Origanum vulgare*
Oriental periwinkle
          Rhazya orientalis
Oriental poppy   *Papaver orientale*
Oriental rocket
          Sisymbrium orientale
Ornamental cabbage
          Brassica oleracea acephala
Orpine   *Sedum telephium*
Oswega tea   *Monarda didyma*
Our Lady's milk thistle
          Silybum marianum
Ox-eye chamomile
          Anthemis tinctoria
Ox-eye daisy
    Chrysanthemum leucanthemum
          Leucanthemum vulgare
Oxford ragwort   *Senecio squalidus*
Oxlip   *Primula elatior*
Oyster plant   *Mertensia maritima*
Ozark sundrops
          Oenothera macrocarpa
          Oenothera missouriensis
Paigles   *Primula elatior*
Painted daisy
          Chrysanthemum carinatum
          Chrysanthemum coccineum
          Pyrethrum hybridum
          Pyrethrum roseum
          Tanacetum coccineum
Painted spurge
          Euphorbia heterophylla
Painted tongue   *Salpiglossis sinuata*
Pale flax   *Linum bienne*
Pale persicaria
          Polygonum lapathifolium
Pale toadflax   *Linaria repens*
Pale willowherb   *Epilobium roseum*
Palm Springs daisy
          Anthemis arabicus
Palsywort   *Primula veris*
Pampas grass   *Cortaderia argentea*
          Cortaderia selloana
Pansy   *Viola tricolor*

Paris daisy
          Argyranthemum frutescens
Parrot's beak   *Lotus berthelotii*
Parrot's bill   *Clianthus puniceus*
Pasque flower   *Pulsatilla vulgaris*
Patience   *Impatiens sultanii*
          Impatiens wallerana
Patient lucy   *Impatiens sultanii*
          Impatiens wallerana
Peach-leaved bellflower
          Campanula grandis
          Campanula latiloba
          Campanula persicifolia
Peacock poppy   *Papaver pavonium*
Pedlar's basket   *Linaria vulgaris*
Pee the bed   *Taraxacum officinale*
Pellitory   *Anacyclus pyrethrum*
Pellitory of Spain
          Anacyclus pyrethrum
Pennycress   *Thlaspi arvense*
Pennywort   *Umbilicus rupestris*
Pepper saxifrage   *Silaum silaus*
Perennial cornflower
          Centaurea dealbata
          Centaurea montana
Perennial honesty   *Lunaria rediviva*
Perennial morning glory
          Ipomoea acuminata
Perennial pea   *Lathyrus latifolius*
Perennial sage   *Salvia superba*
Perennial sweet pea
          Lathyrus latifolius
Perennial wall rocket
          Diplotaxis tenuifolia
Perfoliate pennycress
          Thlaspi perfoliatum
Perforate Saint John's wort
          Hypericum perforatum
Persian buttercup
          Ranunculus asiaticus
Persian everlasting pea
          Lathyrus rotundifolius
Persian speedwell   *Veronica persica*
Persian violet   *Exacum affine*
Peruvian marigold
          Helianthus annuus
Petty spurge   *Euphorbia peplis*
          Euphorbia peplus
Pheasant's eye   *Adonis annua*
Phu   *Valeriana officinalis*
Picotee   *Dianthus caryophyllus*
Pigmy hawk's beard
          Crepis pygmaea
Pigmy sunflower
          Actinea grandiflora
Pilewort   *Erechtites hieracifolia*
          Ranunculus ficaria
Pill-bearing spurge   *Euphorbia hirta*
Pincushion flower
          Scabiosa atropurpurea
          Scabiosa caucasica

Pincushion plant	*Cenia barbata*
	Cotula barbata
	Cotula cenia
Pineapple sage	*Salvia rutilans*
Pink dandelion	*Crepis incana*
Pink evening primrose	
	Oenothera rosea
Pink-head knotweed	
	Polygonum capitatum
Pink oxalis	*Oxalis articulata*
Pink pokers	*Limonium suworowii*
Pink purslane	*Montia sibirica*
Pink rock-jasmine	
	Androsace carnea
Pink sand verbena	
	Abronia umbellata
Pink willowherb	*Epilobium roseum*
Piss the bed	*Taraxacum officinale*
Ploughman's spikenard	
	Inula conyza
Plume flower	*Celosia plumosa*
Plume poppy	*Bocconia cordata*
	Macleaya cordata
Poached egg flower	
	Limnanthes douglasii
Poached egg plant	
	Limnanthes douglasii
Poet's jessamine	
	Jasminum officinale
Poinsettia	*Euphorbia pulcherrima*
Poison potato	*Solanum carolinense*
Policeman's helmet	
	Impatiens glandulifera
Polyanthus	*Primula polyantha*
Poor man's weather glass	
	Anagallis arvensis
Pork and beans	
	Sedum × rubrotinctum
Poroporo	*Solanum laciniatum*
Potato bush	*Solanum rantonnetii*
Potato vine	*Solanum jasminoides*
Pot marigold	*Calendula officinalis*
Prairie blazing star	
	Liatris pycnostachya
Prairie evening primrose	
	Oenothera missouriensis
Prairie flax	*Linum lewisii*
Prairie lupin	*Lupinus lepidus*
Prayer plant	*Maranta leuconeura*
Pretty betsy	*Centranthus ruber*
Prickly phlox	*Gilia californica*
Prickly poppy	
	Argemone grandiflora
Pride of California	
	Lathyrus splendens
Primrose	*Primula acaulis*
	Primula vulgaris
Primrose jasmine	
	Jasminum mesnyi
	Jasminum primulinum

Prince of Wales' feathers	
	Celosia cristata plumosa
	Tanacetum densum amani
Prince's feather	
	Amaranthus hypochondriacus
Privet honeysuckle	*Lonicera pileata*
Procumbent cinquefoil	
	Potentilla anglica
Prophet flower	*Arnebia echioides*
Prostrate vervain	
	Verbena bracteata
Purple archangel	
	Lamium purpureum
Purple avens	*Geum rivale*
Purple boneset	
	Eupatorium purpureum
Purple bugloss	
	Echium plantagineum
Purple coneflower	
	Echinacea purpurea
Purple deadnettle	
	Lamium purpureum
Purple-eyed grass	
	Sisyrinchium douglasii
Purple granadilla	*Passiflora edulis*
Purple loosestrife	
	Lythrum salicaria
Purple mullein	
	verbascum phoeniceum
Purple sage	
	Salvia officinalis purpurascens
Purple saxifrage	
	Saxifraga oppositifolia
Purple spurge	*Euphorbia peplis*
	Euphorbia peplus
Purple toadflax	*Linaria purpurea*
Purple willowherb	
	Lythrum salicaria
Puya	*Puya alpestris*
	Puya berteroniana
Pyramidal bugle	*Ajuga pyramidalis*
Pyramidal saxifrage	
	Saxifraga cotyledon
Pyrenean valerian	
	Valeriana pyrenaica
Pyrethrum	
	Chrysanthemum coccineum
	Pyrethrum hybridum
	Pyrethrum roseum
	Tanacetum cinerariifolium
	Tanacetum coccineum
Queen Anne's thimbles	
	Gilia capitata
Queen of the meadow	
	Eupatorium purpureum
	Filipendula ulmaria
Queen of the night	
	Ipomoea noctiflora
Queen of the prairie	
	Filipendula rubra
	Spiraea lobata
Quicksilver weed	
	Thalictrum dioicum

Ragged robin   *Lychnis flos-cuculi*

Rag paper   *Verbascum thapsus*

Ragwort   *Senecio jacobaea*

Rainbow pink   *Dianthus chinensis*

Rain daisy
    Dimorphotheca pluvialis

Ramping fumitory
    Fumaria capreolata

Rampion   *Campanula rapunculus*

Rampion bellflower
    Campanula rapunculus

Rat's tail statice
    Limonium suworowii
    Psylliostachys suworowii
    Statice suworowii

Rattlesnake master
    Liatris squarrosa

Rattlesnake root   *Polygala senega*

Red baneberry   *Actaea erythocarpa*
    Actaea rubra
    Actaea spicata rubra

Red campion   *Lychnis dioica*
    Melandrium diurnum
    Silene dioica

Red catchfly   *Lychnis viscaria*

Red deadnettle
    Lamium purpureum

Red flax   *Linum grandiflorum*
    Linum rubrum

Red horned poppy
    Glaucium corniculatum
    Glaucium grandiflorum

Red legs   *Polygonum persicaria*

Red maids   *Calandrinia menziesii*

Red morning glory
    Ipomoea coccinea

Red mountain spinach
    Atriplex hortensis rubra

Red orach
    Atriplex hortensis rubra

Red pepper   *Capsicum annuum*

Red rattle   *Pedicularis palustris*

Red ribbons   *Clarkia concinna*
    Euchardium concinna

Red root   *Potentilla erecta*

Red shank   *Polygonum persicaria*

Red star thistle
    Centaurea calcitrapa

Red valerian   *Centranthus ruber*
    Kentranthus ruber

Redwood ivy
    Vancouveria planipetala

Reflexed stonecrop
    Sedum reflexum

Regal geraniums
    Pelargonium × domesticum

Regal pelargoniums
    Pelargonium × domesticum

Rest harrow   *Ononis repens*

Robb's bonnet   *Euphorbia robbiae*

Rock campion   *Silene rupestris*

Rock cinquefoil   *Potentilla rupestris*

Rock cranesbill
    Geranium macrorrhizum

Rockery speedwell
    Veronica prostrata

Rocket   *Hesperis matronalis*

Rocket candytuft   *Iberis amara*
    Iberis coronaria

Rocket larkspur
    Consolida ambigua
    Delphinium ajacis

Rock fringe willowherb
    Epilobium obcordatum

Rock purslane
    Calandrinia umbellata

Rock sea lavender
    Limonium binervosum

Rock soapwort
    Saponaria ocymoides

Rock speedwell   *Veronica fruticans*

Rock stonecrop
    Sedum forsteranum

Rockwood lily   *Ranunculus lyalli*

Roman chamomile
    Anthemis nobilis

Roof houseleek
    Sempervivum tectorum

Rose balsam   *Impatiens balsamina*

Rosebay willowherb
    Epilobium angustifolium

Rose campion
    Agrostemma coronaria
    Lychnis coronaria

Rose mallow   *Lavatera trimestris*

Rose moss   *Portulaca grandiflora*

Rose of heaven   *Lychnis coeli-rosa*
    Silene coeli-rosa
    Silene oculata
    Viscaria elegans

Rose periwinkle   *Amsonia roseus*
    Catharanthus roseus
    Vinca rosea

Roseroot   *Rhodiola rosea*
    Sedum rhodiola
    Sedum rosea

Rose root sedum   *Sedum rosea*

Rose-scented geranium
    Pelargonium capitatum

Rose-scented pelargonium
    Pelargonium capitatum

Rose verbena   *Verbena aubletia*
    Verbena canadensis

Rough hawk's beard   *Crepis biennis*

Round-leaved speedwell
    Veronica filiformis

Rue   *Ruta graveolens*

Rue-leaved saxifrage
    Saxifraga tridactylotes

Running myrtle   *Vinca minor*

Russian aconite
    Aconitum orientale

Russian chamomile
    Anthemis nobilis

Russian comfrey
    Symphytum peregrinum
    Symphytum × uplandicum

Russian knotgrass
    Polygonum erectum

Russian statice
    Limonium suworowii
    Psylliostachys suworowii
    Statice suworowii

Russian vine
    Bilderdykia baldschuanicum
    Polygonum baldschuanicum

Rusty foxglove   *Digitalis ferruginea*

Sage   *Salvia officinalis*

Sage brush   *Artemisia tridentata*

Saint Barnaby's thistle
    Centaurea solstitalis

Saint Benedict's thistle
    Cnicus benedictus

Saint James' wort   *Senecio jacobaea*

Saint John's chamomile
    Anthemis sancti-johannis

Saint Mary's milk thistle
    Silybum marianum

Saint Mary's thistle
    Silybum marianum

Saint Patrick's cabbage
    Saxifraga umbrosa
    Saxifraga × urbium

Saint Peter's wort   *Primula veris*

Salpiglossis   *Salpiglossis sinuata*

Sand catchfly   *Silene conica*

Sand flower   *Ammobium alatum*

Sand phlox   *Phlox bifida*

Sand pink   *Dianthus arenarius*

Sand toadflax   *Linaria arenaria*

Sand verbena   *Abronia latifolia*

Saponaria   *Vaccaria pyramidata*

Satin flower   *Clarkia amoena*
    Godetia grandiflora
    Lunaria annua
    Lunaria biennis
    Sisyrinchium stratiatum

Satin leaf   *Heuchera hispida*

Satin poppy
    Meconopsis napaulensis

Savannah flower   *Echites andrewsii*

Sawwort   *Serratula shawii*

Scabwort   *Inula helenium*

Scarlet bidi-bidi
    Acaena microphylla

Scarlet flax   *Linum grandiflorum*
    Linum rubrum

Scarlet larkspur
    Delphinium cardinale

Scarlet lobelia   *Lobelia cardinalis*

Scarlet monkey flower
    Mimulus cardinalis

Scarlet pimpernel
    Anagallis arvensis

Scarlet plume   *Euphorbia fulgens*

Scarlet sage   *Salvia splendens*

Scarlet trumpet honeysuckle
    Lonicera × brownii

Scarlet wisteria
    Daubentonia tripetii
    Sesbania tripetii

Scotch creeper
    Tropaeolum speciosum

Scotch marigold
    Calendula officinalis

Scotch thistle
    Onopordum acanthium

Scottish bluebell
    Campanula rotundifolia

Scottish flame flower
    Tropaeolum speciosum

Sea alyssum   *Lobularia maritima*

Sea campion   *Silene maritima*
    Silene vulgaris maritima

Sea fig   *Cryophytum crystallinum*
    Mesembryanthemum
    crystallinum

Sea holly   *Eryngium maritimum*

Sea pea   *Lathyrus japonicus*
    Lathyrus maritimus

Sea pearlwort   *Sagina maritima*

Sea pink   *Armeria maritima*

Sea poppy   *Glaucium corniculatum*
    Glaucium grandiflorum

Sea spurge   *Euphorbia paralias*

Senega   *Polygala senega*

Serbian bellflower
    Campanula poscharskyana

Setterwort   *Helleborus foetidus*

Shaggy hawkweed
    Hieracium villosum

Shaggy starfish   *Stapelia hirsuta*

Shamrock   *Medicago lupulina*
    Oxalis acetosella

Shasta daisy
    Chrysanthemum maximum
    Chrysanthemum × superbum
    Leucanthemum × superbum

Sheep's fescue   *Festuca ovina*

Shell flower   *Moluccella laevis*

Shepherd's barometer
    Anagallis arvensis

Shepherd's club   *Verbascum thapsus*

Shepherd's crook
    Lysimachia clethroides

Shepherd's knot
    Potentilla tormentilla

Shield nasturtium
    Tropaeolum lobbianum
    Tropaeolum peltophorum

Shining cranesbill
    Geranium lucidum

Shining scabious   *Scabiosa lucida*

POPULAR GARDEN PLANTS

Shoo-fly — *Nicandra physalodes*
Shooting star
    Dodecatheon alpinum
    Dodecatheon meadia
Short-leaved gentian
    Gentiana brachyphylla
Show geraniums
    Pelargonium × domesticum
Show pelargoniums
    Pelargonium × domesticum
Shrubby musk
    Mimulus aurantiacus
Shrubby penstemon
    Penstemon cordifolius
Siberian bugloss
    Brunnera macrophylla
Siberian wallflower
    Cheiranthus × allionii
    Erysimum asperum
    Erysimum perofskianum
Sicily thyme — *Thymus nitidus*
    Thymus richardii nitidus
Sierra shooting star
    Dodecatheon jeffreyi
Signet marigold — *Tagetes signata*
    Tagetes tenuifolia
Silver lace vine
    Bilderdykia aubertii
    Polygonum aubertii
Silver ragwort — *Senecio bicolor*
Silver sage — *Salvia argentea*
Silver thistle
    Onopordum arabicum
    Onopordum nervosum
Silver weed — *Potentilla anserina*
Silvery cinquefoil
    Potentilla anserina
Simpler's joy — *Verbena hastata*
Skyrocket — *Gilia coronopifolia*
    Gilia rubra
Small balsam — *Impatiens parviflora*
Small catchfly — *Silene gallica*
Small mallow — *Malva pusilla*
Small scabious — *Scabiosa columbaria*
Small yellow foxglove
    Digitalis lutea
Smooth hawk's beard
    Crepis capillaris
Snakeroot — *Polygala senega*
Snakeweed — *Polygonum bistorta*
Snapdragon — *Antirrhinum majus*
Sneezeweed — *Achillea ptarmica*
    Helenium autumnale
Sneezewort — *Achillea ptarmica*
Snow-in-summer
    Helichrysum thyrsoideum
    Ozothamnus thyrsoideum
Snow-on-the-mountain
    Euphorbia marginata
Snow poppy — *Eomecon chionanthe*
Soapwort — *Saponaria officinalis*

Soft cranesbill — *Geranium molle*
Soldiers and sailors
    Pulmonaria officinalis
Southernwood
    Artemisia abrotanum
Sow's ear — *Stachys byzantina*
Spaniard — *Aciphylla colensoi*
Spanish catchfly — *Silene otites*
Spanish pellitory
    Anacyclus pyrethrum
Spanish poppy
    Papaver rupifragum
Spear-leaved willowherb
    Epilobium lanceolatum
Speedwell — *Veronica officinalis*
Spider flower — *Cleome spinosa*
Spider houseleek
    Sempervivum arachnoideum
Spiderworts
    Tradescantia × andersoniana
    Tradescantia virginiana
Spike gayfeather — *Liatris spicata*
Spiked loosestrife
    Lythrum salicaria
Spiked rampion
    Phyteuma spicatum
Spiked speedwell — *Veronica spicata*
Spiny alyssum — *Alyssum spinosum*
Spotted deadnettle
    Lamium maculatum
Spotted dog — *Pulmonaria officinalis*
Spotted gentian — *Gentiana punctata*
Spotted hawkweed
    Hieracium maculatum
Spotted loosestrife
    Lysimachia punctata
Spotted medick — *Medicago arabica*
Spotted touch-me-not
    Impatiens biflora
    Impatiens capensis
Spreading bellflower
    Campanula patula
Spreading globe flower
    Trollius laxus
Spring adonis — *Adonis vernalis*
Spring anemone — *Anemone vernalis*
    Pulsatilla vernalis
Spring beauty — *Montia perfoliata*
Spring bell — *Sisyrinchium douglasii*
Spring cinquefoil
    Potentilla tabernaemontani
    Potentilla verna
Spring gentian — *Gentiana verna*
    Gentiana verna angulosa
Spring sandwort — *Minuartia verna*
Spring speedwell — *Veronica verna*
Spring vetch — *Lathyrus vernuus*
Spurred bellflower
    Campanula alliariifolia
Square-stalked willowherb
    Epilobium tetragonum

Square-stemmed
Saint John's wort

Square-stemmed Saint John's wort	*Hypericum tetrapterum*
Squirrel corn	*Dicentra canadensis*
Staggerweed	*Dicentra canadensis*
Staggerwort	*Senecio jacobaea*
Standing cypress	*Gilia coronopifolia*
	Gilia rubra
Star daisy	*Lindheimera texana*
Stardust	*Gilia hybrida*
	Gilia lutea
	Leptosiphon hybridus
Starfish plant	*Stapelia variegata*
Star ipomoea	*Ipomoea coccinea*
Star of Bethlehem	*Campanula isophylla*
Star of the veldt	*Dimorphotheca aurantiaca*
	Dimorphotheca sinuata
Starry saxifrage	*Saxifraga stellaris*
	Saxifraga stellata
Star thistle	*Centaurea calcitrapa*
Starwort	*Aletris farinosa*
Statice	*Limonium latifolium*
	Limonium sinuatum
	Statice latifolium
Stavesacre	*Delphinium staphisagria*
Steeple bellflower	*Campanula pyramidalis*
Steeple bells	*Cobaea scandens*
Stemless gentian	*Gentiana acaulis*
Sticklewort	*Agrimonia eupatoria*
Sticky catchfly	*Lychnis viscaria*
Sticky groundsel	*Senecio viscosus*
Sticky phacelia	*Eutoca viscida*
	Phacelia viscida
Sticky sage	*Salvia glutinosa*
Sticky sandwort	*Minuartia viscosa*
Stinking bob	*Geranium robertianum*
Stinking hawk's beard	*Crepis foetida*
Stinking hellebore	*Helleborus foetidus*
Stinkweed	*Thlaspi arvense*
Stock	*Matthiola incana*
Stokes's aster	*Stokesia cyanea*
	Stokesia laevis
Stone orpine	*Sedum reflexum*
Strawbells	*Uvularia perfoliata*
Strawberry geranium	*Saxifraga sarmentosa*
	Saxifraga stolonifera
Strawflower	*Helichrysum bracteatum*
Striated catchfly	*Silene conica*
String of beads	*Senecio rowleyanus*
String of pearls	*Senecio rowleyanus*
String of sovereigns	*Lysimachia nummularia*

Sturt's desert pea	*Clianthus speciosus*
Sugar scoop	*Tiarella unifoliata*
Sulphur cinquefoil	*Potentilla recta*
Summer adonis	*Adonis aestivalis*
Summer cypress	*Kochia scoparia trichophylla*
Summer forget-me-not	*Anchusa capensis*
Summer jasmine	*Jasminum officinale*
Summer pheasant's eye	*Adonis aestivalis*
Summer starwort	*Erinus alpinus*
Sundrops	*Oenothera fruticosa*
	Oenothera linearis
Sunflower	*Helianthus annuus*
Sun plant	*Portulaca grandiflora*
Sunray	*Helipterum manglesii*
	Rhodanthe manglesii
Sun spurge	*Euphorbia helioscopia*
Swan river daisy	*Brachycome iberidifolia*
Swan river everlasting	*Helipterum manglesii*
	Rhodanthe manglesii
Sweat root	*Polemonium reptans*
Sweet alison	*Lobularia maritima*
Sweet alyssum	*Alyssum maritimum*
	Lobularia maritima
Sweet bergamot	*Monarda didyma*
Sweet bugle	*Lycopus virginicus*
Sweet four o'clock	*Mirabilis longiflora*
Sweet golden rod	*Solidago odora*
Sweet mace	*Tagetes lucida*
Sweet nancy	*Achillea ageratum*
Sweet pea	*Lathyrus odoratus*
Sweet rocket	*Hesperis matronalis*
Sweet scabious	*Erigeron annuus*
	Scabiosa atropurpurea
Sweet-scented marigold	*Tagetes lucida*
Sweet spurge	*Euphorbia dulcis*
Sweet sultan	*Centaurea imperialis*
	Centaurea moschata
Sweet violet	*Viola odorata*
Sweet william	*Dianthus barbatus*
Sweet william catchfly	*Silene armeria*
Swine's snout	*Taraxacum officinale*
Tagetes	*Tagetes signata*
	Tagetes tenuifolia
Tahoka daisy	*Aster tenacetifolius*
Tall gayfeather	*Liatris scariosa*
Tall meadow rue	*Thalictrum polygamum*
Tall nasturtium	*Tropaeolum majus*
Tall rocket	*Sisymbrium altissimum*

POPULAR GARDEN PLANTS

Tangier pea	*Lathyrus tingitanus*
Tangier scarlet pea	
	Lathyrus tingitanus
Tansy	*Tanacetum vulgare*
Tansy phacelia	
	Phacelia tanacetifolia
Tartarian honeysuckle	
	Lonicera tatarica
Tassel flower	
	Amaranthus caudatus
Telegraph plant	*Desmodium gyrans*
Ten-week stock	
	Matthiola incana annua
Thick-leaved stonecrop	
	Sedum dasyphyllum
Thoroughwort	
	Eupatorium perfoliatum
Thyme-leaved speedwell	
	Veronica serpyllifolia
Texas blue-bonnet	
	Lupinus subcarnosus
	Lupinus texensis
Thousand weed	
	Achillea millefolium
Tickseed	*Coreopsis grandiflora*
Tidy tips	*Layia elegans*
	Layia platyglossa
Toad cactus	*Stapelia variegata*
Toadflax	*Linaria maroccana*
Toad plant	*Stapelia variegata*
Tobacco	*Nicotiana tabacum*
Tobacco plant	*Nicotiana affinis*
	Nicotiana alata
Tormentil	*Potentilla erecta*
Touch-me-not	
	Impatiens balsamina
	Impatiens noli-tangere
Town hall clock	
	Adoxa moschatellina
Trailing bellflower	
	Cyananthus lobatus
Trailing gazania	
	Gazania leucolaena
	Gazania rigens leucolaena
	Gazania uniflora
Trailing lobelia	
	Lobelia erinus pendula
Trailing myrtle	*Vinca minor*
Trailing violet	*Viola hederacea*
Transvaal daisy	*Gerbera jamesonii*
Treacle mustard	
	Erysimum cheiranthoides
Treasure flowers	
	Gazania × hybrids
Tree paeony	*Paeonia arborea*
	Paeonia moutan
	Paeonia suffruticosa
Trinity flowers	
	Tradescantia × andersoniana
	Tradescantia virginiana
Tropaeolum	
	Tropaeolum polyphyllum

True clary	*Salvia sclarea*
True columbine	*Aquilegia vulgaris*
True michaelmas daisy	
	Aster novi-belgii
Trumpet gentian	*Gentiana acaulis*
	Gentiana kochiana
Trumpet honeysuckle	
	Lonicera sempervirens
Trumpet vine	*Campsis radicans*
Tuberous pea	*Lathyrus tuberosus*
Tufted alkanet	*Anchusa caespitosa*
Tufted harebell	
	Wahlenbergia albomarginata
Tufted pansy	*Viola cornuta*
Tufted saxifrage	
	Saxifraga caespitosa
Tulip poppy	*Papaver glaucum*
Tumbleweed	
	Amaranthus graecizans
Tumbling ted	*Saponaria ocymoides*
Turban buttercup	
	Ranunculus asiaticus
Turkey corn	*Dicentra canadensis*
Turkey pea	*Dicentra canadensis*
Turkish tobacco	*Nicotiana rustica*
Tussock bellflower	
	Campanula carpatica
Twiggy mullein	
	Verbascum virgatum
Twinberry	*Lonicera involucrata*
Twinspur	*Diascia barberae*
Umbrella plant	
	Peltiphyllum peltatum
	Saxifraga peltata
Unicorn plant	*Martynia louisiana*
	Proboscidea jussieui
Upright cinquefoil	*Potentilla recta*
Upright spurge	*Euphorbia stricta*
Valley lupin	*Lupinus vallicola*
Vanilla leaf	*Achlys triphylla*
	Liatris odoratissima
Velvet flower	*Salpiglossis sinuata*
Velvet groundsel	*Senecio petasites*
Velvet mullein	*Verbascum thapsus*
Venus's looking glass	
	Legousia speculum-veneris
	Specularia speculum
	Specularia speculum-veneris
Venus's navelwort	
	Omphalodes linifolia
Vernal gentian	*Gentiana verna*
Vernal sandwort	*Minuartia verna*
Vervain	*Verbena lasiostachys*
	Verbena officinalis
	Verbena rigida
Violet cress	*Ionopsidium acaule*
Viper's bugloss	*Echium lycopsis*
	Echium plantagineum
	Echium vulgare
Virginia bluebell	
	Mertensia virginica

Virginia cowslip

Virginia cowslip
    Mertensia virginica
Virginian cowslip
    Mertensia pulmonarioides
Virginia(n) stock
    Cheiranthus maritimus
    Malcolmia maritima
Wall daisy   *Erigeron karkinskianus*
Wall fumitory   *Fumaria muralis*
Wall harebell
    Campanula portenschlagiana
Wall mustard   *Diplotaxis tenuifolia*
Wall pennywort
    Umbilicus rupestris
Wall-pepper   *Sedum acre*
Wall rocket   *Diplotaxis tenuifolia*
Wall speedwell   *Veronica arvensis*
Wandering jenny
    Lysimachia nummularia
Wandering jew
    Tradescantia albiflora
    Tradescantia fluminensis
Water avens   *Geum rivale*
Water bugle   *Lycopus virginicus*
Water cowslip   *Caltha palustris*
Water flower   *Geum urbanum*
Water forget-me-not
    Myosotis palustris
Water pimpernel
    Veronica beccabunga
Waxbells   *Kirengeshoma palmata*
Weasel's snout
    Antirrhunum majus
    Lamium album
    Lamium galeobdolon
Weld   *Reseda luteola*
Welsh poppy   *Meconopsis cambrica*
Western bleeding heart
    Dicentra formosa
Western mugwort
    Artemisia gnaphalodes
    Artemisia ludoviciana
    Artemisia purshiana
White amaranth   *Amaranthus albus*
White bachelor's buttons
    Ranunculus aconitifolius
White baneberry   *Actaea alba*
    Actaea pachypoda
    Actaea spicata alba
White buttercup
    Ranunculus amplexicaulis
White campion   *Silene alba*
    Silene latifolia
White cinquefoil   *Potentilla alba*
White cohosh   *Actaea alba*
White comfrey
    Symphytum orientale
White cup   *Nierembergia repens*
White deadnettle   *Lamium album*
White false hellebore
    Veratrum album

White flax   *Linum catharticum*
White fumitory
    Fumaria capreolata
White helleborine
    Veratrum album
    Veratrum viride
White jasmine   *Jasminum officinale*
White-leaved everlasting
    Helichrysum angustifolium
White lupin   *Lupinus albus*
White mignonette   *Reseda alba*
White mugwort
    Artemisia lactiflora
    Artemisia purshiana
White mullein   *Verbascum lychnitis*
    Verbascum thapsus
White pasque flower
    Pulsatilla alba
White poppy   *Papaver somniferum*
White purslane
    Euphorbia corollata
White rocket   *Diplotaxis erucoides*
White sage   *Artemisia gnaphalodes*
    Artemisia ludoviciana
    Artemisia purshiana
White sails   *Spathiphyllum wallisii*
White sanicle
    Eupatorium ageratoides
    Eupatorium rugosum
White snakeroot
    Eupatorium ageratoides
    Eupatorium rugosum
White stonecrop   *Sedum album*
White vervain   *Verbena urticifolia*
White weed
    Leucanthemum vulgare
Whorlflower   *Morina longifolia*
Wig knapweed   *Centaurea phrygia*
Wild balsam
    Impatiens noli-tangere
Wild candytuft   *Iberis amara*
Wild carnation
    Dianthus caryophyllus
Wild clary   *Salvia verbenacea*
Wild cranesbill
    Geranium maculatum
Wild geranium
    Geranium maculatum
Wild horehound
    Eupatorium teucrifolium
Wild horsehound
    Eupatorium teucrifolium
Wild indigo   *Baptisia tinctoria*
Wild marjoram   *Origanum vulgare*
Wild mignonette   *Reseda lutea*
Wild paeony   *Paeonia officinalis*
Wild pansy   *Viola tricolor*
Wild pea   *Lathyrus sylvestris*
Wild pink   *Dianthus plumarius*
Wild sage   *Salvia horminoides*
    Salvia nemorosa

POPULAR GARDEN PLANTS

Yellow pheasant's eye

Wild spaniard	*Aciphylla colensoi*
Wild sunflower	*Inula helenium*
Wild thyme	*Thymus drucei*
	Thymus serpyllum
Wild vanilla	*Liatris odoratissima*
Wild woodbine	
	Lonicera periclymenum
Willow bellflower	
	Campanula grandis
	Campanula latiloba
	Campanula persicifolia
Willow gentian	
	Gentiana asclepiadea
Willow-leaf ox-eye	
	Buphthalum salicifolium
Wind poppy	
	Stylomecon heterophylla
Winged everlasting	
	Ammobium alatum
Winged pea	*Lotus berthelotii*
Winged statice	
	Limonium sinuatum
	Statice sinuatum
Winter cherry	
	Solanum capsicastrum
Winter jasmine	
	Jasminum nudiflorum
Winter savory	*Satureia montana*
Wishbone flower	*Torenia fournieri*
Witch's gloves	*Digitalis purpurea*
Wolfsbane	*Aconitum lycotonum*
Aconitum lycotonum vulgaria	
	Aconitum vulparia
Wood avens	*Geum urbanum*
Wood betony	*Betonica officinalis*
	Stachys betonica
	Stachys officinalis
Woodbine	*Lonicera periclymenum*
Wood cranesbill	
	Geranium sylvaticum
Wood forget-me-not	
	Myosotis sylvatica
Wood groundsel	*Senecio sylvaticus*
Wood pimpernel	
	Lysimachia nemorum
	Lysimachia vulgaris
Wood pink	*Dianthus inodorus*
	Dianthus sylvestris
Wood poppy	
	Stylophorum diphyllum
Wood ragwort	*Senecio nemorensis*
Wood sorrel	*Oxalis acetosella*
Wood speedwell	
	Veronica montana
Wood spurge	
	Euphorbia angydaloides
Wood woundwort	
	Stachys sylvatica
Woody nightshade	
	Solanum dulcamara

Woolly betony	*Stachys byzantina*
	Stachys lanata
	Stachys olympica
Woolly foxglove	*Digitalis lanata*
Woolly hawkweed	
	Hieracium lanatum
Woolly mullein	*Verbascum thapsus*
Woolly speedwell	
	Veronica candida
	Veronica incana
Woolly thistle	
	Onopordum acanthium
Woolly woundwort	
	Stachys byzantina
Wormwood	*Artemisia absinthium*
Woundwort	*Betonica grandiflora*
	Betonica macrantha
	Solidago virgaurea
	Stachys grandiflora
	Stachys macrantha
Yarrow	*Achillea millefolium*
Yellow aconite	
	Aconitum napellus lycotonum
Yellow archangel	
	Galeobdolon luteum
	Lamiastrum galeobdolon
	Lamium galeobdolon
Yellow bachelor's buttons	
	Ranunculus acris
Yellow bellflower	
	Campanula thyrsoides
Yellow bird's nest	
	Monotropa hypophega
Yellow bugle	*Ajuga chamaepitys*
Yellow bush tobacco	
	Nicotiana glauca
Yellow centaury	*Cicendia filiformis*
Yellow chinese poppy	
	Meconopsis integrifolia
Yellow daisy	*Rudbeckia hirta*
Yellow flax	*Linum flavum*
Yellow forget-me-not	
	Myosotis discolor
Yellow gentian	*Gentiana lutea*
Yellow horned poppy	
	Glaucium flavum
	Glaucium luteum
Yellow loosestrife	
	Lysimachia punctata
	Lysimachia vulgaris
Yellow lupin	*Lupinus luteus*
Yellow meadow rue	
	Thalictrum flavum
Yellow medick	*Medicago falcata*
Yellow monkshood	
	Aconitum anthora
Yellow mountain saxifrage	
	Saxifraga aizoides
Yellow ox-eye	
	Buphthalum salicifolium
Yellow pheasant's eye	
	Adonis vernalis

Yellow pimpernel
  Lysimachia nemorum
Yellow rest harrow   *Ononis natrix*
Yellow scabious
  Scabiosa columbaria ochraleuca
Yellow sorrel   *Oxalis corniculata*
Yellow star thistle
  Centaurea solstitalis
Yellow toadflax   *Linaria vulgaris*
Yellow tuft   *Alyssum argenteum*
  Alyssum murale
Yellow vetchling   *Lathyrus aphaca*

Yellow violet   *Viola saxatilis*
Yellow willowherb
  Lysimachia vulgaris
Yellow woundwort   *Stachys recta*
Yerba mansa
  Anemonopsis californica
Youth and old age   *Zinnia elegans*
Zonal geraniums
  Pelargonium × hortorum
Zonal pelargoniums
  Pelargonium × hortorum
Zulu giant   *Stapelia gigantea*

TREES, BUSHES, AND SHRUBS

It is often said that the difference between trees and shrubs is simple; trees have a single woody stem from which branches grow to form a crown whereas a shrub has several woody stems rising from ground level, forming a crown. But this is over-simplification. The exact shape of a tree can be completely altered by wind action, by a difference in spacing or even by artificial pruning.

A bush is generally defined as a woody plant that is between a shrub and a tree in size.

Silver maple –
Acer saccharinum

Aaron's beard

Aaron's beard	*Hypericum calycinum*
Abaca	*Musa textilis*
Abata cola	*Cola acuminata*
Abele	*Populus alba*
Abelitzia	*Zelkova abelicea*
Abyssinian euphorbia	*Euphorbia trigona*
Abyssinian tea	*Catha edulis*
Aceituno	*Simarouba glauca*
Achiote	*Bixa orellana*
Achocha	*Cyclanthera pedata*
Adam's apple	*Tabernaemontana coronaria*
	Tabernaemontana divaricata
Adam's laburnum	*+ Laburnocytisus adami*
Adam's needle	*Yucca filamentosa*
	Yucca gloriosa
Adventure bay pine	*Phyllocladus asplenifolius*
Afara	*Terminalia superba*
Afghan ash	*Fraxinus xanthoxyloides*
Afghan cherry	*Prunus jacquemontii*
African boxwood	*Myrsine africana*
African bread tree	*Treculia africana*
African cypress	*Widdringtonia cupressoides*
African fern pine	*Podocarpus gracilior*
African hemp	*Sparmannia africana*
African juniper	*Juniperus procera*
African locust	*Parkia biglobosa*
African locust bean	*Parkia filicoidea*
African mahogany	*Khaya nysasica*
	Khaya senegalensis
African milk barrel	*Euphorbia horrida*
African milkbush	*Synadenium grantii*
African milk tree	*Euphorbia trigona*
African nutmeg	*Monodora myristica*
African oil palm	*Elaeis guineensis*
African peach	*Nauclea latifolia*
African red alder	*Cunonia capensis*
African tulip tree	*Spathodea campanulata*
African walnut	*Coula edulis*
African yellowwood	*Podocarpus elongatus*
Aguacate	*Persea americana*
Ague tree	*Sassafras albidum*
Ahuehuete	*Taxodium mucronatum*
Ailanto	*Ailanthus glandulosa*

Aino mulberry	*Morus australis*
Akee	*Blighia sapida*
Alabama snow wreath	*Neviusia alabamensis*
Alaska cedar	*Chamaecyparis nootkatensis*
Alaska yellow cedar	*Chamaecyparis nootkatensis*
Albany bottlebrush	*Callistemon speciosus*
Alberta white spruce	*Picea glaucaalbertiana*
Alcock's spruce	*Picea alcoquiana*
	Picea bicolor
Alder buckthorn	*Frangula alnus*
	Rhamnus frangula
Alder-leaved rowan	*Sorbus alnifolia*
Aleppo pine	*Pinus halepensis*
Alerce	*Fitzroya cupressoides*
	Tetraclinis articulata
Alexander palm	*Ptychosperma elegans*
Alexandra palm	*Archontophoenix alexandrae*
Alexandrian laurel	*Calophyllum inophyllum*
	Danae racemosa
Alexandrian senna	*Cassia acutifolia*
	Cassia senna
Algarroba bean	*Ceratonia siliqua*
Algarrobo	*Prosopis chilensis*
	Prosopis juliflora
Algerian ash	*Fraxinus xanthoxyloides dimorpha*
Algerian fir	*Abies numidica*
Algerian oak	*Quercus canariensis*
	Quercus mirbeckii
Allamanda	*Allamanda cathartica*
Alleghany plum	*Prunus alleghaniensis*
Alleghany service berry	*Amelanchier laevis*
Alleghany spurge	*Pachysandra procumbens*
Alligator apple	*Annona glabra*
	Annona palustris
Alligator juniper	*Juniperus deppeana pachyphlaea*
	Juniperus pachyphloea
Alligator pear	*Persea americana*
All saint's cherry	*Prunus cerasus semperflorens*
Allspice	*Pimenta dioica*
	Pimenta officinalis
Almond	*Prunus dulcis*
Almond-leaved pear	*Pyrus amygdaliformis*
Almond willow	*Salix amygdaloides*
	Salix triandra

TREES, BUSHES, AND SHRUBS

Alpen rose
    Rhododendron ferrugineum

Alpine ash   *Eucalyptus delegatensis*
    Eucalyptus gigantea

Alpine azalea   *Azalea procumbens*
    Chamaecistus procumbens
    Loiseleuria procumbens

Alpine buckthorn   *Rhamnus alpina*

Alpine celery-topped pine
    Phyllocladus alpinus

Alpine currant   *Ribes alpinum*

Alpine elder   *Sambucus racemosa*

Alpine fir   *Abies lasiocarpa*
    Abies amabilis

Alpine gum   *Eucalyptus archeri*

Alpine laburnum
    Laburnum alpinum

Alpine larch   *Larix lyalli*

Alpine laurel   *Kalmia microphylla*

Alpine rose   *Rosa alpina*

Alpine totara   *Podocarpus nivalis*

Alpine whitebeam
    Sorbus chamaemespilus

Altai mountain thorn
    Crataegus altaica

Althaea   *Althaea frutex*
    Hibiscus syriacus

Amatungulu   *Carissa grandiflora*
    Carissa macrocarpa

Ambarella   *Spondias cytherea*

Ambash   *Herminiera elaphroxylon*

Amboina pine   *Agathis dammara*

Amboina pitch tree   *Agathis alba*

American arbor-vitae
    Thuya occidentalis

American ash   *Fraxinus americana*

American aspen
    Populus tremuloides

American basswood
    Tilia americana

American beautybush
    Kolkwitzia amabilis

American beech   *Fagus americana*
    Fagus grandifolia

American black cherry
    Prunus serotina

American blackcurrant
    Ribes americanum

American bladdernut
    Staphylea trifolia

American boxwood   *Cornus florida*

American chestnut
    Castanea dentata

American colombo
    Frasera carolinensis

American crab apple
    Malus angustifolia

American cranberry
    Oxycoccus macrocarpus
    Vaccinium macrocarpon

American dogwood   *Cornus sericea*

American elder
    Sambucus canadensis

American elm   *Ulmus americana*

American filbert
    Corylus americana

American fringe tree
    Chionanthus virginicus

American green alder
    Alnus crispa mollis

American hazel   *Corylus americana*

American holly   *Ilex opaca*

American hop-hornbeam
    Ostrya virginiana

American hornbeam
    Carpinus caroliniana

American judas tree
    Cercis canadensis

American larch   *Larix laricina*

American laurel   *Kalmia latifolia*

American lime   *Tilia americana*

American mangrove
    Rhizophora mangle

American mastic tree   *Schinus molle*

American mock orange
    Prunus caroliniana

American mountain ash
    Sorbus americana

American mulberry   *Morus rubra*

American oil palm
    Elaeis melanococca
    Elaeis oleifera

American olive
    Osmanthus americanus

American persimmon
    Diospyros virginiana

American pistachio   *Pistacia texana*

American plane
    Platanus occidentalis

American red buckeye
    Aesculus pavia

American redbud   *Cercis canadensis*

American red elder
    Sambucus pubens

American red oak   *Quercus rubra*

American red pine   *Pinus resinosa*

American red plum
    Prunus americana

American red spruce   *Picea rubens*

American sarsaparilla
    Aralia niducaulis

American silverberry
    Elaeagnus argentea
    Elaeagnus commutata

American sloe
    Prunus alleghaniensis

American smoke tree
    Cotinus americanus
    Cotinus obovatus

American spikenard
    Aralia racemosa

American storax
Styrax americanus

American swamp laurel
Kalmia glauca

American sweet gum
Liquidambar styraciflua

American sycamore
Platanus occidentalis

American wayfaring tree
Viburnum lantanoides

American white ash   *Fraxinus alba*
Fraxinus americana

American white oak
Quercus bicolor

American wild plum
Prunus americana

American yellowwood
Cladrastis lutea
Cladrastis tinctoria

American yew   *Taxus brevifolia*
Taxus canadensis

Ammoniacum
Dorema ammoniacum

Amur cork tree
Phellodendron amurense

Amur lilac   *Syringa amurensis*

Amur lime   *Tilia amurensis*

Amur maple   *Acer ginnala*
Acer tataricum ginnala

Amur privet   *Ligustrum amurense*

Anacahuita   *Cordia boissieri*

Anaconoa   *Cordia sebestena*

Ana tree   *Acacia albida*

Andaman marble   *Diospyros kurzii*

Angelica tree   *Aralia elata*
Aralia spinosa

Angel's trumpet   *Datura suaveolens*

Angelwing jasmine
Jasminum nitidum

Aniseed tree   *Illicium floridanum*

Anglo-Japanese yew   *Taxus media*

Anime resin   *Hymenaea courbaril*

Aniseed tree   *Illicium anisatum*

Annatto   *Bixa orellana*

Annual mallow   *Lavatera trimestris*

Annual poinsettia
Euphorbia heterophylla

Anona blanca   *Annona diversifolia*

Antarctic beech
Nothofagus antarctica

Anthony nut   *Staphylea pinnata*

Ant tree   *Triplaris americana*

Apache pine   *Pinus engelmannii*

Apache plume   *Fallugia paradoxa*

Apamata   *Tabebuia serratifolia*

Apiesdoring   *Acacia galpinii*

Apothecary's rose
Rosa gallica officinalis
Rosa officinalis

Appalachian tea   *Ilex glabra*

Appleblossom cassia
Cassia javanica

Appleblosson senna
Cassia javanica

Apple box   *Eucalyptus bridgesiana*

Apple guava   *Psidium guajava*

Apple gum   *Eucalyptus clavigera*

Apple-ring acacia   *Acacia albida*

Apple rose   *Rosa villosa*

Apricot   *Prunus armeniaca*

Apricot plum   *Prunus simonii*

Arabian coffee   *Coffea arabica*

Arabian jasmine   *Jasminum sambac*

Arabian tea   *Catha edulis*

Araroba   *Andira araroba*

Arar tree   *Tetraclinis articulata*

Arbor-vitae   *Thuya occidentalis*

Arching forsythia
Forsythia suspensa fortunei

Arctic willow   *Salix arctica*

Areca-nut palm   *Areca catechu*

Areca palm
Chrysalidocarpus lutescens

Areng palm   *Arenga pinnata*

Argan tree   *Argania spinosa*

Argyle apple   *Eucalyptus cinerea*

Arizona ash   *Fraxinus velutina*

Arizona cork fir
Abies lasiocarpa arizonica

Arizona cypress
Cupressus arizonica

Arizona pine
Pinus ponderosa arizonica

Arizona walnut   *Juglans major*

Arkansas rose   *Rosa arkansana*

Armand's pine   *Pinus armandii*

Armenian oak   *Quercus pontica*

Arolla pine   *Pinus cembra*

Aronia prunifolia
Purple chokeberry

Arrow broom   *Genista sagittalis*

Arrowwood
Viburnum acerifolium
Viburnum dentatum

Arroyo willow   *Salix lasiolepis*

Asgara   *Pterostyrax hispida*

Ashe juniper   *Juniperus ashei*

Ashe magnolia   *Magnolia ashei*

Ashleaf maple   *Acer negundo*

Asian pear   *Pyrus pyrifolia*

Asiatic sweetleaf
Symplocos paniculata

Asoka tree   *Saraca indica*

Aspen   *Populus tremula*

Assai palm   *Euterpe edulis*
Euterpe oleracea

Assam rubber   *Ficus belgica*
Ficus elastica

TREES, BUSHES, AND SHRUBS

Balsam willow

Assam tea
Camellia sinensis assamensis

Assyrian plum   Cordia myxa

Athel   Tamarix aphylla

Atlantic white cedar
Chamaecyparis thyoides

Atlas cedar   Cedrus atlantica

August plum   Prunus americana

Australian banyan   Ficus macrocarpa
Ficus macrophylla

Australian beech
Eucalyptus polyanthemos
Nothofagus moorei

Australian beefwood
Casuarina equisetifolia

Australian blackwood
Acacia melanoxylon

Australian bottle plant
Jatropha podagrica

Australian bower plant
Pandorea jasminoides

Australian brush cherry
Eugenia myrtifolia
Syzygium paniculatum

Australian cabbage palm
Corypha australis
Livistona australis

Australian cherry
Exocarpus cupressiformis

Australian fan palm
Corypha australis
Livistona australis

Australian fever bush
Alstonia scholaris

Australian finger lime
Microcitrus australasica

Australian forest oak
Casuarina torulosa

Australian heath   Epacris impressa

Australian honeysuckle
Banksia grandis

Australian ivy palm
Brassaia actinophylla

Australian laurel
Pittosporum tobira

Australian lilac
Hardenbergia monophylla

Australian mountain ash
Eucalyptus regnans

Australian nut
Macadamia integrifolia

Australian pepper tree
Schinus molle

Australian pine
Araucaria heterophylla
Casuarina equisetifolia

Australian river oak
Casuarina cunninghamiana

Australian rosemary
Westringia fruticosa

Australian round lime
Microcitrus australis

Australian sarsaparilla
Hardenbergia violacea

Australian smokebush
Conospermum stoechadis

Australian tea tree
Leptospermum laevigatum

Australian umbrella tree
Brassaia actinophylla

Australian willow myrtle
Agonis flexuosa

Austrian briar   Rosa foetida

Austrian copper rose
Rosa foetida bicolor

Austrian pear   Pyrus austriaca

Austrian pine   Pinus nigra
Pinus nigra caramanica
Pinus nigra nigra

Austrian whitebeam
Sorbus austriaca

Austrian yellow rose   Rosa foetida

Autumn cherry
Prunus subhirtella autumnalis

Autumn olive   Elaeagnus umbellata

Avaram   Cassia auriculata

Avignon berry   Rhamnus infectoria

Avocado pear   Persea americana

Awl tree   Morinda citrifolia

Ayrshire rose   Rosa arvensis

Azarole   Crataegus azarolus

Azediracta   Melia azedarach

Azorean holly   Ilex perado

Azure ceanothus
Ceanothus caeruleus

Babassu   Orbignya barbosiana

Babassu palm   Orbignya speciosa

Babul   Acacia nilotica

Bachelor's buttons
Kerria japonica plena

Bael tree   Aegle marmelos

Bag-flower
Clerodendrum thomsoniae

Balata   Manilkara bidentata

Bald cypress   Taxodium distichum

Baldhip rose   Rosa gymnocarpa

Balearic box   Buxus balearica

Balfour's aralia
Polyscias balfouriana

Balkan gorse   Genista lydia

Balkan maple   Acer heldreichii
Acer hyrcanum

Ball tree   Aegle marmelos

Balm of gilead   Abies balsamea
Cedronella canariensis
Populus candicans
Populus gileadensis

Balsam fir   Abies balsamea

Balsam of Peru   Myroxylon pereirae

Balsam poplar   Populus balsamifera
Populus tacamahaca

Balsam willow   Salix pyrifolia

Balsa wood	*Ochroma pyramidale*
Bamboo briar	*Acacia nudicaulis*
Bamboo-leaved oak	
	Quercus myrsinaefolia
	Quercus prinus
Bamboo palm	
	Chamaedorea erumpens
	Rhapis excelsa
Bamenda cola	*Cola anomala*
Banana	*Musa acuminata*
Banana shrub	*Michelia figo*
Banana yucca	*Yucca baccata*
Banksian rose	*Rosa banksiae*
Banyan	*Ficus benghalensis*
	Ficus indica
Baobab	*Adansonia digitata*
Barbados cedar	*Cedrela odorata*
Barbados cherry	*Eugenia michelii*
	Eugenia uniflora
	Malpighia glabra
Barbados nut	*Jatropha curcus*
Barbados pride	
	Adenanthera pavonina
	Caesalpinia pulcherrima
Barbary gum	*Acacia gummifera*
Barbasco	*Jacquinia armillaris*
	Jacquinia barbasco
Barberry	*Berberis vulgaris*
Barrel palm	*Colpothrinax wrightii*
Barren's clawflower	
	Calothamnus validus
Bartram oak	
	Quercus × heterophylla
Barwood	*Baphia nitida*
	Pterocarpus erinaceus
Basford willow	*Salix basfordiana*
	Salix × rubens basfordiana
Basket oak	*Quercus prinus*
Basket willow	*Salix purpurea*
	Salix viminalis
Basswood	*Tilia americana*
Bastard bullet tree	
	Houmiria floribunda
Bastard cedar	*Guazuma ulmifolia*
Bastard cherry	*Eretia tinifola*
Bastard gimlet	*Eucalyptus diptera*
Bastard indigo	*Amorpha fruticosa*
Bastard jute	*Hibiscus cannabinus*
Bastard logwood	*Acacia berteriana*
Bastard sandalwood	
	Myoporum sandwicense
Bastard service tree	
	Sorbus × thuringiaca
Bastard teak	
	Pterocarpus marsupium
Batoko plum	*Flacourtia indica*
	Flacourtia ramontchi
Bat willow	*Salix alba coerulea*
Bayberry	*Myrica pennsylvanica*

Bay gall bush	*Ilex coriacea*
	Ilex lucida
Bay laurel	*Laurus nobilis*
Bay-rum tree	*Pimenta racemosa*
Bay tree	*Pimenta racemosa*
Bay willow	*Salix pentandra*
Beach plum	*Prunus maritima*
Beach pine	*Pinus contorta*
Bead tree	*Melia azedarach*
Beaked filbert	*Corylus cornuta*
Beaked hazel	*Corylus cornuta*
Beak willow	*Salix bebbiana*
Bearberry	
	Arctostaphylos manzanita
	Rhamnus purshiana
Bearberry willow	*Salix uva-ursi*
Bear huckleberry	
	Gaylussacia ursina
Bear oak	*Quercus ilicifolia*
Bear's grape	
	Arctostaphylos uva-ursi
Beautiful fir	*Abies amabilis*
Beautyberry	*Callicarpa americana*
Beauty bush	*Kolkwitzia amabilis*
Bechtel crab apple	
	Malus ioensis plena
Beechwood	
	Casuarina equisetifolia
Beefsteak plant	
	Acalypha wilkesiana
Beefwood	*Mimusops balata*
Bela tree	*Aegle marmelos*
Belgian evergreen	
	Dracaena sanderana
Bell-flowered cherry	
	Prunus campanulata
Bellflower heather	*Erica cinerea*
Bell-fruited mallee	
	Eucalyptus preissiana
Bell-fruit tree	
	Codonocarpus cotinifolius
Bell heather	*Erica cinerea*
Belmore sentry palm	
	Howea belmoreana
	Kentia belmoreana
Bengal quince	*Aegle marmelos*
Benguet pine	*Pinus insularis*
Benjamin bush	*Lindera benzoin*
Benjamin tree	*Ficus benjamina*
Ben oil tree	*Moringa oleifera*
Bentham's cornel	*Cornus capitata*
Benzoin	*Lindera benzoin*
	Styrax benzoin
Bergamot	
	Citrus aurantium bergamia
Bergamot orange	
	Citrus aurantium bergamia
	Citrus bergamia
Berg cypress	
	Widdringtonia cupressoides

Black fibre palm

Berlin poplar
　　　　　Populus × berolinensis
Bermuda cedar
　　　　　Juniperus bermudiana
Bermuda olivewood bark
　　　　　Cassine laneana
Bermuda palmetto
　　　　　Sabal bermudana
Berry heath　　　Erica baccans
Besom heath　　　Erica scoparia
Be-still tree　　Thevetia peruviana
Betel　　　　　　Piper betel
　　　　　　　　Piper betle
Betel-nut palm　　Areca aleracea
　　　　　　　　Areca catechu
Betel palm　　　Areca catechu
Betel pepper　　　Piper betel
　　　　　　　　Piper betle
Bhendi tree　　Thespesia populnea
Bhutan cypress
　　　　　Cupressus duclouxiana
　　　　　Cupressus torulosa
Bhutan pine　　　Pinus excelsa
　　　　　　　Pinus wallichiana
Bible fig　　　Ficus sycomorus
Bible leaf
　　　Hypericum androsaemum
Bigarade　　Citrus aurantium
Big-berry manzanita
　　　　Arctostaphylos glauca
Bigbud hickory　Carya tomentosa
Big-cone pine　　Pinus coulteri
Big-cone spruce
　　　Pseudotsuga macrocarpa
Big-leaved maple
　　　　Acer macrophyllum
Big-leaved storax　Styrax obassia
Bignay　　　Antidesma bunius
Big-toothed aspen
　　　Populus grandidentata
Big-tooth euphorbia
　　　Euphorbia grandidens
Bilberry　　Vaccinium myrtillus
Bilimbi　　　Averrhoa bilimbi
Bilsted gum
　　　Liquidambar styraciflua
Bimble box　Eucalyptus populifolia
　　　　　Eucalyptus populnea
Bimli jute　Hibiscus cannabinus
Bimlipatum　Hibiscus cannabinus
Birch bark cherry　Prunus serrula
Birch leaf maple　Acer tetramerum
Bird-catcher tree
　　　　Pisonia umbellifera
Bird cherry ,　　　Prunus avium
　　　　　　　　Prunus padus
　　　　Prunus pennsylvanica
Bird of paradise
　　　　Caesalpinia gilliesii
　　　　Poinciana gilliesii

Bird's eye bush　　Ochna japonica
　　　　　　　Ochna serrulata
Bird's eye maple　Acer saccharinum
Biscochito　Ruprechtia coriacea
Bishop pine　　Pinus muricata
Bitter almond　Prunus dulcis amara
Bitter apple　Citrullus colocynthis
Bitter ash　　　Picraena excelsa
Bitter bark　　Alstonia scholaris
　　　　　　Pinckneya pubens
Bitterbush　Picramnia pentandra
Bitter cassava　Manihot esculenta
　　　　　Manihot utilissima
Bitter cherry　Prunus emarginata
Bitter damson　Simarouba amara
Bitter gallberry　　Ilex glabra
Bitternut　　Carya cordiformis
Bitter oak　　Quercus cerris
Bitter orange　Citrus aurantium
Bitter pecan
　　　　　Carya aquatica
Bitterwood　　Quassia amara
　　　　Simarouba glauca
Black alder　　Alnus glutinosa
　　　　　Ilex verticillata
　　　　　Viburnum molle
Black apricot　Prunus × dasycarpa
Black ash　　　Fraxinus nigra
Black bark　Diospyros whyteana
Blackbead
　　Pithecellobium guadalupense
　　Pithecellobium unguis-cati
Black bean tree
　　　Castanospermum australe
Black bearberry　Arctous alpinus
Black beech　Nothofagus solandri
Blackberry　　Robus fruticosus
Black birch　　Betula lenta
　　　　　　Betula nigra
Black box　Eucalyptus largiflorens
　　　　Eucalyptus × rariflora
Black boy　Xanthorrhoea preissii
Blackbutt　Eucalyptus pilularis
Blackbutt peppermint
　　　　Eucalyptus smithii
Black calabash　Enallagma latifolia
Black cherry　　Prunus serotina
Black chokeberry
　　　　Aronia melanocarpa
Black cottonwood
　　　Populus heterophylla
　　　Populus trichocarpa
Blackcurrant　　Ribes nigrum
Black cutch tree　Acacia catechu
Black cypress pine
　　　　Callitris calcarata
　　　　Callitris endlicheri
Black dogwood　Rhamnus frangula
Black elder　　Sambucus nigra
Black fibre palm　Arenga pinnata

Black gum	*Eucalyptus aggregata*
	Nyssa sylvatica
Black haw	*Bumelia lanuginosa*
	Viburnum lentago
	Viburnum prunifolium
Black hawthorn	
	Crataegus douglasii
Black highbush blueberry	
	Vaccinium atrococcum
Black huckleberry	
	Gaylussacia baccata
	Gaylussacia resinosa
Blacking plant	*Hibiscus sinensis*
Black ironwood	*Olea laurifolia*
Black Italian poplar	
	Populus serotina
Blackjack oak	*Quercus marilandica*
Black juniper	*Juniperus wallichiana*
Black kauri pine	
	Agathis microstachys
Black larch	*Larix laricina*
Black laurel	*Gordonia lasianthus*
Black-leaved plum	
	Prunus cerasifera nigra
Black locust tree	
	Robinia pseudoacacia
Black mangrove	*Avicennia nitida*
Black maple	*Acer nigrum*
Black mountain ash	
	Eucalyptus sieberi
Black mulberry	*Morus nigra*
Black oak	*Quercus velutina*
Black olive	*Bucida buceras*
Black palm	*Normanbya normanbyi*
Black peppermint	
	Eucalyptus amygdalina
	Eucalyptus salicifolia
Black persimmon	*Diospyros nigra*
	Diospyros texana
Black pine	*Pinus thunbergii*
	Podocarpus amarus
Black plum	*Syzygium cumini*
Black poplar	*Populus nigra*
Black pussy willow	
	Salix melanostachys
Black sally	*Eucalyptus stellulata*
Black sapote	*Diospyros digyna*
	Diospyros nigra
Black sassafras	
	Atherosperma moschatum
Black spruce	*Picea mariana*
	Picea nigra
Black stinkwood	*Ocotea bullata*
Blackthorn	
	Crataegus calpodendron
	Prunus spinosa
Blackthorn oak	
	Quercus marilandica
Black titi	*Cliftonia monophylla*
	Cyrilla racemiflora
Black tupelo	*Nyssa sylvatica*
Black walnut	*Juglans nigra*
Black wattle	*Acacia mearnsii*
Black willow	*Salix nigra*
Blackwood	*Acacia melanoxylon*
	Acacia penninervis
Blackwood acacia	
	Acacia melanoxylon
Black wood	*Dalbergia latifolia*
Bladder ketmia	*Hibiscus trionum*
Bladdernut	*Diospyros whyteana*
	Staphylea holocarpa
Blaeberry	*Vaccinium myrtillus*
Blakely's red gum	
	Eucalyptus blakelyi
Blaxland's stringybark	
	Eucalyptus blaxlandii
Blaze	*Prunus cerasifera nigra*
Bleeding glory flower	
	Clerodendrum thomsoniae
Bleeding heart vine	
	Clerodendrum thomsoniae
Blood-leaf Japanese maple	
	Acer palmatum atropurpureum
Blood-twig dogwood	
	Cornus sanguinea
Bloodwood	
	Haematoxylum campeachianum
Blue ash	*Fraxinus quadrangulata*
Blue atlas cedar	
	Cedrus atlantica glauca
Blue barberry	*Mahonia aquifolium*
Blue beech	*Carpinus caroliniana*
Blueberry	*Vaccinium corymbosum*
Blueberry ash	
	Elaeocarpus reticulatus
Blue birch	*Betula coerulea-grandis*
Blueblossom	
	Ceanothus thyrsiflorus
Blue broom	*Erinacea anthyllis*
	Erinacea pungens
Bluebush	*Eucalyptus macrocarpa*
Blue cedar	*Cedrus atlantica glauca*
Blue Chinese juniper	
	Juniperus chinensis
	columnaris glauca
Blue douglas fir	
	Pseudotsuga menziesii glauca
Blue dracaena	*Cordyline indivisa*
	Dracaena indivisa
Blue elder	*Sambucus caerulea*
Blue fan palm	*Brahea armata*
	Erythea armata
Blue gum	*Eucalyptus globulus*
Blue haw	*Viburnum rufidulum*
Blue hesper palm	*Brahea armata*
	Erythea armata
Blue holly	*Ilex × meservae*
Bluejack oak	*Quercus incana*
Blue latan	*Latania loddigesii*
Blue-leaf wattle	*Acacia cyanophylla*

Blue-leaved stringybark
    Eucalyptus agglomerata
Blue lilly-pilly
    Syzygium coolminianum
Blue magnolia *Magnolia acuminata*
Blue mountain mallee
    Eucalyptus stricta
Blue oak   *Quercus douglasii*
Blue palm   *Erythea armata*
Blue palmetto
    Rhapidophyllum hystrix
    Sabal palmetto
Blue passion flower
    Passiflora caerulea
Blue pine   *Pinus wallichiana*
Blue Spanish fir
    Abies pinsapo glauca
Blue spruce   *Picea pungens*
    Picea pungens glauca
Blue tangle *Gaylussacia frondosa*
Blue wattle *Acacia cyanophylla*
Blue weeping gum
    Eucalyptus sepulcralis
Blue willow *Salix alba caerulea*
    Salix caesia
Blue yucca   *Yucca baccata*
Bog bilberry *Vaccinium uliginosum*
Bog heather   *Erica tetralix*
Bog kalmia   *Kalmia poliifolia*
Bog laurel   *Kalmia poliifolia*
Bog myrtle   *Myrica gale*
Bog rosemary
    Andromeda polifolia
Bog spruce   *Picea mariana*
    Picea nigra
Bog whortleberry
    Vaccinium uliginosum
Boldo   *Peumus boldus*
Bolinus ridge ceanothus
    Ceanothus masonii
Boobyalla *Myoporum insulare*
Boojum tree *Idria columnaris*
Bornmüller's fir
    Abies bornmuellerana
Bosisto's box
    Eucalyptus bosistoana
Bosnian pine *Pinus leucodermis*
Botany Bay gum
    Xanthorrhoea arborea
Botany Bay tea tree *Correa alba*
Bo tree   *Ficus religiosa*
Bottlebrush buckeye
    Aesculus parviflora
Bottle brush *Callistemon citrinus*
Bottle palm *Colpothrinax wrightii*
    Hyophorbe lagencaulis
Bottle tree *Adansonia gregori*
Bourbon palm *Latania borbonica*
Bourbon rose *Rosa × borboniana*
Bournemouth pine *Pinus pinaster*

Bourtree   *Sambucus nigra*
Bower plant *Pandorea jasminoides*
Bowstring hemp
    Calotropis gigantea
Bow-wood tree *Maclura pomifera*
Boxberry *Gaultheria procumbens*
Box blueberry *Vaccinium ovatum*
Box elder   *Acer negundo*
Box huckleberry
    Gaylussacia brachycera
Box-leaved holly   *Ilex crenata*
Box sand myrtle
    Leiophyllum buxifolium
Box thorn   *Bursaria spinosa*
Boxwood *Buxus sempervirens*
Bracelet honey myrtle
    Melaleuca armillaris
Bracelet wood *Jacquinia armillaris*
Bracted fir   *Abies procera*
Bramble   *Rubus fruticosus*
Bramble acacia *Acacia victoriae*
Bramble wattle *Acacia victoriae*
Branch thorn *Erinacea anthyllis*
    Erinacea pungens
Brasiletto *Caesalpinia vesicaria*
Brazilian araucaria
    Araucaria angustifolium
Brazil cherry *Eugenia brasiliensis*
    Eugenia dombeyi
    Eugenia michelii
    Eugenia uniflora
Brazilian ironwood
    Caesalpinia ferrea
Brazilian passion flower
    Passiflora caerulea
Brazilian pepper tree
    Schinus terebinthifolia
Brazilian pine
    Araucaria angustifolia
Brazilian skyflower *Duranta ellisia*
    Duranta repens
Brazil nut *Bertholletia excelsa*
Brazilwood
    Caesalpinia braziliensis
    Caesalpinia echinata
    Caesalpinia sappan
Bread and cheese
    Crataegus monogyna
Breadfruit *Artocarpus altilis*
    Artocarpus incisus
    Pandanus odoratissimus
Breadnut *Brosimum alicastrum*
Bread tree
    Encephalartos altensteinii
Brewer's weeping spruce
    Picea brewerana
Briançon apricot
    Prunus brigantina
Briar   *Erica arborea*
Briar rose   *Rosa canina*
Bridewort *Spiraea salicifolia*

COMMON NAMES

Brisbane box

Common name	Botanical name
Brisbane box	*Tristania conferta*
Bristle-cone fir	*Abies bracteata*
	Abies venusta
Bristle-cone pine	*Pinus aristata engelm*
Bristly locust	*Robinia hispida*
Bristly sarsaparilla	*Aralia hispida*
Brittle-leaf manzanita	*Arctostaphylos crustacea*
Brittle willow	*Salix fragilis*
Broadleaf cockspur thorn	*Crataegus × prunifolia*
Broadleaf podocarpus	*Podocarpus nagi*
Broad-leaved bottle tree	*Brachychiton australis*
Broad-leaved ironbark	*Eucalyptus fibrosa*
Broad-leaved kindling bark	*Eucalyptus dalrympleana*
Broad-leaved lime	*Tilia platyphyllos*
Broad-leaved mahogany	*Swietenia macrophylla*
Broad-leaved peppermint	*Eucalyptus dives*
Broad-leaved sally	*Eucalyptus camphora*
Broad-leaved spindle	*Euonymus latifolius*
Broad-leaved whitebeam	*Sorbus latifolia*
Bronvaux medlar	+ *Crataegomespilus dardarii*
Bronze shower	*Cassia moschata*
Broom hickory	*Carya glabra*
Broom palm	*Coccothrinax argentea*
	Thrinax argentea
Broom wattle	*Acacia calamifolia*
Brown barrel	*Eucalyptus fastigiata*
Brown cabbage tree	*Pisonia grandis*
Brown dogwood	*Cornus glabrata*
Brown mallee	*Eucalyptus astringens*
Brown pine	*Podocarpus elatus*
Brown stringybark	*Eucalyptus baxteri*
	Eucalyptus capitellata
Brush box	*Tristania conferta*
Brush cherry	*Syzygium paniculatum*
Buckberry	*Gaylussacia ursina*
Buckbrush	*Ceanothus cuneatus*
Buckthorn	*Bumelia lycioides*
Buckwheat brush	*Cliftonia monophylla*
Buckwheat tree	*Cliftonia monophylla*

Common name	Botanical name
Buddhist pine	*Podocarpus macrophyllus*
Buddleia	*Buddleia davidii*
Buffalo berry	*Shepherdia argentea*
	Shepherdia canadensis
Buffalo currant	*Ribes aureum*
	Ribes odoratum
Buffalo thorn	*Ziziphus mucronata*
Buffalo wattle	*Acacia kettlewelliae*
Buisson ardent	*Pyracantha coccinea*
Bulgarian fir	*Abies borisii-regis*
Bullace	*Prunus domestica institia*
	Prunus insititia
Bull banksia	*Banksia grandis*
Bull bay	*Magnolia grandiflora*
Bull-hoof tree	*Bauhinia purpurea*
Bull-horn acacia	*Acacia cornigera*
Bullick	*Eucalyptus megacarpa*
Bullock's heart	*Annona reticulata*
Bull's horn acacia	*Acacia spadicigera*
Bull thatch	*Sabal jamaicensis*
Bunchberry	*Cornus canadensis*
Bundy	*Eucalyptus elaeophora*
	Eucalyptus goniocalyx
Bungalay	*Eucalyptus botryoides*
Bunya-bunya	*Araucaria bidwillii*
Bunya pine	*Araucaria bidwillii*
Burdekin plum	*Pleiogynium cerasiferum*
Burmese fishtail palm	*Caryota mitis*
Burmese pride	*Amhertsia nobilis*
Burmese rosewood	*Pterocarpus indicus*
Burnet rose	*Rosa pimpinellifolia*
Burning bush	*Combretum macrophyllum*
	Cotinus coggygria purpureus
	Euonymus atropurpurea
Burracoppin mallee	*Eucalyptus burracoppinensis*
Burr oak	*Quercus macrocarpa*
Burr rose	*Rosa roxburghii*
Bursting heart	*Euonymus americana*
Bush chinquapin	*Castanea alnifolia*
	Castanea sempervirens
	Castanopsis sempervirens
	Chrysolepis sempervirens
Bush fig	*Ficus capensis*
Bush hollyhock	*Althaea frutex*
	Hibiscus syriacus
Bush honeysuckle	*Lonicera nitida*
Bush mallow	*Althaea frutex*
	Lavatera olbia
	Hibiscus syriacus
Bush palmetto	*Sabal adansonii*
	Sabal minor
Bush willow	*Combretum erythrophyllum*

Californian sassafras

Bush yate   *Eucalyptus lehmanii*

Butcher's broom   *Ruscus aculeatus*

Buttercup bush   *Cassia corymbosa*

Buttercup flower
          Allamanda cathartica

Buttercup tree
          Cochlospermum religiosum
          Cochlospermum vitifolium

Buttercup winter hazel
          Corylopsis paucifolia

Butterfly bush
          Buddleia alternifolia
          Buddleia davidii

Butterfly lavender
          Lavandula stoechas

Butterfly palm   *Areca lutescens*
          Chrysalidocarpus lutescens

Butterfly pea   *Clitoria ternatea*

Butterfly tree   *Bauhinia purpurea*

Butter nut   *Juglans cinerea*

Buttonball   *Platanus occidentalis*

Button bush
          Cephalanthus occidentalis

Buttonwood   *Conocarpus erectus*
          Platanus occidentalis

Cabbage gum
          Eucalyptus amplifolia
          Eucalyptus clavigera
          Eucalyptus paucifolia

Cabbage palm   *Areca aleracea*
          Corypha australis
          Livistonia australis
          Roystonea oleracea

Cabbage palmetto   *Sabal palmetto*

Cabbage rose   *Rosa centifolia*

Cabbage tree   *Andira araroba*
          Cordyline australis
          Cussonia paniculata
          Cussonia spicata
          Sabal palmetto

Cacao   *Theobroma cacao*

Cade   *Juniperus oxycedrus*

Cafta   *Catha edulis*

Caimito   *Chrysophyllum cainito*

Cajuput   *Melaleuca leucadendron*

Calaba tree
          Calophyllum brasiliense

Calabash nutmeg
          Monodora myristica

Calabash tree   *Crescenta cujete*

Calabrian pine   *Pinus brutia*

Calabur   *Muntingia calabura*

Calamondin   × *Citrofortunella mitis*

Calappa palm
          Actinorhytis calapparia

Calico bush   *Kalmia latifolia*

Calif of Persia willow
          Salix aegyptiaca

California big tree
          Sequoiadendron giganteum

California huckleberry
          Vaccinium ovatum

California incense cedar
          Calocedrus decurrens

California juniper
          Juniperus occidentalis

Californian allspice
          Calycanthus occidentalis

Californian anemone bush
          Carpenteria californica

Californian bay
          Umbellularia californica

Californian bayberry
          Myrica californica

Californian blackcurrant
          Ribes bracteosum

Californian black oak
          Quercus kelloggii

Californian black walnut
          Juglans hindsii

Californian buckeye chestnut
          Aesculus californica

Californian buckthorn
          Rhamnus purshiana

Californian coast redwood
          Sequoia sempervirens

Californian cypress
          Cupressus goveniana
          Cupressus stephensonii

Californian fan palm
          Washingtonia filifera

Californian field oak
          Quercus agrifolia

Californian fuchsia
          Zauschneria californica

Californian horse chestnut
          Aesculus californica

Californian juniper
          Juniperus californica

Californian laurel
          Umbellularia californica

Californian lilac
          Ceanothus thyrsiflorus

Californian live oak
          Quercus agrifolia
          Quercus chrysolepis

Californian mock orange
          Carpenteria californica

Californian mountain pine
          Pinus monticola

Californian nutmeg
          Torreya californica

Californian olive
          Umbellularia californica

Californian pepper tree
          Schinus molle

Californian privet
          Ligustrum ovalifolium

Californian red fir   *Abies magnifica*

Californian redwood
          Sequoia sempervirens

Californian sassafras
          Umbellularia californica

Californian scrub oak

Californian scrub oak
Quercus dumosa

Californian tree mallow
Lavatera assurgentiflora

California wax myrtle
Myrica californica

Californian white oak
Quercus lobata

Californian walnut
Juglans californica

Californian yew *Taxus brevifolia*

Calisaya *Cinchona calisaya*

Calotropis *Calotropis procera*

Calumba *Jateorhiza calumba*

Cambridge oak *Quercus warburgii*

Camden woollybutt
Eucalyptus macarthurii

Camel's foot *Bauhinia purpurea*

Camel thorn *Acacia giraffae*
Alhagi camelorum

Campbell's magnolia
Magnolia campbellii

Campeachy-wood
Haematoxylum campeachianum

Camperdown elm
Ulmus glabra camperdown

Camphor tree
Cinnamomum camphora

Camwood *Baphia nitida*

Canada pitch *Tsuga canadensis*

Canada plum *Prunus nigra*

Canadian dwarf juniper
Juniperus depressa

Canadian hemlock
Tsuga canadensis

Canadian juniper
Juniperus communis depressa

Canadian maple *Acer rubrum*

Canadian red pine *Pinus resinosa*

Canadian tea
Gaultheria procumbens

Canadian yew *Taxus canadensis*

Canary balm
Cedronella canariensis

Canary date palm
Phoenix canariensis

Canary holly *Ilex perado*

Canary Island date palm
Phoenix canariensis

Canary Island juniper
Juniperus cedrus

Canary Island pine
Pinus canariensis

Canary Island holly
Ilex platyphylla

Canary Island laurel
Laurus azorica
Laurus canariensis
Laurus maderensis

Canary palm *Phoenix canariensis*

Candelabra cactus
Euphorbia lactea

Candelabra spruce
Picea montigena

Candelabra tree
Araucaria angustifolia
Euphorbia candelabrum
Euphorbia ingens

Candelilla
Euphorbia antisyphilitica

Candle-bark gum
Eucalyptus rubida

Candleberry
Myrica pennsylvanica

Candleberry myrtle
Myrica carolinensis
Myrica cerifera
Myrica faya

Candleberry tree
Aleurites moluccana

Candle bush *Cassia didymobotrya*

Candlenut tree
Aleurites moluccana

Candlestick senna *Cassia alata*

Candle tree *Parmentiera cereifera*

Cane apple *Arbutus unedo*

Canistel *Pouteria campechiana*

Cannonball tree
Couroupita guianensis

Canoe birch *Betula papyrifera*

Canoe cedar
Chamaecyparis nootkatensis

Canyon live oak
Quercus chrysolepis

Canyon maple *Acer macrophyllum*

Caoutchouc *Hevea brasiliensis*

Cape box *Buxus macowani*

Cape chestnut
Calodendrum capense

Cape fig *Ficus capensis*

Cape figwort *Phygelius capensis*

Cape gardenia
Tabernaemontana divaricata

Cape gum *Acacia horrida*
Acacia senegal

Cape heath *Erica hyemalis*

Cape hibiscus
Hibiscus diversifolius

Cape jasmine *Acacia senegal*
Gardenia florida
Gardenia grandiflora
Gardenia jasminoides
Tabernaemontana coronaria
Tabernaemontana divaricata

Cape myrtle *Lagerstroemia indica*
Myrsine africana

Cape pittosporum
Pittosporum viridiflorum

Caper bush *Capparis spinosa*

Caper spurge *Euphorbia lathyris*

Cappadocian maple
Acer cappadocicum

Capulin cherry	*Prunus salicifolia*
Cardinal spear	*Erythrina arborea*
	Erythrina herbacea
Cardinal willow	
	Salix fragilis decipiens
Caribbean pine	*Pinus caribaea*
Caribbean royal palm	
	Roystonea oleracea
Carib wood	*Sabinea carinalis*
Caricature plant	
	Graptophyllum hortense
	Graptophyllum pictum
Carmel ceanothus	
	Ceanothus griseus
Carmel creeper	
	Ceanothus griseus horizontalis
Carnanba palm	*Copernica cerifera*
Carnauba palm	*Copernica cerifera*
Carnauba wax palm	
	Copernica prunifera
Carob	*Ceratonia siliqua*
Carob tree	*Jacaranda procera*
Carolina allspice	
	Calycanthus fertilis
	Calycanthus floridus
Carolina ash	*Fraxinus caroliniana*
Carolina buckthorn	
	Rhamnus carolinian
Carolina hemlock	
	Tsuga caroliniana
Carolina ipecac	
	Euphorbia ipecacuanhae
Carolina poplar	*Populus eugenei*
Carolina silverbell	*Halesia carolina*
Carolina spurge	
	Euphorbia ipecacuanhae
Caroline poplar	*Populus angulata*
	Populus × canadensis
Carpathian spruce	
	Picea abies carpathica
Carrion tree	
	Couroupita guianensis
Cartagena bark	
	Cinchona cordifolia
Cascade fir	*Abies amabilis*
Cascara	*Picramnia antidesma*
Cascara sagrada	
	Rhamnus purshiana
Cascarilla	*Croton cascarilla*
Cashew	*Anacardium occidentale*
Caspian locust	*Gleditsia caspica*
Caspian willow	*Salix acutifolia*
Cassa-banana	*Sicana odorifera*
Cassada wood	
	Turpinia occidentalis
Cassandra	
	Chamaedaphne calyculata
Cassava	*Manihot esculenta*
Cassava wood	
	Turpinia occidentalis
Cassia	*Cinnamomum cassia*

Cassia bark tree	
	Cinnamomum cassia
Cassia flower tree	
	Cinnamomum loureirii
Cassie	*Acacia farnesiana*
Cassina	*Ilex cassine*
	Ilex vomitoria
Castilla rubber tree	
	Castilla elastica
Castor aralia	*Kalopanax pictus*
Castor bean	*Ricinus communis*
Castor oil palm	*Aralia japonica*
	Fatsia japonica
Castor oil plant	*Fatsia japonica*
	Ricinus communis
Catalina ceanothus	
	Ceanothus arboreus
Catalina cherry	*Prunus lyonii*
Catalina ironwood	
	Lyonothamnus floribundus
Catalina mountain lilac	
	Ceanothus arboreus
Catalonian jasmine	
	Jasminum grandiflorum
Catalpa	*Catalpa bignonioides*
Catawba	*Catalpa speciosa*
Catberry	
	Nemopanthus mucronatus
Catclaw acacia	*Acacia greggii*
Catechu	*Acacia catechu*
	Areca catechu
Catesby oak	*Quercus laevis*
Cat's claw	*Pithecellobium unguis-cati*
Cat spruce	*Picea glauca*
Cat thyme	*Teucrium marum*
Caucasian alder	*Alnus subcordata*
Caucasian ash	*Fraxinus oxycarpa*
Caucasian elm	
	Zelkova carpinifolia
	Zelkova crenata
Caucasian fir	*Abies nordmanniana*
Caucasian lime	*Tilia × euchlora*
Caucasian maple	
	Acer cappadocicum
Caucasian nettle tree	
	Celtis caucasica
Caucasian oak	
	Quercus macranthera
Caucasian pear	*Pyrus caucasica*
Caucasian wing-nut	
	Pterocarya fraxinifolia
Caucasian whortleberry	
	Vaccinium arctostaphylos
Cayenne cherry	*Eugenia michelii*
	Eugenia uniflora
Ceara rubber	*Manihot glaziovii*
Cedar elm	*Ulmus crassifolia*
Cedar of Goa	*Cupressus lusitanica*
Cedar of Lebanon	*Cedrus libani*
Cedar pine	*Pinus glabra*

Cedar wattle

Cedar wattle	*Acacia alata*
	Acacia terminalis
Cedrat lemon	*Citrus medica cedra*
Cedron	*Simaba cedron*
Celery pine	*Phyllocladus alpinus*
	Phyllocladus asplenifolius
	Phyllocladus trichomanoides
Celery-top pine	
	Phyllocladus rhomboidalis
Celery-topped pine	
	Phyllocladus asplenifolius
Century plant	*Yucca recurvifolia*
Cevennes pine	
	Pinus nigra cebennensis
Ceylon cinnamon	
	Cinnamomum zeylanicum
Ceylon date palm	*Phoenix zelanica*
Ceylon ebony	*Diospyros ebenaster*
	Diospyros ebenum
Ceylon gooseberry	
	Aberia gardneri
	Dovyalis hebecarpa
Ceylon mahogany	*Melia dubium*
Ceylon oak	*Schleichera oleosa*
Ceylon olive	*Elaeocarpus serratus*
Ceylon rosewood	
	Albizzia odoratissima
Ceylon tea	*Cassine glauca*
Chaconia	*Warszewiczia coccinea*
Chalk maple	*Acer leucoderme*
Champaca	*Michelia champaca*
Champion's oak	
	Quercus velutina rubrifolia
Champney rose	*Rosa × noisettiana*
Charcoal tree	
	Byrsonima crassifolia
Chaste tree	*Vitex agnus-castus*
Chat	*Catha edulis*
Chattamwood	
	Bumelia lanuginosa
Cheken	*Eugenia cheken*
	Luma chequen
Checkerberry	
	Gaultheria procumbens
	Mitchella repens
Cheddar whitebeam	*Sorbus anglica*
Chenille plant	*Acalypha hispida*
Chennar tree	*Platanus orientalis*
Chequered juniper	
	Juniperus deppeana pachyphlaea
Chequer tree	*Sorbus torminalis*
Cherimalla	*Annona cherimola*
Cherimoya	*Annona cherimola*
Cherokee bean	*Erythrina arborea*
	Erythrina herbacea
Cherokee rose	*Rosa laevigata*
Cherry birch	*Betula lenta*
Cherry elaeagnus	*Elaeagnus edulis*
	Elaeagnus multiflora

Cherry laurel	*Prunus caroliniana*
	Prunus laurocerasus
Cherry plum	*Prunus cerasifera*
Cherrystone juniper	
	Juniperus monosperma
Chess apple	*Sorbus aria*
Chestnut dioon	*Dioon edule*
Chestnut-leaved oak	
	Quercus castaneifolia
Chestnut oak	*Quercus acutissima*
	Quercus muehlenbergii
	Quercus prinus
Chestnut rose	*Rosa roxburghii*
Chichester elm	*Ulmus × vegeta*
Chickasaw plum	
	Prunus angustifolia
Chicle	*Achras sapota*
Chicot	*Gymnocladus dioica*
Chicozapote	*Manilkara zapota*
Chilean boldo tree	*Peumus boldus*
Chilean cedar	
	Austrocedrus chilensis
Chilean fire bush	
	Embothrium coccineum
Chilean fire tree	
	Embothrium coccineum
Chilean guava	*Eugenia ugni*
	Myrtus ugni
	Ugni molinae
Chilean hazel	*Gevuina avellana*
Chilean incense cedar	
	Austrocedrus chilensis
	Calocedrus chilensis
Chilean jasmine	
	Mandevilla suaveolens
Chilean myrtle	*Myrtus chequen*
Chilean nut	*Gevuina avellana*
Chilean tea tree	*Lycium chilense*
	Lycium gracillianum
Chilean totara	
	Podocarpus nubigenus
Chilean wine palm	*Jubaea chilensis*
	Jubaea spectabilis
Chilean yew	*Podocarpus andinus*
Chile pine	*Araucaria araucana*
Chilghoza pine	*Pinus gerardiana*
Chinaberry	*Melia azedarach*
China cane	*Rhapis excelsa*
China rose	*Hibiscus sinensis*
	Rosa chinensis
China tree	*Koelreuteria paniculata*
	Melia azedarach
China turpentine tree	
	Pistacia terebinthus
Chinese angelica tree	
	Aralia chinensis
Chinese anise	*Illicium anisatum*
Chinese arbor-vitae	
	Thuya orientalis
Chinese ash	*Fraxinus chinensis*
Chinese aspen	*Populus adenopoda*

TREES, BUSHES, AND SHRUBS

Chinese rubber tree

Chinese banyan	*Ficus retusa*
Chinese beech	*Fagus englerana*
	Fagus longipetiolata
Chinese bottle tree	
	Firmiana simplex
Chinese box	*Murraya paniculata*
Chinese box orange	
	Severinia buxifolia
Chinese box thorn	
	Lycium barbarum
	Lycium chinense
Chinese bramble	
	Rubus cockburnianus
Chinese bush cherry	
	Prunus glandulosa
	Prunus tomentosa
Chinese butternut	
	Juglans cathayensis
Chinese catalpa	*Catalpa ovata*
Chinese cedar	*Cedrela sinensis*
	Toona sinensis
Chinese cherry	*Prunus conradinae*
Chinese chestnut	
	Castanea mollissima
Chinese cinnamon	
	Cinnamomum cassia
Chinese coffee tree	
	Gymnocladus chinensis
Chinese cork oak	
	Quercus variabilis
Chinese cork tree	
	Phellodendron chinensis
Chinese cowtail pine	
	Cephalotaxus sinensis
Chinese crab apple	
	Malus hupehensis
	Malus spectabilis
Chinese date	*Ziziphus jujuba*
Chinese dogwood	
	Cornus kousa chinensis
Chinese douglas fir	
	Pseudotsuga sinensis
Chinese elm	*Ulmus parvifolia*
Chinese evergreen magnolia	
	Magnolia delavayi
Chinese fan palm	
	Livistona chinensis
Chinese filbert	*Corylus chinensis*
Chinese fir	
	Cunninghamia lanceolata
Chinese flowering apple	
	Malus spectabilis
Chinese flowering ash	
	Fraxinus mariesii
Chinese fountain palm	
	Livistona chinensis
Chinese fringe tree	
	Chionanthus retusa
Chinese hackberry	*Celtis japonica*
	Celtis sinensis
Chinese hawthorn	
	Crateagus pinnatifida
	Photinia serrulata
Chinese hazel	*Corylus chinensis*
Chinese hemlock	*Tsuga chinensis*
Chinese hibiscus	
	Hibiscus rosa-sinensis
	Hibiscus sinensis
Chinese hickory	*Carya cathayensis*
Chinese hill cherry	
	Prunus mutabilis stricta
	Prunus serrulata hupehensis
Chinese holly	*Ilex cornuta*
	Osmanthus heterophyllus
Chinese honey locust	
	Gleditsia chinensis
	Gleditsia sinensis
Chinese incense cedar	
	Calocedrus macrolepis
Chinese ixora	*Ixora chinensis*
Chinese jujube tree	
	Ziziphus jujuba
Chinese juniper	*Juniperus chinensis*
Chinese kidney bean	
	Wisteria sinensis
Chinese larch	*Larix potaninii*
Chinese lilac	*Syringa × chinensis*
Chinese lime	*Tilia oliveri*
Chinese laurel	*Antidesma bunius*
Chinese magnolia	
	Magnolia sinensis
Chinese mint bush	
	Elsholtzia stauntonii
Chinese mountain ash	
	Sorbus hupehensis
Chinese necklace poplar	
	Populus lasiocarpa
Chinese parasol tree	
	Firmiana simplex
Chinese peach	*Prunus davidiana*
Chinese pear	*Pyrus ussuriensis*
Chinese persimmon	
	Diospyros chinensis
	Diospyros kaki
Chinese pine	*Keteleeria davidiana*
	Pinus tabuliformis
Chinese pistachio	*Pistacia chinensis*
Chinese plum-yew	
	Cephalotaxus fortuni
Chinese poplar	*Populus lasiocarpa*
Chinese privet	*Ligustrum lucidum*
	Ligustrum sinense
Chinese quince	
	Pseudocydonia sinensis
Chinese red-barked birch	
	Betula albosinensis
Chinese redbud	*Cercis chinensis*
	Cercis racemosa
Chinese rowan	*Sorbus hupehensis*
Chinese rubber tree	
	Eucommia ulmoides

Chinese sacred bamboo
Nandina domestica

Chinese sand pear   *Pyrus pyrifolia*

Chinese scarlet rowan
Sorbus commixta embley
Sorbus discolor

Chinese scholar tree
Sophora japonica

Chinese silk thread tree
Eucommia ulmoides

Chinese silkworm thorn
Cudrania tricuspidata

Chinese snakebark maple
Acer davidii

Chinese snowball
Viburnum macrocephalum

Chinese soapberry
Sapindus mukorossi

Chinese spruce   *Picea asperata*

Chinese star anise   *Illicium verum*

Chinese stewartia
Stewartia sinensis

Chinese stuartia   *Stewartia sinensis*

Chinese swamp cypress
Glyptostrobus lineatus

Chinese tallow tree
Sapium sebiferum

China tea plant   *Lycium chinense*

Chinese thuya   *Thuya orientalis*

Chinese sweet gum
Liquidambar formosana

Chinese tulip tree
Liriodendron chinense

Chinese varnish tree
Rhus potaninii

Chinese walnut   *Juglans cathayensis*

Chinese water pine
Glyptostrobus lineatus

Chinese weeping cypress
Cupressus funebris

Chinese whitebeam   *Sorbus folgneri*

Chinese white cedar
Thuya orientalis

Chinese white pine   *Pinus armandii*

Chinese wing-nut
Pteryocarya stenoptera

Chinese witch-hazel
Hamamelis mollis

Chinese wood-oil tree
Aleurites fordii

Chinese yellow wood
Cladrastis chinensis
Cladrastris sinensis

Chinese yew   *Taxus celebica*
Taxus chinensis

Chinese zelkova   *Zelkova sinica*

Chinquapin chestnut
Castanea pumila

Chinquapin oak   *Quercus prinoides*

Chinquapin rose   *Rosa roxburghii*

Chios mastic tree   *Pistacia lentiscus*

Chittamwood   *Cotinus americanus*
Cotinus obovatus

Chocolate tree   *Theobroma cacao*

Chokeberry   *Aronia arbutifolia*

Choke cherry   *Prunus virginiana*

Christmas berry
Heteromeles arbutifolia

Christmas berry tree
Schinus terebinthifolia

Christmas box
Sarcococca buxaceae

Christmas candle   *Cassia alata*

Christmas heather
Erica canaliculata

Christmas palm   *Veitchia merrillii*

Christmas star
Euphorbia pulcherrima

Christmas tree   *Nuytsia floribunda*
Picea abies

Christ's thorn   *Paliurus aculeatus*
Paliurus australis
Paliurus spina-christi
Paliurus virgatus

Chusan palm
Trachycarpus fortunei

Cider gum   *Eucalyptus gunnii*

Cigar-box cedar   *Cedrela odorata*

Cigar tree   *Catalpa speciosa*

Ciliate heath   *Erica ciliaris*

Cilician fir   *Abies cilicica*

Cinnamon
Cinnamomum zeylanicum

Cinnamon rose   *Rosa majalis*

Cinnamon tree
Cinnamomum zeylanicum

Cinnamon wattle   *Acacia leprosa*

Citrange   × *Citroncirus webberi*

Citron   *Citrus medica*

Clammy azalea
Rhododendron viscosum

Clammy locust   *Robinia viscosa*

Clanwilliam cedar
Widdringtonia juniperoides

Claret ash
Fraxinus oxycarpa raywood

Clearing nut   *Strychnos potatorum*

Clementine   *Citrus reticulata*

Cliff date palm   *Phoenix rupicola*

Cliff whitebeam   *Sorbus rupicola*

Climbing fig   *Ficus pumila*
Ficus repens
Ficus stipulata

Climbing hydrangea
Pileostegia viburnoides
Schizophragma viburnoides

Clove tree   *Eugenia caryophyllus*
Syzygium aromaticum

Clown fig   *Ficus aspera*

Clustered fishtail palm
Caryota mitis

Cluster fig   *Ficus glomerata*
                        Ficus racemosa
Cluster pine   *Pinus pinaster*
Coach-whip   *Fouquieria splendens*
Coarse-leaved mallee
                        Eucalyptus grossa
Coastal myall   *Acacia binervia*
Coast banksia   *Banksia integrifolia*
Coast ceanothus
                     Ceanothus ramulosus
Coast redwood
                  Sequoia sempervirens
Cobana   *Stahlia monosperma*
Cobnut   *Corylus avellana*
Coca   *Erythroxylum coca*
Cocaine plant   *Erythroxylum coca*
Cock's comb   *Erythrina crista-galli*
Cockscomb beech
                 Fagus sylvatica cristata
Cockspur coral tree
                   Erythrina crista-galli
Cockspur thorn
                   Crataegus crus-galli
Cocoa   *Theobroma cacao*
Coco-de-mer   *Lodoicea maldavica*
Cocona   *Solanum topiro*
Coconut palm   *Cocos nucifera*
Coco palm   *Chrysobalanus icaco*
Coco plum   *Chrysobalanus icaco*
Coffee   *Coffea arabica*
Coffee berry   *Rhamnus californica*
Coffee senna   *Cassia occidentalis*
Coffee tree   *Polyscias guilfoylei*
Coffin juniper
                  Juniperus recurva coxii
Coffin tree   *Taiwania flousiana*
Cohune palm   *Orbignya cohune*
Coigue   *Nothofagus dombeyi*
Coigue de magellanes
                  Nothofagus betuloides
Cola   *Cola nitida*
Cola tree   *Cola acuminata*
Colorado blue spruce
                  Picea pungens glauca
Colorado red cedar
                 Juniperus scopulorum
Colorado white fir   *Abies concolor*
Colorado spruce   *Picea pungens*
Columnar spruce
            Juniperus communis stricta
Commercial apple   *Malus pumila*
Common acacia
                 Robinia pseudoacacia
Common alder   *Alnus glutinosa*
Common almond
                     Prunus amygdalus
                     Prunus communis
                        Prunus dulcis
Common apricot
                     Prunus armeniaca

Common ash   *Fraxinus excelsior*
Common barberry
                       Berberis vulgaris
Common beech   *Fagus sylvatica*
Common bearberry
                Arctostaphylos uva-ursi
Common briar   *Rosa canina*
Common broom   *Cytisus genista*
Common box   *Buxus sempervirens*
Common broom   *Cytisus scoparius*
Common buckthorn
                   Rhamnus catharticus
Common camellia
                     Camellia japonica
Common coral bean
             Erythrina corallodendron
Common crab apple
                       Malus sylvestris
Common cypress pine
                       Callitris preissii
                       Callitris robusta
Common dogwood
                     Cornus sanguinea
Common elder   *Sambucus nigra*
Common elm   *Ulmus procera*
Common fig   *Ficus carica*
Common gardenia
                        Gardenia florida
                     Gardenia grandiflora
                    Gardenia jasminoides
Common goosberry
                     Ribes uva-crispa
Common gorse   *Ulex europaeus*
Common guava   *Psidium guajava*
Common hawthorn
                   Crataegus monogyna
Common hazel   *Corylus avellana*
Common holly   *Ilex aquifolium*
Common hornbeam
                      Carpinus betulus
Common horse chestnut
                 Aesculus hippocastanum
Common jujube tree
                       Ziziphus jujuba
Common juniper
                    Juniperus communis
Common laburnum
                 Laburnum anagyroides
                     Laburnum vulgare
Common larch   *Larix decidua*
Common laurel
                   Prunus laurocerasus
Common lavender
                 Lavandula angustifolium
Common lilac   *Syringa vulgaris*
Common lime   *Tilia × europaea*
                      Tilia × vulgaris
Common maple   *Acer campestre*
Common mountain ash
                     Sorbus aucuparia
Common mulberry   *Morus nigra*

Common myrtle   *Myrtus communis*
Common oak   *Quercus robur*
Common oleander
    Nerium oleander
Common olive   *Olea europaea*
Common osier   *Salix viminalis*
Common papaw   *Carica papaya*
Common pawpaw   *Carica papaya*
Common peach   *Prunus persica*
Common pear   *Pyrus communis*
Common persimmon
    Diospyros virginiana
Common privet
    Ligustrum vulgare
Common quince   *Cydonia oblonga*
Common rose mallow
    Hibiscus moscheutos
Common rowan   *Sorbus aucuparia*
Common sage   *Salvia officinalis*
Common sallow   *Salix cinerea*
    Salix cinerea oleifolia
Common screw pine
    Pandanus utilis
Common silver birch
    Betula pendula
Common silver fir   *Abies alba*
Common spindle tree
    Euonymus europaea
Common spruce   *Picea abies*
Common walnut   *Juglans regia*
Common white birch
    Betula pubescens
Common white jasmine
    Jasminum officinale
Common witch hazel
    Hamamelis virginiana
Common yellow azalea
    Rhododendron luteum
Common yellowwood
    Podocarpus falcatus
Common yew   *Taxus baccata*
Confederate jasmine
    Jasminum nitidum
Confederate rose
    Hibiscus mutabilis
Confetti bush
    Coleonema pulchrum
Congo fig   *Ficus dryepondtiana*
Congo mallee   *Eucalyptus dumosa*
Connemara heath
    Daboecia cantabrica
Contorted hazel
    Corylus avellana contorta
Contorted pagoda tree
    Sophora japonica + pendula
Contorted willow
    *Salix babylonica pekinensis
    tortuosa*
    Salix matsudana tortuosa
Cooba   *Acacia salicina*
Coolibah   *Eucalyptus microtheca*

Cootamundra wattle
    Acacia baileyana
    Acacia saileyana
Copal tree   *Ailanthus altissima*
Copper beech
    Fagus sylvatica purpurea
Copper-leaf   *Acalypha wilkesiana*
Copper-pod tree
    Peltophorum pterocarpum
Coquito palm   *Jubaea chilensis*
    Jubaea spectabilis
Coralbark maple
    Acer japonicum aconitifolium
    Acer palmatum senkaki
Coralbark willow
    Salix alba britzensis
    Salix alba chermesina
    Salix caprea chermesina
Coral bean   *Erythrina arborea*
    Erythrina herbacea
Coralberry   *Ardisia crenata*
    Symphoricarpus orbiculatus
    Symphoricarpus rubra vulgaris
    Symphoricarpus vulgaris
Coral gum   *Eucalyptus torquata*
Coral hibiscus
    Hibiscus schizopetalus
Coral pea   *Adenanthera pavonina*
Coral plant   *Jatropha multifida*
    Russelia equisetiformis
    Russelia juncea
Coral tree
    Erythrina corallodendron
    Erythrina crista-galli
    Macaranga grandifolia
Coralwood   *Adenanthera pavonina*
Cordia   *Cordia sebestena*
Cork elm   *Ulmus thomasii*
Cork fir   *Abies lasiocarpa arizonica*
Cork hopbush
    Kallstroemia platyptera
Cork oak   *Quercus suber*
Corkscrew hazel
    Corylus avellana contorta
Corkscrew wattle   *Acacia tortuosa*
Corkscrew willow
    *Salix babylonica pekinensis
    tortuosa*
    Salix matsudana tortuosa
Corkwood   *Leitneria floridana*
    Ochroma pyramidale
Corkwood tree
    Duboisia myoporoides
Cornel   *Cornus mas*
Cornelian cherry
    Cornus capitata mas
    Cornus mas
Cornish elm   *Ulmus angustifolia*
    Ulmus angustifolia cornubiensis
    Ulmus carpinifolia cornubiensis
    Ulmus minor stricta cornubiensis
    Ulmus stricta

TREES, BUSHES, AND SHRUBS

Cuban belly palm

Cornish heath *Erica vagans*
Coromandel ebony
Diospyros melanoxylon
Correosa *Rhus microphylla*
Corsican heath *Erica terminalis*
Corsican pine
Pinus nigra maritima
Corylopsis *Corylopsis spicata*
Cosmetic bark tree
Murraya paniculata
Costa Rican guava
Psidium friedrichsthalianum
Costa Rican holly
Olmediella betschlerana
Costorphine plane
Acer pseudoplatanus costorphinense
Cotoneaster *Cotoneaster frigidus*
Cottage mezereon
Daphne mezereum
Cottage tea tree *Lycium barbarum*
Lycium chinense
Cotton gum *Nyssa aquatica*
Cotton rose *Hibiscus mutabilis*
Cotton tree
Bombax malabaricum
Cottonwood *Populus deltoides*
Cottony jujube
Ziziphus mauritania
Council tree *Ficus altissima*
Country walnut
Aleurites moluccana
Cowberry *Viburnum lentago*
Vaccinium vitisidaea
Cow-itch *Rhus radicans*
Cow-itch cherry *Malpighia urens*
Cow-itch tree
Lagunaria patersonii
Cow's tail pine
Cephalotaxus harringtonia
Cowtail pine *Cephalotaxus fortunii*
Cow tree *Brosimum alicastrum*
Coyoli palm *Acrocomia mexicana*
Coyote willow *Salix exigua*
Cox's juniper
Juniperus recurva coxii
Crackerberry *Cornus canadensis*
Crack willow *Salix fragilis*
Crampbark *Viburnum trilobum*
Viburnum opulus
Cranberry
Vaccinium macrocarpon
Vaccinium oxycoccus
Vaccinium vitis-idaea
Cranberry bush *Viburnum opulus*
Viburnum trilobum
Cranberry cotoneaster
Cotoneaster apiculatus
Cranberry tree
Viburnum trilobum

Cranesbill myrtle
Lagerstroemia indica
Cream bush *Holodiscus ariifolius*
Holodiscus discolor
Cream nut *Bertholleta excelsa*
Creeping barberry *Mahonia repens*
Creeping blue blossom
Ceanothus thyrsiflorus repens
Creeping boobialla
Mysporum parvifolium
Creeping cedar
Juniperus horizontalis
Creeping dogwood
Cornus canadensis
Creeping fig *Ficus pumila*
Ficus repens
Ficus stipulata
Creeping juniper
Juniperus horizontalis
Juniperus procumbens
Creeping mountain ash
Sorbus reducta
Creeping rubber plant *Ficus pumila*
Ficus repens
Ficus stipulata
Creeping savin juniper
Juniperus horizontalis
Creeping willow *Salix repens*
Creeping winterberry
Gaultheria procumbens
Creosote bush *Larrea divaricata*
Larrea tridentata
Crested moss rose
Rosa centifolia cristata
Cretan maple *Acer sempervirens*
Cretan zelcova *Zelcova abelicea*
Cricket-bat willow
Salix alba caerulea
Crimean lime *Tilia × euchlora*
Crimean pine
Pinus nigra caramanica
Crimson bottle brush
Callistemon citrinus
Crimson dwarf cherry
Prunus × cistena
Crimson mallee box
Eucalyptus lansdowneana
Cross-leaved heath *Erica tetralix*
Croton *Codiaeum variegatum*
Crowberry *Empetrum nigrum*
Crown of thorns
Paliurus spina-christi
Crown plant *Calotropis gigantea*
Crucifixion thorn
Holacantha emoryi
Cry-baby tree
Erythrina crista-galli
Cuachilote *Parmentiera edulis*
Cuban bast *Hibiscus elatus*
Cuban belly palm
Colpothrinax wrightii

Cuban manac

Cuban manac *Calyptronoma dulcis*
Cuban pine *Pinus caribaea*
Pinus occidentalis
Cuban pink trumpet tree
Tabebuia pallida
Cuban royal palm *Roystonea regia*
Cucumber tree
Magnolia acuminata
Cucurite palm
Maximiliana caribaea
Maximiliana maripa
Maximiliana regia
Cuipo *Cavanillesia platanifolia*
Cultivated apple *Malus domestica*
Cup gum *Eucalyptus cosmophylla*
Curare
Chondrodendron tomentosum
Strychnos toxifera
Curly palm *Howea belmoreana*
Kentia belmoreana
Curly sentry palm
Howea belmoreana
Curry-leaf tree *Murraya koenigii*
Murraya paniculata
Curry plant
Helichrysum serotinum
Custard apple *Annona cherimola*
Annona reticulata
Annona squamosa
Cutch *Acacia catechu*
Cut-leaf birch
Betula pendula dalecarlica
Cut-leaf lilac *Syringa laciniata*
Cut-leaf oak
Quercus robur filicifolia
Cut-leaf purple beech
Fagus sylvatica rohanii
Cut-leaf walnut
Juglans regia laciniata
Cutleaf zelkova
Zelkova sinica verschaffeltii
Cut-leaved alder
Alnus glutinosa imperialis
Cut-leaved beech
Fagus sylvatica heterophylla
Cut-leaved elder
Sambucus nigra laciniata
Cut-leaved hazel
Corylus avellana heterophylla
Cut-leaved mountain ash
Sorbus aucuparia asplenifolia
Cut-tail *Eucalyptus fastigiata*
Cuyamaca cypress
Cupressus stephensonii
Cypress golden oak
Quercus alnifolia
Cypress hebe *Hebe cupressoides*
Cypress oak
Quercus robur fastigiata
Cypress spurge
Euborbia cyparissias
Cyprian cedar *Cedrus brevifolia*

Cyprian plane
Platanus orientalis insularis
Cyprus cedar *Cedrus brevifolia*
Cyprus strawberry tree
Arbutus andrachne
Cyprus turpentine tree
Pistacia terebinthus
Dagger plant *Yucca aloifolia*
Dahoon holly *Ilex cassine*
Dahurian buckthorn
Rhamnus davurica
Dahurian larch *Larix gmelinii*
Daimyo oak *Quercus dentata*
Daisy bush
Olearia nummularifolia
Olearia phlogopappa
Dalmatian broom
Genista dalmatica
Genista sylvestris pungens
Damask rose *Rosa damascena*
Dammar *Agathis dammara*
Damson *Prunus domestica*
Prunus domestica institia
Prunus insititia
Dane's elder *Sambucus ebulus*
Danewort *Sambucus ebulus*
Dangleberry *Gaylussacia frondosa*
Daniell's euodia *Euodia daniellii*
Daphne lilac
Syringa microphylla superba
Dark-leaved willow
Salix myrsinifolia
Salix nigricans
Darlington oak *Quercus laurifolia*
Darwin stringybark
Eucalyptus tetradonta
Darwin woollybutt
Eucalyptus miniata
Date palm *Phoenix dactylifera*
Date plum *Diospyros chinensis*
Diospyros kaki
Diospyros lotus
Diospyros virginiana
Datil *Yucca baccata*
Dattock tree
Detarium senegalense
David's pine *Pinus armandii*
Dawick beech
Fagus sylvatica dawyck
Dawn redwood
Metasequoia glyptostroboides
Day cestrum *Cestrum diurnum*
Day jessamine *Cestrum diurnum*
Dead-rat tree *Adansonia digitata*
Deal pine *Pinus strobus*
Deane's gum *Eucalyptus deanei*
Deccan hemp *Hibiscus cannabinus*
Deciduous cypress
Taxodium distichum
Deckaner hemp
Hibiscus cannabinus

Dragon's blood

Deciduous camellia
    Stewartia pseudocamellia
    Stuartia pseudocamellia

Deep-veined maple   *Acer argutum*

Deerberry   *Vaccinium stamineum*

Deerbrush
    Ceanothus integerrimus

Deer bush *Ceanothus integerrimus*

Deer oak   *Quercus sadlerana*

Degame
    Calycophyllum candissimum

Delavay's silver fir *Abies delavayi*

Del norte manzanita
    Arctostaphylos cinerea

Deodar   *Cedrus deodara*

Derris   *Derris elliptica*

Desert almond *Prunus fasciculata*

Desert apricot *Prunus fremontii*

Desert cassia   *Cassia covesii*
    Cassia eremophila

Desert fan palm
    Washingtonia filifera

Desert gum   *Eucalyptus rudis*

Desert ironwood *Olneya tesota*

Desert juniper *Juniperus utahensis*

Desert kurrajong
    Brachychiton gregorii

Desert olive
    Forestiera neomexicana

Desert peach *Prunus andersonii*

Desert rose *Adenium obesum*

Desert rose mallow
    Hibiscus farragei

Desert sumac *Rhus microphylla*

Desert tea *Ephedera vulgaris*

Desert willow *Chilopsis linearis*

Desmond mallee
    Eucalyptus desmondensis

Deutzia   *Deutzia gracilis*

Devil's bit *Alstonia scholaris*

Devil's club *Oplopanax horridus*

Devi's fig *Argemone mexicana*

Devil's maple *Acer diabolicum*

Devil's shoestrings
    Viburnum alnifolium

Devil's walking stick *Aralia spinosa*

Devil tree *Alstonia scholaris*

Devil wood
    Osmanthus americanus

Dhobi's nut
    Semecarpus anacardium

Digger pine *Pinus sabiniana*

Dita bark *Alstonia scholaris*

Divi-divi *Caesalpinia coriaria*

Dockmackie
    Viburnum acerifolium

Dogberry *Cornus sanguinea*
    Sorbus americana
    Viburnum alnifolium

Dog briar   *Rosa canina*

Dog hobble *Viburnum alnifolium*

Dog rose   *Rosa canina*

Dogwood   *Cornus sanguinea*

Dombey's southern beech
    Nothofagus dombeyi

Doom palm *Hyphaene thebaica*

Dorset heath   *Erica ciliaris*

Doub pine *Borassus flabellifer*

Double almond
    Prunus dulcis roseoplena

Double cherry-plum
    Prunus × blireana

Double coconut
    Lodoicea maldavica

Doble crimson thorn
    Crataegus laevigata

Double flowering gorse
    Ulex europaeus plenus

Double gean *Prunus avium plena*

Double pink thorn
    Crataegus laevigata

Double spruce *Picea mariana*
    Picea nigra

Double white cherry
    Prunus avium plena

Double white thorn
    Crataegus laevigata

Douglas fir *Pseudotsuga menziesii*

Douglas maple
    Acer glabrum douglasii

Doum palm *Hyphaene thebaica*
    Palmae hyphaene

Dove tree *Davidia involucrata*

Downton elm
    Ulmus × hollandica smithii

Downy birch *Betula pubescens*

Downy black poplar
    Populus nigra betulifolia

Downy cherry *Prunus tomentosa*

Downy chestnut *Castanea alnifolia*

Downy clethra *Clethra tomentosa*

Downy hawthorn *Crataegus mollis*

Downy Japanese maple
    Acer japonicum

Downy myrtle
    Rhodomyrtus tomentosa

Downy oak *Quercus pubescens*

Downy poplar
    Populus heterophylla

Downy rose *Rosa tomentosa*

Downy tree of heaven
    Ailanthus vilmoriniana

Downy willow *Salix lapponum*

Dracaena fig *Ficus pseudopalma*

Dragon bones *Euphorbia lactea*

Dragon-eye palm
    Pinus densiflora oculus-draconis

Dragon's blood
    Daemonorops draco

Dragon's claw willow
    Salix babylonica pekinensis tortuosa
    Salix matsudana tortuosa
Dragon spruce   *Picea asperata*
Dragon tree   *Dracaena draco*
Drooping cowtail pine
    Cephalotaxus harringtonia drupacea
Drooping juniper
    Juniperus recurva
Drooping she-oak
    Casuarina stricta
Duck's foot tree   *Ginkgo biloba*
Dudgeon   *Buxus sempervirens*
Duke cherry   *Prunus × effusus*
    Prunus × gondouinii
Duke of Argyll's tea tree
    Lycium barbarum
    Lycium chinense
Dunkeld larch   *Larix × eurolepis*
Durian   *Durio zibethinus*
Durmast oak   *Quercus petraea*
Dutch elm   *Ulmus × hollandica*
    Ulmus major
Dutch lavender
    Lavandula angustifolia vera
    Lavandula vera
Dwarf Alberta spruce
    Picea glauca albertiana nana
Dwarf American cherry
    Prunus pumila
Dwarf banana   *Musa nana*
Dwarf bilberry
    Vaccinium caespitosum
Dwarf birch   *Betula glandulosa*
    Betula nana
Dwarf broom   *Cytisus demissus*
Dwarf buckeye   *Aesculus parviflora*
Dwarf buckthorn   *Rhamnus pumila*
Dwarf cherry   *Prunus fruticosa*
    Prunus pumila
Dwarf chestnut oak
    Quercus prinoides
Dwarf cornel   *Cornus canadensis*
    Cornus suecica
Dwarf elder   *Sambucus ebulus*
Dwarf elm   *Ulmus pumila*
Dwarf fan palm
    Chamaerops humilis
Dwarf furze   *Ulex minor*
Dwarf gorse   *Ulex minor*
Dwarf hawthorn
    Crataegus monogyna compacta
Dwarf holly   *Malpighia coccigera*
Dwarf horse chestnut
    Aesculus parviflora
Dwarf huckleberry
    Gaylussacia dumosa
    Gaylussacia frondosa
Dwarf laurel   *Kalmia angustifolia*

Dwarf mountain pine   *Pinus mugo*
Dwarf nealie   *Acacia bynoeana*
Dwarf palmetto   *Sabal adansonii*
    Sabal minor
Dwarf pea tree   *Caragana pygmaea*
Dwarf poinciana
    Caesalpinia pulcherrima
Dwarf pomegranate
    Punica granatum nana
Dwarf quince
    Chaenomeles japonica
Dwarf russian almond
    Prunus tenella
Dwarf she-oak   *Casuarina nana*
Dwarf Siberian pine   *Pinus pumila*
Dwarf stone pine   *Pinus pumila*
Dwarf sumac   *Rhus copallina*
Dwarf willow   *Salix herbacea*
Dyer's broom   *Genista tinctoria*
Dyer's greenweed
    Genista tinctoria
Dysentery bark   *Simarouba amara*
Eagle's claw maple
    Acer platanoides laciniatum
Eared willow   *Salix aurita*
Ear-leaved umbrella tree
    Magnolia fraseri
East African juniper
    Juniperus procera
Eastern cottonwood
    Populus deltoides
Eastern hemlock   *Tsuga canadensis*
Eastern hop-hornbeam
    Ostrya virginiana
Eastern hornbeam
    Carpinus orientalis
Eastern larch   *Larix laricina*
Eastern red cedar
    Juniperus virginiana
Eastern strawberry tree
    Arbutus andrachne
Eastern sycamore
    Platanus occidentalis
Eastern white pine   *Pinus strobus*
East Himalayan fir   *Abies spectabilis*
    Abies webbiana
East Himalayan spruce
    Picea spinulosa
East Indian ebony
    Diospyros ebenaster
    Diospyros ebenum
East Indian fig tree
    Ficus benghalensis
    Ficus indica
East Indian satinwood
    Chloroxylon swietenia
East Indian walnut   *Albizzia lebbeck*
East Indian wine palm
    Phoenix rupicola
East Siberian fir   *Abies nephrolepis*

Evergreen currant

Eastwood manzanita *Arctostaphylos glandulosa*

Ebony *Diospyros ebenaster*
Diospyros ebenum

Ebonywood *Bauhinia variegata*

Ecuador laurel *Cordia alliodora*

Edging box *Buxus suffruticosa*

Edible banana *Musa acuminata*
Musa × paradisiaca

Eggfruit *Pouteria campechiana*

Egg-yolk willow *Salix alba vitellina*

Eglantine rose *Rosa eglanteria*
Rosa rubiginosa

Egyptian doom palm *Hyphaene thebaica*

Egyptian privet *Lawsonia inermis*

Egyptian sycamore *Ficus sycomorus*

Egyptian thorn *Acacia senegal*

Elephant apple *Feronia elephantum*
Feronia limonia

Elephant apple tree *Dillenia indica*

Elephant bush *Portulacaria afra*

Elephant's ear *Enterolobium cyclocarpum*

Elephant hedge bean tree *Schotia latifolia*

Elephant's ear wattle *Acacia dunnii*

Elephant's foot *Beaucarnea recurvata*

Elephant tree *Bursera microphylla*

Elliott's blueberry *Vaccinium elliottii*

Elliott's pine *Pinus elliottii*

Elm-leaved sumac *Rhus coriaria*

Emblic *Phyllanthus emblica*

Emmerson's thorn *Crataegus submollis*

Emory oak *Quercus emoryi*

Empress candle plant *Cassia alata*

Empress tree *Paulownia tomentosa*

Emu bush *Eremophila maculata*

Encina *Quercus agrifolia*

Engelmann's oak *Quercus engelmannii*

Engelmann's spruce *Picea engelmannii*

Engler's beech *Fagus englerana*

English elm *Ulmus procera*

English hawthorn *Crataegus laevigata*

English holly *Ilex aquifolium*

English laurel *Prunus laurocerasus*

English oak *Quercus rober*

English walnut *Juglans regia*

English yew *Taxus baccata*

Epaulette tree *Pterostyrax hispida*

Erect Japanese cherry *Prunus amanogawa*

Erman's birch *Betula ermanii*

Escabon *Cytisus proliferus*

Escallonia *Escallonia macrantha*

Ethiopian acacia *Acacia abyssinica*

Ethiopian date palm *Phoenix abyssinica*

Eucryphia *Eucryphia glutinosa*
Eucryphia pinnatifolia

Euodia *Euodia hupehensis*

European alder *Alnus incana*

European ash *Fraxinus excelsior*

European aspen *Populus tremula*

European bird cherry *Prunus padus*

European bladdernut *Staphylea pinnata*

European chestnut *Castanea sativa*

European cranberry bush *Viburnum opulus*

European elder *Sambucus nigra*

European fan palm *Chamaerops humilis*

European field elm *Ulmus carpinifolia*

European golden ball *Forsythia europaea*

European green alder *Alnus viridis*

European hop-hornbeam *Ostrya carpinifolia*

European hornbeam *Carpinus betulus*

European larch *Larix decidua*

European mountain ash *Sorbus aucuparia*

European red elder *Sambucus racemosa*

European scrub pine *Pinus mugo pumilo*

European silver fir *Abies alba*

European spindle tree *Euonymus europaea*

European whitebeam *Sorbus aria wilfred fox*

European white birch *Betula pendula*

European white elm *Ulmus laevis*

European wild cranberry *Oxycoccus oxycoccus*
Oxycoccus palustris
Vaccinium oxycoccus

Everblooming French heather *Erica doliiformis*

Everglades palm *Acoelorrhaphe wrightii*

Evergreen ash *Fraxinus uhdei*

Evergreen blueberry *Vaccinium myrsinites*

Evergreen cherry *Prunus ilicifolia*

Evergreen currant *Ribes viburnumifolium*

Evergreen laburnum
    Piptanthus laburnifolius
    Piptanthus nepalensis
Evergreen magnolia
    Magnolia grandiflora
Evergreen oak   *Quercus ilex*
Evergreen pear   *Pyrus kawakamii*
Evergreen rose   *Rosa sempervirens*
Evergreen spindle tree
    Euonymus japonica
Evergreen sumac   *Rhus virens*
Ewart's mallee
    Eucalyptus ewartiana
Exeter oak   *Quercus* × *hispanica*
    lucombeana
Fairy rose   *Rosa chinensis minima*
False acacia   *Robinia pseudoacacia*
False buckeye   *Ungnadia speciosa*
False buckthorn
    Bumelia lanuginosa
False brazilwood
    Caesalpinia peltophoroides
False cactus   *Euphorbia lactea*
False castor oil plant
    Fatsia japonica
False dogwood *Sapindus saponaria*
False heath   *Fabiana imbricata*
False holly
    Osmanthus heterophyllus
False indigo   *Amorpha fruticosa*
False ipecac   *Psychotria emetica*
False lombardy poplar
    Populus robusta
False medlar
    Sorbus chamaemespilus
False olive   *Cassine orientalis*
False winter's bark
    Cinnamodendron corticosum
Fancy annie   *Delonix regia*
Farges catalpa   *Catalpa fargesii*
Farges fir   *Abies fargesii*
    Abies sutchuenensis
Farkleberry *Vaccinium arboreum*
Fastigiate beech
    Fagus sylvatica fastigiata
Fearn tree   *Jacaranda acutifolia*
Feather-cone fir   *Abies procera*
Feather-duster palm
    Rhopalostylis sapida
Feathery cassia
    Cassia artemisioides
February daphne
    Daphne mezereum
Fehi banana   *Musa fehi*
    Musa troglodytarum
Feijoa   *Feijoa sellowiana*
Felt-leaf ceanothus
    Ceanothus arboreus
Female lombardy poplar
    Populus nigra italica foemina
Fern-leaf aralia *Polyscias filicifolia*

Fernleaf beech
    Fagus sylvatica asplenifolia
Fern-leaved beech
    Fagus sylvatica heterophylla
Fern-leaved bramble
    Rubus laciniatus
Fern-leaved elder
    Sambucus nigra laciniata
Fern-leaved oak
    Quercus robur asplenifolia
Fern palm   *Cycas circinalis*
Fern podocarpus
    Podocarpus elongatus
Fern rhapis   *Rhapis excelsa*
Fetterbush *Leucothoe fontanesiana*
    Lyonia lucida
Fever bush   *Garrya fremontii*
    Lindera benzoin
Fever tree   *Acacia xanthophloea*
    Pinckneya pubens
Fiddle-leaf fig   *Ficus lyrata*
    Ficus pandurata
Fiddler's spurge
    Euphorbia cyathophora
Fiddlewood
    Citharexylum fruticosum
    Citharexylum spinosum
Field briar   *Rosa agrestis*
Field maple   *Acer campestre*
Field rose   *Rosa arvensis*
Fig-leaf palm   *Aralia japonica*
    Aralia sieboldii
    Fatsia japonica
Fijian kauri pine *Agathis vitiensis*
Fiji fan palm *Pritchardia pacifica*
Filbert   *Corylus avellana*
    Corylus maxima
Finger tree *Euphorbia tirucalli*
Finnish whitebeam *Sorbus fennica*
    Sorbus × *hybrida*
Fire birch   *Betula populifolia*
Firebush   *Hamelia patens*
Firecracker   *Russelia juncea*
Fire cherry *Prunus pennsylvanica*
Fire-dragon *Acalypha wilkesiana*
Fire-on-the-mountain
    Euphorbia cyathophora
    Euphorbia heterophylla
Firethorn *Pyracantha coccinea*
Fire tree   *Nuytsia floribunda*
Firewheel   *Grevillea wilsonii*
Firewheel tree
    Stenocarpus sinuatus
Fish-poison tree *Piscidia erythrina*
Fishtail camellia
    Camellia × *williamsii c.f.coates*
Five fingers tree
    Neopanax arboreus
Five-seeded hawthorn
    Crataegus pentagyna
Flaky fir   *Abies squamata*

Fragrant olive

Flamboyant tree *Delonix regia*

Flame bottle tree
Brachychiton acerifolius

Flame creeper
Combretum microphyllum

Flamegold *Koelreuteria elegans*

Flame of the forest *Delonix regia*
Spathodea campanulata

Flame-of-the-woods *Ixora coccinea*
Ixora incarnata

Flame tree
Brachychiton acerifolius
Brachychiton australis
Delonix regia
Sterculia acerfolia

Flat-topped yate
Eucalyptus occidentalis

Flooded box
Eucalyptus microtheca

Flooded gum *Eucalyptus grandis*

Florida corkwood
Leitneria floridana

Florida mahogany *Persea borbonia*

Florida maple *Acer barbatum*

Florida royal palm *Roystonea elata*

Florida silver palm
Coccothrinax argentata

Florida strangler fig *Ficus aurea*

Florida thatch palm
Thrinax parviflora

Florida yew *Taxus floridana*

Florist's genista *Genista canariensis*

Florist's willow *Salix caprea*

Floss silk tree *Chorisia speciosa*

Flower fence
Caesalpinia pulcherrima

Flowering almond
Prunus glandulosa
Prunus jacquemontii
Prunus japonica
Prunus triloba

Flowering ash *Fraxinus dipetala*
Fraxinus ornus

Flowering banana *Musa ornata*

Flowering currant
Ribes sanguineum

Flowering dogwood
Cornus capitata florida
Cornus florida

Flowering nutmeg
Leycesteria formosa

Flowering plum *Prunus cerasifera*

Flowering quince
Chaenomeles speciosa

Flowering spurge
Euphorbia corollata

Flowering willow *Chilopsis linearis*

Flower-of-an-hour
Hibiscus trionum

Flower-of-love
Tabernaemontana divaricata

Fly honeysuckle
Lonicera xylosteum

Foetid yew *Torreya taxifolia*

Folgner's whitebeam
Sorbus folgneri

Folhado *Clethra arborea*

Foliage flower *Breynia disticha*

Fontainebleau service tree
Sorbus × latifolia

Forest fever tree
Anthockistazam besiaca

Forest red gum
Eucalyptus tereticornis

Forest wild medlar
Vangueria esculenta

Formosa incense cedar
Calocedrus formosana

Formosan azalea *Azalea oldhamii*
Rhododendron oldhamii

Formosan cedar
Chamaecyparis formosensis

Formosan cherry
Prunus campanulata

Formosan cypress
Chamaecyparis formosensis

Formosan gum
Liquidambar formosana

Formosan hemlock
Tsuga formosana

Formosa rice tree *Aralia japonica*
Fatsia japonica

Forrest's fir *Abies delavayi forrestii*

Forrest's maple *Acer forrestii*

Forrest's marlock
Eucalyptus forrestiana

Forrest's silver fir
Abies delavayi forrestii

Forster's sentry palm
Howea forsterana
Kentia forsterana

Forsythia *Forsythia × intermedia*

Fountain buddleia
Buddleia alternifolia

Fountain bush
Russelia equisetiformis
Russelia juncea

Fountain dracaena
Cordyline australis

Fountain tree
Spathodea campanulata

Four-winged mallee
Eucalyptus tetraptera

Foxberry *Vaccinium vitis-idaea*

Foxglove tree
Paulownia tomentosa

Foxtail *Acalypha hispida*

Foxtail pine *Pinus balfouriana*

Fragrant champaca
Michelia champaca

Fragrant myall
Acacia homalophylla

Fragrant olive *Osmanthus fragrans*

Fragrant snowbell

Fragrant snowbell	*Styrax obassia*
Fragrant sumac	*Rhus aromatica*
Franceschi palm	*Brahea elegans*
Frangipani	*Plumeria acuminata*
Frangipani tree	*Plumeria rubra*
Frankincense	*Boswellia thurifera*
Frankincense pine	*Pinus taeda*
Franklin tree	*Franklinia alatamaha*
Fraser's balsam fir	*Abies fraseri*
Frazer river douglas fir	
	Pseudotsuga menziesii caesia
Fremont's box thorn	
	Lycium pallidum
Fremont poplar	*Populus fremontii*
French hales	*Sorbus devoniensis*
French heather	*Erica hyemalis*
French lavender	
	Lavandula stoechas
French mulberry	
	Callicarpa americana
French physic nut	*Jatropha curcus*
French rose	*Rosa gallica*
French tamarisk	*Tamarix gallica*
French willow	*Salix triandra*
Fringed heath	*Erica ciliaris*
Fringed hibiscus	
	Hibiscus schizopetalus
Fringed lavender	
	Lavandula dentata
Fringe tree	*Chionanthus virginicus*
Frosted thorn	
	Crataegus × prunifolia
Frosty wattle	*Acacia pruinosa*
Fuchsia-flowered gooseberry	
	Ribes speciosum
Fuchsia gum	
	Eucalyptus forrestiana
Fuji cherry	*Prunus incisa*
Full-moon maple	*Acer japonicum*
Furin holly	*Ilex geniculata*
Furry willow	*Salix adenophylla*
	Salix cordata
Furze	*Ulex europaeus*
Fustic	*Chlorophora tinctoria*
Gaboon	*Aucomea klainiana*
Galapee tree	
	Sciadophyllum brownii
Gale	*Myrica gale*
Gallberry	*Ilex glabra*
Gambel's oak	*Quercus gambelii*
Gander's oak	*Quercus × ganderi*
Garland crab apple	
	Malus coronaria
Garland flower bush	
	Daphne cneorum
Garlic pear	*Crataera gynandra*
Gbanja cola	*Cola nitida*
Gean	*Prunus avium*
	Prunus avium sylvestris
Gebang palm	*Corypha elata*
Geiger tree	*Cordia sebestena*
Genipap	*Genipa americana*
Genipe	*Melicoccus bijugatus*
Genista	*Cytisus canariensis*
Genoa broom	*Genista januensis*
Georgia bark tree	
	Pinckneya pubens
Georgia pine	*Pinus palustris*
Geraldton wax	
	Chamaelaucium uncinatum
Geranium-leaf aralia	
	Polyscias guilfoylei
Geranium tree	*Cordia sebestena*
Gerard's pine	*Pinus gerardiana*
German greenweed	
	Genista germanica
Ghost gum	*Eucalyptus papuana*
Ghost tree	*Davidia involucrata*
Ghostweed	*Euphorbia marginata*
Giant cedar	*Thuya plicata*
Giant chinquapin	
	Castanopsis chrysophylla
Giant dogwood	
	Cornus controversa
Giant dracaena	*Cordyline australis*
Giant filbert	*Corylus maxima*
Giant fir	*Abies grandis*
Giant gum	*Eucalyptus regnans*
Giant pine	*Pinus lambertiana*
Giant protea	*Protea cynaroides*
Giant sequoia	
	Sequoiadendron giganteum
Giant woolly protea	
	Protea barbigera
Gibb's firethorn	
	Pyracantha atalantoides
Gidgee myall	*Acacia homalophylla*
Giles' netbush	*Calothamnus gilesii*
Gimlet gum	*Eucalyptus salubris*
Gingerbread palm	
	Hyphaene thebaica
	Palmae hyphaene
Gingerbread plum	
	Parinari macrophylla
Gingerbread tree	
	Parinari macrophylla
Gippsland fountain palm	
	Livistona australis
Gippsland palm	*Corypha australis*
	Livistona australis
Gippsland waratah	*Telopea oreades*
Glastonbury thorn	
	Crataegus monogyna biflora
Crataegus monogyna praecox	
	Crataegus praecox
Glory bower	
	Clerodendrum fargesii
Glory pea	*Clianthus formosus*

Glory tree *Clerodendrum fargesii*
Clerodendrum thomsoniae
Clerodendrum trichotomum

Glory wattle *Acacia spectabilis*

Glossy-leaf fig *Ficus retusa*

Glossy privet *Ligustrum lucidum*

Goat willow *Salix caprea*

Goddess magnolia
Magnolia sprengeri diva

Godsberry *Diospyros lotus*

Gold coast bombax
Bombax buonopozense

Gold dust *Acacia acinacea*

Gold-dust dracaena
Dracaena surculosa

Golden acacia
Robinia pseudoacacia frisia

Golden American elder
Sambucus canadensis aurea

Golden apple *Aegle marmelos*
Spondias lutea
Spondias cytherea

Golden ash
Fraxinus excelsior jaspidea

Golden ashleaf maple
Acer negundo auratum

Golden-bark ash
Fraxinus excelsior jaspidea

Golden bay *Laurus nobilis aureus*

Golden beech
Fagus sylvatica zlatia

Golden bell *Forsythia suspensa*

Golden bush *Cassinia fulvida*

Golden-blotched hedgehog holly
Ilex aquifolium ferox aurea

Golden cassia *Cassia fasciculata*

Golden chain
Laburnum anagyroides
Laburnum vulgare

Golden chain tree
Laburnum × watereri

Golden chestnut
Chrysolepis chrysophylla

Golden Chinese juniper
Juniperus chinensis aurea

Golden-cup oak
Quercus chrysolepis

Golden curls tree
Salix babylonica pekinensis
tortuosa aurea

Golden curls willow
Salix xerythroflexuosa

Golden currant *Ribes aureum*
Ribes odoratum

Golden deodar cedar
Cedrus deodara aurea

Golden dewdrop *Duranta ellisia*
Duranta repens

Golden dogwood
Cornus alba aurea

Golden elder
Sambucus nigra aurea

Golden-feather palm
Chrysalidocarpus lutescens

Golden fig *Ficus aurea*

Golden-flowered daphne
Daphne aurantiaca

Golden hazel
Corylus avellana aurea

Golden heather *Cassinia fulvida*

Golden hinoki cypress
Chamaecyparis obtusa crippsii

Golden holly
Ilex aquifolium aurea

Golden Irish yew
Taxus baccata aureovariegata
Taxus baccata fastigiata aurea

Golden larch *Pseudolarix amabilis*
Pseudolarix kaempferi

Golden lawson's cypress
Chamaecyparis lawsoniana
stewartii

Golden-leaved barberry
Berberis thunbergii aurea

Golden-leaved catalpa
Catalpa bignonioides aurea

Golden-leaved Japanese maple
Acer shirasawanum aureum

Golden-leaved laburnum
Laburnum anagyroides aureum

Golden locust
Robinia pseudoacacia frisia

Golden mimosa *Acacia baileyana*

Golden moon maple
Acer japonicum aureum

Golden oak *Quercus alnifolia*
Quercus robur concordia

Golden poplar
Populus serotina aurea

Golden privet
Ligustrum ovalifolium aureum

Golden rain *Cassia fistula*
Laburnum anagyroides
Laburnum vulgare

Golden rain tree
Koelreuteria paniculata

Golden rain wattle
Acacia prominens

Golden shower *Cassia fistula*

Golden sycamore
Acer pseudoilatanus worlei

Golden trumpet
Allamander cathartica

Golden trumpet tree
Tabebuia chrysantha

Golden variegated dogwood
Cornus alba spaethii

Golden wattle *Acacia pycnantha*

Golden weeping ash
Fraxinus excelsior aurea pendula

Golden weeping holly
Ilex aquifolium aurea pendula

Golden weeping willow
Salix sepulcralis chrysocoma

Golden willow   *Acacia cyanophylla*
     Salix alba vitellina

Golden wonder
     Cassia didymobotrya
     Cassia splendida

Golden wreath   *Acacia saligna*

Golden yew   *Taxus baccata aurea*

Goldilocks   *Helichrysum stoechas*

Goldspire   *Azara integrifolia*

Gomuti palm   *Arenga pinnata*

Good-luck palm
     Chamaedorea elegans

Good-luck plant
     Cordyline fruticosa
     Cordyline terminalis
     Dracaena terminalis

Goodyer elm   *Ulmus angustifolia*

Goora nut   *Cola acuminata*

Gooseberry   *Ribes uva-crispa*

Goosegog   *Ribes uva-crispa*

Gooseberry tree
     Phyllanthus acidus

Goose plum   *Prunus americana*

Gorse   *Ulex europaeus*

Governor's plum   *Flacourtia indica*
     Flacourtia ramontchi

Gowen's cypress
     Cupressus goveniana

Graceful wattle   *Acacia decora*

Grampian stringybark
     Eucalyptus alpina

Grand fir   *Abies grandis*

Granite bottlebrush
     Melaleuca armillaris

Granjeno   *Celtis iguanaea*

Grapefruit   *Citrus × paradisi*

Grass palm   *Cordyline australis*

Grass tree   *Xanthorrhoea australis*

Greater malayan chestnut
     Chrysolepis megacarpa

Great laurel
     Rhododendron maximum
     Rhododendron ponticum

Great Malayan chestnut
     Castanopsis megacarpa

Great plains cottonwood
     Populus sargentii

Great rose mallow
     Hibiscus grandiflorus

Great sallow   *Salix caprea*

Great white cherry
     Prunus tai-haku

Grecian fir   *Abies cephalonica*

Grecian juniper   *Juniperus excelsa*

Grecian strawberry tree
     Arbutus andrachne

Greek fir   *Abies cephalonica*

Greek myrtle   *Myrtus communis*

Greek whitebeam   *Sorbus graeca*

Green alder   *Alnus crispa mollis*
     Alnus viridis

Green almond   *Pistacia vera*

Green ash   *Fraxinus pennsylvanica*

Green-bark ceanothus
     Ceanothus spinosus

Green briar   *Smilax rotundifolia*

Green ebony   *Tabebuia flavescens*

Greengage
     Prunus domestica italica

Greenglow plum   *Prunus cerasifera*

Greenheart   *Nectandra rodiaei*
     Ocotea bullata
     Ocotea radiaei

Greenleaf manzanita
     Arctostaphylos patula

Green manzanita
     Arctostaphylos patula

Green mountain sallow
     Salix andersoniana

Green oak   *Quercus pubescens*

Green osier   *Cornus alternifolia*

Green rose
     Rosa chinensis viridiflora

Green wattle   *Acacia decurrens*
     Acacia farnesiana

Greenweed   *Genista tinctoria*

Gregg's pine   *Pinus greggii*

Grey alder   *Alnus incana*

Grey birch   *Betula alleghaniensis*
     Betula populifolia

Grey box   *Eucalyptus moluccana*

Grey-budded snakebark maple
     Acer rufinerve

Grey corkwood
     Erythrina vespertilio

Grey dogwood   *Cornus racemosa*

Grey douglas fir
     Pseudotsuga menziesii caesia

Grey goddess   *Brahea armata*

Grey gum   *Eucalyptus punctata*
     Eucalyptus tereticornis

Grey heath   *Erica cinerea*

Grey ironbark
     Eucalyptus paniculata

Greyleaf cherry   *Prunus canescens*

Grey mulga   *Acacia brachybotrya*

Grey peppermint
     Eucalyptus radiata

Grey pine   *Pinus banksiana*

Grey poplar   *Populus canescens*

Grey sage brush   *Atriplex canescens*

Grey sallow   *Salix cinerea*
     Salix cinerea atrocinerea

Grey willow   *Salix cinerea*
     Salix humilis

Ground cherry   *Prunus fruticosa*

Ground rattan cane   *Rhapis excelsa*

Ground senna   *Cassia chamaecrista*

Grouseberry   *Viburnum trilobum*

Grumichama *Eugenia brasiliensis*
            Eugenia dombeyi
Grumixameira *Eugenia brasiliensis*
              Eugenia dombeyi
Guadalupe cypress
            Cupressus guadalupensis
Guadalupe palm *Brahea edulis*
Guajilote *Parmentiera edulis*
Guanabana *Annona muricata*
Guatemalan rhubarb
            Jatropha integerrima
Guava *Psidium guajava*
       Psidium guineense
Guayabo hormiguero
            Triplaris surinamensis
Guelder rose *Viburnum opulus*
Guernsey elm
    Ulmus carpinifolia sarniensis
    Ulmus minor stricta sarniensis
Guiana chestnut *Pachira aquatica*
Guindo *Nothofagus antarctica*
Guinea pepper *Xylopia aethiopica*
Guinea plum *Parinari excelsa*
Gully ash *Eucalyptus smithii*
Gully gum *Eucalyptus smithii*
Gulmohur *Delonix regia*
Gum acacia *Acacia nilotica*
            Acacia senegal
Gum ammoniac
            Dorema ammoniacum
Gum-arabic tree *Acacia nilotica*
                 Acacia senegal
                 Acacia seyal
Gumbo-limbo *Bursera simaruba*
Gum cistus *Cistus ladanifer*
Gum elastic *Bumelia lanuginosa*
Gum elemi *Bursera simaruba*
Gumi *Elaeagnus edulis*
      Elaeagnus multiflora
Gum-lac *Schleichera oleosa*
Gum-top stringybark
            Eucalyptus delegatensis
Gum tragacanth
            Astragalus gummiferi
Gungurru *Eucalyptus caesia*
Gutta percha *Palaquim gutta*
Gutta-percha tree
            Eucommia ulmoides
Habbel *Juniperus drupacea*
Hackberry *Celtis occidentalis*
Hackmatack *Larix laricina*
            Populus balsamifera
Hagberry *Prunus padus*
Hag briar *Smilax hispida*
Hairy alpen rose
            Rhododendron hirsutum
Hairy birch *Betula pubescens*
Hairy greenweed *Genista pilosa*
Hairy greenwood *Genista pilosa*

Hairy huckleberry
            Vaccinium hirsutum
Hairy manzanita
            Arctostaphylos columbiana
Hairy wattle *Acacia pubescens*
Hai-tung crab apple
            Malus spectabilis
Halberd-leaved willow *Salix hastata*
Hall's crab apple *Malus halliana*
Handflower tree
            Chiranthodendron
            pentadactylon
Handkerchief tree
            Davidia involucrata
Hansen's cherry *Prunus tomentosa*
Hard beech *Nothofagus truncata*
Hard maple *Acer saccharum*
Hardy fuchsia *Fuchsia conica*
               Fuchsia magellanica
Hardy orange *Citrus trifoliata*
              Poncirus trifoliata
Harrington plum yew
            Cephalotaxus harringtonia
Harry Lauder's walking stick
            Corylus avellana contorta
Hat-rack cactus *Euphorbia lactea*
Hat tree *Brachychiton discolor*
Havard oak *Quercus havardii*
Hawaiian good-luck plant
            Cordyline terminalis
            Dracaena terminalis
Hawaiian hibiscus *Hibiscus sinensis*
Hawthorn *Crataegus monogyna*
Hawthorn-leaf crab apple
            Malus florentina
Hawthorn-leaved maple
            Acer crataegifolium
Hazel alder *Alnus rugosa*
Heart-leaf manzanita
            Arctostaphylos andersonii
Heart-leaved silver gum
            Eucalyptus cordata
Heart-leaved willow *Salix cordata*
Heartnut *Juglans ailantifolia*
Heather *Calluna erica*
Heavenly bamboo
            Nandina domestica
Hedge euphorbia
            Euphorbia neriifolia
Hedgehog broom
            Erinacea anthyllis
            Erinacea pungens
Hedgehog fir *Abies pinsapo*
Hedgehog holly
            Ilex aquifolium ferox
Hedgehog rose *Rosa rugosa*
Hedge maple *Acer campestre*
Hedge thorn *Carissa bispinosa*
Hedge wattle *Acacia armata*
Heldreich's maple *Acer heldreichii*
Hemlock *Conium maculatum*

Hemp palm	*Trachycarpus fortunei*
Hemp tree	*Vitex agnus-castus*
Hemsley's storax	*Styrax hemsleyana*
Henna	*Lawsonia alba*
	Lawsonia inermis
Hercules' club	*Aralia spinosa*
	Zanthoxylum clava-herculis
Hers' maple	*Acer hersii*
Hiba	*Thujopsis dolabrata*
	Thuyopsis dolabrata
Hiba arbor-vitae	*Thujopsis dolabrata*
	Thuyopsis dolabrata
Hibiscus	*Althaea frutex*
	Hibiscus syriacus
Hickory pine	*Pinus aristata*
	Pinus pungens
Hicks' yew	*Taxus media hicksii*
Higan cherry	*Prunus subhirtella*
	Prunus subhirtella autumnalis
High-bush blueberry	*Vaccinium corymbosum*
Highbush cranberry	*Viburnum trilobum*
Highclere holly	*Ilex × altaclarensis golden king*
High-ground willow oak	*Quercus incana*
Highland pine	*Pinus sylvestris rubra*
Hill cherry	*Prunus serrulata*
	Prunus serrulata spontanea
Hill gooseberry	*Rhodomyrtus tomentosa*
Hill guava	*Rhodomyrtus tomentosa*
Himalayan alder	*Alnus nitida*
Himalayan birch	*Betula jacquemontii*
	Betula utilis
Himalayan bird cherry	*Prunus cornuta*
Himalayan cedar	*Cedrus deodara*
Himalayan cherry	*Prunus rufa*
Himalayan cotoneaster	*Cotoneaster simonsii*
Himalayan cypress	*Cupressus torulosa*
Himalayan fir	*Abies spectabilis*
	Abies webbiana
Himalayan hemlock	*Tsuga dumosa*
Himalayan holly	*Ilex dipyrena*
Himalayan honeysuckle	*Leycesteria formosa*
Himalayan jasmine	*Jasminum humile*
Himalayan juniper	*Juniperus recurva*
Himalayan larch	*Larix griffithii*
Himalayan lilac	*Syringa emodi*

Himalayan musk rose	*Rosa brunonii*
Himalayan pine	*Pinus wallichiana*
Himalayan spruce	*Picea smithiana*
Himalayan tree-cotoneaster	*Cotoneaster frigidus*
Himalayan whitebeam	*Sorbus cuspidata*
Himalayan white pine	*Pinus wallichiana*
Himalayan yew	*Taxus wallichiana*
Hinoki cypress	*Chamaecyparis obtusa*
Hoarwithy	*Viburnum lantana*
Hoaryleaf ceanothus	*Ceanothus crassifolius*
Hoary manzanita	*Arctostaphylos canescens*
Hoary willow	*Salix elaeagnus*
Hobblebush	*Viburnum alnifolium*
	Viburnum lantanoides
Hobble marsh	*Viburnum alnifolium*
Hog cranberry	*Arctostaphylos uva-ursi*
Hogg's double yellow rose	*Rosa lutea hoggii*
Hognut	*Carya glabra*
Hog plum	*Prunus americana*
	Prunus reverchonii
	Spondias mombin
	Ximenia americana
Holford pine	*Pinus × holfordiana*
Holly barberry	*Mahonia aquifolium*
Holly-leaf ceanothus	*Ceanothus purpureus*
Hollyleaf sweetspire	*Itea illicifolia*
Holly-leaved barberry	*Berberis ilicifolia*
Holly-leaved cherry	*Prunus ilicifolia*
Holly-leaved grevillea	*Grevillea aquifolium*
Holly-leaved olive	*Osmanthus heterophyllus*
Holly mahonia	*Mahonia aquifolium*
Holly oak	*Quercus ilex*
Hollywood juniper	*Juniperus chinensis kaizuka*
Holm oak	*Quercus ilex*
Holy flax	*Santolina rosmarinifolia*
Hondapara	*Dillenia indica*
Hondo spruce	*Picea jezoensis hondoensis*
Honduras mahogany	*Swietenia macrophylla*
Honeyberry	*Celtis australis*
	Melicoccus bijugatus
Honey bush	*Melianthus major*

Honey flower — *Protea mellifera*
Honey locust — *Gleditsia triacanthos*
— *Gleditschia triacanthos*
Honey myrtle — *Melaleuca huegelii*
Honey palm — *Jubaea chilensis*
— *Jubaea spectabilis*
Honey protea — *Protea mellifera*
Honeyshuck — *Gleditsia triacanthos*
Hoop pine
— *Araucaria cunninghamii*
Hop hornbeam — *Ostrya carpinifolia*
Hop tree — *Ptelea trifoliata*
Horizontal box — *Buxus prostrata*
Hornbeam — *Carpinus betulus*
Hornbeam-leaved maple
— *Acer carpinifolium*
Hornbeam maple
— *Acer carpinifolium*
Horned holly — *Ilex cornuta*
Horned mallee
— *Eucalyptus eremophila*
Horned maple — *Acer diabolicum*
Horse briar — *Smilax rotundifolia*
Horse cassia — *Cassia grandis*
Horse chestnut
— *Aesculus hippocastanum*
Horseradish tree — *Moringa oleifera*
— *Moringa pterygosperma*
Horse sugar — *Symplocos tinctoria*
Horsetail she-oak
— *Casuarina equisetifolia*
Horsetail tree
— *Casuarina equisetifolia*
Hottentot's bean — *Schotia afra*
House lime — *Sparmannia africana*
House pine
— *Araucaria heterophylla*
Hualo — *Nothofagus glauca*
Huamuchii — *Pithecellobium dulce*
Huanuco — *Cinchona micrantha*
Huckleberry — *Cyrilla racemiflora*
— *Vaccinium myrtillus*
Huckleberry oak
— *Quercus vaccinifolia*
Huisache — *Acacia farnesiana*
Hulver bush — *Ilex aquifolium*
Hungarian hawthorn
— *Crataegus nigra*
Hungarian lilac — *Syringa josikaea*
Hungarian oak — *Quercus conferta*
— *Quercus frainetto*
Hungarian thorn — *Crataegus nigra*
Huntingdon elm
— *Ulmus × hollandica vegeta*
— *Ulmus × vegeta*
Huon pine — *Dacrydium franklinii*
— *Lagarostrobus franklinii*
Hupeh cherry
— *Prunus serrulata hupehensis*

Hupeh crab apple
— *Malus hupehensis*
Hupeh rowan — *Sorbus hupehensis*
Hurricane palm
— *Ptychosperma macarthurii*
Hybrid black poplar
— *Populus × canadensis*
Hybrid buckeye — *Aesculus × hybrida*
Hybrid catalpa
— *Catalpa × erubescens*
— *Catalpa × hybrida*
Hybrid cockspur thorn
— *Crataegus × lavallei*
Hybrid larch — *Larix × eurolepis*
Hybrid magnolia
— *Magnolia ×soulangeana*
Hybrid rowan — *Sorbus × thuringiaca*
Hybrid strawberry tree
— *Arbutus × andrachnoides*
Hybrid wing nut
— *Pterocarya × rehderana*
Icaco — *Chrysobalanus icaco*
Ichang lemon — *Citrus ichangensis*
Idesia — *Idesia polycarpa*
Ignatius bean — *Strychnos ignatii*
Iigiri tree — *Idesia polycarpa*
Ilama — *Annona diversifolia*
Ilang-ilang — *Cananga odorata*
Illawarra flame tree
— *Brachychiton acerifolius*
Illyarie — *Eucalyptus erythrocorys*
Illyarri — *Eucalyptus erythrocorys*
Imou pine — *Dacrydium cupressinum*
Inaja palm — *Maximiliana caribaea*
— *Maximiliana maripa*
— *Maximiliana regia*
Incense cedar
— *Calocedrus decurrens*
— *Libocedrus decurrens*
Incense juniper — *Juniperus thurifera*
Incense rose — *Rosa primula*
India date palm — *Phoenix rupicola*
— *Phoenix sylvestris*
Indian almond — *Sterculia foetida*
— *Terminalia catappa*
Indian azalea — *Rhododendron simsii*
Indian bael — *Aegle marmelos*
Indian banyan — *Ficus benghalensis*
— *Ficus indica*
Indian bean — *Catalpa speciosa*
Indian bean tree
— *Catalpa bignonioides*
Indian cedar — *Cedrus deodara*
Indian cher pine — *Pinus roxburghii*
Indian cherry — *Rhamnus caroliniana*
Indian currant
— *Symphoricarpus orbiculatus*
— *Symphoricarpus rubra vulgaris*
— *Symphoricarpus vulgaris*
Indian elm — *Ulmus rubra*

Indian gooseberry
Phyllanthus acidus
Indian gum *Acacia arabica*
Indian hawthorn *Raphiolepis indica*
Raphiolepis umbellata
Indian hemp *Hibiscus cannabinus*
Indian horse chestnut
Aesculus indica
Indian jalap *Ipomoea turpethum*
Indian jujube *Ziziphus mauritania*
Indian laburnum *Cassia fistula*
Indian laurel
Calophyllum inophyllum
Ficus retusa
Persea indica
Terminalia alata
Indian lilac *Melia azedarach*
Indian mulberry *Morinda citrifolia*
Indian neem tree *Melia indica*
Indian night jasmine
Nyctanthes arbor-tristis
Indian oak
Barringtonia acutangula
Indian olive *Olea ferruninea*
Indian plum *Oemleria cerasiformis*
Indian rhododendron
Melastoma malabathricum
Indian sarsaparilla
Hemidesmus indica
Indian senna *Cassia angustifolia*
Indian silk-cotton tree
Bombax malabaricum
Indian sorrell *Hibiscus sabdariffa*
Indian spice tree *Vitex agnus-castus*
Indian spruce *Picea smithiana*
Indian tree spurge
Euphorbia tirucalli
Indian walnut *Aleurites moluccana*
Indian rubber tree *Ficus belgica*
Ficus elastica
Indigbo *Terminalia ivorensis*
Inkberry *Ilex glabra*
Inland lodgepole pine
Pinus contorta latifolia
Interior live oak *Quercus wislizenii*
Ipecac spurge
Euphorbia ipecacuanhae
Ipecacuanha
Cephaelis ipecacuanha
Uragoga ipecacuanha
Irish gorse *Ulex europaeus strictus*
Irish heath *Daboecia cantabrica*
Erica erigena
Erica mediterranea
Irish juniper
Juniperus communis hibernica
Juniperus communis stricta
Irish yew *Taxus baccata fastigiata*
Iroko *Chlorophora excelsa*
Iron tree *Metrosideros robusta*
Parrotia persica

Ironwood *Bumelia lycioides*
Cliftonia monophylla
Cyrilla racemiflora
Mesurea ferrea
Ostrya virginiana
Ironwood tree *Eugenia confusa*
Eugenia garberi
Island manzanita
Arctostaphylos insularis
Island oak *Quercus tomentella*
Islay plum *Prunus ilicifolia*
Isu tree *Distylium racemosum*
Italian alder *Alnus cordata*
Italian buckthorn
Rhamnus alaternus
Italian cypress
Cupressus sempervirens
Italian jasmine *Jasminum humile*
Italian maple *Acer opalus*
Italian oak *Quercus frainetto*
Italian poplar *Populus nigra italica*
Italian stone pine *Pinus pinea*
Ita palm *Mauritia flexuosa*
Mauritia setigera
Ivory nut palm
Phytelephas macrocarpa
Ivory palm
Phytelephas macrocarpa
Ivybush *Kalmia latifolia*
Jaboncillo *Sapindus saponaria*
Jaborand *Pilocarpus jaborandi*
Jaborandi *Pilocarpus jaborandi*
Jaboticaba *Myrciaria cauliflora*
Jacaranda *Jacaranda acutifolia*
Jacaranda mimosifolia
Jacareuba *Calophyllum brasiliense*
Jack fruit *Artocarpus heterophyllus*
Jack oak *Quercus ellipsoidalis*
Quercus marilandica
Jack pine *Pinus banksiana*
Jackwood *Cordia dentata*
Jacobite rose *Rosa × alba*
Jacob's coat *Acalypha wilkesiana*
Jacob's staff *Fouquieria splendens*
Jacquemont's birch
Betula jacquemontii
Jaggery palm *Caryota urens*
Jalap *Ipomoea purga*
Jamaica caper tree
Capparis cynophallophora
Jamaica dogwood
Piscidia erythrina
Jamaica mandarin orange
Glycosmis pentaphylla
Jamaican damson plum
Chrysophyllum oliviforme
Jamaican dogwood
Piscidia piscipula
Jamaican manac
Calyptronoma occidentalis

Jamaican trumpet tree
    Tabebuia riparia

Jamaica nutmeg
    Monodora myristica

Jamaica palmetto   *Sabal jamaicensis*

Jamaica pepper   *Pimenta officinalis*

Jamaica plum   *Spondias lutea*

Jamaica quassia   *Picraena excelsa*

Jamaica sarsaparilla   *Smilax ornata*

Jamaica sorrell   *Hibiscus sabdariffa*

Jambolan   *Eugenia jambolana*

Jambolan plum   *Syzygium cumini*

Jambool   *Syzygium cumini*
    Syzygium samarangense

Jambosa   *Syzygium samarangense*

Jambu   *Syzygium cumini*

Jambul   *Eugenia jambolana*

Japanese alder   *Alnus japonica*

Japanese alpine cherry
    Prunus nipponica

Japanese angelica tree   *Aralia elata*

Japanese anise   *Illicium anisatum*

Japanese apricot   *Prunus mume*

Japanese arbor-vitae
    Thuya standishii

Japanese aspen   *Populus sieboldii*

Japanese barberry
    Berberis thunbergii

Japanese bead tree
    Melia azedarach

Japanese beech   *Fagus crenata*
    Fagus japonica
    Fagus sieboldii

Japanese bigleaf magnolia
    Magnolia hypoleuca

Japanese bitter orange
    Poncirus trifoliata

Japanese black pine
    Pinus thunbergii

Japanese bush cherry
    Prunus japonica

Japanese cedar
    Cryptomeria japonica

Japanese cherry   *Prunus serrulata*

Japanese cherry birch
    Betula grossa

Japanese chestnut
    Castanea crenata
    Chrysolepis cuspidata

Japanese chestnut oak
    Quercus acutissima

Japanese chinquapin
    Castanopsis cuspidata

Japanese cork tree
    Phellodendron japonicum

Japanese cornel   *Cornus officinalis*

Japanese cornelian cherry
    Cornus officinalis

Japanese cowtail pine
    Cephalotaxus harringtonia

Japanese crab apple
    Malus floribunda

Japanese cucumber tree
    Magnolia hypoleuca

Japanese dogwood
    Cornus controversa
    Cornus kousa

Japanese double pink cherry
    Prunus serrulata kanzan

Japanese douglas fir
    Pseudotsuga japonica

Japanese elm
    Ulmus davidiana japonica
    Ulmus japonica

Japanese euonymus
    Euonymus japonica

Japanese euptelea
    Euptelea polyandra

Japanese evergreen oak
    Quercus acuta

Japanese false cypress
    Chamaecyparis obtusa

Japanese fatsia   *Aralia japonica*
    Fatsia japonica

Japanese fern palm   *Cycas revoluta*

Japanese figleaf palm
    Fatsia japonica

Japanese fir   *Abies firma*

Japanese flowering cherry
    Prunus yedoensis

Japanese hackberry   *Celtis japonica*
    Celtis sinensis

Japanese hairy alder   *Alnus hirsuta*

Japanese hazel   *Corylus sieboldiana*

Japanese hemlock   *Tsuga sieboldii*

Japanese hill cherry
    Prunus sargentii

Japanese hibiscus
    Hibiscus schizopetalus

Japanese holly   *Ilex crenata*
    Ilex integra

Japanese honey locust
    Gleditsia japonica

Japanese hop-hornbeam
    Ostrya japonica

Japanese hornbeam
    Carpinus japonica

Japanese horse chestnut
    Aesculus turbinata

Japanese jasmine   *Jasminum mesnyi*

Japanese lacquer tree
    Rhus verniciflua

Japanese lantern
    Hibiscus schizopetalus

Japanese larch   *Larix kaempferi*

Japanese large-leaved birch
    Betula maximowicziana

Japanese laurel   *Aucuba japonica*

Japanese lime   *Tilia japonica*

Japanese linden   *Tilia japonica*

Japanese loquat
    Eriobotrya japonica

Japanese magnolia
Magnolia obovata

Japanese mahonia
Mahonia japonica

Japanese maple *Acer japonicum*
Acer palmatum

Japanese medlar
Eriobotrya japonica

Japanese nutmeg *Torreya nucifera*

Japanese pagoda tree
Sophora japonica

Japanese pear *Pyrus pyrifolia*

Japanese persimmon
Diospyros chinensis
Diospyros kaki

Japanese photinia *Photinia glabra*

Japanese pittosporum
Pittosporum tobira

Japanese plum *Eriobotrya japonica*
Prunus japonica
Prunus salicina
Prunus triflora

Japanese plum-fruited yew
Cephalotaxus harringtonia

Japanese plum yew
Cephalotaxus harringtonia drupacea

Japanese poplar
Populus maximowiczii

Japanese privet
Ligustrum japonicum
Ligustrum ovalifolium

Japanese pussy willow
Salix gracilistyla

Japanese quince
Chaenomeles speciosa

Japanese raisin tree *Hovenia dulcis*

Japanese red birch
Betula maximowicziana

Japanese red cedar
Cryptomeria japonica

Japanese red oak *Quercus acuta*

Japanese red pine *Pinus densiflora*

Japanese rose *Rosa rugosa*

Japanese rowan *Sorbus commixta*

Japanese sago palm *Cycas revoluta*

Japanese snakebark maple
Acer capillipes

Japanese snowball
Viburnum plicatum

Japanese snowbell *Styrax japonica*

Japanese snow flower
Deutzia gracilis

Japanese spindle tree
Euonymus hamiltonianus yedoensis
Euonymus japonica

Japanese spruce
Picea maximowiczii
Picea polita

Japanese stewartia
Stewartia pseudocamellia

Japanese stone pine *Pinus pumila*

Japanese sweetspike *Itea japonica*

Japanese thuya *Thuya standishii*

Japanese torreya *Torreya nucifera*

Japanese tree lilac
Syringa reticulata

Japanese umbrella pine
Sciadopitys verticillata

Japanese varnish tree
Firmiana simplex
Rhus verniciflua

Japanese viburnum
Viburnum japonicum

Japanese walnut *Juglans ailantifolia*

Japanese white birch
Betula japonica
Betula mandschurica
Betula platyphylla japonica

Japanese white pine
Pinus parviflora

Japanese willow *Salix urbaniana*

Japanese wing nut
Pterocarya rhoifolia

Japanese winterberry *Ilex serrata*

Japanese witch hazel
Hamamelis japonica

Japanese woodoil tree
Aleurites cordata

Japanese yellowwood
Cladrastis platycarpa

Japanese yew
Podocarpus macrophyllus
Taxus cuspidata

Japanese zelkova *Zelkova serrata*

Japan pepper
Zanthoxylum piperitum

Jarrah *Eucalyptus marginata*

Jasmine box *Phillyrea decora*
Phyllirea vilmoriniana

Java apple *Syzygium samarangense*

Java fig *Ficus benjamina*

Java glory bean
Clerodendrum × speciosum

Java plum *Eugenia jambolana*
Syzygium cumini

Javillo *Hura crepitans*

Jeffrey's hemlock *Tsuga × jeffreyi*

Jeffrey's pine *Pinus jeffreyi*

Jelecote pine *Pinus patula*

Jelly palm *Butia capitata*
Butia yatay

Jersey elm
Ulmus minor stricta sarniensis
Ulmus × sarniensis
Ulmus stricta sarniensis

Jersey pine *Pinus virginiana*

Jerusalem pine *Pinus halepensis*

Jerusalem sage *Phlomis fruticosa*

Jerusalem thorn *Paliurus aculeatus*
Paliurus spina-christi
Parkinsonia aculeata

Jesuit's bark	*Cinchona calisaya*
Jew's mallow	*Corchorus olitorius*
	Kerria japonica
Jim brush	*Ceanothus sorediatus*
Jim bush	*Ceanothus sorediatus*
Jinbul	*Eucalyptus microtheca*
Jobo tree	*Spondias mombin*
Jocote	*Spondias purpurea*
John Downie crab apple	
	Malus sylvestris john downie
Jointwood	*Cassia nodosa*
Joseph's coat	*Alternant ficoidea*
Joshua tree	*Yucca brevifolia*
Jove's fruit	*Lindera melissifolia*
Judas tree	*Cercis siliquastrum*
June berry	*Amelanchier arborea*
	Amelanchier laevis
	Amelanchier lamarckii
Jungle flame	*Ixora coccinea*
	Ixora incarnata
Jungle geranium	*Ixora coccinea*
	Ixora incarnata
Juniper rush	*Genista raetam*
Jupiter's beard	
	Anthyllis barba-jovis
Jute	*Corchorus capsularis*
	Corchorus olitorius
Kaffia plum	
	Acokanthera oblongifolia
Kaffir bean	*Schotia afra*
Kaffir fig	*Ficus nekbudu*
	Ficus utilis
Kaffir plum	
	Harpephyllum caffrum
Kaffir thorn	*Lycium afrum*
Kahika	*Podocarpus dacrydioides*
Kahikatea	
	Dacrycarpus dacrydioides
	Podocarpus dacrydioides
Kai apple	*Aberia caffra*
	Dovyalis caffra
Kaki	*Diospyros chinensis*
	Diospyros kaki
Kamahi tree	
	Weinmannia racemosa
Kamala tree	
	Mallotus philippinensis
Kamani	*Terminalia catappa*
Kamila tree	*Mallotus philippinensis*
Kangaroo apple	*Solanum aviculare*
	Solanum laciniatum
Kangaroo thorn	*Acacia armata*
Kanooka	*Tristania laurina*
Kapok bush	
	Cochlospermum frazeri
Kapok tree	*Ceiba pentandra*
Karaka	*Corynocarpus laevigata*
Karanda	*Carissa carandas*
Karapincha	*Murraya koenigii*
Karo	*Pittosporum crassifolium*

Karri	*Eucalyptus diversicolor*
Karri gum	*Eucalyptus diversicolor*
Karri tree	*Paulownia tomentosa*
Karroo thorn	*Acacia karroo*
Karum tree	*Pongamia pinnata*
Kashi holly	*Ilex chinensis*
	Ilex purpurea
Kashmir cypress	
	Cupressus cashmeriana
Kashmir rowan	
	Sorbus cashmiriana
Kassod tree	*Cassia siamea*
Katsura	
	Cercidiphyllum japonicum
Kau apple	*Aberia caffra*
	Dovyalis caffra
Kauri pine	*Agathis australis*
Kava	*Piper methysticum*
Kawaka	*Calocedrus plumosa*
Kawaka tree	*Libocedrus plumosa*
Kawa-kawa	*Macropiper excelsum*
Kaya	*Torreya nucifera*
Kazanlik rose	
	Rosa damascena trigintipetala
Keaki	*Zelkova serrata*
Keg fig	*Diospyros chinensis*
	Diospyros kaki
Kei apple	*Aberia caffra*
	Dovyalis caffra
Kellogg oak	*Quercus kelloggii*
Kenaf	*Hibiscus cannabinus*
Kenari	*Canarium commune*
Kentia palm	*Howea forsterana*
	Kentia forsterana
Kentish cob	*Corylus maxima*
Kentucky coffee tree	
	Gymnocladus dioica
Kermes oak	*Quercus coccifera*
Key lime	*Citrus aurantifolia*
Key palm	*Thrinax morrisii*
Khair	*Acacia catechu*
Khasya pine	*Pinus insularis*
	Pinus khasya
Khat	*Catha edulis*
Khingan fir	*Abies nephrolepis*
Killarney strawberry tree	
	Arbutus unedo
Kilmarnock willow	
	Salix caprea pendula
Kimberley grey box	
	Eucalyptus argillacea
	Eucalyptus leucophylla
King Boris' fir	*Abies borisii-regis*
Kingdon Ward's carmine	
cherry	*Prunus cerasoides rubea*
King mandarin	
	Citrus × nobilis king
King nut	*Carya laciniosa*
King of Siam	*Citrus × nobilis king*

King orange

King orange	*Citrus × nobilis king*
King protea	*Protea cynaroides*
King William pine	*Athrotaxis selaginoides*
Kinnikinick	*Arctostaphylos uva-ursi*
Kino	*Coccoloba uvifera*
Kinos	*Pterocarpus marsupium*
Kirishima azalea	*Azalea obtusa*
	Rhododendron obtusum
Kitembilla	*Aberia gardneri*
	Dovyalis hebecarpa
Kittul tree	*Caryota urens*
Klinki pine	*Araucaria hunsteinii*
Knicker tree	*Gymnocladus dioica*
Knife acacia	*Acacia cultriformis*
Knife-leaf wattle	*Acacia cultriformis*
Knobcone pine	*Pinus attenuata*
Koa	*Acacia koa*
Kohuhu	*Pittosporum tenuifolium*
Kokio	*Kokia drynarioides*
Kokoon tree	*Kokoona zeylanica*
Kola	*Cola acuminata*
	Cola nitida
Kola nut	*Cola acuminata*
Kola nut tree	*Cola vera*
Konara oak	*Quercus glandulifera*
Korean azalea	*Rhododendron poukhanense*
Korean fir	*Abies koreana*
Korean forsythia	*Forsythia ovata*
Korean hill cherry	*Prunus leveilleana*
	Prunus serrulata pubescens
Korean lilac	*Syringa meyeri palibin*
	Syringa velutina
Korean lime	*Tilia insularis*
Korean maple	*Acer pseudosieboldianum*
Korean pine	*Pinus koraiensis*
Korean thuya	*Thuya koraiensis*
Kousa	*Cornus kousa*
Kousso	*Hagenia abyssinica*
Kowhai	*Sophora tetraptera*
Koyama spruce	*Picea koyamai*
Kruse's mallee	*Eucalyptus kruseana*
Kumaon pear	*Pyrus pashia*
Kumoi	*Pyrus pyrifolia*
Kumquat	*Fortunella margarita*
Kurogane holly	*Ilex rotunda*
Kurrajong	*Brachychiton populneus*
Kusamaki tree	*Podocarpus macrophyllus*
Labrador tea plant	*Ledum groenlandicum*
	Ledum latifolium
Lacebark	*Hoheria lyallii*
	Hoheria populnea
Lacebark pine	*Pinus bungeana*
Lace-cap hydrangea	*Hydrangea macrophylla*
Lace-cap viburnum	*Viburnum plicatum tomentosum*
Lace-cup viburnum	*Viburnum plicatum tomentosum*
Lacquer tree	*Rhus verniciflua*
Lac tree	*Schleichera oleosa*
Lady banks' rose	*Rosa banksiae*
Lady-of-the-night	*Brunfelsia americana*
Lady-of-the-woods	*Betula pendula*
Lady palm	*Rhapis excelsa*
Lady's handkerchief tree	*Davidia involucrata*
Lagos ebony	*Diospyros mespiliformis*
Lambkill	*Kalmia angustifolia*
Lancashire whitebeam	*Sorbus lancastriensis*
Lancaster red rose	*Rosa gallica officinalis*
Lancaster rose	*Rosa gallica officinalis*
Lancewood	*Oxandra lanceolata*
	Pseudopanax crassifolius
Lantern tree	*Crinodendron hookerianum*
Lapland willow	*Salix lapponum*
Large-coned douglas fir	*Pseudotsuga macrocarpa*
Large gallberry	*Ilex coriacea*
	Ilex lucida
Large-leaved banyan	*Ficus wightiana*
Large-leaved cucumber tree	*Magnolia macrophylla*
Large-leaved lime	*Tilia platyphyllos*
Large-leaved magnolia	*Magnolia macrophylla*
Large-leaved podocarp	*Podocarpus macrophyllus*
Large pussy willow	*Salix discolor*
Large tupelo	*Nyssa aquatica*
Late lilac	*Syringa villosa*
Laurel	*Cordia alliodora*
	Prunus laurocerasus
Laurel magnolia	*Magnolia grandiflora*
Laurel negro	*Cordia alliodora*
Laurel oak	*Quercus imbricaria*
	Quercus laurifolia
Laurel sumac	*Rhus laurina*
Laurel tree	*Persea borbonia*
Laurel willow	*Salix pentandra*
Laurelwood	*Calophyllum inophyllum*
Laurustinus	*Viburnum tinus*

Lodgepole pine

Lavender cotton
Santolina chamaecyparissus
Santolina incana

Lavender tree
Heteropyxis natalensis

Lawson's cypress
Chamaecyparis lawsoniana

Lawson's holly
Ilex × altaclarensis lawsoniana

Lead plant *Amorpha canescens*

Least whitebeam *Sorbus minima*

Least willow *Salix herbacea*

Leatherleaf
Chamaedaphne calyculata
Viburnum wrightii

Leather oak *Quercus durata*

Leatherwood *Cyrilla racemiflora*
Dirca palustris
Eucryphia lucida

Lebanon oak *Quercus libani*

Lebbek tree *Albizzia lebbeck*

Leechee *Litchi chinensis*

Lehmann's gum
Eucalyptus lehmannii

Lemandarin *Citrus × limonia*

Lemon *Citrus limon*

Lemonade berry *Rhus integrifolia*

Lemonade sumac *Rhus integrifolia*

Lemonade tree *Adensonia digitata*

Lemon bottlebrush
Callistemon citrinus

Lemonleaf *Gaultheria shallon*

Lemon plant *Aloysia triphylla*
Lippia citriodora

Lemon-scented gum
Eucalyptus citriodora

Lemon-scented ironbark
Eucalyptus staigeriana

Lemon-scented spotted gum
Eucalyptus citriodora

Lemon-scented verbena
Aloysia triphylla

Lemon sumac *Rhus aromatica*

Lemon verbena *Aloysia citriodora*
Aloysia triphylla
Lippia citriodora
Verbena triphylla

Lemonwood
Pittosporum eugenioides

Lenga *Nothofagus pumilo*

Lentisco *Pistacia texana*
Rhus virens

Lerp mallee *Eucalyptus incrassata*

Lesser flowering quince
Chaenomeles japonica

Lettuce tree *Pisonia alba*

Leverwood *Ostrya virginiana*

Leyland cypress
× cupressocyparis leylandii

Liberian coffee *Coffea liberica*

Lichfield crab apple
Malus sylvestris john downie

Lichi *Litchi chinensis*

Licuri palm *Syagrus coronata*

Life-of-man *Aralia racemosa*

Lightwood *Acacia implexa*
Ceratopetalum apetalum

Ligiri tree *Idesia polycarpa*

Lignum vitae *Guaiacum officinale*

Likiang spruce *Picea likiangensis*

Lilac broom *Carmichaelia odorata*

Lilac *Syringa oblata*

Lillypilly *Acmena smithii*
Eugenia smithii

Lily-flowered magnolia
Magnolia liliiflora

Lily-of-the-valley tree
Clethra arborea

Lily thorn *Catesbaea spinosa*

Lily tree *Magnolia denudata*

Limber pine *Pinus flexilis*

Limeberry *Triphasia trifolia*

Lime *Citrus aurantifolia*

Lime-leaf maple *Acer distylum*

Limequat
× Citrofortunella floridana
× Citrofortunella swinglei
Citrus aurantifolia × fortunella

Linden *Tilia cordata*
Tilia × europaea

Linden viburnum
Viburnum dilatatum

Ling *Calluna vulgaris*

Lion's ear *Leonotis leonurus*

Lipstick tree *Bixa orellana*

Litchi *Litchi chinensis*

Little cokernut palm
Jubaea chilensis
Jubaea spectabilis

Little lady palm *Rhapis excelsa*

Little-leaf fig *Ficus australis*
Ficus rubiginosa

Little tree willow *Salix arbuscula*

Little walnut *Juglans microcarpa*
Juglans rupestris

Live oak *Quercus virginiana*

Lobel's maple *Acer lobelii*

Loblolly bay *Gordonia lasianthus*

Loblolly magnolia
Magnolia grandiflora

Loblolly pine *Pinus taeda*

Lobster claw *Clianthus puniceus*

Lobster plant
Euphorbia pulcherrima

Locust bean *Ceratonia siliqua*

Locust tree *Hymenaea courbaril*
Robinia pseudoacacia

Lodgepole pine *Pinus contorta*
Pinus contorta latifolia

Logwood
Haematoxylum campeachianum
Lombardy poplar *Populus nigra*
Populus nigra italica
Lombardy poplar cherry
Prunus amanagawa
London plane *Platanus × acerifolia*
Platanus × hispanicus
Platanus hybrida
Long buchu *Agathosma crenulata*
Long-flowered marlock
Eucalyptus macrandra
Long john *Triplaris americana*
Long-leaf pine *Pinus palustris*
Long-leaved argyle apple
Eucalyptus cephalocarpa
Long-leaved Indian pine
Pinus roxburghii
Long-thatch palm
Calyptronoma occidentalis
Looking-glass bush
Coprosma baueri
Coprosma repens
Looking-glass tree
Heritiera macrophylla
Loquat *Eriobotrya japonica*
Lord's candlestick *Yucca gloriosa*
Lote tree *Celtis australis*
Lotus tree *Ziziphus lotus*
Love tree *Cercis siliquastrum*
Low-bush blueberry
Vaccinium angustifolium
Lowland fir *Abies grandis*
Low's white fir
Abies concolor lowiana
Luccombe oak *Quercus × hispanica*
Luchu pine *Pinus luchuensis*
Lucky bean tree
Erythrina lysistemon
Lucky nut *Thevetia peruviana*
Ludwig's oak
Quercus × ludoviciana
Luster-leaf holly *Ilex latifolia*
Lyall's larch *Larix lyalli*
Lychee *Litchi chinensis*
Nephelium litchi
Macadamia nut
Macadamia integrifolia
Macadamia tetraphylla
Macarthur palm
Ptychosperma macarthurii
Macartney rose *Rosa bracteata*
Macassar ebony
Diospyros ebenaster
Diospyros ebenum
Macaw-fat *Elaeis guineensis*
Macdonald oak
Quercus macdonaldii
Macedonian oak *Quercus frainetto*
Quercus macedonica
Quercus trojana

Macedonian pine *Pinus peuce*
Mackay's heath *Erica mackaiana*
McNab's cypress
Cupressus macnabiana
Madagascar nutmeg
Ravensara aromatica
Madagascar palm
Chrysalidocarpus lutescens
Madagascar plum
Flacourtia indica
Flacourtia ramontchi
Madagascar tamarind
Vangueria edulis
Madar *Calotropis gigantea*
Madden cherry
Maddenia hypoleuca
Madeira broom *Genista virgata*
Madeira holly *Ilex perado*
Madeira redwood
Swietenia mahagoni
Madeira walnut *Juglans regia*
Madeira whortleberry
Vaccinium padifolium
Madre *Gliricidia maculata*
Gliricidia sepium
Madrona *Arbutus menziesii*
Magdeburg apple
Malus × magdeburgensis
Magic flower *Cantua buxifolia*
Magnolia *Magnolia × soulangiana*
Magnolia-leaved willow
Salix magnifica
Mahala-mat *Ceanothus prostratus*
Mahaleb *Prunus mahaleb*
Mahoe *Hibiscus elatus*
Hibiscus ramiflorus
Hibiscus tiliaceus
Melicytus ramiflorus
Thespesia populnea
Mahogany *Swietenia macrophylla*
Mahogany birch *Betula lenta*
Mahogany pine *Podocarpus totara*
Ma huang *Ephedra vulgaris*
Maidenhair tree *Ginkgo biloba*
Maiden's gum *Eucalyptus maidenii*
Eucalyptus globus maidenii
Maingay's oak *Quercus maingayi*
Malabar nut *Adhatoda vasica*
Malabar plum *Syzygium jambos*
Malay apple *Eugenia malaccensis*
Syzygium malaccense
Malay banyan *Ficus retusa*
Malay ixora *Ixora chinensis*
Malay jewel vine *Derris scandens*
Malay palm *Daemonorops grandis*
Male berry *Lyonia ligustrina*
Male blueberry *Lyonia ligustrina*
Mallee *Eucalyptus dumosa*
Mallee box *Eucalyptus bakeri*
Mamey *Mammea americana*

Mamey colorado — *Pouteria sapota*
Mamey sapote — *Pouteria sapota*
Mammee — *Mammea americana*
Mammee apple
— *Mammea americana*
Mammee sapote — *Pouteria sapota*
Mammoth tree
— *Sequoiadendron giganteum*
Mamoncillo — *Melicoccus bijugatus*
Manaca — *Brunfelsia uniflora*
Manchester poplar
— *Populus nigra betulifolia*
Manchineel
— *Hippomane mancinella*
Manchurian alder — *Alnus hirsuta*
Manchurian apricot
— *Prunus mandshurica*
Manchurian ash
— *Fraxinus mandshurica*
Manchurian cherry — *Prunus maackii*
Manchurian fir — *Abies holophylla*
Manchurian lime — *Tilia mandshurica*
Manchurian walnut
— *Juglans mandshurica*
Mandarin lime — *Citrus × limonia*
Mandarin orange — *Citrus reticulata*
Mandioca — *Manihot esculenta*
Mango — *Mangifera indica*
Mangosteen — *Garcinia mangostana*
Mangrove date palm
— *Phoenix paludosa*
Mangrove palm — *Nypa fruticans*
Manila hemp — *Musa textilis*
Manila palm — *Veitchia merrillii*
Manila tamarind
— *Pithecellobium dulce*
Manio — *Podocarpus nubigenus*
Manioc — *Manihot esculenta*
Manna ash — *Fraxinus ornus*
Manna bush — *Tamarix gallica*
Manna gum — *Eucalyptus viminalis*
Manna tree — *Alhagi maurorum*
Manuka — *Leptospermum scoparium*
Manzanita
— *Arctostaphyols manzanita*
Manzanote
— *Olmediella betschlerana*
Maori holly — *Olearia illicifolia*
Maple-leaved viburnum
— *Viburnum acerifolium*
Maraja palm — *Bactris gasipaes*
Maranon — *Anacardium occidentale*
Maria — *Calophyllum brasiliense*
Maries fir — *Abies mariesii*
Mariposa manzanita
— *Arctostaphylos mariposa*
Maritime pine — *Pinus pinaster*
Marking-nut tree
— *Semecarpus anacardium*

Markry — *Rhus radicans*
Marlberry — *Ardisia escallonioides*
Marmalade box — *Genipa americana*
Marmalade plum — *Manilkara zapota*
— *Pouteria sapota*
Maroc fir — *Abies marocana*
Marri — *Eucalyptus calophylla*
— *Eucalyptus camaldulensis*
Marsh ledum — *Ledum palustre*
Marsh rosemary
— *Andromeda polifolia*
Martin's maple — *Acer martinii*
Marumi kumquat
— *Fortunella japonica*
Mastic tree — *Pistacia lentiscus*
— *Schinus molle*
Mata kuching
— *Nephelium malaiense*
Matai — *Podocarpus spicatus*
— *Prumnopitys taxifolia*
Matasano — *Casimiroa tetrameria*
Match-me-if-you-can
— *Acalypha wilkesiana*
Maté — *Ilex paraguariensis*
Matico — *Piper angustifolium*
Matrimony thorn
— *Lycium balbarum*
Matrimony vine — *Lycium balbarum*
Maul oak — *Quercus chrysolepis*
Maxwell spruce
— *Picea abies maxwellii*
May — *Crataegus laevigata*
— *Crataegus monogyna*
May flower — *Epigaea repens*
May rose — *Rosa majalis*
Mayten — *Maytenus boaria*
Mazari palm
— *Nannorrhops ritchiana*
Mazzard — *Prunus avium*
Meadow fern — *Myrica gale*
Meadow rose — *Rosa blanda*
Mealberry
— *Arctostaphylos uva-ursi*
Meal tree — *Viburnum lantana*
Mealy stringybark
— *Eucalyptus cinerea*
Mediterranean cypress
— *Cupressus sempervirens*
Mediterranean hackberry
— *Celtis australis*
Mediterranean heather
— *Erica mediterranea*
Mediterranean storax
— *Styrax officinalis*
Medlar — *Mespilus germanica*
— *Mimusops elengi*
Medusa's head
— *Euphorbia caput-medusae*
Meloncillo — *Passiflora suberosa*
Melon tree — *Carica papaya*

Melukhie	*Corchorus olitorius*
Memorial rose	*Rosa wichuraina*
Mercury	*Rhus radicans*
Merrit gum	*Eucalyptus flocktoniae*
Mescal bean	*Sophora secundiflora*
Mesquite	*Prosopis glandulosa*
	Prosopis juliflora
Messmate stringybark	
	Eucalyptus obliqua
Mexican apple	*Casimiroa edulis*
Mexican blue palm	*Brahea armata*
Mexican buckeye	
	Ungnadia speciosa
Mexican cherry	–*Prunus capuli*
Mexican cypress	
	Taxodium mucronatum
Mexican cypress	
	Cupressus lusitanica
Mexican elm	*Ulmus mexicana*
Mexican fan palm	
	Washingtonia robusta
Mexican fire plant	
	Euphorbia cyathophora
	Euphorbia heterophylla
Mexican flameleaf	
	Euphorbia pulcherrima
Mexican hand plant	
	Chiranthodendron
	pentadactylon
Mexican incense bush	
	Eupatorium micranthum
	Eupatorium weinmannianum
Mexican juniper	*Juniperus flaccida*
Mexican lime	*Citrus aurantifolia*
Mexican manzanita	
	Arctostaphylos pungens
Mexican mulberry	
	Morus microphylla
Mexican nut pine	*Pinus cembroides*
Mexican orange blossom	
	Choisya ternata
Mexican palo verde	
	Parkinsonia aculeata
Mexican pine	*Pinus patula*
Mexican poppy	
	Argemone mexicana
Mexican red sage	*Salvia fulgens*
Mexican stone pine	
	Pinus cembroides
Mexican swamp cypress	
	Taxodium mucronatum
Mexican tassel bush	
	Garrya macrophylla
Mexican Washington palm	
	Washingtonia robusta
Mexican weeping pine	*Pinus patula*
Mexican white pine	
	Pinus ayacahuite
Meyer's blue juniper	
	Juniperus squamata meyeri
Mezereon	*Daphne mezereum*

Mickey-mouse plant	
	Ochna japonica
	Ochna serrulata
Midland hawthorn	
	Crataegus laevigata
	Crataegus oxyacanthoides
Mignonette tree	*Lawsonia alba*
	Lawsonia inermis
Mile tree	*Casuarina equisetifolia*
Milk barrel	*Euphorbia cereiformis*
	Euphorbia heptagona
	Euphorbia leviana
Milk bush	*Euphorbia tirucalli*
Milk tree	*Sapium hippomane*
Mimosa	*Acacia dealbata*
Mimosa tree	*Albizzia julibrissin*
Min fir	*Abies recurvata*
Ming aralia	*Polyscias fruticosa*
Miniature date palm	
	Phoenix roebelenii
Miniature fan palm	*Rhapis excelsa*
Miniature holly	
	Malpighia coccigera
Miraculous berry	
	Synsepalum dulcificum
Miraculous fruit	
	Synsepalum dulcificum
Mirbeck's oak	*Quercus canariensis*
Miro	*Podocarpus ferrugineus*
	Prumnopitys ferruginea
Mirror plant	*Coprosma repens*
Missey-mooney	*Sorbus americana*
Mississippi hackberry	
	Celtis laevigata
Mistletoe fig	*Ficus deltoides*
	Ficus diversifolia
Mistletoe rubber plant	
	Ficus deltoides
	Ficus diversifolia
Mitcham lavender	
	Lavender officinalis
	Lavender spica
Miyabe's maple	*Acer miyabei*
Mlanje cedar	
	Widdringtonia whytei
Mobala plum	
	Parinari curatellifolia
Mochi tree	*Ilex integra*
Mockernut	*Carya tomentosa*
Mock orange	*Bumelia lycioides*
	Philadelphus coronarius
	Pittosporum tobira
	Pittosporum undulatum
	Prunus caroliniana
	Styrax americanus
Modoc cypress	*Cupressus bakeri*
Mogadore gum	*Acacia gummifera*
Mole plant	*Euphorbia lathyris*
Molle	*Schinus molle*
Momi fir	*Abies firma*

TREES, BUSHES, AND SHRUBS

Monarch birch
    Betula maximowicziana

Money pine   *Larix kaempferi*

Money tree
    Eucalyptus pulverulenta

Mongolian lime   *Tilia mongolica*

Mongul oak   *Quercus mongolica*

Monkey-bread tree
    Adansonia digitata

Monkey jack   *Artocarpus lakoocha*

Monkey nut   *Lecythis zabucayo*

Monkey-pistol   *Hura crepitans*

Monkeypod tree   *Samanea saman*

Monkey pot tree   *Leycythis usitata*

Monkey puzzle   *Araucaria araucana*

Monkey's dinner bell
    Hura crepitans

Monk's pepper tree
    Vitex agnus-castus

Monterey ceanothus
    Ceanothus rigidus

Monterey cypress
    Cupressus macrocarpa

Monterey pine   *Pinus radiata*

Montezuma pine
    Pinus montezumae

Monthly rose   *Rosa chinensis*

Montpelier broom
    Cytisus monspessulanus

Montpelier maple
    Acer monspessulanum

Montpelier rock rose
    Cistus monspeliensis

Moolar   *Eucalyptus microtheca*

Moonah   *Melaleuca lanceolata*
    Melaleuca pubescens

Moonlight holly
    Ilex aquifolium flavescens

Moon trefoil   *Medicago arborea*

Moorberry   *Vaccinium uliginosum*

Moose bark
    Acer pennsylvanicum

Mooseberry   *Viburnum alnifolium*
    Viburnum pauciflorum

Moose elm   *Ulmus rubra*

Moosewood   *Acer pennsylvanicum*
    Dirca palustris
    Viburnum alnifolium

Mop-head acacia
    Robinia pseudoacacia inermis
    Robinia pseudoacacia
    umbraculifera

Mop-head hydrangea
    Hydrangea hortensia

Moreton Bay chestnut
    Castanospermum australe

Moreton Bay fig   *Ficus macrocarpa*
    Ficus macrophylla

Moreton Bay pine
    Araucaria cunninghamii

Morinda spruce   *Picea smithiana*

Morning glory bush
    Ipomoea fistulosa

Morning glory tree
    Ipomoea arborescens

Morning, noon and night
    Brunfelsia australis

Moroccan broom
    Cytisus battandieri

Moroccan fir   *Abies macrocana*

Morocco gum   *Acacia gummifera*

Morocco ironwood
    Argania spinosa

Morro manzanita
    Arctostaphylos morroensis

Moss rose   *Rosa centifolia mucosa*

Mossy cup oak   *Quercus cerris*
    Quercus macrocarpa

Mossy locust   *Robinia hispida*

Motillo   *Soleanea berteriana*

Mottled spurge   *Euphorbia lactea*

Mountain agathis   *Agathis alba*

Mountain alder   *Alnus crispa mollis*
    Alnus tenuifolia

Mountain ash   *Eucalyptus regnans*
    Sorbus aucuparia

Mountain avens   *Dryas octopetala*

Mountain azalea
    Loiseleuria procumbens

Mountain beech
    Nothofagus solandri
    cliffortioides

Mountain box   *Arctostaphylos*
    uva-ursi

Mountain camellia   *Stewartia ovata*

Mountain cedar   *Juniperus ashei*

Mountain cherry   *Prunus prostrata*

Mountain currant   *Ribes alpinum*

Mountain cypress
    Widdringtonia cupressoides

Mountain dogwood
    Cornus nuttallii

Mountain grape
    Mahonia aquifolium

Mountain guava
    Psidium montanum

Mountain gum
    Eucalyptus dalrympleana

Mountain hemlock
    Tsuga mertensiana

Mountain hickory
    Acacia penninervis

Mountain holly
    Nemopanthus mucronatus
    Olearia illicifolia
    Prunus ilicifolia

Mountain immortelle
    Erythrina poeppigiana

Mountain kiepersol
    Cussonia paniculata

Mountain laurel   *Kalmia latifolia*

Mountain mahoe   *Hibiscus elatus*

Mountain mahogany   *Betula lenta*
                     Cercocarpus montanus
Mountain maple   *Acer spicatum*
Mountain papaw
                Carica cundinamarcensis
Mountain pawpaw
                Carica cundinamarcensis
Mountain papaya   *Carica pubescens*
Mountain pepper
                Drimys lanceolata
Mountain pine
                Dacrydium bidwillii
                Pinus mugo
                Pinus uncinata
Mountain rimu
                Dacrydium laxifolium
Mountain rose bay
                Rhododendron catawbiense
Mountain silver bell
                Halesia monticola
Mountain snowdrop tree
                Halesia monticola
Mountain soursop
                Annona montana
Mountain spurge
                Pachysandra terminalis
Mountain sumac   *Rhus copallina*
Mountain-sweet
                Ceanothus americanus
Mountain white pine
                Pinus monticola
Mountain willow   *Salix arbuscula*
Mount Atlas mastic tree
                Pistacia atlantica
Mount Etna broom
                Genista aethnensis
Mount Morgan wattle
                Acacia podalyriifolia
Mount Morrison barberry
                Berberis morrisonensis
Mount Morrison juniper
                Juniperus morrisonicola
Mount Morrison spruce
                Picea morrisonicola
Mount Omei rose   *Rosa omeiensis*
                   Rosa sericea
Mount wellington peppermint
                Eucalyptus coccifera
Mourning cypress
                Cupressus funebris
Moutan   *Paeonia suffruticosa*
Moutan paeony   *Paeonia moutan*
                 Paeonia suffruticosa
Mudar   *Calotropis procera*
Mudgee wattle   *Acacia spectabilis*
Mueller's cypress pine
                Callitris muelleri
Mugga   *Eucalyptus sideroxylon*
Muku tree   *Aphananthe aspera*
Mulberry fig   *Ficus sycomorus*
Mulga   *Acacia aneura*

Murray red gum
                Eucalyptus camaldulensis
Murray river pine
                Callitris columellaris
                Callitris glauca
Murtillo   *Ugni molinae*
Musk rose   *Rosa moschata*
Musk willow   *Salix aegyptiaca*
Muskwood   *Olearia argophylla*
Mu tree   *Aleurites montana*
Myallwood   *Acacia homalophylla*
Myrobalan   *Phyllanthus emblica*
             Terminalia bellirica
             Terminalia catappa
Myrobalan plum   *Prunus cerasifera*
Myrrh   *Commifera abyssinica*
         Commifera molmol
         Commifera myrrha
Myrtle   *Cyrilla racemiflora*
          Myrtus communis
          Umbellularia californica
Myrtle beech
                Nothofagus cunninghamii
Myrtle spurge   *Euphorbia lathyris*
Mysore fig   *Ficus mysorensis*
Nagami kumquat
                Fortunella margarita
Nagi   *Podocarpus nagi*
Naio   *Myoporum sandwicense*
Naked coral tree
                Erythrina coralloides
Nanking cherry   *Prunus tomentosa*
Nannyberry   *Viburnum lentago*
Nanny plum   *Viburnum lentago*
Nan-shan bush
                Cotoneaster adpressus praecox
Nanten   *Nandina domestica*
Napolean's button
                Napoleona heudottii
Narrow-leaved ash
                Fraxinus angustifolia
Narrow-leaved black
peppermint
                Eucalyptus nicholii
Narrow-leaved bottle tree
                Brachychiton rupestris
Narrow-leaved ironbark
                Eucalyptus crebra
Narrow-leaved peppermint
                Eucalyptus radiata
Narrow-leaved pittosporum
                Pittosporum phillyraeoides
Nashi   *Pyrus pyrifolia*
Natal fig   *Ficus natalensis*
Natal orange   *Strychnos spinosa*
Natal plum   *Carissa grandiflora*
Native frangipani
                Hymenosporum flavum
Nazeberry   *Manilkara zapote*
Necklace poplar   *Populus deltoides*
                   Populus lasiocarpa

Nectarine
    Prunus persica nectarina
    Prunus persica nucipersica

Needle bush   *Azim tetracantha*

Needle-bush wattle   *Acacia rigens*

Needle fir   *Abies holophylla*

Needleflower tree
    Posoqueria latifolia

Needle furze   *Genista anglica*

Needle juniper   *Juniperus rigida*

Needle palm
    Rhapidophyllum hystrix
    Yucca filamentosa

Needle rose   *Rosa acicularis*

Nepal barberry   *Berberis aristata*

Nepalese alder   *Alnus nepalensis*

Nepalese white thorn
    Pyracantha crenulata

Nepal laburnum
    Piptanthus laburnifolius

Nepal nut pine   *Pinus gerardiana*

Nepal privet   *Ligustrum lucidum*

Nero's crown
    Tabernaemontana coronaria

Nest spruce   *Picea abies nidiformis*

Netted willow   *Salix reticulata*

Nettle tree   *Celtis australis*
    Celtis occidentalis

New Caledonia pine
    Araucaria columnaris

New England boxwood
    Cornus florida

New Jersey tea plant
    Ceanothus americanus

New Zealand black pine
    Podocarpus spicatus
    Prumnopitys taxifolia

New Zealand christmas tree
    Metrosideros excelsa
    Metrosideros robusta
    Metrosideros tomentosa

New Zealand flax   *Phormium tenax*

New Zealand holly
    Olearia macrodonta

New Zealand honeysuckle tree
    Knightia excelsa

New Zealand laburnum
    Sophora tetraptera

New Zealand laurel
    Corynocarpus laevigata

New Zealand lilac   *Hebe hulkeana*

New Zealand pittosporum
    Pittosporum tenuifolium

New Zealand red beech
    Nothofagus fusca

New Zealand sophora
    Sophora tetraptera

New Zealand tea tree
    Leptospermum scoparium

New Zealand tree fuchsia
    Fuchsia excorticata

New Zealand white pine
    Dacrycarpus dacrydioides

New Zealand wineberry
    Aristotelia racemosa

Ngaio   *Myoporum laetum*

Nibung palm
    Oncosperma tigillarium

Nicaraguan cocoa-shade
    Gliricidia maculata
    Gliricidia sepium

Nichol's willow-leaved
peppermint
    Eucalyptus nicholii

Nicker bean   *Entada gigas*

Nicker tree   *Gymnocladus dioica*

Nicobar breadfruit   *Pandanis leram*

Niger seed   *Guizotia abyssinica*

Night jasmine   *Cestrum nocturnum*
    Nyctanthes arbor-tristis

Night jessamine
    Cestrum nocturnum

Nikau palm   *Rhopalostylis sapida*

Nikko fir   *Abies homolepis*

Nikko maple
    Acer maximowiczianum
    Acer nikoense

Nine bark   *Physocarpus opulifolius*

Niobe willow   *Salix × blanda*

Nipa palm   *Nypa fruticans*

Nirre   *Nothofagus antarctica*

Nish broom   *Genista nyssana*

Nispero   *Manilkara zapota*

Noah's ark juniper
    Juniperus communis compressa

Noble fir   *Abies procera*

Nodding pond cypress
    Taxodium ascendens nutans

Noisette rose   *Rosa × noisettiana*

Nootka cypress
    Chamaecyparis lawsoniana
    nootkatensis
    Chamaecyparis nootkatensis

Norfolk Island hibiscus
    Lagunaria patersonii

Norfolk Island pine
    Araucaria excelsa
    Araucaria heterophylla

North American honeysuckle
    Lonicera ledebourii

North American salal
    Gaultheria shallon

North American snakebark
maple
    Acer pennsylvanicum

Northern bayberry
    Myrica pennsylvanica

Northern bungalow palm
    Archontophoenix alexandrae

Northern catalpa   *Catalpa speciosa*

Northern cypress pine
    Callitris intratropica

Northern downy rose
Rosa sherardii

Northern Japanese hemlock
Tsuga diversifolia

Northern Japanese magnolia
Magnolia kobus

Northern pin oak
Quercus ellipsoidalis

Northern pitch pine   *Pinus rigida*

Northern prickly ash
Zanthoxylum americanum

Northern red oak   *Quercus rubra*

Northern white cedar
Thuya occidentalis

Norway maple   *Acer platanoides*

Norway pine   *Pinus resinosa*

Norway spruce   *Picea abies*

Nosegay tree   *Plumeria rubra*

Nutgall tree   *Rhus chinensis*

Nutmeg   *Myristica fragrans*

Nutmeg hickory
Carya myristiciformis

Nut palm   *Cycas media*

Nut pine   *Picea cembroides*
Picea monophylla
Pinus cembroides
Pinus edulis
Pinus monophylla

Nuttall's dogwood   *Cornus nuttallii*

Nux regia   *Juglans regia*

Nux vomica tree
Styrchnos nux-vomica

Nyasaland mahogany
Khaya nyasica

Nyman's hybrid eucryphia
*Eucryphia × nymansensis
nymansay*

Nypa palm   *Nypa fruticans*

Oak-leaf fig   *Ficus montana*
Ficus quercifolia

Oak-leaf hydrangea
Hydrangea quercifolia

Obeche   *Triplochiton scleroxylon*

Ocean spray   *Holodiscus ariifolius*
Holodiscus discolor

Ocotillo   *Fouquieria splendens*

Octopus tree   *Brassaia actinophylla*

Ogechee lime   *Nyssa candicans*

Ohio buckeye   *Aesculus glabra*

Oil palm   *Elaeis guineensis*

Okinawa pine   *Pinus leucodermis*

Oklohoma plum   *Prunus gracilis*

Old English lavender
Lavandula angustifolia
Lavandula officinalis
Lavandula spica

Oldfield birch   *Betula populifolia*

Oldfield pine   *Pinus taeda*

Oldfield's mallee
Eucalyptus oldfieldii

Old man's beard
Chionanthus virginicus

Old red damask rose
Rosa officinalis

Oleander-leaved euphorbia
Euphorbia neriifolia

Oleander   *Nerium olean*

Oleaster
Elaeagnus angustifolia
Elaeagnus latifolia

Olibanus tree   *Boswellia thurifera*

Olive   *Olea europaea*

Olive bark tree   *Terminalia catappa*

Oliver's lime   *Tilia oliveri*

Omeo round-leaved gum
Eucalyptus neglecta

One-leaved nut pine
Pinus cembroides monophylla

Ontario poplar   *Populus candicans*

Opiuma   *Pithecellobium dulce*

Opopanax   *Acacia farnesiana*

Oppossumwood   *Halesia carolina*

Orange   *Citrus sinensis*

Orange ball tree   *Buddleia globosa*

Orange-bark myrtle   *Myrtus luma*
Luma apiculata

Orange champaca
Michelia champaca

Orange eye   *Buddleia davidii*

Orange jasmine
Murraya paniculata

Orange jessamine   *Murraya exotica*

Orange-twig willow
Salix alba chermesina

Orange wattle   *Acacia cyanophylla*

Orchard apple   *Malus domestica*

Orchid tree   *Bauhinia purpurea*

Oregon alder   *Alnus oregona*
Alnus rebra

Oregon ash   *Fraxinus latifolia*
Fraxinus oregona

Oregon cherry   *Prunus emarginata*

Oregon crab apple   *Malus fusca*

Oregon douglas fir
Pseudotsuga menziesii

Oregon grape
Mahonia aquifolium
Mahonia nervosa

Oregon holly   *Ilex aquifolium*

Oregon maple
Acer macrophyllum

Oregon myrtle
Umbellularia californica

Oregon oak   *Quercus garryana*

Oregon plum   *Prunus subcordata*

Oregon tea   *Ceanothus sanguineus*

Oregon white oak
Quercus garryana

Oriental alder   *Alnus orientalis*

Oriental arborvitae *Platycladus orientalis*
Oriental beech *Fagus orientalis*
Oriental cherry *Prunus serrulata*
Oriental cork oak *Quercus variabilis*
Oriental hornbeam *Carpinus orientalis*
Oriental pear *Pyrus pyrifolia*
Oriental plane *Platanus orientalis*
Oriental spruce *Picea orientalis*
Oriental sweet gum *Liquidambar orientalis*
Oriental thorn *Crataegus laciniata*
Crataegus orientalis
Oriental white oak *Quercus aliena*
Ornamental quince *Chaenomeles speciosa nivalis*
Osage orange *Maclura aurantiaca*
Maclura pomifera
Oshima cherry *Prunus speciosa*
Osier *Salix viminalis*
Oso-berry *Nuttallia cerasiformis*
Oemleria cerasiformis
Osmaronia cerasiformis
Otaheite apple *Spondias dulcis*
Otaheite chestnut *Inocarpus edulis*
Otaheite gooseberry *Phyllanthus acidus*
Otaheite apple *Spondias cytherea*
Otaheite orange *Citrus otaitensis*
Otaheite walnut *Aleurites moluccana*
Otay manzanita *Arctostaphylos otayensis*
Oteniqua yellowwood *Podocarpus falcatus*
Our Lord's candle *Yucca whipplei*
Ouricuri palm *Syagrus coronata*
Oval kumquat *Fortunella margarita*
Oval-leaved privet *Ligustrum ovalifolium*
Oval-leaved southern beech *Nothofagus betuloides*
Ovens' acacia *Acacia pravissima*
Ovens' wattle *Acacia pravissima*
Overcup oak *Quercus lyrata*
Overtop palm *Rhyticocos amara*
Owe cola *Cola verticilliata*
Ox-hoof tree *Bauhinia purpurea*
Oyster bay pine *Callitris rhomboides*
Ozark chestnut *Castanea ozarkensis*
Ozark white cedar *Juniperus ashei*
Ozark witch hazel *Hamamelis vernalis*
Pacaya *Chamaedorea tepejilote*

Pacific dogwood *Cornus nuttallii*
Pacific madrone *Arbutus menziesii*
Pacific plum *Prunus subcordata*
Pacific silver fir *Abies amabilis*
Pacific white fir *Abies concolor lowiana*
Pacific willow *Salix lasiandra*
Pacific yew *Taxus brevifolia*
Padang cassia *Cinnamomum burmanii*
Padauk *Pterocarpus indicus*
Paddy river box *Eucalyptus macarthurii*
Pagoda dogwood *Cornus alternifolia*
Pagoda flower *Clerodendrum paniculatum*
Clerodendrum × speciosum
Pagoda tree *Plueria rubra*
Sophora japonica
Pahautea *Calocedrus bidwillii*
Pahautea tree *Libocedrus bidwillii*
Painted leaf *Euphorbia cyathophora*
Euphorbia pulcherrima
Painted maple *Acer pictum*
Painted spurge *Euphorbia heterophylla*
Pajaro manzanita *Arctostaphylos pajaroensis*
Pale hickory *Carya pallida*
Pale laurel *Kalmia poliifolia*
Palestine oak *Quercus calliprinos*
Pali-mara *Alstonia scholaris*
Palma christi *Ricinus communis*
Palma corcho *Microcyas calocoma*
Palma pita *Yucca treculeana*
Palmella *Yucca elata*
Palmetto *Sabal palmetto*
Palmetto thatch *Thrinax parviflora*
Palm lily *Cordyline australis*
Yucca gloriosa
Palm willow *Salix caprea*
Palmyra palm *Borassus flabellifer*
Palo santo *Triplaris americana*
Palo verde *Cercidium floridum*
Palta *Persea americana*
Panama candle tree *Parmentiera cereifera*
Panama orange *× Citrofortunella mitis*
Panama rubber tree *Castilla elastica*
Pandang *Pandanus odoratissimus*
Pandanus palm *Pandanus tectorius*
Panicled dogwood *Cornus racemosa*
Papaw *Carica papaya*
Papaya *Carica papaya*
Paperbark cherry *Prunus serrula*

Paperbark maple

Paperbark maple	*Acer griseum*
Paperbark thorn	*Acacia woodii*
Paperbark tree	
	Melaleuca quinquenervia
Paper birch	*Betula papyrifera*
Paper flower	*Bougainvillea glabra*
Paper mulberry	
	Broussonetia papyrifera
Paper plant	*Aralia japonica*
	Fatsia japonica
Paradise apple	*Malus pumila*
Paradise nut	*Lecythis zabucayo*
Paradise palm	*Howea forsterana*
Paradise tree	*Melia azedarach*
	Simarouba glauca
Paraguayan trumpet tree	
	Tabebuia argentea
Paraguay jasmine	
	Brunfelsia australis
Paraguay tea	*Ilex paraguariensis*
Para nut	*Bertholleta excelsa*
Para-para	*Pisonia umbellifera*
Para rubber tree	
	Hevea brasiliensis
Parasol de Saint Julien	
	Populus tremuloides pendula
Parasol pine	
	Sciadopitys verticillata
Pareira	
	Chondrodendron tomentosum
Parlour palm	
	Chamaedorea elegans
Parrot bill	*Clianthus puniceus*
Parrot leaf	*Alternant fioidea*
Parrot's bill	*Clianthus puniceus*
Parry manzanita	
	Arctostaphylos manzanita
Parry pine	*Pinus quadrifolia*
Parsley-leaved elder	
	Sambucus nigra laciniata
Parsley-leaved thorn	
	Crataegus apiifolia
Partridge berry	
	Gaultheria procumbens
	Mitchella repens
Partridge cane	*Rhapis excelsa*
Partridge pea	*Cassia fasciculata*
Passion flower	*Passiflora caerulea*
Patagonian cypress	
	Fitzroya cupressoides
Patana palm	*Oenocarpus batava*
Paulownia	*Paulownia tomentosa*
Pawpaw	*Asimina triloba*
	Carica papaya
Peach	*Prunus persica*
Peach-leaved willow	
	Salix amygdaloides
Peach palm	*Bactris gasipaes*
	Gulielma gassipaes
Peach protea	*Protea grandiceps*

Peachwood	*Caesalpinia echinata*
	Haematoxylum campeachianum
Peacock flower	
	Caesalpinia pulcherrima
	Delonix regia
Peacock flower-fence	
	Adenanthera pavonina
Peacock thorn	
	Crataegus calpodendron
	Crataegus tomentosa
Pear-fruited mallee	
	Eucalyptus pyriformis
Pear fruit	*Margyricarpus pinnatus*
	Margyricarpus setosus
Pearl acacia	*Acacia podalyriifolia*
Pearl berry	
	Margyricarpus pinnatus
	Margyricarpus setosus
Pearl bush	*Exochorda grandiflora*
	Exochorda racemosa
Pearl fruit	*Margyricarpus pinnatus*
	Margyricarpus setosus
Pear thorn	
	Crataegus calpodendron
	Crataegus tomentosa
Pecan	*Carya illinoensis*
	Carya pecan
Peepul	*Ficus religiosa*
Pegwood	*Cornus sanguinea*
Pejibaye	*Bactris gasipaes*
Peking willow	
	Salix babylonica pekinensis
	Salix matsudana
Pencil cedar	*Juniperus virginiana*
Pencil juniper	*Juniperus virginiana*
Pencil tree	*Euphorbia tirucalli*
Pendant silver lime	*Tilia petiolaris*
Pendunculate oak	*Quercus rober*
Pennsylvania maple	
	Acer pennsylvanicum
Pepper hibiscus	
	Malaviscus arboreus
Peppermint tree	*Agonis flexuosa*
Peppermint wattle	
	Acacia terminalis
Pepperidge	*Nyssa sylvatica*
Pepper tree	*Drimys lanceolata*
	Macropiper excelsum
	Schinus molle
Pepperwood	
	Umbellularia californica
	Zanthoxylum clava-herculis
Père David's maple	*Acer davidii*
Peregrina	*Jatropha integerrima*
Perfumed cherry	*Prunus mahaleb*
Perny's holly	*Ilex pernyi*
Persian albizzia	*Albizzia julibrissin*
Persian ironwood	*Parrotia persica*
Persian lilac	*Melia azedarach*
	Syringa × persica

Plantain

Persian maple	*Acer velutinum*
Persian walnut	*Juglans regia*
Persian yellow rose	
	Rosa foetida persiana
Persimmon	*Diospyros kaki*
	Diospyros virginiana
Peruvian balsam	
	Myroxylon pereirae
Peruvian bark	*Cinchona succirubra*
Peruvian fuchsia tree	
	Fuchsia boliviana
Peruvian mastic tree	*Schinus molle*
Peruvian pepper tree	*Schinus molle*
Petai	*Parkia speciosa*
Petticoat palm	
	Copernica macroglossa
	Washingtonia filifera
Petty morel	*Aralia racemosa*
Petty whin	*Genista anglica*
Pfitzer juniper	
	Juniperus × media pfitzerana
Pharoah's fig	*Ficus sycomorus*
Pheasant berry	
	Leycesteria formosa
Philippine fig	*Ficus pseudopalma*
Philippine medusa	*Acalypha hispida*
Phillyrea	*Phillyrea latifolia*
Phoenician juniper	
	Juniperus phoenicea
Phoenix tree	*Firmiana simplex*
Physic nut	*Jatropha curcus*
	Jatropha multifida
Phytolacca	*Phytolacca dioica*
Piassaba	*Attalea funifira*
	Leopoldinia piassaba
Piassava palm	*Vonitra fibrosa*
Piccabeen bungalow palm	
	Archontophoenix cunninghamiana
Piccabeen palm	*Archontophoenix cunninghamiana*
Pichi	*Fabiana imbricata*
Picrasma	*Picrasma quassioides*
Pie cherry	*Prunus cerasus*
Pigeon berry	*Duranta ellisia*
	Duranta repens
Pigeon plum	
	Coccoloba diversifolia
Pig laurel	*Kalmia angustifolia*
Pigmy juniper	
	Juniperus squamata pygmaea
Pig nut	*Carya cordiformis*
Pignut hickory	*Carya glabra*
Pignut palm	
	Hyophorbe lagenicaulis
Pillar apple	*Malus tschonoskii*
Pimbina	*Viburnum trilobum*
Pimento	*Pimenta dioica*
	Pimenta officinalis
Pinang	*Areca catechu*

Pin cherry	*Prunus pennsylvanica*
Pindo palm	*Butia capitata*
Pineapple broom	
	Cytisus battandieri
Pineapple bush	
	Dasypogon bromeliaefolius
Pineapple guava	*Feijoa sellowiana*
Pineapple sage	*Salvia rutilans*
Pineapple shrub	
	Calycanthus floridus
Pink-and-white shower	
	Cassia nodosa
Pink ball tree	*Dombeya wallichii*
Pink broom	
	Notospartium carmicheliae
Pink cassia	*Cassia javanica*
Pink cedar	*Acrocarpus fraxinifolius*
Pink dogwood	
	Cornus florida rubra
Pink horse chestnut	
	Aesculus × carnea
Pink pearl lilac	*Syringa swegiflexa*
Pink poui	*Tabebuia rosea*
	Tabebuia pentaphylla
Pink sand verbena	
	Abromia umbellata
Pink shower	*Cassia grandis*
Pink siris	*Albizzia julibrissin*
Pink snowball tree	
	Dombeya × cayeuxii
Pink tecoma	*Tabebuia pentaphylla*
Pink trumpet tree	*Tabebuia rosea*
Pink tulip tree	
	Magnolia campbellii
Pin oak	*Quercus palustris*
Pinwheel jasmine	
	Jasminum gracillimum
Pinyon pine	*Pinus cembroides*
	Pinus edulis
Pipe tree	*Sambucus nigra*
Pipperidge	*Berberis vulgaris*
Pirana pine	*Aruacaria angustifolia*
Pirul	*Schinus molle*
Pistacia nut	*Pistacia vera*
Pistachio	*Pistacia vera*
Pistol bush	
	Duvernoia adhatodioides
Pitanga	*Eugenia michelii*
	Eugenia pitanga
	Eugenia uniflora
Pitch pine	*Pinus palustris*
	Pinus rigida
Pith tree	*Herminiera elaphroxylon*
Pitomba	*Eugenia luschnathiana*
Pittosporum	
	Pittosporum tenuifolium
Pituri	*Duboisia myoporoides*
Piute cypress	*Cupressus nevadensis*
Plantain	*Musa acuminata*
	Musa × paradisiaca

Platterleaf

Platterleaf	*Coccoloba uvifera*
Plough breaker	*Erythrina zeyheri*
Plum	*Prunus domestica*
Plume albizzia	*Albizzia distachya*
	Albizzia lophantha
Plume nutmeg	
	Atherosperma moschata
Plum fir	*Podocarpus andinus*
Plum-fruited yew	
	Podocarpus andinus
Plum juniper	*Juniperus drupacea*
Plum-leaved apple	
	Malus prunifolia
Plumwood	*Eucryphia moorai*
Plum yew	*Cephalotaxus fortunii*
Plymouth crowberry	
	Corema conradii
Plymouth pear	*Pyrus cordata*
Pocket handkerchief tree	
	Davidia involucrata
Poet's jasmine	*Jasminum officinale*
Poet's laurel	*Laurus nobilis*
Pohutukawa	*Metrosideros excelsa*
Poinciana	*Delonix regia*
Poinsettia	*Euphorbia pulcherrima*
Point Reyes ceanothus	
	Ceanothus gloriosus
Poison ash	*Chionanthus virginicus*
Poison bay	*Illicium floridanum*
Poison dogwood	*Rhus vernix*
Poison elder	*Rhus vernix*
Poison haw	*Viburnum molle*
Poison ivy	*Rhus radicans*
Poison oak	*Rhus radicans*
Poison sumac	*Rhus vernix*
Polecat bush	*Rhus aromatica*
Polisandro	*Stahlia monosperma*
Polished willow	*Salix laevigata*
Pollard's almond	
	Prunus × amygdalopersica pollardii
Pomegranate	*Punica granatum*
Pomelo	*Citrus grandis*
	Citrus maxima
Pomerac jambos	
	Syzygium malaccense
Pomette bleue	
	Crataegus brachyacantha
Pompelmous	*Citrus maxima*
Pond apple	*Annona glabra*
Pond cypress	*Taxodium ascendens*
Ponderosa pine	*Pinus ponderosa*
Pontine oak	*Quercus pontica*
Pony tail	*Beaucarnea recurvata*
Poonga oil tree	*Pongamia pinnata*
Pop ash	*Fraxinus caroliniana*
Popinac	*Acacia farnesiana*
Porcupine palm	
	Rhapidophyllum hystrix
Portia tree	*Thespesia populnea*
Port Jackson fig	*Ficus australis*
	Ficus rubiginosa
Port Jackson pine	
	Callitris rhomboides
Port Jackson willow	
	Acacia cyanophylla
Port Macquarie pine	
	Callitris macleayana
Port Orford cedar	
	Chamaecyparis lawsoniana
Portugal heath	*Erica lusitanica*
Portugal laurel	*Prunus lusitanica*
Portuguese broom	*Cytisus albus*
Portuguese crowberry	
	Corema album
Portuguese cypress	
	Cupressus lusitanica
Portuguese oak	*Quercus faginea*
	Quercus lusitanica
Possum apple	
	Diospyros virginiana
Possumhaw holly	*Ilex decidua*
Possum haw	
	Viburnum acerifolium
Possum oak	*Quercus nigra*
Possumwood	
	Diospyros virginiana
Post oak	*Quercus stellata*
Posy bush	*Dais cotonifolia*
Potato tree	*Solanum crispum*
Potato vine	*Solanum jasminoides*
Potomac cherry	*Prunus yedoensis*
Pottery tree	*Moquila utilis*
Poverty pine	*Pinus virginiana*
Prairie cherry	*Prunus gracilis*
Prairie crab apple	*Malus ioensis*
Prairie mimosa	
	Desmanthus illinoensis
Prairie rose	*Rosa setigera*
Prairie senna	*Cassia fasciculata*
Prairie tea	*Croton monanthogynus*
Prairie willow	*Salix humilis*
Pretty whin	*Genista anglica*
Prickleweed	
	Desmanthus illinoensis
Prickly ash	*Aralia spinosa*
	Xanthoxylum americanum
	Zanthoxylum americanum
Prickly broom	*Ulex europaeus*
Prickly cardinal	*Erythrina zeyheri*
Prickly castor oil tree	
	Kalopanax pictus
Prickly custard apple	
	Annona muricata
Prickly cyad	
	Encephalartos altensteinii
Prickly cypress	
	Juniperus formosana
Prickly juniper	*Juniperus oxycedrus*

Prickly moses	*Acacia verticillata*
Prickly palm	*Bactris major*
Prickly pine	*Pinus pungens*
Prickly pole	*Bactris guineensis*
Prickly wattle	*Acacia juniperina*
Pride of Barbados	
	Caesalpinia pulcherrima
Pride of Bolivia	*Tipuana tipu*
Pride of Burma	*Amhertsia nobilis*
Pride of China	*Melia azedarach*
Pride of India	
	Koelreuteria paniculata
	Lagerstroemia speciosa
	Melia azedarach
Prim privet	*Ligustrum vulgare*
Primrose jasmine	
	Jasminum mesnyi
Primrose tree	
	Lagunaria patersonii
Prince Albert's yew	
	Saxegothaea conspicua
Prince Rupprecht larch	
	Larix gmelini
	principis-rupprechtii
Princess palm	*Dictyosperma album*
Princess tree	
	Paulownia tomentosa
Prostrate broom	
	Cytisus decumbens
Provence rose	*Rosa centifolia*
	Rosa gallica
Provision tree	*Pachira aquatica*
Prune	*Prunus domestica*
Puerto Rican hat palm	
	Palmetto causiarum
	Sabal causiarum
Puerto Rican holly	
	Olmediella betschlerana
Puerto Rican royal palm	
	Roystonea borinquena
Pudding berry	*Cornus canadensis*
Pudding-pipe tree	*Cassia fistula*
Puka	*Meryta sinclairii*
Pulasan	*Nephelium lappaceum*
	Nephelium mutabile
Pumelo	*Citrus grandis*
Pummelo	*Citrus maxima*
Pumpkin ash	*Fraxinus tomentosa*
Punk tree	
	Melaleuca quinquenervia
Purging buckthorn	
	Rhamnus catharticus
Purging cassia	*Cassia fistula*
Purging fistula	*Cassia fistula*
Purging nut	*Jatropha curcus*
Purple anise	*Illicium floridanum*
Purple apricot	*Prunus × dasycarpa*
Purple beech	
	Fagus sylvatica purpurea

Purple birch	
	Betula pendula purpurea
Purple broom	*Cytisus purpurea*
Purple chokeberry	
	Aronia prunifolia
	Malus floribunda
Purple crab apple	
	Malus × purpurea
Purple English oak	
	Quercus rober purpurescens
Purple fern-leaved beech	
	Fagus sylvatica rohanii
Purple flash	
	Prunus cerasifera pissardii
Purple heather	*Erica cinerea*
Purple-leaf barberry	*Berberis*
	thunbergii atropurpurea
	Berberis vulgaris atropurpurea
Purple-leaf birch	
	Betula pendula purpurea
Purple-leaf bird cherry	
	Prunus padus colorata
Purple-leaf sand cherry	
	Prunus × cistena
Purple-leaved cob	
	Corylus maxima purpurea
Purple-leaved daphne	
	Daphne × houtteana
Purple-leaved filbert	
	Corylus maxima purpurea
Purple-leaved plum	
	Prunus cerasifera pissardii
Purple-leaved sycamore	
	Acer pseudoplatanus
	altropurpureum
Purple mombin	*Spondias purpurea*
Purple orchid tree	
	Bauhinia variegata
Purple osier	*Salix purpurea*
Purple plum	*Prunus cerasifera*
	atropurpurea
Purple smoke tree	
	Cotinus coggygria purpureus
Purple spruce	
	Picea likiangensis purpurea
Purple-twig willow	*Salix acutifolia*
Purple weigela	
	Weigela florida purpurea
Pururi tree	*Vitex lucens*
Pussy willow	*Salix caprea*
	Salix discolor
Puzzle willow	*Salix × ambigua*
	Salix mutabilis
Pygmy date palm	
	Phoenix roebelenii
Pygmy rowan	*Sorbus reducta*
Pyrenean oak	*Quercus pyrenaica*
Pyrenean pine	
	Pinus nigra cebennensis
Pyrenean whitebeam	
	Sorbus mougeotii
Qat	*Catha edulis*

Quaking aspen	
	Populus tremuloides
Quandong	*Fusanus acuminatus*
Queen palm	
	Arecastrum romanzoffianum
Queen sago	*Cycas circinalis*
Queen's cape myrtle	
	Lagerstroemia speciosa
Queensland bottle tree	
	Brachychiton rupestris
	Sterculia rupestris
Queensland firewheel tree	
	Stenocarpus sinuatus
Queensland hog plum	
	Pleiogynium cerasiferum
Queensland kauri	*Agathis brownii*
Queensland lacebark	
	Brachychiton discolor
Queensland nut	
	Macadamia integrifolia
	Macadamia ternifolia
Queensland pittosporum	
	Pittosporum rhombifolium
Queensland poplar	
	Homalanthus populifolius
Queensland pyramidal tree	
	Lagunaria patersonii
Queensland silver wattle	
	Acacia podalyriifolia
Queensland umbrella tree	
	Brassaia actinophylla
Queensland wattle	
	Acacia podalyriifolia
Queensland yellowwood	
	Rhodosphaera rhodanthema
Queen's umbrella tree	
	Brassaia actinophylla
Quercitron	*Quercus velutina*
Quick	*Crataegus monogyna*
Quickbeam	*Sorbus aucuparia*
Quicken tree	*Sorbus aucuparia*
Quick-set thorn	
	Crataegus laevigata
Quickthorn	*Crataegus monogyna*
Quince	*Cydonia oblonga*
Quinine	*Cinchona officinalis*
Quinine bush	*Garrya elliptica*
Quiverleaf	
	Populus tremuloides
Rabbit-eye blueberry	
	Vaccinium virgatum
Rabbit root	*Aralia nidicaulis*
Raffia palm	*Raphia farinifera*
	Raphia ruffia
Railway poplar	*Populus regenerata*
Rainbow dogwood	
	Cornus florida rainbow
Rain tree	*Brunfelsia undulata*
	Samanea saman
Ramanas rose	*Rosa rugosa*
Ramarama	*Myrtus bullata*

Rambutan	*Nephelium lappaceum*
Ramin	*Gonystylus bancanus*
Ramontchi	*Flacourtia indica*
	Flacourtia ramontchi
Ramsthorn	*Rhamnus catharticus*
Rangpur lime	*Citrus × limonia*
Rantry	*Sorbus aucuparia*
Raoul beech	*Nothofagus procera*
Raspberry-jam tree	
	Acacia acuminata
Rata	*Metrosideros robusta*
Rattan cane	*Calamus rotang*
Rauli beech	*Nothofagus procera*
Real yellowwood	
	Podocarpus latifolius
Red alder	*Alnus oregona*
	Alnus rubra
Red ash	*Fraxinus pennsylvanica*
Red-barked dogwood	*Cornus alba*
Red bay	*Persea borbonia*
Red bearberry	
	Arctostaphylos uva-ursi
Red beech	*Nothofagus fusca*
Red beefwood	
	Casuarina equisetifolia
Red-berried elder	
	Sambucus pubens
	Sambucus racemosa
Redberry buckthorn	
	Rhamnus crocea
Red-berry juniper	
	Juniperus pinchotii
Red bilberry	
	Vaccinium parvifolium
	Vaccinium vitis-idaea
Red birch	*Betula nigra*
Red bloodwood	
	Eucalyptus gummifera
Red broad-leaved lime	
	Tilia platyphyllos rubra
Red buckeye	*Aesculus pavia*
Redbud	*Cercis canadensis*
Red-bud maple	*Acer trautvetteri*
Red canella	
	Cinnamodendron corticosum
Red cedar	
	Acrocarpus fraxinifolius
	Juniperus virginiana
Red chokeberry	
	Aronia arbutifolia
Red cinchona	*Cinchona succirula*
Red cypress pine	
	Callitris endlicheri
Red elder	*Viburnum opulus*
Red elm	*Ulmus rubra*
	Ulmus serotina
Red fir	*Abies magnifica*
Red flag bush	
	Mussaenda eythrophylla
Red-flowered mallee	
	Eucalyptus ewartiana

Red-flowering gum
    Eucalyptus ficifolia

Red gum   *Eucalyptus calophylla*
    Eucalyptus camaldulensis
    Ceratopetalum gummiferum
    Liquidambar styraciflua

Red haw   *Crataegus mollis*

Red hawthorn *Crataegus laevigata*

Red heart   *Ceanothus spinosus*

Red hickory   *Carya ovalis*

Red horse-chestnut
    Aesculus × carnea

Red-hot cat-tail *Acalypha hispida*

Red huckleberry
    Vaccinium parvifolium

Red ironbark
    Eucalyptus sideroxylon

Red jasmine   *Plumeria rubra*

Red latan   *Latania borbonica*
    Latania lontaroides

Red-leaved acacia *Acacia rubida*

Red-leaved wattle *Acacia rubida*

Red mahogany
    Eucalyptus resinifer

Red mallee
    Eucalyptus erythronema

Red mangrove
    Rhizophora mangle

Red manjack   *Cordia nitida*

Red maple   *Acer rubrum*

Red may   *Crataegus laevigata*

Red mombin   *Spondias purpurea*

Red moort   *Eucalyptus nutans*

Red morell *Eucalyptus longicornis*

Red mulberry   *Morus rubra*

Red oak   *Quercus borealis*
    Quercus rubra

Red-osier dogwood *Cornus sericea*
    Cornus stolonifera

Red-paper tree *Albizzia rhodesica*

Red pine *Dacrydium cupressinum*
    Pinus resinosa
    Podocarpus dacrydioides

Red pokers *Hakea bucculenta*

Red root *Ceanothus americanus*

Red river gum
    Eucalyptus camaldulensis

Red rose   *Rosa gallica*

Red sandalwood
    Pterocarpus santalinus
    Santalum rubrum

Red sandalwood tree
    Adenanthera pavonina

Red saunders
    Pterocarpus santalinus

Red signal heath *Erica mammosa*

Red silk-cotton tree *Bombax ceiba*
    Bombax malabaricum

Red silver fir   *Abies ambilis*

Red siris   *Albizzia toona*

Red snakebark maple
    Acer capillipes

Red spotted gum
    Eucalyptus mannifera

Red spruce   *Picea rubens*
    Picea rubra

Red-stemmed acacia
    Acacia rubida

Red stopper   *Eugenia confusa*
    Eugenia garberi

Red stringybark
    Eucalyptus macrorhyncha

Red titi   *Cyrilla racemiflora*

Redwater tree
    Erythrophleum guineese
    Erythrophleum suaveolens

Red willow   *Cornus amomum*
    Salix laevigata

Redwood *Adenanthera pavonina*
    Sequoia sempervirens

Redwood rose *Rosa gymnocarpa*

Reed palm *Chamaedorea seifrizii*

Reed rhapis   *Rhapis humilis*

Reefwood *Stenocarpus salignus*

Remarkable-cone pine
    Pinus radiata

Rewa rewa *Knightia excelsa*

Rhododendron
    Rhododendron arboreum
    Rhododendron ponticum

Ribbon gum *Eucalyptus viminalis*

Ribbon plant *Dracaena sanderana*

Ribbonwood   *Hoheria lyallii*
    Hoheria populnea
    Hoheria sexstylosa

Ribbonwood tree
    Plagianthus regius

Rice-paper tree
    Tetrapanax papyriferus

Ridge-fruited mallee
    Eucalyptus incrassata

Rimu *Dacrydium cupressinum*

Ring-cupped oak *Quercus glauca*

Ringworm cassia *Cassia alata*

Ringworm powder tree
    Andira araroba

Ringworm senna *Cassia elata*

Rio grande cherry
    Eugenia aggregata

River birch   *Betula nigra*

River maple *Acer saccharinum*

River oak
    Casuarina cunninghamiana

River peppermint
    Eucalyptus andreana
    Eucalyptus elata
    Eucalyptus longifolia

River tea tree
    Melaleuca leucadendron

River walnut *Juglans microcarpa*
    Juglans rupestris

Robertson's peppermint
        Eucalyptus robertsonii
Robinia   *Robinia pseudoacacia*
Robin redbreast bush
        Melaleuca lateritia
Roblé beech   *Nothofagus obliqua*
Roblé blanco   *Nothofagus pumilo*
Roblé de maule   *Nothofagus glauca*
Roblé pellin   *Nothofagus obliqua*
Robusta coffee   *Coffea canephora*
Rock birch   *Betula nana*
Rock buckthorn   *Rhamnus saxitilis*
Rock cherry   *Prunus prostrata*
Rock chestnut oak   *Quercus prinus*
Rock cotoneaster
        Cotoneaster horizontalis
Rock elm   *Ulmus thomasi*
Rock maple   *Acer glabrum*
        Acer saccharum
Rock sallow   *Salix petraea*
Rock whitebeam   *Sorbus rupicola*
Rocky mountain cherry
        Prunus besseyi
Rocky mountain fir
        Abies lasiocarpa
Rocky mountain maple
        Acer glabrum
Rocky mountains bramble
        Rubus deliciosus
Rocky mountain scrub oak
        Quercus undulata
Rocky mountains juniper
        Juniperus jackii
        Juniperus scopulorum
Roebelin palm   *Phoenix roebelenii*
Roman candle   *Yucca gloriosa*
Roman laurel   *Laurus nobilis*
Rooiels   *Cunonia capensis*
Ropebark   *Dirca palustris*
Rosa mundi   *Rosa gallica versicolor*
Rose acacia   *Robinia hispida*
Rose apple   *Eugenia aquea*
        Syzygium jambos
        Syzygium malaccense
Rose bay oleander
        Nerium oleander
Rose bay
        Rhododendron maximum
        Rhododendron ponticum
Rose box
        Cotoneaster microphyllus
Rosebud cherry   *Prunus subhirtella*
        Prunus subhirtella ascendens
Rose gum   *Eucalyptus grandis*
Rose imperial
        Cochlospermum vitifolium
Roselle   *Hibiscus sabdariffa*
Rose mallee
        Eucalyptus × rhodantha
Rose mallow   *Lavatera trimestris*

Rosemary   *Rosmarinus officinalis*
Rose of sharon   *Althaea frutex*
        Hibiscus syriacus
        Hypericum calycinum
Rose of Venezuela
        Brownea grandiceps
Rosewood   *Tipuana tipu*
Rosy trumpet tree   *Tabebuia rosea*
Rottnest Island pine
        Callitris preissii
Rouen lilac   *Syringa × chinensis*
Rough-barked maple
        Acer triflorum
Rough-barked Mexican pine
        Pinus montezumae
Rough-leaved hydrangea
        Hydrangea aspera
Rough netbush   *Calothamnus asper*
Rough-shell macadamia nut
        Macadamia tetraphylla
Round buchu   *Agathosma betulina*
Round-eared willow   *Salix aurita*
Round kumquat
        Fortunella japonica
Round-leaved dogwood
        Cornus rugosa
Round-leaved mallee
        Eucalyptus orbifolia
Round-leaved moort
        Eucalyptus platypus
Round-leaved snow gum
        Eucalyptus perriniana
Roundwood   *Sorbus americana*
Rowan   *Sorbus aucuparia*
Royal bay   *Laurus nobilis*
Royal fern   *Osmunda regalis*
Royal hakea   *Hakea victoriae*
Royal jasmine
        Jasminum grandiflorum
Royal lime   *Tilia × europaea pallida*
Royal palm   *Roystonea regia*
Royal poinciana   *Delonix regia*
Royoc   *Morinda royoc*
Rubber euphorbia
        Euphorbia tirucalli
Rubber plant   *Ficus belgica*
        Ficus elastica
Rubber tree   *Hevea brasiliensis*
Ruby saltbush
        Enchylaena tomentosa
Rue   *Ruta graveolens*
Ruffled fan palm   *Licula grandis*
Ruffle pine   *Aiphanes caryotifolia*
Rum cherry   *Prunus serotina*
Russian almond   *Prunus tenella*
Russian cedar   *Pinus cembra*
Russian elm   *Zelkova carpinifolia*
Russian mountain ash
        Sorbus aucuparia rossica-major

TREES, BUSHES, AND SHRUBS

Sarvis holly

Russian olive
    Elaeagnus angustifolia
Russian pea shrub   *Caragana frutex*
Russian rock birch
    Betula ermanii
Russian vine
    Polygonum baldschuanicum
Rusty fig   *Ficus australis*
    Ficus rubiginosa
Rusty nannyberry
    Viburnum rufidulum
Rusty podocarp
    Podocarpus ferrugineus
Sabal   *Serenoa repens*
Sabicu   *Lysiloma latisiliqua*
Sacramento rose
    Rosa stellata mirifica
Sacred bamboo
    Nandina domestica
Sacred fig   *Ficus religiosa*
Sacred fir   *Abies religiosa*
Sacred flower   *Cantua buxifolia*
Sacred garlic pear
    Crateva religiosa
Sage-leaved pear   *Pyrus salvifolia*
Sage-leaved rock rose
    Cistus salviifolius
Sage-leaved willow
    Salix salviaefolia
Sage tree   *Vitex agnus-castus*
Sage willow   *Salix candida*
Saghalin spruce   *Picea glehnii*
Sago palm   *Caryota urens*
    Cycas circinalis
    Cycas revoluta
    Metroxylon sagu
Saigon cinnamon
    Cinnamomum loureirii
Saint Daboec's heath
    Daboecia cantabrica
Saint John's bread
    Ceratonia siliqua
Saint Lucie cherry   *Prunus mahaleb*
Saint Mary's wood
    Calophyllum brasiliense
Sakhalin fir   *Abies sachalinensis*
Sakhalin spruce   *Picea glehnii*
Sal   *Shorea robusta*
Salac   *Salacea edulis*
Sallow   *Salix caprea*
Sallow thorn
    Hippophae rhamnoides
Sally wattle   *Acacia binervia*
Salmon berry   *Rubus parviflorus*
Salmon gum
    Eucalyptus salmonophloia
Salmon white gum
    Eucalyptus lane-poolei
Salt cedar   *Tamarix gallica*
Salt tree
    Halimodendron halodendron

Saman tree   *Samanea saman*
Sandal tree   *Sandoricum indicum*
Sandalwood   *Santalum album*
Sandalwood tree
    Adenanthera pavonina
Sandbar willow   *Salix interior*
Sandberry   *Arctostaphylos uva-ursi*
Sandbox tree   *Hura crepitans*
Sand cherry   *Prunus besseyi*
    Prunus depressa
    Prunus pumila
Sanderswood
    Pterocarpus santalinus
Sandfly bush   *Zieria smithii*
Sand hickory   *Carya pallida*
San Diego ceanothus
    Ceanothus cyaneus
Sand jack   *Quercus incana*
Sand myrtle
    Leiophyllum buxifolium
Sand pear   *Pyrus pyrifolia*
    Pyrus ussuriensis
Sand pine   *Pinus clausa*
Sand plum   *Prunus angustifolia*
Sand willow   *Salix arenaria*
San José hesper palm
    Brahea brandegeei
Santa Barbara ceanothus
    Ceanothus dentatus
    Ceanothus impressus
Santa Cruz cypress
    Cupressus abramsiana
Santa Lucia fir   *Abies bracteata*
    Abies venusta
Santa Maria
    Calophyllum brasiliense
Sapele
    Entandrophragma cylindricum
Sapodilla   *Manilkara zapota*
Sapodilla plum   *Achras sapota*
    Manilkara zapota
Sapote   *Pouteria sapota*
Sapote amarillo
    Pouteria campechiana
Sapote borracho
    Pouteria campechiana
Sappanwood   *Caesalpinia sappan*
Sapphire berry
    Symplocos paniculata
Sapree wood
    Widdringtonia cupressoides
Sapucia nut   *Lecythis zabucayo*
Sargent's cherry   *Prunus sargentii*
Sargent's cypress
    Cupressus sargentii
Sargent spruce   *Picea brachytyla*
Sargent's rowan
    Sorbus sargentiana
Sarsaparilla   *Aralia nudicaulis*
Sarvis holly   *Ilex amelanchier*

Sasanqua camellia

Sasanqua camellia	
	Camellia sasanqua
Sassafras	*Sassafras albidum*
Sassy bark	
	Erythrophleum guineese
Satin bush	*Podalyria sericea*
Satin hibiscus	*Hibiscus huegelii*
Satinleaf	
	Chrysophyllum oliviforme
Satinwood	*Chloroxylon swietenia*
	Fagara flava
	Liquidambar styraciflua
	Murraya paniculata
Satiny willow	*Salix pellita*
Satsuma orange	*Citrus reticulata*
Saucer magnolia	
	Magnolia × soulangiana
Sausage tree	*Kigelia africana*
	Kigelia pinnata
Savin	*Juniperus sabina*
Sawara cypress	
	Chamaecyparis pisifera
Saw briar	*Smilax glauca*
Saw cabbage palm	
	Acoelorrhaphe wrightii
Saw-leaf zelkova	*Zelkova serrata*
Saw palmetto	*Serenoa repens*
Sawtooth oak	*Quercus acutissima*
Scaly-leaved Nepal juniper	
	Juniperus squamata
Scandinavian juniper	
	Juniperus suecica
Scarab cypress	
	Chamaecyparis lawsoniana
	allumii
Scarlet bush	*Hamelia patens*
Scarlet gum	*Eucalyptus phoenicia*
Scarlet haw	*Crataegus pedicellata*
Scarlet maple	*Acer rubrum*
Scarlet oak	*Quercus coccinea*
Scarlet sumac	*Rhus glabra*
Scarlet willow	
	Salix alba chermesina
Scented paperbark	
	Melaleuca squarrosa
Scholar's tree	*Sophora japonica*
Schrenk's spruce	
	Picea schrenkiana
Scorpion senna	*Coronilla emerus*
Scotch briar	*Rosa pimpinellifolia*
Scotch broom	*Cytisus scoparius*
Scotch elm	*Ulmus glabra*
Scotch fir	*Pinus sylvestris*
Scotch heath	*Erica cinerea*
Scotch heather	*Calluna vulgaris*
Scotch laburnum	
	Laburnum alpinum
Scotch rose	*Rosa pimpinellifolia*
Scots pine	*Pinus sylvestris*
Scouler willow	*Salix scoulerana*

Screw bean	*Prosopis pubescens*
Screw pine	*Pandanus leram*
Screw-pod wattle	*Acacia implexa*
Scribbly gum	
	Eucalyptus haemastoma
Scrub bottle-tree	
	Brachychiton discolor
Scrub oak	*Quercus dumosa*
	Quercus illicifolia
Scrub palmetto	*Sabal adansonii*
	Sabal etonia
	Sabal minor
	Sabal repens
	Serenoa repens
Scrub pine	*Pinus banksiana*
	Pinus virginiana
Scrub sumac	*Rhus microphylla*
Sea apple	*Syzygium grande*
Sea ash	
	Zanthoxylum clava-herculis
	Xanthoxlyum clava-herculis
Sea buckthorn	
	Hippophae rhamnoides
Sea fig	*Ficus superba*
Sea grape	*Coccoloba uvifera*
Sealing-wax palm	
	Cyrostachys lakka
	Cyrostachys renda
Sea orach	*Atriplex halinus*
Sea purslane	
	Atriplex portulacoides
Seaside alder	*Alnus maritima*
Sea-urchin tree	*Hakea laurina*
Selu	*Cordia myxa*
Senegal date palm	
	Phoenix reclinata
Senegal gum	*Acacia senegal*
Senegal mahogany	
	Khaya senegalensis
Senegal rosewood	
	Pterocarpus erinaceus
Senna	*Cassia acutifolia*
Sentol	*Sandoricum koetjapa*
Sentry palm	*Howea forsterana*
	Kentia forsterana
September elm	*Ulmus serotina*
Serbian laurel	
	Prunus laurocerasus serbica
Serbian spruce	*Picea omorika*
Serpentine manzanita	
	Arctostaphylos obispoensis
Service tree	*Sorbus domestica*
Sessile oak	*Quercus petraea*
Seven barks	
	Hydrangea arborescens
Seven sisters rose	
	Rosa multiflora grevillei
Seville orange	*Citrus aurantium*
Shad berry	*Amelanchier laevis*
Shad blow	*Amelanchier laevis*

TREES, BUSHES, AND SHRUBS

Silver birch

Shad bush   *Amelanchier laevis*
Shaddock   *Citrus maxima*
Shagbark hickory   *Carya ovata*
Shaggy-bark manzanita
          Arctostaphylos tomentosa
Shag-spine pea shrub
          Carangana jubata
Shallon   *Gaultheria shallon*
Shamel ash   *Fraxinus uhdei*
Shantung maple   *Acer truncatum*
Sharp cedar   *Juniperus oxycedrus*
Shaving-brush tree
          Pseudobombax ellipticum
Shea butter tree
          Butyrospermum paradoxum
          Butyrospermum parkii
She balsam   *Abies fraseri*
Sheepberry   *Viburnum lentago*
          Viburnum prunifolium
Sheep laurel   *Kalmia angustifolia*
Shell-bark hickory   *Carya laciniosa*
          Carya ovata
Shensi fir   *Abies chensiensis*
She-oak   *Casuarina stricta*
She pine   *Podocarpus elatus*
Shingle oak   *Quercus imbricaria*
Shingle tree
          Acrocarpus fraxinifolius
Shining honeysuckle
          Lonicera nitida
Shining privet   *Ligustrum lucidum*
Shining sumach   *Rhus copallina*
Shining willow   *Salix lucida*
Shinnery oak   *Quercus havardii*
Shin oak   *Quercus gambelii*
Shipmast acacia
          Robinia pseudoacacia
          appalachia
Shittimwood   *Acacia nilotica*
          Bumelia lanuginosa
          Bumelia lycioides
          Halesia carolina
Shore juniper   *Juniperus conferta*
Shore pine   *Pinus contorta*
Shore plum   *Prunus maritima*
Short-leaf pine   *Pinus echinata*
Shot huckleberry
          Vaccinium ovatum
Showy crab apple
          Malus floribunda
Shrubby dogwood
          Cornus sanguinea
Shrubby germander
          Teucrium fruticans
Shrubby horsetail
          Ephedra distachya
Shrubby mallow   *Lavatera olbia*
Shrubby musk
          Mimulus aurantiacus
Shrubby pavia   *Aesculus parviflora*

Shrubby trefoil   *Ptelea trifoliata*
Shrub tobacco   *Nicotiana glauca*
Shui-hsa
          Metasequoia glyptostroboides
Shumard's oak   *Quercus shumardii*
Shumard's red oak
          Quercus shumardii
Siberian apricot   *Prunus sibirica*
Siberian balsam poplar
          Populus laurifolia
Siberian crab apple   *Malus baccata*
          Malus × robusta
Siberian dogwood   *Cornus alba*
Siberian elm   *Ulmus pumila*
Siberian fir   *Abies sibirica*
Siberian larch   *Larix russica*
          Larix sibirica
Siberian pea tree
          Caragana arborescens
Siberian spruce   *Picea obovata*
Sibipiruna
          Caesalpinia peltophoroides
Sicilian sumac   *Rhus coriaria*
Sickle senna   *Cassia tora*
Sicklepod   *Cassia tora*
Siebold's beech   *Fagus sieboldii*
Siebold's hemlock   *Tsuga sieboldii*
Sierra fir   *Abies concolor lowiana*
Sierra juniper
          Juniperus occidentalis
Sierra Leone tamarind
          Dialium guineense
Sierra plum   *Prunus subcordata*
Sierra redwood
          Sequoiadendron giganteum
Sikkim larch   *Larix griffithiana*
          Larix griffithii
Sikkim spruce   *Picea spinulosa*
Silk-cotton tree   *Ceiba pentandra*
          Cochlospermun religiosum
Silk fruit tree
          Seriocarpus conyzoides
Silk grass   *Yucca filamentosa*
Silk tassel bush   *Garrya elliptica*
Silk tree   *Albizzia julibrissin*
Silkworm tree   *Morus alba*
Silky camellia
          Stuartia malacodendron
          Stewartia malacodendron
Silky dogwood   *Cornus amomum*
          Cornus purpusii
Silky oak   *Grevillea robusta*
Silky willow   *Salix sericea*
Silver beech   *Nothofagus menziesii*
Silver bell   *Halesia carolina*
          Halesia tetraptera
Silver berry   *Elaeagnus angustifolia*
          Elaeagnus argentea
          Elaeagnus commutata
Silver birch   *Betula pendula*

Silver broom palm
Thrinax argentea

Silver buffalo berry
Shepherdia argentea

Silverbush *Sophora tomentosa*

Silver dogwood
Cornus alba elegantissima

Silver dollar gum
Eucalyptus polyanthemos

Silver dollar tree
Eucalyptus cinerea

Silver fir *Abies alba*

Silver heather
Cassinia vauvilliersii albida

Silver hedgehog holly
Ilex aquifolium argentea

Silver holly *Berberis hypokerina*

Silver-leaf manzanita
Arctostaphylos silvicola

Silver-leaved mountain gum
Eucalyptus pulverulenta

Silver-leaved poplar *Populus alba*

Silver lime *Tilia tomentosa*

Silver mallee scrub
Eucalyptus polybractea

Silver maple *Acer saccharinum*

Silver palm *Coccothrinax argentata*

Silver pear *Pyrus salicifolia*

Silver peppermint
Eucalyptus risdonii
Eucalyptus tenuiramis

Silver privet *Ligustrum
ovalifolium argenteum*

Silver saw palm
Acoelorrhaphe wrightii

Silver spruce *Picea sitchensis*

Silver thatch
Coccothrinax argentea

Silver top *Eucalyptus nitens*

Silver tree
Leucodendron argenteum

Silver trumpet tree
Tabebuia argentia

Silver variegated dogwood
Cornus alba elegantissima

Silver wattle *Acacia dealbata*

Silver willow *Salix alba argentea*
Salix alba sericea

Simarouba *Simarouba amara*

Simon's plum *Prunus simonii*

Singapore holly
Malpighia coccigera

Singapore oak *Quercus conocarpa*

Single-leaf ash
Fraxinus excelsior diversifolia

Single-leaf pinyon pine
Picea monophylla

Single-leaf pine *Pinus monophylla*

Single-leaved ash
Fraxinus angustifolia veltheimii
Fraxinus anomala

Siris tree *Albizzia lebbeck*

Siskiyou cypress
Cupressus bakeri mathewsii

Siskiyou-mat *Ceanothus pumilus*

Siskiyou spruce *Picea brewerana*

Sissoo *Dalbergia sissoo*

Sitka alder *Alnus sinuata*

Sitka spruce *Picea sitchensis*

Skewerwood *Euonymus europaea*

Skunk bush *Rhus trilobata*

Skyflower *Duranta ellisia*
Duranta repens

Slash pine *Pinus elliotii*

Sleeping hibiscus
Malvaviscus arboreus

Sleepy mallow
Malaviscus arboreus

Slender lady palm *Rhapis excelsa*
Rhapis humilis

Slender smokebush
Conospermum huegelii

Slippery elm *Ulmus rubra*

Sloe *Prunus alleghaniensis*
Prunus spinosa

Small-fruited grey gum
Eucalyptus propinqua

Small-fruited Queensland nut
Macadamia ternifolia

Small-leaved elm *Ulmus alata*

Small-leaved gum
Eucalyptus parvifolia

Small-leaved lime *Tilia cordata*

Small-leaved rubber plant
Ficus benjamina

Small-leaved sumac
Rhus microphylla

Small pussy willow *Salix humilis*

Smoke bush
Conospermum huegelii
Cotinus coggygria

Smoke tree *Cotinus coggygria*
Rhus cotinus

Smooth alder *Alnus serrulata*
Alnus rugosa

Smooth Arizona cypress
Cupressus glabra

Smooth-barked Arizona cypress
Cupressus arizonica
Cupressus glabra

Smooth Japanese maple
Acer japonicum
Acer palmatum

Smooth-leaved elm
Ulmus carpinifolia
Ulmus carpinifolia sarniensis
Ulmus minor

Smooth rambutan
Alectryon subcinereus

Smooth rose *Rosa blanda*

Smooth senna *Cassia laevigata*

Smooth sumac *Rhus glabra*

Smooth Tasmanian cedar
Athrotaxis cupressoides

Smooth winterberry
Ilex laevigata

Smooth withe-rod
Viburnum nudum

Smooth withy-rod
Viburnum nudum

Snailseed *Coccoloba diversifolia*

Snakebark maple *Acer capillipes*
Acer crataegifolium
Acer forrestii
Acer hersii
Acer laxiflorum
Acer pennsylvanicam
Acer rufinerve

Snake bush
Duvernoia adhatodioides

Snakewood tree *Cecropia palmata*

Snappy gum *Eucalyptus micrantha*
Eucalyptus racemosa

Snowball tree
Viburnum opulus sterile

Snowbell tree *Styrax japonica*

Snowberry *Gaultheria hispida*
Symphoricarpus albus
Symphoricarpus rivularis

Snowbush *Breynia disticha*
Ceanothus cordulatus

Snow camellia
Camellia japonica rusticana

Snowdrop tree *Halesia carolina*
Halesia tetraptera

Snowflake aralia *Trevesia palmata*

Snowflake tree *Trevesia palmata*

Snowflower
Chionanthus virginicus

Snow gum *Eucalyptus niphophila*

Snow heath *Erica herbacea*

Snow heather *Erica carnea*

Snow in summer
Helichrysum rosmarinifolium
Ozothamnus rosmarinifolius
Ozothamnus thyrsoideus

Snow pear *Pyrus nivalis*

Snowy mespil *Amelanchier laevis*
Amelanchier lamarckii
Amelanchier ovalis

Snuffbox tree *Oncoba spinosa*

Soap bark tree *Quillaia saponaria*

Soapberry *Sapindus saponaria*
Shepherdia canadensis

Soapberry tree
Sapindus drummondii

Soap bush *Noltea africana*

Soap tree *Quillaia saponaria*
Yucca elata

Soapweed *Yucca elata*
Yucca glauca

Soapwell *Yucca glauca*

Socotra cucumber tree
Dendrosicyos socotrana

Soft-leaved rose *Rosa villosa*

Soft maple *Acer rubrum*
Acer saccharinum

Soldier rose mallow
Hibiscus militaris

Soledad pine *Pinus torreyana*

Solitary palm
Ptychosperma elegans

Somali tea *Catha edulis*

Sonoma manzanita
Arctostaphylos densiflora

Sonoran palmetto *Sabal uresana*

Sorbet *Cornus mas*

Sorrel tree
Oxydendrum arboreum

Sorrowless tree *Saraca indica*

Soulard crab apple
Malus × soulardii

Sourberry *Rhus integrifolia*

Sour cherry *Prunus cerasus*

Sour gum *Nyssa sylvatica*

Sour orange *Citrus aurantium*

Soursop *Annona muricata*

Sour top *Vaccinium canadanse*

Sourwood tree
Oxydendrum arboreum

South African sagewood
Buddleia salviifolia

South American apricot
Mammea americana

South American crowberry
Empetrum rubrum

South American pine
Dacrydium fonkii

South American royal palm
Roystonea oleracea

Southern arrowwood
Viburnum dentatum

Southern balsam fir *Abies fraseri*

Southern black haw
Viburnum rufidulum

Southern blue gum
Eucalyptus globulus

Southern buckthorn
Bumelia lycioides

Southern catalpa
Catalpa bignonioides

Southern cross silver mallee
Eucalyptus crucis

Southern Japanese hemlock
Tsuga sieboldii

Southern live oak
Quercus virginiana

Southern magnolia
Magnolia grandiflora

Southern nettle tree
Celtis australis

Southern pine *Pinus palustris*

Southern pitch pine *Pinus palustris*

Southern prickly ash	
	Xanthoxylum clava-herculis
	Zanthoxylum clava-herculis
Southern red cedar	
	Juniperus silicicola
Southern sassafras	
	Atherosperma moschatum
Southern sugar maple	
	Acer barbatum
Southern white cedar	
	Chamaecyparis thyoides
Southern wild crab apple	
	Malus angustifolia
Southern yellow pine	
	Pinus palustris
Southern yew	
	Podocarpus macrophyllus
South Queensland kauri	
	Agathis robusta
South sea ironwood	
	Casuarina equisetifolia
Spadic	*Erythroxylum coca*
Spaeth's flowering ash	
	Fraxinus spaethiana
Spanish bayonet	*Yucca aloifolia*
	Yucca baccata
Spanish broom	*Genista hispanica*
	Spartium junceum
Spanish buckeye	
	Ungnadia speciosa
Spanish cedar	*Cedrela odorata*
	Toona odorata
Spanish cherry	*Mimusops elengi*
Spanish chestnut	*Castanea sativa*
Spanish cordia	*Cordia sebestena*
Spanish dagger	*Yucca aloifolia*
	Yucca carnerosana
Spanish fir	*Abies pinsapo*
Spanish gorse	*Genista hispanica*
Spanish heath	*Erica australis*
	Erica carnea
	Erica lusitanica
Spanish jasmine	
	Jasminum grandiflorum
Spanish juniper	*Juniperus thurifera*
Spanish lime	*Melicoccus bijugatus*
Spanish mahogany	
	Swietenia mahagoni
Spanish oak	*Quercus falcata*
	Quercus palustris
Spanish plum	*Spondias purpurea*
Spanish red oak	*Quercus falcata*
Spanish royal palm	
	Roystonea hispaniolana
Spanish savin	
	Juniperus sabina tamariscifolia
Spanish stopper	*Eugenia buxifolia*
	Eugenia foetida
	Eugenia myrtoides
Speckled alder	*Alnus rugosa*
Spiceberry	*Ardisia crenata*

Spice bush	*Benzoin aestivale*
	Lindera benzoin
Spice guava	*Psidium montanum*
Spicy jatropha	
	Jatropha integerrima
Spikenard	*Aralia nudicaulis*
	Aralia racemosa
Spike winter hazel	
	Corylopsis spicata
Spindle palm	
	Hyophorbe verschaffeltii
Spindle tree	*Euonymus europaea*
Spineless broad-leaved holly	
	Ilex × altaclarensis camelliifolia
Spineless yucca	*Yucca elephantipes*
Spiny pine	*Aiphanes caryotifolia*
Spinning gum	
	Eucalyptus perriniana
Spiral eucalyptus	
	Eucalyptus cinerea
Sponge tree	*Acacia farnesiana*
Spoonleaf yucca	*Yucca filamentosa*
Spoon tree	*Cunonia capensis*
Spoonwood	*Kalmia latifolia*
Spotted dracaena	
	Dracaena surculosa
Spotted emu bush	
	Eremophila maculata
Spotted fig	*Ficus infectoria*
	Ficus lacor
	Ficus virens
Spotted gum	*Eucalyptus maculata*
Spotted laurel	*Aucuba japonica*
Spotted mountain gum	
	Eucalyptus goniocalyx
Spray bush	*Holodiscus discolor*
Spread-leaved pine	*Pinus patula*
Spring cherry	*Prunus subhirtella*
Spring heath	*Erica herbacea*
Spruce pine	*Pinus glabra*
	Pinus virginiana
Spurge laurel	*Daphne laureola*
Spurge olive	*Cneorum tricoccum*
Spur-leaf tree	*Tetracentron sinense*
Square-fruited mallee	
	Eucalyptus tetraptera
Squarenut	*Carya tomentosa*
Squaw berry	*Mitchella repens*
Squaw bush	*Viburnum trilobum*
Squaw carpet	*Ceanothus prostratus*
Squaw huckleberry	
	Vaccinium stamineum
Squaw vine	*Mitchella repens*
Stagbush	*Viburnum prunifolium*
Stagger bush	*Lyonia mariana*
Stag's horn sumac	*Rhus hirta*
	Rhus typhina
Star acacia	*Acacia verticillata*

TREES, BUSHES, AND SHRUBS

Swamp immortelle

Star anise	Illicium anisatum
	Illicium religiosum
	Illicium verum
Star apple	Chrysophyllum cainito
Star fruit	Damasonium alisma
Star jasmine	
	Jasminum gracillimum
	Jasminum multiflorum
Starleaf	Brassaia actinophylla
Starry magnolia	Magnolia stellata
Star wattle	Acacia verticillata
Steedman's gum	
	Eucalyptus steedmanii
Sticky wattle	Acacia howittii
Stiff dogwood	Cornus stricta
Stiff-leaved juniper	Juniperus rigida
Stinking ash	Ptelea trifoliata
Stinking cedar	Torreya taxifolia
Stinking elder	Sambucus pubens
Stinking juniper	
	Juniperus foetidissima
Stinking tutsan	
	Hypericum hircinum
Stinking weed	Cassia occidentalis
Stinkwood	Celtis kraussiana
	Gustavia augusta
Stinking yew	Torreya californica
Stone pine	Picea monophylla
	Pinus pinea
	Pinus monophylla
Storax	Liquidambar orientalis
	Styrax officinalis
Stoward's mallee	
	Eucalyptus stowardii
Strangler fig	Ficus aurea
Strawberry bush	
	Euonymus americana
Strawberry guava	
	Psidium cattleianum
Strawberry-raspberry	
	Rubus illecebrosus
Strawberry shrub	
	Calycanthus floridus
Strawberry tree	Cornus capitata
	Euonymus americana
Strickland's gum	
	Eucalyptus stricklandii
Stringybark tree	
	Eucalyptus globulus
Striped maple	
	Acer pennsylvanicum
Strongback	Bourreria ovata
Strychnine	Strychnos nux-vomica
Sturt's desert pea	
	Clianthus formosus
Styptic weed	Cassia occidentalis
Subalpine fir	Abies lasiocarpa
Sudan gum-arabic	Acacia senegal
Sugar apple	Annona squamosa
Sugarberry	Celtis laevigata
	Celtis occidentalis

Sugarbush	Protea mellifera
	Protea repens
Sugar gum	Eucalyptus cladocalyx
	Eucalyptus corynocalyx
Sugar maple	Acer saccharum
Sugar palm	Arenga pinnata
Sugar pine	Pinus lambertiana
Sugar tree	Acer barbatum
Sulfer rose	Rosa hemisphaerica
	Rosa sulfurea
Sulphur rose	Rosa hemisphaerica
	Rosa sulfurea
Summerberry	Viburnum trilobum
Summer haw	Crataegus flava
Summer lilac	Buddleia davidii
Summer sweet	Clethra alnifolia
Summit cedar	Athrotaxis laxifolia
Sun fruit	Heliocarpus americanus
Sunrise horsechestnut	
	Aesculus neglecta erythroblastos
Sunset hibiscus	
	Abelmoschus manihot
Sunshine wattle	
	Acacia botrycephala
	Acacia paniculata
Suntwood	Acacia nilotica
Surinam cherry	Eugenia uniflora
Surinam quassia	Quassia amara
Sutchuen fir	Abies fargesii
	Abies sutchuenensis
Swamp azalea	
	Rhododendron viscosum
Swamp banksia	Banksia littoralis
Swamp bay	Magnolia virginiana
Swamp bloodwood	
	Eucalyptus ptychocarpa
Swamp blueberry	
	Vaccinium corymbosum
Swamp candleberry	
	Myrica pennsylvanica
Swamp chestnut oak	
	Quercus michauxii
	Quercus prinus
Swamp cottonwood	
	Populus heterophylla
Swamp currant	Ribes lacustre
Swamp cypress	
	Taxodium distichum
Swamp dogwood	Ptelea trifoliata
Swamp elder	Viburnum aopulus
Swamp gum	Eucalyptus ovata
	Eucalyptus rudis
Swamp haw	Viburnum cassinoides
Swamp hickory	Carya cordiformis
Swamp holly	Ilex amelanchier
Swamp honeysuckle	
	Azalea viscosa
	Rhododendron viscosum
Swamp immortelle	Erythrina fusca
	Erythrina glauca
	Erythrina ovalifolia

Swamp laurel

Swamp laurel	*Kalmia glauca*
Swamp locust	*Gleditsia aquatica*
Swamp mahogany	*Eucalyptus robustus*
Swamp mallee	*Eucalyptus spathulata*
Swamp maple	*Acer rubrum*
Swamp paperbark	*Melaleuca ericifolia*
	Melaleuca rhaphiophylla
Swamp post oak	*Quercus lyrata*
Swamp privet	*Forestiera acuminata*
Swamp red bay	*Persea borbonia*
Swamp rose mallow	*Hibiscus moscheutos*
Swamp rose	*Rosa palustris*
Swamp she oak	*Casuarina equisitifolia*
Swamp sumac	*Rhus vernix*
Swamp tea tree	*Melaleuca quinquenervia*
Swamp wattle	*Acacia elongata*
Swamp white cedar	*Chamaecyparis thyoides*
Swamp white oak	*Quercus bicolor*
Swedish birch	*Betula alba dalecarlia*
	Betula pendula dalecarlica
Swedish juniper	*Juniperus communis pyramidalis*
	Juniperus communis suecica
Swedish myrtle	*Myrtus communis*
Swedish whitebeam	*Sorbus intermedia*
	Sorbus suecica
Sweet acacia	*Acacia farnesiana*
	Acacia suaveolens
Sweet almond	*Prunus dulcis*
Sweet bay	*Laurus nobilis*
	Magnolia virginiana
	Persea borbonia
Sweet berry	*Viburnum lentago*
Sweet birch	*Betula lenta*
Sweet briar	*Rosa eglanteria*
	Rosa rubiginosa
Sweet buckeye	*Aesculus flava*
	Aesculus octandra
Sweet cassava	*Manihot dulcis*
Sweet cherry	*Prunus avium*
Sweet chestnut	*Castanea sativa*
Sweet cistin	*Halimium lasianthum formosum*
Sweet crab apple	*Malus coronaria*
Sweet elder	*Sambucus canadensis*
Sweet elm	*Ulmus fulva*
Sweet gale	*Myrica gale*
Sweet gallberry	*Ilex coriacea*
	Ilex lucida

Sweet gum	*Liquidambar styraciflua*
Sweet haw	*Viburnum prunifolium*
Sweetleaf	*Symplocos tinctoria*
Sweet lemon	*Citrus lumia*
Sweet lime	*Citrus limetta*
Sweet locust	*Gleditsia triacanthos*
Sweet olive	*Osmanthus fragrans*
Sweet orange	*Citrus sinensis*
Sweet pea bush	*Podalyrica calptrata*
Sweet pepperbush	*Clethra alnifolia*
Sweet-potato tree	*Manihot esculenta*
Sweet-scented crab apple	*Malus coronaria*
Sweet-scented sumac	*Rhus aromatica*
Sweetsop	*Annona squamosa*
Sweetspire	*Itea virginica*
Sweet sumac	*Rhus aromatica*
Sweet thorn	*Acacia karroo*
Sweet verbena tree	*Backhousia citriodora*
Sweet viburnum	*Viburnum odoratissimum*
	Viburnum prunifolium
Swiss stone pine	*Pinus cembra*
Swiss mountain pine	*Pinus mugo*
Swiss willow	*Salix helvetica*
Swollen-thorn acacia	*Acacia cornigera*
Sword bean	*Entada gigas*
Sugarbush	*Rhus ovata*
Sugar sumac	*Rhus ovata*
Sulphur rose	*Rosa hemisphaerica*
	Rosa sulphurea
Sycamore	*Acer pseudoplatanus*
Sycamore fig	*Ficus sycomorus*
Sydney blue gum	*Eucalyptus saligna*
Sydney golden wattle	*Acacia longifolia*
Sydney peppermint	*Eucalyptus piperita*
Syrian ash	*Fraxinus syriaca*
Syrian bead tree	*Melia azedarach*
Syrian juniper	*Juniperus drupacea*
Syringa	*Philadelphus coronarius*
Szechuan birch	*Betula platyphylla*
Szechwan fir	*Abies sutchuenensis*
Table dogwood	*Cornus controversa*
Table mountain pine	*Pinus pungens*
Tabletop elm	*Ulmus glabra pendula*
Tacamahak	*Populus balsamifera*
Tachibana orange	*Citrus tachibana*
Tag alder	*Alnus serrulata*

Tagasaste	*Cytisus palmensis*
Tagua	*Phytelephas macrocarpa*
Taiwan cherry	
	Prunus campanulata
Taiwan fir	*Cunninghamia konishii*
Taiwan spruce	*Picea morrisonicola*
Tala palm	*Borassus flabellifer*
Talipot palm	
	Corypha umbraculifera
Tall conebush	
	Isopogon anemonifolius
Tallow shrub	*Myrica cerifera*
Tallow tree	*Detarium senegalense*
	Sapium salicifolium
Tallow-wood	
	Eucalyptus microcorys
	Ximenia americana
Tall tutsan	*Hypericum × inodorum*
Tamarack	*Larix laricina*
Tamarind	*Tamarindus indica*
Tamarindo	*Tamarindus indica*
Tamarind-of-the-Indies	
	Vangueria edulis
Tamarisk	*Tamarix anglica*
Tanbark oak	
	Lithocarpus densiflorus
Tanekaha tree	
	Phyllocladus trichomanoides
Tangelo	*Citrus × tangelo*
Tangerine	*Citrus reticulata*
Tanglefoot	*Viburnum alnifolium*
Tanglefoot beech	
	Nothofagus gunnii
Tangor	*Citrus × nobilis*
Tanner's cassia	*Cassia auriculata*
Tanner's sumac	*Rhus coriaria*
Tanner's tree	*Coriaria nepalensis*
Tanoak	*Lithocarpus densiflorus*
Tansy-leaved thorn	
	Crataegus tanacetifolia
Tapa-cloth tree	
	Broussonetia papyrifera
Tapioca	*Manihot esculenta*
Tara	*Caesalpinia spinosa*
Tarajo	*Ilex latifolia*
Tarata	*Pittosporum eugenioides*
Tarentum myrtle	
	Myrtus communis tarentina
Tartar dogwood	*Cornus alba*
Tartar maple	*Acer tataricum*
Tartoga	*Jatropha podagrica*
Tasmanian beech	
	Nothofagus cunninghamii
Tasmanian blue gum	
	Eucalyptus globulus
Tasmanian brown gum	
	Eucalyptus johnstonii
Tasmanian cypress pine	
	Callitris oblonga

Tasmanian daisy bush	
	Olearia phlogopappa
Tasmanian laurel	
	Anopterus glandulosus
Tasmanian snow gum	
	Eucalyptus coccifera
Tasmanian waratah	
	Telopea truncata
Tasmanian waterbush	
	Myoporum tetrandrum
Tasmanian waxberry	
	Gaultheria hispida
Tassel cherry	*Prunus litigiosa*
Tassel-white	*Itea virginica*
Tatarian maple	*Acer tataricum*
Tawhiwhi	
	Pittosporum tenuifolium
Tea-berry	*Gaultheria procumbens*
	Viburnum cassinoides
Tea olive	*Osmanthus fragrans*
Tea plant	*Viburnum lentago*
Tea tree	
	Leptospermum scoparium
	Melaleuca quinquenervia
Tea viburnum	*Viburnum setigerum*
Tecate cypress	*Cupressus forbesii*
	Cupressus guadalupensis
Teak	*Tectona grandis*
Tea-leaf willow	*Salix phylicifolia*
Tea-oil plant	*Camellia oleifera*
Tea plant	*Camellia sinensis*
	Camellia thea
Tea rose	*Rosa × odorata*
Temple juniper	*Juniperus rigida*
Temple orange	
	Citrus × nobilis temple
Temple tree	*Plumeria rubra*
Tenasserim pine	*Pinus kerkusii*
	Pinus merkusii
Terebinth	*Pistacia terebinthus*
Tetterbush	*Lyonia lucida*
Texan buckeye	*Ungnadia speciosa*
Texan walnut	*Juglans microcarpa*
	Juglans rupestris
Texas ash	*Fraxinus texensis*
Texas ebony	
	Pithecellobium flexicaule
Texas mimosa	*Acacia greggii*
Texas mountain laurel	
	Sophora secundiflora
Texas palmetto	*Sabal mexicana*
Texas red oak	*Quercus texana*
Thatch-leaf palm	
	Howea forsterana
	Kentia forsterana
Thatch palm	*Coccothrinax crinata*
Thatch screw pine	
	Pandanus tectorius
Thick-leaved sallow	*Salix crassifolia*
Thirty thorn	*Acacia seyal*

Thorny elaeagnus

Thorny elaeagnus	*Elaeagnus pungens*
Thread palm	*Washingtonia robusta*
Thurlow weeping willow	*Salix elegantissima*
Tibetan cherry	*Prunus mugus*
	Prunus serrula
Tibetan hazel	*Corylus tibetica*
Ti-es	*Pouteria campechiana*
Tiger-tail spruce	*Picea polita*
	Picea torana
Timor white gum	*Eucalyptus alba*
	Eucalyptus platyphylla
Tinnevelly senna	*Cassia angustifolia*
Tingiringi gum	*Eucalyptus glaucescens*
Tipu tree	*Tipuana tipu*
Tisswood	*Persea borbonia*
Titi	*Cliftonia monophylla*
	Cyrilla racemiflora
	Oxydendrum arboreum
Titoki	*Alectryon excelsum*
Toatoa tree	*Phyllocladus glaucus*
Tobacco sumac	*Rhus virens*
Tobago cane	*Bactris guineensis*
Tobira	*Pittosporum tobira*
Toddy palm	*Borassus flabellifer*
	Caryota urens
Tollon	*Heteromeles arbutifolia*
Tolu balsam	*Myroxylon pereirae*
Tolu tree	*Myroxylon balsamum*
Tonka bean	*Dipteryx odorata*
Toog	*Bischofia javanica*
Toon	*Toona sinensis*
Toothache tree	*Xanthoxylum americanum*
	Zanthoxylum americanum
Toothbrush tree	*Salvadora persica*
Topal holly	*Ilex × attenuata*
Toringo crab apple	*Malus sieboldii*
Tornillo	*Prosopis pubescens*
Torote	*Bursera microphylla*
Torrey pine	*Pinus torreyana*
Tortuguero	*Polygala cowellii*
Tossa jute	*Corchorus olitorius*
Totara	*Podocarpus totara*
Towai tree	*Weinmannia racemosa*
Toyon	*Heteromeles arbitifolia*
Trailing rose	*Rosa arvensis*
Trailing sallow	*Salix aurita*
Tramp's spurge	*Euphorbia corollata*
Transcaucasian birch	*Betula medwediewii*
Transvaal hard pear	*Olinia emarginata*

Transvaal teak	*Pterocarpus angolensis*
Trautvetter's maple	*Acer trautvetteri*
Traveller's palm	*Ravenala madagascariensis*
Traveller's tree	*Ravenala madagascariensis*
Tree anemone	*Carpenteria californica*
Tree dracaena	*Dracaena arborea*
Tree flax	*Linum arboreum*
Tree fuchsia	*Fuchsia arborescens*
	Schotia brachypetala
Tree hazel	*Corylus colurna*
Tree heath	*Erica arborea*
Tree hibiscus	*Hibiscus elatus*
Tree hollyhock	*Althaea frutex*
	Hibiscus syriacus
Tree lupin	*Lupinus arboreus*
Tree mallow	*Lavatera arborea*
	Lavatera olbia
Tree of heaven	*Ailanthus altissima*
Tree of gold	*Tabebuia argentea*
Tree of kings	*Cordyline fruticosa*
	Cordyline terminalis
	Dracaena terminalis
Tree of life	*Guaiacum officinale*
	Mauritia flexuosa
	Mauritia setigera
Tree of sadness	*Nyctanthes arbor-tristis*
Tree of the sun	*Chamaecyparis obtusa*
Tree paeony	*Paeonia delavayi*
	Paeonia suffruticosa
Tree poppy	*Dendromecon rigidum*
Tree purslane	*Atriplex halimus*
Tree rhododendron	*Rhododendron arboreum*
Tree tomato	*Cyphomandra betacea*
Tree wisteria	*Bolusanthus speciosus*
Trembling aspen	*Populus tremula*
	Populus tremuloides
Trident maple	*Acer buergeranum*
Trifoliate orange	*Poncirus trifoliata*
Trinidad palm	*Sabal mauritiiformis*
Trip-toe	*Viburnum alnifolium*
Tropical almond	*Terminalia catappa*
Tropical snowflake	*Trevesia palmata*
Tropic laurel	*Ficus benjamina*
True laurel	*Laurus nobilis*
True service tree	*Sorbus domestica*
Truffle oak	*Quercus robur*
Trumpet tree	*Cecropia peltata*
Tuart gum	*Eucalyptus gomphocephalus*

TREES, BUSHES, AND SHRUBS

Venezuelan mahogany

Tuba root	*Derris elliptica*
Tubeflower	
	Clerodendrum indicum
Tucuma	*Astrocaryum aculeatum*
Tufted fishtail palm	*Caryota mitis*
Tufted willow	*Salix nivalis*
Tulipan	*Spathodea campanulata*
Tulip poplar	
	Liriodendron tulipifera
Tulip tree	*Liriodendron tulipifera*
	Spathodea campanulata
Tumbledown gum	
	Eucalyptus dealbata
Tung	*Aleurites montana*
Tung-oil tree	*Aleurites fordii*
Tung tree	*Aleurites fordii*
Tupelo	*Nyssa sylvatica*
Tupelo gum	*Nyssa aquatica*
Turkestan rose	*Rosa rugosa*
Turkey oak	*Quercus cerris*
	Quercus incana
	Quercus laevis
Turkish filbert	*Corylus colurna*
Turkish fir	*Abies bornmuellerana*
Turkish hazel	*Corylus colurna*
Turk's cap	*Malaviscus arboreus*
Turk's turban	
	Clerodendrum indicum
Turner's oak	*Quercus × turneri*
Turpentine pine	*Callitris verrucosa*
Turpentine tree	*Coprifera mopane*
	Pistacia terebinthus
	Siliphium terebinthaceum
	Syncarpia glomulifera
Turpeth	*Ipomoea turpethum*
Tutsan	*Hypericum androsaemum*
Twin flower	*Linnaea borealis*
Twin-flowered daphne	
	Daphne pontica
Twisted heath	*Erica cinerea*
Twisted-leaf pine	*Pinus teocote*
Twisted wattle	*Acacia tortuosa*
Twisted willow	
	Salix babylonica pekinensis tortuosa
Twistwood	*Viburnum lantana*
Two-leaved nut pine	
	Pinus cembroides edulis
	Pinus edulis
Two-styled hawthorn	
	Crataegus laevigata
Two-winged gimlet	
	Eucalyptus diptera
Ubame oak	*Quercus phillyraeoides*
Udo	*Aralia cordata*
Ugli fruit	
	Citrus paradisi × citrus reticulata
Ulmo	*Eucryphia cordifolia*
Umbrella pine	
	Hedyscepe canterburyana
	Pinus pinea
	Sciadopitys verticillata
Umbrella thorn	*Acacia tortilis*
Umbrella tree	*Magnolia tripetala*
	Musanga cecropioides
	Schefflera arboricola
Umkokolo	*Aberia caffra*
	Dovyalis caffra
Uni	*Ugni molinae*
Upas tree	*Antiaris toxicaria*
Upland cypress	
	Taxodium ascendens
Upland sumach	*Rhus glabra*
Upland tupelo	*Nyssa sylvatica*
Upside-down tree	
	Adansonia digitata
Urn gum	*Eucalyptus urnigera*
Utah ash	*Fraxinus anomala*
Utah juniper	
	Juniperus osteosperma
Valley oak	*Quercus lobata*
Valonia oak	*Quercus ithaburensis*
	Quercus macrolepis
Van Moltke's lime	*Tilia × moltkei*
Van Volxem's maple	
	Acer velutinum vanvolxemii
Variegated croton	
	Codiaeum variegatum
Variegated Japanese dogwood	
	Cornus controversa variegata
Variegated sweet chestnut	
	Castanea sativa albomarginata
	Castanea sativa aureomarginata
Varnish-leaved gum	
	Eucalyptus vernicosa
	Fraxinus xanthoxyloides
Varnish tree	*Ailanthus altissima*
	Aleurites moluccana
	Koelreuteria paniculata
	Rhus verniciflua
	Semecarpus anacardium
Varnish wattle	*Acacia verniciflua*
Vegetable mercury	
	Brunfelsia uniflora
Vegetable tallow tree	
	Sapium sebiferum
Veitch's magnolia	
	Magnolia × veitchii
Veitch's screw pine	
	Pandanus veitchii
Veitch's silver fir	*Abies veitchii*
Velvet ash	*Fraxinus velutina*
Velvet leaf	*Vaccinium canadense*
Velvet sumac	*Rhus typhina*
Velvet tamarind	*Dialium guineense*
Venetian sumac	*Cotinus coggygria*
	Rhus cotinus
Venezuelan mahogany	
	Swietenia candollea

Victoria rosemary
Westringia rosmariniformis

Victorian box
Pittosporum undulatum

Vilmorin's rowan   Sorbus vilmorinii

Vilmori's sorbus   Sorbus vilmorinii

Vine cactus   Fouquieria splendens

Vinegar tree   Rhus glabra

Vineleaf maple   Acer cissifolium

Vine maple   Acer circinatum

Violet tree   Polygala cowellii

Violet willow   Salix daphnoides

Violetwood   Acacia homalophylla

Virgilia   Cladrastis lutea

Virginia dogwood   Cornus florida

Virginian bird cherry
Prunus virginiana

Virginian snowflower
Chionanthus virginicus

Virginian sumac   Rhus typhina

Virginia sweetspire   Itea virginica

Virginia willow   Itea virginica

Virgin's palm   Dioon edule

Voss's laburnum   Laburnum vossii
Laburnum × watereri

Wadadura   Leycythis grandiflora

Wadalee gum tree   Acacia catechu

Wahoo elm
Euonymus atropurpurea
Ulmus alata

Walking-stick ebony
Diospyros monbuttensis

Walking-stick palm
Linospadix monostachya

Wallaba   Eperua falcata

Wallangarra   Acacia accola

Wall germander
Teucrium chamaedrys

Wallowa   Acacia calamifolia

Wallwort   Sambucus ebulus

Wampee   Clausena lansium

Wampi   Clausena lansium

Warminster broom
Cytisus × praecox

Warted yate
Eucalyptus megacornata

Wartleaf ceanothus
Ceanothus papillosus

Warty birch   Betula pendula

Washington palm
Washingtonia filifera

Washington thorn
Crataegus phaenopyrum

Water ash   Acer negundo
Fraxinus caroliniana
Ptelea trifoliata

Water beech   Carpinus caroliniana

Water birch   Betula occidentalis

Water blossom pea
Podalyria calyptrata

Waterbush
Myoporum tenuifolium

Water chestnut   Pachira aquatica

Water elder   Viburnum opulus

Water elm   Planera aquatica
Planera ulmifolia
Ulmus americana

Water-filter nut
Strychnos potatorum

Water fir
Metasequoia glyptostroboides

Water hickory   Carya aquatica

Water holly   Mahonia nervosa

Waterlily protea   Protea aurea

Water locust   Gleditsia aquatica

Water oak   Quercus nigra

Water rose-apple tree
Syzygium aqueum

Water sallow   Salix aquatica

Water tupelo   Nyssa aquatica

Waterwell tree
Warszewiczia coccinea

Wattle-leaved peppermint
Eucalyptus acacae formis

Waukegan juniper
Juniperus horizontalis douglasii

Wavyleaf ceanothus
Ceanothus foliosus

Wavy Saint John's wort
Hypericum undulatum

Wax-leaf privet
Ligustrum lucidum

Wax mallow   Malvaviscus arboreus

Wax apple
Syzygium samarangense

Waxberry   Gaultheria hispida
Myrica cerifera

Wax-leaf privet
Ligustrum japonicum

Wax myrtle   Myrica carolinensis
Myrica cerifera

Wax palm   Ceroxylon alpinum

Wax tree   Rhus succedanea

Wayfaring tree
Viburnum alnifolium
Viburnum lantana

Weaver's broom
Spartium junceum

Weddel palm
Microcoelum weddelianum

Wedding-cake tree
Cornus controversa

Weeping alder
Alnus incana pendula

Weeping ash
Fraxinus excelsior pendula

Weeping aspen
Populus tremula pendula

Weeping atlas cedar
Cedrus atlantica glauca

Weeping beech
Fagus sylvatica pendula
Weeping birch
Betula pendula tristis
Betula pendula youngii
Weeping boree *Acacia vestita*
Weeping bottlebrush
Callistemon viminalis
Weeping cherry *Prunus ivensii*
Prunus subhirtella pendula
Weeping cotoneaster
Cotoneaster hybridus pendulus
Cotoneaster × watereri
Weeping cypress
Cupressus funebris
Weeping elm
Ulmus glabra camperdown
Ulmus glabra pendula
Weeping european larch
Larix decidua pendula
Weeping fig *Ficus benjamina*
Weeping forsythia
Forsythia sieboldii
Forsythia suspensa
Weeping golden box
Buxus aurea pendula
Weeping hawthorn
Crataegus monogyna pendula
Weeping holly
Ilex aquifolium pendula
Weeping larch *Larix × pendula*
Weeping laurel *Ficus benjamina*
Weeping mountain ash
Sorbus aucuparia pendula
Weeping mulberry
Morus alba pendula
Weeping myall *Acacia pendula*
Weeping oak
Quercus robur pendula
Weeping pear
Pyrus salicifolia pendula
Weeping pea tree
Caragana arborescens pendula
Weeping peking willow
Salix matsudana pendula
Weeping podocarpus
Podocarpus elongatus
Weeping purple osier
Salix purpurea pendula
Weeping purple willow
Salix purpurea pendula
Weeping rosebud cherry
Prunus subhirtella pendula
Weeping sally
Eucalyptus mitchelliana
Salix caprea pendula
Weeping Scotch laburnum
Laburnum alpinum pendulum
Weeping silver lime
Tilia petiolaris
Weeping spring cherry
Prunus subhirtella pendula
rosea

Weeping spruce *Picea brewerana*
Weeping swamp cypress
Taxodium distichum pendens
Weeping Swedish birch
Betula pendula dalecarlica
Weeping tea tree
Melaleuca leucadendron
Weeping wattle *Acacia saligna*
Weeping willow *Salix babylonica*
Salix babylonica pekinensis
pendula
Salix caprea pendula
Salix × chrysocoma
Salix sepulcralis chrysocoma
Weeping willow-leaved pear
Pyrus salicifolia pendula
Weeping wych elm
Ulmus glabra horizontalis
Weigela *Weigela florida*
Wellingtonia
Sequoiadendron giganteum
West African barwood
Pterocarpus angolensis
West African ebony
Diospyros mespiliformis
West African kino
Pterocarpus erinaceus
West African rubber tree
Ficus vogelii
West Asian plane
Platanus orientalis
West Australian mahogany
Eucalyptus marginata
Western arbor-vitae *Thuya plicata*
Western balsam poplar
Populus trichocarpa
Western burning bush
Euonymus occidentalis
Western catalpa *Catalpa speciosa*
Western choke cherry
Prunus virginiana demissa
Western gorse *Ulex gallii*
Western hemlock
Tsuga heterophylla
Western Himalayan cedar
Cedrus deodara
Western juniper
Juniperus occidentalis
Western larch *Larix occidentalis*
Western laurel
Kalmia microphylla
Western lombardy poplar
Populus nigra plantierensis
Western oak *Quercus garryana*
Western plane
Platanus occidentalis
Western redbud *Cercis occidentalis*
Western red cedar *Thuya plicata*
Western sand cherry
Prunus besseyi
Western tea myrtle
Melaleuca nesophylla

Western white pine
Pinus monticola
Western yellow pine
Pinus ponderosa
Western yew *Taxus brevifolia*
Westfelton yew
Taxus baccata dovastoniana
West Himalayan fir *Abies pindrow*
West Himalayan spruce
Picea smithiana
West Indian birch
Bursera simaruba
West Indian blackthorn
Acacia farnesiana
West Indian boxwood
Gossypiospermum praecox
West Indian cedar
Cedrela odorata
Toona odorata
West Indian cherry *Cordia nitida*
West Indian dogwood
Piscidia piscipula
West Indian holly *Leea coccinea*
West Indian jasmine
Plumeria rubra
West Indian laurel fig
Ficus perforata
West Indian lime
Citrus aurantifolia
West Indian mahogany
Swietenia mahagoni
West Indian silkwood
Zanthoxylum flavum
West Indies walnut
Juglans jamaicensis
Westland pine
Dacrydium colensoi
Westonbirt dogwood
Cornus alba sibirica
Weymouth pine *Pinus strobus*
Wheatley elm
Ulmus carpinifolia sarniensis
Ulmus minor stricta sarniensis
Ulmus × sarniensis
Wheel of fire *Stenocarpus sinuatus*
Wheel tree
Trochodendron aralioides
Whin *Ulex europaeus*
Whinberry *Vaccinium myrtillus*
Whistlewood
Acer pennsylvanicum
Whistling tree *Acacia seyal*
White alder *Alnus incana*
Alnus rhombifolia
Clethra acuminata
White ash *Eucalyptus fraxinoides*
Fraxinus alba
Fraxinus americana
White-barked birch
Betula jacquemontii
White-barked himalayan birch
Betula jacquemontii

White bark pine *Pinus albicaulis*
White basswood
Tilia heterophylla
Whitebeam *Sorbus aria*
White birch *Betula papyrifera*
Betula pendula
Betula populifolia
Betula pubescens
White box *Eucalyptus albens*
White broom *Genista raetam*
White butternut *Juglans cinerea*
White cedar
Chamaecyparis thyoides
Melia dubium
Tabebuia pallida
Thuya occidentalis
White cinnamon *Canella alba*
Canella winterana
White cypress
Chamaecyparis thyoides
White cypress pine
Callitris columellaris
Callitris glauca
White dogwood *Cornus florida*
White dragon tree
Sesbania formosa
White elm *Ulmus americana*
White fir *Abies amabilis*
Abies concolor
White gum *Eucalyptus rossii*
White-heart hickory
Carya tomentosa
White heather
Calluna vulgaris alba
White ironbark
Eucalyptus leucoxylon
White jasmine
Jasminum officinale
White jute *Corchorus capsularis*
White-leaf manzanita
Arctostaphylos viscida
White-leaved marlock
Eucalyptus tetragona
White mahogany
Eucalyptus acmenioides
Eucalyptus umbra
White maple *Acer saccharinum*
White mountain dogwood
Viburnum alnifolium
White mulberry *Morus alba*
White oak *Quercus alba*
White peppermint
Eucalyptus pulchella
White pine *Pinus strobus*
Podocarpus elatus
Podocarpus dacrydioides
White popinac *Leucaena glauca*
White poplar *Populus alba*
White Portuguese broom
Cytisus albus
Cytisus multiflorus albus
White rose of York *Rosa × alba*

TREES, BUSHES, AND SHRUBS

White sally   *Eucalyptus pauciflora*

White sandalwood
    Santalum album

White sapote   *Casimiroa edulis*

White silk-cotton tree
    Ceiba pentandra

White Spanish broom
    Cytisus multiflorus

White spruce   *Picea glauca*

White stinkwood   *Celtis africanus*

White stopper   *Eugenia axillaris*

White stringybark
    Eucalyptus globoidea

White swamp azalea
    Rhododendron viscosum

White tea tree
    Melaleuca leucadendron

Whitethorn   *Crataegus laevigata*
    Crataegus monogyna

White titi   *Cyrilla racemiflora*

White-top peppermint
    Eucalyptus radiata

White walnut   *Juglans cinerea*

White wax tree
    Ligustrum lucidum

White wicky   *Kalmia cuneata*

White willow   *Salix alba*

White winter heather
    Erica hyemalis

Whitewood
    Liriodendron tulipifera
    Tabebuia riparia
    Tilia americana

Whiteywood
    Melicytus ramiflorus

Whortleberry
    Vaccinium corymbosum
    Vaccinium myrtillus

Whortle-leaved willow
    Salix myrsinites

Whortle willow   *Salix myrsinites*

Wickson plum   *Prunus × sultana*

Wicky   *Kalmia angustifolia*

Wicopy   *Dirca palustris*

Wig tree   *Cotinus coggygria*

Wild almond   *Prunus fasciculata*

Wild cherry   *Prunus avium*
    Prunus ilicifolia

Wild chestnut   *Pachira insignis*

Wild China soapberry
    Sapindus marginatus

Wild China tea
    Sapindus drummondii

Wild cinnamon   *Canella winterana*

Wild cocoa tree   *Pachira aquatica*

Wild coffee   *Polyscias guilfoylei*
    Psychotria nervosa
    Psychotria sulzneri

Wild cotton   *Hibiscus diversifolius*
    Hibiscus moscheutos

Wild crab apple
    Malus angustifolia
    Malus sylvestris

Wild custard-apple
    Annona senegalensis

Wild date   *Yucca baccata*

Wild date palm   *Phoenix rupicola*
    Phoenix sylvestris

Wild-goose plum
    Prunus munsoniana

Wild hippo   *Euphorbia corollata*

Wild holly   *Ilex aquifolium*

Wild ipecac
    Euphorbia ipecacuanhae

Wild irishman   *Discaria toumatou*

Wild lilac   *Ceanothus sanguineus*

Wild lime   *Xanthoxylum fagara*
    Zanthoxylum fagara

Wild olive   *Elaeagnus angustifolia*
    Elaeagnus latifolia
    Halesia carolina
    Nyssa aquatica
    Olea africana
    Osmanthus americanus

Wild orange   *Prunus caroliniana*

Wild peach   *Kigelia africana*
    Prunus fasciculata

Wild pear   *Pyrus communis*
    Pyrus pyraster

Wild pepper tree
    Vitex agnus-castus

Wild plum   *Prunus domestica*

Wild poinsettia
    Warszewiczia coccinea

Wild pomegranate
    Burchelia bubalina

Wild pride of India
    Galpinia transvaalica

Wild raisin   *Viburnum cassinoides*
    Viburnum lentago

Wild red cherry
    Prunus pennsylvanica

Wild robusta coffee
    Coffea canephora

Wild rose-apple
    Syzygium pyenanthum

Wild rosemary   *Croton cascarilla*
    Ledum palustre

Wild sarsaparilla   *Aralia nudicaulis*

Wild senna   *Cassia hebecarpa*
    Cassia marilandica

Wild sensitive plant
    Cassia nictitans

Wild service tree   *Sorbus torminalis*

Wild snowball
    Ceanothus americanus

Wild soursop   *Annona montana*

Wild sweet crab apple
    Malus coronaria

Willimore cedar
    Widdringtonia schwarzii

Willow acacia   *Acacia salicina*

Willow cherry	*Prunus incana*
Willow-leaf bay	
	Laurus nobilis angustifolia
Willow-leaf magnolia	
	Magnolia salicifolia
Willow-leaved bottlebrush	
	Callistemon salignus
Willow-leaved jessamine	
	Cestrum parqui
Willow-leaved pear	
	Pyrus salicifolia
Willow-leaved poplar	
	Populus angustifolia
Willowmore cedar	
	Widdringtonia schwarzii
Willow myrtle	*Agonis flexuosa*
Willow oak	*Quercus phellos*
Willow pittosporum	
	Pittosporum phillyraeoides
Willow podocarp	
	Podocarpus chilinus
	Podocarpus salignus
Wilson's barberry	
	Berberis wilsoniae
Wilson's beauty bush	
	Kolkwitzia amabilis
Wilson's berberis	
	Berberis wilsoniae
Wilson's douglas fir	
	Pseudotsuga wilsoniana
Wilson's yellowwood	
	Cladrastis wilsonii
Wilwilli	*Erythrina monosperma*
	Erythrina tahitiensis
Windmill jasmine	
	Jasminum nitidum
Windmill palm	
	Trachycarpus fortunei
Wineberry	*Rubus phoenicolasius*
Wine palm	*Borassus flabellifer*
	Caryota urens
	Jubaea chilensis
	Jubaea spectabilis
Winged elm	*Ulmus alata*
Winged spindle tree	
	Euonymus alata
Winged wattle	*Acacia alata*
Wing-rib sumac	*Rhus copellina*
Wingseed	*Ptelea trifoliata*
Winterberry	*Ilex glabra*
	Ilex verticillata
Winter cherry	
	Prunus subhirtella autumnalis
Winter daphne	*Daphne odora*
Winter-flowering cherry	
	Prunus subhirtella autumnalis
Winter-flowering jasmine	
	Jasminum nudiflorum
Wintergreen	
	Gaultheria procumbens
Winter heath	*Erica carnea*

Winter savory	*Satureia montana*
Winter's bark	*Drimys winteri*
Wintersweet	
	Acokanthera oblongifolia
	Chimonanthus praecox
Winter thorn	*Acacia albida*
Wirilda	*Acacia retinoides*
Wiry honey myrtle	
	Melaleuca nematophylla
Wisconsin weeping willow	
	Salix × blanda
Wisteria tree	*Sesbania tripetii*
Witch alder	
	Fothergilla monticola
Witch hazel	*Corylus hamamelis*
Witch hobble	
	Viburnum alnifolium
	Viburnum lantanoides
Withe-rod	*Viburnum cassinoides*
Wi tree	*Spondias cytherea*
Wire-netting bush	
	Corokia cotoneaster
Woadwaxen	*Genista tinctoria*
Wolfberry	
	Symphoricarpus occidentalis
Wolley-dod's rose	
	Rosa villosa duplex
Wonderboom	*Ficus pretoriae*
Wonder tree	*Ricinus communis*
Wood apple	*Feronia elephantum*
	Feronia limonia
Wood rose	*Rosa gymnocarpa*
Woodward's blackbutt	
	Eucalyptus woodwardii
Woodwaxen	*Genista tinctoria*
Woolly butia palm	
	Butia eriospatha
Woollybutt	*Eucalyptus longifolia*
Woolly lavender	*Lavandula lanata*
Woolly-leaf ceanothus	
	Ceanothus tomentosus
Woolly netbush	
	Calothamnus villosus
Woolly-podded broom	
	Cytisus grandiflorus
Woolly tea tree	
	Leptospermum lanigerum
Woolly willow	*Salix lanata*
Woman's tongue tree	
	Albizzia lebbeck
Wormwood senna	
	Cassia artemisioides
Wyalong wattle	
	Acacia cardiophylla
Wych elm	*Ulmus glabra*
Yankee point ceanothus	
	Ceanothus griseus horizontalis
Yarran	*Acacia homalophylla*
Yatay palm	*Butia yatay*
Yate tree	*Eucalyptus cornuta*

TREES, BUSHES, AND SHRUBS

Zanzibar coffee

Yaupon *Ilex cassine*
Ilex vomitoria
Yeddo hawthorn
Raphiolepis japonica
Raphiolepis umbellata
Yeddo spruce *Picea jezoensis*
Yellow azalea
Rhododendron luteum
Yellow banksian rose
Rosa banksiae lutea
Yellow bark *Cinchona calisaya*
Yellow-bark ash
Fraxinus excelsior jaspidea
Yellow-bark oak *Quercus velutina*
Yellow-bark thorn *Acacia woodii*
Yellow bell *Allamanda cathartica*
Yellowbells *Tecoma stans*
Yellow-berried holly
Ilex aquifolium fructuluteo
Yellow-berried mountain ash
Sorbus aucuparia xanthocarpa
Yellow-berried yew
Taxus baccata fructo-luteo
Yellow bignonia *Tecoma stans*
Yellow birch *Betula alleghaniensis*
Betula lutea
Yellow bloodwood
Eucalyptus eximia
Yellow box *Eucalyptus melliodora*
Yellow buckeye *Aesculus flava*
Aesculus octandra
Yellow butterfly palm
Chrysalidocarpus lutescens
Yellow catalpa *Catalpa ovata*
Yellow cedar *Thuya occidentalis*
Yellow chestnut oak
Quercus muehlenbergii
Yellow-cotton tree
Cochlospermum religiosum
Yellow cucumber tree
Magnolia cordata
Yellow cypress
Chamaecyparis nootkatensis
Yellow elder *Tecoma stans*
Yellow flame tree
Peltophorum pterocarpum
Yellow-flowered gum
Eucalyptus woodwardii
Yellow-fruited holly
Ilex aquifolium bacciflava
Yellow-fruited thorn
Crataegus flava
Yellow guava *Psidium guajava*
Yellow gum *Eucalyptus johnstonii*
Yellow haw *Crataegus flava*
Yellow jasmine *Jasminum mesnyi*
Yellow latan *Latania verschaffeltii*
Yellow locust tree
Robinia pseudoacacia
Yellow mombin *Spondias lutea*
Spondias mombin

Yellow oak
Quercus muehlenbergii
Yellow oleander
Thevetia peruviana
Yellow palm
Chrysalidocarpus lutescens
Yellow pine *Pinus echinata*
Yellow poplar
Liriodendron tulipifera
Yellow poui *Tabebuia serratifolia*
Yellow princess palm
Dictyosperma aureum
Yellow root
Xanthorrhiza simplicissima
Yellow silver pine
Dacrydium intermedium
Yellow-stemmed dogwood
Cornus stolonifera flaviramea
Yellow stringybark
Eucalyptus muellerana
Yellow-topped mallee ash
Eucalyptus luehmanniana
Yellow tree-lupin
Lupinus arboreus
Yellow trumpet tree *Tecoma stans*
Yellowwood *Cladrastis lutea*
Rhodosphaera rhodanthema
Zanthoxylum americanum
Yerba-de-maté
Ilex paraguariensis
Yerba mansa
Anemopsis californica
Yertchuk
Eucalyptus consideniana
Yesterday and today
Brunfelsia australis
Yew-leaved torreya
Torreya taxifolia
Yezo spruce *Picea jezoensis*
York and Lancaster rose
Rosa damascena versicolor
Yoruba ebony
Diospyros monbuttensis
Yoshino cherry *Prunus yedoensis*
Young's golden juniper
Juniperus chinensis aurea
Young's weeping birch
Betula pendula youngii
Yuca *Manihot esculenta*
Yucca *Yucca gloriosa*
Yulan *Magnolia denudata*
Magnolia heptapeta
Magnolia yulan
Yunnan lilac *Syringa yunnanensis*
Yunnan mahonia
Mahonia lomariifolia
Zamang tree *Samanea saman*
Zambak *Jasminum sambac*
Zanona palm *Socratea exorhiza*
Zanzibar coffee
Coffea zanguebariae

Zapote blanco

Zapote blanco	*Casimiroa edulis*
Zebra wood	*Connarus guianensis*
	Diospyros kurzii
Zitherwood	
	Citharexylum spinosum

Zoeschen maple	
	Acer × zoeschense
Zombi palm	*Zombia antillarum*
Zulu fig tree	*Ficus nekbudu*
	Ficus utilis

WILD FLOWERS

It is not possible in a volume such as this to include every species of wild flower; even one limited to European or North American floras would run to a substantial number of books, therefore preference has been given to the more common and widely distributed plants.

It is now illegal to dig up wild plants in many parts of the world, and the gathering of wild flowers by collectors and amateur 'flower lovers' coupled with the widespread use of modern herbicides has resulted in the near-extinction of many species. Wild flowers should be left undisturbed to be enjoyed by all.

Common poppy –
Papaver rhoeas

Acrid lettuce

Acrid lettuce	*Lactuca virosa*
Agrimony	*Agrimonia eupatoria*
Alexanders	*Smyrnium olustratum*
Allseed	*Radiola linoides*
Alpine blue sowthistle	
	Cicerbita alpina
Alpine cinquefoil	
	Potentilla crantzii
Alpine clematis	*Clematis alpina*
Alpine enchanter's nightshade	
	Circaea alpina
Alpine fleabane	*Erigeron boreatis*
Alpine forget-me-not	
	Myosotis alpestris
Alpine lady's mantle	
	Alchemilla alpina
Alpine lettuce	*Cicerbita alpina*
Alpine meadow rue	
	Thalictrum alpinum
Alpine nightshade	*Circaea alpina*
Alpine penny cress	*Thlaspi alpestre*
Alpine rivulet saxifrage	
	Saxifraga rivularis
Alpine sawwort	*Saussurea alpina*
Alpine speedwell	*Veronica alpina*
Alpine willow herb	
	Epilobium anagallidifolium
Alpine woundwort	*Stachys alpina*
Alsike clover	*Trifolium hybridum*
Alternate-flowered water milfoil	
	Myriophyllum alterniflorum
Alternate-leaved golden saxifrage	
	Chrysosplenium alternifolium
American pondweed	
	Potamogeton epihydrus
American speedwell	
	Veronica peregrina
Amphibious bistort	
	Polygonum amphibium
Angular solomon's seal	
	Polygonatum odoratum
Annual knawel	
	Scleranthus annuus
Annual mercury	
	Mercurialis annua
Annual sea-blite	*Suaeda maritima*
Arrowhead	*Sagittaria sagittifolia*
Arrowhead orache	
	Atriplex prostrata
Asarabacca	*Asarum europaeum*
Asarina	*Asarina procumbens*
Astrantia	*Astrantia major*
Autumn felwort	
	Gentiana amarella
	Gentianella amarella
Autumn gentian	
	Gentiana amarella
	Gentianella amarella

Autumn hawkbit	
	Leontodon autumnalis
Autumn lady's tresses	
	Spiranthes autumnalis
	Spiranthes spiralis
Autumn squill	*Scilla autumnalis*
Awlwort	*Subularia aquatica*
Bachelor's buttons	
	Chrysanthemum parthenium
	Tanacetum parthenium
Balm-leaved figwort	
	Scrophularia scorodonia
Baneberry	*Actaea spicata*
Barren strawberry	
	Potentilla sterilis
Basil thyme	*Acinos arvensis*
Bath asparagus	
	Ornithogalum pyrenaicum
Beach orache	*Atriplex littoralis*
Beaked hawk's beard	
	Crepis vesicaria
Beautiful Saint John's wort	
	Hypericum pulchrum
Bee orchid	*Ophrys apifera*
Beggarticks	*Bidens frondosa*
Bell heather	*Erica cinera*
Bermuda buttercup	
	Oxalis pes-caprae
Betony	*Betonica officinalis*
	Stachys betonica
Bilberry	*Vaccinium myrtillis*
Bindweed	*Calystegia sepium*
Bird's eye primrose	
	Primula farinosa
Bird's foot trefoil	
	Lotus corniculatus
Bird's nest orchid	
	Neottia nidus-avis
Birthwort	*Aristolochia clematitis*
Birdsfoot	*Ornithopus perpusillus*
Bithynian vetch	*Vicia bithynica*
Biting stonecrop	*Sedum acre*
Bittersweet	*Solanum dulcamara*
Bitter vetch	*Lathyrus montanus*
Black bindweed	
	Fallopia convolvulus
Black bitter vetch	*Lathyrus niger*
Black bryony	*Tamus communis*
Black horehound	*Ballota nigra*
Black medick	*Medicago lupulina*
Black mullein	*Verbascum nigrum*
Black mustard	*Brassica nigra*
Black nightshade	*Solanum nigrum*
Black vetch	*Lathyrus niger*
Bladder campion	*Silene vulgaris*
Blood-drop emlets	
	Mimulus luteus
Blood-red geranium	
	Geranium sanguiuneum

Bloody geranium
Geranium sanguiuneum

Blotched monkey-flower
Mimulus luteus

Blue anemone   *Anemone apennina*

Bluebell   *Endymion non-scriptus*
Hyacinthoides non-scripta

Blue comfrey
Symphytum peregrinum

Blue flax   *Linum perenne*

Blue fleabane   *Erigeron acer*

Blue gromwell   *Lithospermum*
purpurocaeruleum

Blue iris   *Iris spuria*

Blue lettuce   *Lactuca perennis*

Blue lobelia   *Lobelia urens*

Blue pimpernel   *Anagallis foemina*

Blue sowthistle
Cicerbita macrophylla
Lactuca macrophylla

Blue water speedwell
Veronica anagallis-aquatica

Bog arum   *Calla palustris*

Bog asphodel
Narthecium ossifragum

Bogbean   *Menyanthes trifoliata*

Bog bilberry   *Vaccinium uliginosum*

Bog dandelion
Taraxacum spectabile

Bog pimpernel   *Anagallis tenella*

Bog pondweed
Potamogeton oblongus
Potamogeton polygonifolius

Bog Saint John's wort
Hypericum elodes

Bog stitchwort   *Stellaria alsine*

Bog violet   *Viola palustris*

Bog water starwort
Callitriche stagnalis

Bog whortleberry
Vaccinium uliginosum

Bog willow herb
Epilobium palustre

Branched broomrape
Orobanche ramosa

Branched plantain
Plantago arenaria

Breckland wormwood
Artemisia campestris

Bristly oxtongue
Hemintia echioides
Picris echioides

Broad-fruited cornsalad
Valerianella rimosa

Broad-leaved dock
Rumex obtusifolius

Broad-leaved everlasting pea
Lathyrus latifolius

Broad-leaved eyebright
Euphrasia occidentalis
Euphrasia tetraquetra

Broad-leaved helleborine
Epipactis helleborine
Epipactis latifolia

Broad-leaved pondweed
Potamogeton natans

Broad-leaved ragwort
Senecio fluviatilis

Broad-leaved spurge
Euphorbia platyphyllos

Broad-leaved willow herb
Epilobium montanum

Brooklime   *Veronica beccabunga*

Brookweed   *Samolus valerandi*

Brown knapweed
Centaurea nemoralis

Buck's horn plantain
Plantago coronopus

Bugle   *Ajuga reptans*

Bulbous buttercup
Ranunculus bulbosus

Bulbous saxifrage
Saxifraga granulata

Bur chervil   *Anthriscus caucalis*

Burdock   *Arctium pubens*

Bur-marigold   *Bidens tripartita*

Bur medick   *Medicago minima*

Burnet rose   *Rosa pimpinellifolia*

Burnet saxifrage
Pimpinella saxifraga

Burnt orchid   *Orchis ustulata*

Bush vetch   *Vicia sepium*

Butcher's broom   *Ruscus aculeatus*

Butterbur   *Petasites hybridus*

Butterfly orchid
Habenaria chlorantha
Platanthera chlorantha

Calamint   *Calamintha sylvatica*

Canadian fleabane
Conyza canadensis
Erigeron canadensis

Canadian golden rod
Solidago canadensis

Canadian waterweed
Elodea canadensis

Canterbury bell
Campanula medium

Caper spurge   *Euphorbia lathyrus*

Carline thistle   *Carlina vulgaris*

Carrot broomrape
Orobanche amethystea
Orobanche maritima

Catmint   *Nepeta cataria*

Cat's ear   *Hypochoeris radicata*

Celery-leaved crowsfoot
Ranuculus sceleratus

Chaff weed   *Anagallis minima*

Chalk hill eyebright
Euphrasia pseudokerneri

Chalk milkwort   *Polygala calcarea*

Chamomile   *Chamaemelum nobile*

Changing forget-me-not
    Myosotis discolor

Charlock   *Sinapis arvensis*

Cheddar pink
    Dianthus gratianopolitanus

Chickweed   *Stellaria media*

Chickweed willow herb
    Epilobium alsinifolium

Chickweed wintergreen
    Trientalis europaea

Chicory   *Cichorum intybus*

Chinese mugwort
    Artemisia verlotorum

Chives   *Allium schoenoprasum*

Ciliate heath   *Erica ciliaris*

Circular-leaved crowfoot
    Ranunculus circinatus

Clary   *Salvia horminoides*

Cleavers   *Galium aparine*

Climbing corydalis
    Corydalis claviculta

Cloudberry   *Rubus chamaemorus*

Clove-scented broomrape
    Orobanche caryophyllacea

Clustered alpine saxifrage
    Saxifraga nivalis

Clustered bellflower
    Campanula glomerata

Clustered dock
    Rumex conglomeratus

Coastal broomrape
    Orobanche amethystea
    Orobanche maritima

Coastal holly   *Eryngium campestre*

Coltsfoot   *Tussilago farfara*

Columbine   *Aquilegia vulgaris*

Comfrey   *Symphytum officinale*

Common agrimony
    Agrimonia eupatoria

Common bindweed
    Calystegia sepium

Common bistort
    Polygonum bistorta

Common bladderwort
    Utricularia vulgaris

Common burdock
    Arctium pubens

Common butterwort
    Pinguicula vulgaris

Common calamint
    Calamintha sylvatica

Common cat's ear
    Hypochoeris radicata

Common centaury
    Centaurium erythraea

Common chickweed
    Stellaria media

Common comfrey
    Symphytum officinale

Common cornsalad
    Valerianella locusta

Common cow wheat
    Melampyrum pratense

Common dog violet
    Viola riviniana

Common duckweed   *Lemna minor*

Common evening primrose
    Oenothera biennis

Common eyebright
    Euphrasia nemorosa

Common field speedwell
    Veronica persica

Common figwort
    Scrophularia nodosa

Common fleabane
    Pulicaria dysenterica

Common forget-me-not
    Myosotis arvensis

Common fumitory
    Fumaria officinalis

Common gromwell
    Lithospermum officinale

Common hawkweed
    Hieracium vulgatum

Common heather
    Calluna vulgaris

Common hemp nettle
    Galeopsis tetrahit

Common knapweed
    Centaurea nigra

Common mallow   *Malva sylvestris*

Common meadow
buttercup   *Ranunculus acris*

Common meadow rue
    Thalictrum flavum

Common milkwort
    Polygala vulgaris

Common mullein
    Verbascum thapsus

Common nettle   *Urtica dioica*

Common orache   *Atriplex patula*

Common poppy   *Papaver rhoeas*

Common ragwort
    Senecio jacobaea

Common red poppy
    Papaver rhoeas

Common restharrow
    Ononis repens

Common rockrose
    Helianthemum nummularium

Common sea lavender
    Limonium vulgare

Commom slender eyebright
    Euphrasia micrantha

Common sorrel   *Rumex acetosa*

Common sowthistle
    Sonchus oleraceus

Common spotted orchid
    Dactylorhiza fuchsii

Common storksbill
    Erodium cicutarium

Common teasel   *Dipsacus fullonum*
    Dipsacus sylvestris

Downy rose

Common toadflax *Linaria vulgaris*

Common tormentil
　　　　　Potentilla erecta

Common twayblade *Listera ovata*

Common valerian
　　　　　Valeriana officinalis

Common vetch *Vicia sativa*

Common violet *Viola riviniana*

Common water crowfoot
　　　　　Ranunculus aquatilis

Common water dropwort
　　　　　Oenanthe fistulosa

Common water starwort
　　　　　Callitriche stagnalis

Common whitlowgrass
　　　　　Erophila verna

Common wild thyme
　　　　　Thymus drucei

Common wintergreen
　　　　　Pyrola minor

Coralroot orchid
　　　　　Corallorhiza trifida

Corky-fruited water dropwort
　　　　　Oenanthe pimpinelloides

Corn bedstraw
　　　　　Galium tricornutum

Corn buttercup
　　　　　Ranunculus arvensis

Corn chamomile
　　　　　Anthemis arvensis

Cornflower *Centaurea cyanus*

Corn gromwell
　　　　　Lithospermum arvense

Corn marigold
　　　　　Chrysanthemum segetum

Corn mignonette
　　　　　Reseda phyteuma

Corn mint *Mentha arvensis*

Corn sowthistle *Sonchus arvensis*

Corn spurrey *Spergula arvensis*

Cornish heath *Erica vagans*

Corsican speedwell
　　　　　Veronica repens

Cowberry *Vaccinium vitis-idaea*

Cow parsley *Anthriscus sylvestris*

Cow parsnip
　　　　　Heracleum sphondylium

Cowslip *Primula veris*

Cranberry *Oxycoccus palustris*
　　　　　Vaccinium oxycoccus

Creamy clover
　　　　　Trifolium ochroleucon

Creeping bellflower
　　　　　Campanula rapunculoides

Creeping buttercup
　　　　　Ranunculus repens

Creeping cinquefoil
　　　　　Potentilla reptans

Creeping forget-me-not
　　　　　Myosotis secunda

Creeping jenny
　　　　　Lysimachia nummularia

Creeping lady's tresses
　　　　　Goodyera repens

Creeping Saint John's wort
　　　　　Hypericum humifusum

Creeping thistle *Cirsium arvense*

Crimson clover
　　　　　Trifolium incarnatum

Cross-leaved heath *Erica tetralix*

Crosswort *Galium cruciata*

Crowberry *Empetrum nigrum*

Crow garlic *Allium vineale*

Crown vetch *Coronilla varia*

Cuckoo flower
　　　　　Cardamine pratensis

Cuckoo-pint *Arum maculatum*

Curled pondweed
　　　　　Potamogeton crispus

Curly dock *Rumex crispus*

Cut-leaved cranesbill
　　　　　Geranium dissectum

Cut-leaved dead nettle
　　　　　Lamium hybridum

Cut-leaved germander
　　　　　Teucrium botrys

Cut-leaved self heal
　　　　　Prunella laciniata

Cyclamen *Cyclamen hederifolium*

Cypress spurge
　　　　　Euphorbia cyparissias

Daisy *Bellis perennis*

Dame's violet *Hesperis matronalis*

Dandelion *Taraxacum officinale*

Dark burdock
　　　　　Arctium nemorosum

Dark hair crowfoot
　　　　　Ranunculus trichophyllus

Dark mullein *Verbascum nigrum*

Dark red helleborine
　　　　　Epipactis atropurpurea
　　　　　Epipactis atrorubens

Deadly nightshade *Atropa bella
donna*

Deptford pink *Dianthus armeria*

Devil's bit scabious *Scabiosa succisa*
　　　　　Succisa pratensis

Dewberry *Rubus caesius*

Dittander *Lepidium latifolium*

Dodder *Cuscuta epithymum*

Dog rose *Rosa canina*

Dog's mercury
　　　　　Mercurialis perennis

Dog violet *Viola canina*

Dove's foot cranesbill
　　　　　Geranium columbinum
　　　　　Geranium molle

Downy pepperwort
　　　　　Lepidium heterophyllum

Downy rose *Rosa tomentosa*

Downy woundwort
    Stachys germanica

Dragon's teeth
    Tetragonolobus maritimus

Dropwort   *Filipendula vulgaris*

Dune storksbill   *Erodium dunense*

Dusky cranesbill
    Geranium phaeum

Dusty miller   *Artemisia stellerana*

Dutch clover   *Trifolium repens*

Dwarf mallow   *Malva neglecta*

Dwarf orchid   *Orchis ustulata*

Dwarf pansy   *Viola kitaibeliana*

Dwarf spurge   *Euphorbia exigua*

Dwarf thistle   *Cirsium acaule*

Dwarf Welsh eyebright
    Euphrasia cambrica

Dyer's rocket   *Reseda luteola*

Early forget-me-not
    Myosotis ramosissima

Early purple orchid   *Orchis mascula*

Early spider orchid
    Ophrys aranifera
    Ophrys sphegodes

Elecampane   *Inula helenium*

Enchanter's nightshade
    Circaea lutetiana

English sticky eyebright
    Euphrasia anglica

English stonecrop
    Sedum anglicum

Erect evening primrose
    Oenothera stricta

Erect hedge parsley   *Torilis japonica*

Evening catchfly   *Silene noctiflora*

Evening primrose
    Oenothera biennis

Everlasting   *Anaphalis margaritacea*
    Helichrysum arenarium

Everlasting pea   *Lathyrus latifolius*

Fairy flax   *Linum catharticum*

False cleavers   *Galium spurium*

False thorow-wax
    Bupleurum subovatum

Fat duckweed   *Lemna gibba*

Fat hen   *Chenopodium album*

Fen bedstraw   *Galium uliginosum*

Fennel   *Foeniculum vulgare*

Fennel pondweed
    Potamogeton pectinatus

Fen orchid   *Liparis loeselii*

Fen pondweed
    Potamogeton coloratus

Fen sowthistle   *Sonchus palustris*

Fenugreek
    Trifolium ornithopodioides

Fen violet   *Viola stagnina*

Feverfew
    Chrysanthemum parthenium
    Tanacetum parthenium

Field bindweed
    Convolvulus arvensis

Field eryngo   *Eryngium campestre*

Field felwort   *Gentiana campestris*
    Gentianella campestris

Field fleawort   *Senecio integrifolius*

Field forget-me-not
    Myosotis arvensis

Field garlic   *Allium oleraceum*

Field gentian   *Gentiana campestris*
    Gentianella campestris

Field gromwell
    Lithospermum arvense

Field pansy   *Viola arvensis*

Field penny cress   *Thlaspi arvense*

Field pepperwort
    Lepidium campestre

Field poppy   *Papaver rhoeas*

Field rose   *Rosa arvensis*

Field scabious   *Knautia arvensis*
    Scabiosa arvensis

Field speedwell   *Veronica agrestis*

Field toadflax   *Linaria arvensis*

Field woundwort   *Stachys arvensis*

Figwort   *Scrophularia nodosa*

Fimbriate medick
    Medicago polymorpha

Fine-leaved heath   *Erica cinera*

Fine-leaved water dropwort
    Oenanthe aquatica

Fingered speedwell
    Veronica triphyllos

Flat-head clover
    Trifolium glomeratum

Fleabane   *Pulicaria dysenterica*

Floating pondweed
    Potamogeton natans

Floating water plantain
    Alisma natans
    Luronium natans

Flowering nutmeg
    Leycesteria formosa

Flowering rush
    Butomus umbellatus

Fluellen   *Kickxia elatine*

Fly honeysuckle
    Lonicera xylosteum

Fly orchid   *Ophrys insectifera*
    Ophrys muscifera

Fool's parsley   *Aethusa cynapium*

Fool's watercress
    Apium nodiflorum

Forget-me-not   *Myosotis arvensis*

Forked larkspur   *Consolida regalis*

Fox-and-cubs
    Hieracium aurantiacum

Foxglove   *Digitalis purpurea*

Fragrant agrimony
    Agrimonia procera

Fragrant evening primrose
    Oenothera stricta

Ground thistle

Fragrant orchid
    Gymnadenia conopsea
French hawk's beard
        Crepis nicaeensis
French mint
    Mentha spicata × suaveolens
Fringed waterlily
        Nymphoides peltata
Fritillary   *Fritillaria meleagris*
Frogbit
    Hydrocharis morsus-ranae
Frog orchid   *Coeloglossum viride*
Frosted orache   *Atriplex laciniata*
Fumitory   *Fumaria officinale*
Garden chervil
        Anthriscus cerefolium
Garden cress   *Lepidium sativum*
Garlic mustard   *Alliaria petiolata*
Germander speedwell
        Veronica chamaedrys
Ghost orchid
    Epipogium aphyllum
Giant bellflower
        Campanula lactiflora
        Campanula latifolia
Giant hogweed
    Heracleum mantegazzianum
Gibbous duckweed   *Lemna gibba*
Gipsywort   *Lycopus europaeus*
Glasswort   *Salicornia europaea*
Globeflower   *Trollius europaeus*
Goat's beard   *Tragopogon pratensis*
Goat's rue   *Galega officinalis*
Golden rod   *Solidago virgaurea*
Golden samphire
        Inula crithmoides
Golden saxifrage
    Chrysosplenium oppositifolium
Goldilocks   *Ranunculus auricomus*
Goldilocks buttercup
        Ranunculus auricomus
Good King Henry
    Chenopodium bonus-henricus
Goose grass   *Galium aparine*
Goutweed
    Aegopodium podagraria
Grape hyacinth
        Muscari atlanticum
        Muscari neglectum
        Muscari racemosum
Grass of parnassus
        Parnassia palustris
Grass-poly   *Lythrum hyssopifolia*
Grass vetchling   *Lathyrus nissolia*
Grassy orach   *Atriplex littoralis*
Great bellflower
        Campanula latifolia
Great bindweed
        Calystegia silvatica
Great burdock   *Arctium lappa*

Great burnet
    Sanguisorba officinalis
Great duckweed
    Spirodela polyrhiza
Greater bird's foot trefoil
    Lotus uliginosus
Greater bladderwort
    Utricularia vulgaris
Greater broomrape
    Orobanche rapum-genistae
Greater burnet saxifrage
        Pimpinella major
Greater butterfly orchid
        Platanthera chlorantha
Greater butterwort
        Pinguicula grandiflora
Greater celandine
        Chelidonium majus
Greater chickweed
        Stellaria neglecta
Greater dodder   *Cuscuta europaea*
Greater eyebright
        Euphrasia arctica
Greater hayrattle
        Rhinanthus serotinus
Greater periwinkle   *Vinca major*
Greater sea spurrey
        Spergularia media
Greater stitchwort
        Stellaria holostea
Greater yellow rattle
        Rhinanthus angustifolius
Great fen ragwort
        Senecio paludodus
Great knapweed
        Centaurea scabiosa
Great lettuce   *Lactuca virosa*
Great mullein   *Verbascum thapsus*
Great pignut
        Bunium bulbocastanum
Great plantain   *Plantago major*
Great sundew   *Drosera anglica*
Great willow herb
        Epilobium hirsutum
Green field speedwell
        Veronica agrestis
Green figwort
        Scrophularia umbrosa
Green hellebore   *Helleborus viridis*
Green nightshade
        Solanum sarrachoides
Green-winged orchid   *Orchis morio*
Grey speedwell   *Veronica polita*
Gromwell
    Lithospermum officinale
Ground elder
    Aegopodium podagraria
Ground ivy   *Glechoma hederacea*
Groundsel   *Senecio vulgaris*
Ground thistle   *Cirsium acaute*
        Cirsium acautonauct

Hairy bindweed   *Calystegia pulchra*
Hairy bird's foot trefoil
         Lotus hispidus
Hairy bitter cress
         Cardamine hirsuta
Hairy-leaved eyebright
         Euphrasia curta
Hairy nightshade   *Solanum luteum*
Hairy rock cress   *Arabis hirsuta*
Hairy Saint John's wort
         Hypericum hirsutum
Hairy spurge   *Euphorbia pilosa*
Hairy stonecrop   *Sedum villosum*
Hairy tare   *Vicia hirsuta*
Hairy vetchling   *Lathyrus hirsutus*
Hairy violet   *Viola hirta*
Hairy willow herb
         Epilobium parviflorum
Hardheads   *Centaurea nigra*
Harebell   *Campanula rotundifolia*
Hare's-foot clover
         Trifolium arvense
Harsh downy-rose   *Rosa tomentosa*
Hautbois strawberry
         Fragaria moschata
Hawkbit   *Leontodon leysseri*
         Leontodon taraxacoides
Hawkweed
         Hieracium umbellatum
Hawkweed ox-tongue
         Picris hieracioides
Hayrattle   *Rhinanthus minor*
Heath bedstraw   *Galium saxatile*
Heath groundsel   *Senecio sylvaticus*
Heath milkwort
         Polygala serpyllifolia
Heath speedwell
         Veronica officinalis
Heath spotted orchid
         Dactylorhiza maculata
Hedge bedstraw   *Galium album*
Hedge bindweed
         Calystegia sepium
Hedge parsley   *Torilis japonica*
Hedgerow cranesbill
         Geranium pyrenaicum
Hedge woundwort
         Stachys sylvatica
Hemlock   *Conium maculatum*
Hemlock water dropwort
         Oenanthe crocata
Hemp agrimony
         Eupatorium cannabinum
Hemp-nettle   *Galeopsis tetrahit*
Henbane   *Hyoscyamus niger*
Henbit   *Lamium amplexicaule*
Herb bennet   *Geum urbanum*
Herb paris   *Paris quadrifolia*
Herb robert
         Geranium robertianum

Hispid marsh mallow
         Althaea hirsuta
Hoary cinquefoil
         Potentilla argentea
Hoary cress   *Cardaria draba*
Hoary mullein
         Verbascum pulverulentum
Hoary plantain   *Plantago media*
Hoary ragwort   *Senecio erucifolius*
Hoary rockrose
         Helianthemum canum
Hogweed   *Heracleum sphondylium*
Honesty   *Lunaria annua*
Honeysuckle
         Lonicera periclymenum
Hop   *Humulus lupulus*
Hop trefoil   *Trifolium campestre*
Horse mint   *Mentha longifolia*
Horehound   *Ballota nigra*
Horseshoe vetch
         Hippocrepis comosa
Hyssop-leaved loosestrife
         Lythrum hyssopifolia
Indian balsam
         Impatiens glandulifera
Intermediate bladderwort
         Utricularia intermedia
Intermediate dead-nettle
         Lamium moluccellifolium
Intermediate wintergreen
         Pyrola media
Irish eyebright
         Euphrasia salisburgensis
Irish heath   *Erica hibernica*
Irish saxifrage   *Saxifraga rosacea*
Irish spurge   *Euphorbia hyberna*
Italian catchfly   *Silene italica*
Italian cuckoo-pint   *Arum italicum*
Italian lords-and-ladies
         Arum italicum
Ivy broomrape   *Orobanche hederae*
Ivy duckweed   *Lemna trisulca*
Ivy   *Hedera helix*
Ivy-leaved bellflower
         Campanula hederacea
         Wahlenbergia hederacea
Ivy-leaved crowfoot
         Ranunculus hederaceus
Ivy-leaved duckweed
         Lemna trisulca
Ivy-leaved speedwell
         Veronica hederifolia
Ivy-leaved toadflax
         Cymbalaria muralis
Japanese knotweed
         Reynoutria japonica
Jersey buttercup
         Ranunculus paludosus
Jersey forget-me-not
         Myosotis sicula
Jersey orchid   *Orchis laxiflora*

Little marsh dandelion

Jersey toadflax   *Linaria pelisseriana*

Keeled-fruited cornsalad
  Valerianella carinata

Kidney vetch   *Anthyllis vulneraria*

Kingcup   *Caltha palustris*

Knapweed   *Centaurea nigra*

Knotgrass   *Polygonum aviculare*

Knotted clover   *Trifolium striatum*

Knotted hedge parsley
  Torilis nodosa

Knotted pearlwort   *Sagina nodosa*

Lady orchid   *Orchis purpurea*

Lady's bedstraw   *Galium verum*

Lady's slipper
  Cypripedium calceolus

Lady's smock   *Cardamine pratensis*

Lady's tresses
  Spiranthes autumnalis
  Spiranthes spiralis

Lamb's lettuce   *Valerianella locusta*

Lamb's tongue   *Plantago media*

Lanceolate water plantain
  Alisma lanceolatum

Large bindweed
  Calystegia silvatica

Large bitter cress
  Cardamine amara

Large-flowered butterwort
  Pinguicula grandiflora

Large-flowered evening primrose
  Oenothera erythrosepala

Large-flowered hemp-nettle
  Galeopsis speciosa

Large-flowered sticky eyebright
  Euphrasia rostkoviana

Large hop trefoil   *Trifolium aureum*

Large lizard clover
  Trifolium molinerii

Larger wild thyme
  Thymus pulegioides

Large thyme   *Thymus pulegioides*

Large wintergreen
  Pyrola rotundifolia

Large yellow foxglove
  Digitalis grandiflora

Large yellow restharrow
  Ononis natrix

Larkspur   *Consolida ambigua*

Late spider orchid
  Ophrys arachnites
  Ophrys fuciflora

Leafy lousewort   *Pedicularis foliosa*

Least birdsfoot
  Ornithopus perpusillus

Least lettuce   *Lactuca saligna*

Least waterlily   *Nuphar pumila*

Least yellow trefoil
  Trifolium micranthum

Leopard's bane
  Doronicum pardalianches

Lesser bindweed
  Convolvulus arvensis

Lesser bladderwort
  Utricularia minor

Lesser broomrape
  Orobanche apiculata
  Orobanche minor

Lesser burdock   *Arctium minus*

Lesser burnet   *Sanguisorba minor*

Lesser butterfly orchid
  Habenaria bifolia
  Platanthera bifolia

Lesser celandine
  Ranunculus ficaria

Lesser centaury
  Centaurium pulchellum

Lesser dandelion
  Taraxacum erythrospermum
  Taraxacum laevigatum

Lesser dodder   *Cuscuta epithymum*

Lesser duckweed   *Lemna minor*

Lesser fleabane   *Pulicaria vulgaris*

Lesser goat's beard
  Tragopogon minor

Lesser herb robert
  Geranium purpureum

Lesser marshwort
  Apium inundatum

Lesser meadow rue
  Thalictrum minus

Lesser periwinkle   *Vinca minor*

Lesser sea spurrey
  Spergularia marina

Lesser skullcap   *Scutellaria minor*

Lesser snapdragon
  Misopates orontium

Lesser solomon's seal
  Polygonatum odoratum

Lesser stitchwort
  Stellaria graminea

Lesser sweet briar   *Rosa micrantha*

Lesser twayblade   *Listera cordata*

Lesser valerian   *Valeriana dioica*

Lesser water forget-me-not
  Myosotis caespitosa

Lesser waterlily   *Nuphar pumila*

Lesser water parsnip   *Berula erecta*

Lesser water plantain
  Alisma ranunculoides
  Baldellia ranunculoides

Lesser wintergreen   *Pyrola minor*

Lesser yellow trefoil
  Trifolium dubium

Lily-of-the-valley
  Convallaria majalis

Ling   *Calluna vulgaris*

Little kneeling eyebright
  Euphrasia confusa

Little marsh dandelion
  Taraxacum palustre

Little three-lobed crowfoot

Little three-lobed crowfoot	
	Ranunculus tripartitus
Livelong	*Sedum telephium*
Lizard orchid	
	Himantoglossum hircinum
	Orchis hircina
Loddon pondweed	
	Potamogeton nodosus
Long-fruited bird's foot trefoil	
	Lotus angustissimus
Long-leaved sundew	
	Drosera intermedia
Long rough-headed poppy	
	Papaver agremone
Long smooth-headed poppy	
	Papaver dubium
Long-stalked pondweed	
	Potamogeton praelongus
Long-stemmed cranesbill	
	Geranium columbinum
Long-styled rose	*Rosa stylosa*
Lords-and-ladies	
	Arum maculatum
Lousewort	*Pedicularis sylvatica*
Lovage	*Levisticum officinale*
Lucerne	*Medicago sativa*
Lupin	*Lupinus nootkatensis*
Mackay's heath	*Erica mackaiana*
Maiden pink	*Dianthus deltoides*
Man orchid	
	Aceras anthropophorum
Mare's tail	*Hippuris vulgaris*
Marjoram	*Origanum vulgare*
Marsh bedstraw	*Galium palustre*
Marsh bird's foot trefoil	
	Lotus uliginosus
Marsh cinquefoil	
	Potentilla palustris
Marsh fleawort	*Senecio palustris*
Marsh forget-me-not	
	Myosotis secunda
Marsh gentian	
	Gentiana pneumonanthe
Marsh hawk's beard	
	Crepis paludosa
Marsh helleborine	
	Epipactis palustris
Marsh lousewort	
	Pedicularis palustris
Marsh mallow	*Althaea officinalis*
Marsh marigold	*Caltha palustris*
Marsh pea	*Lathyrus palustris*
Marsh ragwort	*Senecio aquaticus*
Marsh Saint John's wort	
	Hypericum elodes
Marsh speedwell	
	Veronica scutellata
Marsh stitchwort	*Stellaria palustris*
Marsh thistle	*Carduus palustris*
	Cirsium palustre
Marsh valerian	*Valeriana dioica*

Marsh willow herb	
	Epilobium palustre
Marsh woundwort	
	Stachys palustris
May lily	*Maianthemum bifolium*
Meadow buttercup	
	Ranunculus acris
Meadow clary	*Salvia pratensis*
Meadow cranesbill	
	Geranium pratense
Meadow rue	*Thalictrum flavum*
Meadow saffron	
	Colchicum autumnale
Meadow saxifrage	
	Saxifraga granulata
Meadowsweet	*Filipendula ulmaria*
Meadow thistle	*Carduus dissectum*
	Carduus pratensis
	Cirsium dissectum
Meadow water dropwort	
	Oenanthe silaifolia
Medium wintergreen	*Pyrola media*
Melancholy thistle	
	Carduus helenioides
	Cirsium heterophyllum
Mexican fleabane	
	Erigeron mucronatus
Milfoil	*Achillea millefolium*
Milk thistle	*Silybum marianum*
Mistletoe	*Viscum album*
Mollyblobs	*Caltha palustris*
Monkshood	*Aconitum napellus*
Monkey flower	*Mimulus guttatus*
Monkey orchid	*Orchis simia*
Moorland crowfoot	
	Ranunculus omiophyllus
Mortar pellitory	*Parietaria judaica*
Moschatel	*Adoxa moschatellina*
Moss campion	*Silene acaulis*
Mossy saxifrage	
	Saxifraga hypnoides
Moth mullein	*Verbascum blattaria*
Mountain avens	*Dryas octopetala*
Mountain campion	*Silene acaulis*
Mountain everlasting	
	Antennaria dioica
Mountain pansy	*Viola lutea*
Mountain Saint John's wort	
	Hypericum montanum
Mountain sorrel	*Oxyria digyna*
Mountain sticky eyebright	
	Euphrasia montana
Mouse-ear hawkweed	
	Hieracium pilosella
	Pilosella officinarum
Mud water starwort	
	Callitriche stagnalis
Mudwort	*Limosella aquatica*
Mugwort	*Artemisia vulgaris*
Musk	*Mimulus moschatus*

Ploughman's spikenard

Musk mallow *Malva moschata*
Musk orchid
  Herminium monorchis
Musk thistle *Carduus nutans*
Musky storksbill
  Erodium moschatum
Naked autumn crocus
  Crocus nudiflorus
Narrow-fruited cornsalad
  Valerianella dentata
Narrow-leaved eyebright
  Euphrasia salisburgensis
Narrow-leaved helleborine
  Cephalanthera longifolia
Narrow-leaved marsh dandelion
  Taraxacum palustre
Narrow-leaved pepperwort
  Lepidium ruderale
Narrow-leaved Saint John's wort
  Hypericum linarifolium
Narrow-leaved sweet briar
  Rosa agrestis
Narrow-leaved vetch
  Vicia angustifolia
Narrow-lipped helleborine
  Epipactis leptochila
Nettle-cure dock
  Rumex obtusifolius
Nettle-leaved bellflower
  Campanula trachelium
Night-flowering catchfly
  Silene noctiflora
Nipplewort *Lapsana communis*
Nodding bur marigold
  Bidens cernua
Nodding catchfly *Silene nutans*
Nodding thistle *Carduus nutans*
Nodding water avens *Geum rivale*
Northern bedstraw
  Galium boreale
Northern downy rose
  Rosa sherardii
Norwegian mugwort
  Artemisia norvegica
Nottingham catchfly *Silene nutans*
Nuttall's waterweed
  Elodea nuttallii
Oblong-leaved sundew
  Drosera intermedia
Old man's beard *Clematis vitalba*
One-flowered wintergreen
  Moneses uniflora
  Pyrola uniflora
Opium poppy
  Papaver somniferum
Orach *Atriplex patula*
Orange balsam *Impatiens capensis*
Orange hawkweed
  Hieracium aurantiacum
  Pilosella aurantiaca
Orpine *Sedum telephium*

Ox-eye daisy
  Chrysanthemum leucanthemum
  Leucanthemum vulgare
Oxford ragwort *Senecio squalidus*
Oxlip *Primula elatior*
Pale butterwort
  Pinguicula lusitanica
Pale flax *Linum bienne*
Pale hairy buttercup
  Ranunculus sardous
Pale heath violet *Viola lactea*
Pale persicaria
  Polygonum lapathifolium
Pale toadflax *Linaria repens*
Pale willow herb
  Epilobium Roseum
Parsley piert *Aphanes arvensis*
Parsley water dropwort
  Oenanthe lachenalii
Pasque flower *Pulsatilla vernalis*
  Pulsatilla vulgaris
Pearly everlasting
  Anaphalis margaritacea
Pedicelled willow herb
  Epilobium roseum
Pellitory-of-the-wall
  Parietaria judaica
Penny royal *Mentha pulegium*
Pepper mint *Mentha × piperita*
Pepper saxifrage *Silaum silaus*
Perennial flax *Linum perenne*
Perennial glasswort
  Arthrocnemum perenne
Perennial honesty
  Lunaria rediviva
Perennial sowthistle
  Sonchus arvensis
Perfoliate honeysuckle
  Lonicera caprifolium
Perfoliate penny cress
  Thlaspi perfoliatum
Perfoliate pondweed
  Potamogeton perfoliatus
Perforate alexanders
  Smyrnium perfoliatum
Perforate Saint John's wort
  Hypericum perforatum
Persian speedwell *Veronica persica*
Petty spurge *Euphorbia peplus*
Pheasant's eye *Narcissus majalis*
Picris broomrape
  Orobanche picridis
Pignut *Conopodium majus*
Pineapple weed
  Matricaria matricarioides
Pink butterwort
  Pinguicula lusitanica
Pink water speedwell
  Veronica catenata
Ploughman's spikenard
  Inula conyza

Plymouth thistle

Plymouth thistle	
	Carduus pycnocephalus
Pond bedstraw	*Galium debile*
Pond chickweed	
	Myosoton aquaticum
Portland spurge	
	Euphorbia portlandica
Prickly lettuce	*Lactuca scariola*
	Lactuca serriola
Prickly poppy	*Papaver agremone*
Prickly restharrow	*Ononis spinosa*
Prickly saltwort	*Salsola kali*
Prickly sowthistle	*Sonchus asper*
Primrose	*Primula vulgaris*
Primrose peerless	
	Narcissus × bifloris
Procumbent cinquefoil	
	Potentilla anglica
Procumbent pearlwort	
	Sagina procumbens
Procumbent yellow sorrel	
	Oxalis corniculata
Prostrate pearlwort	
	Sagina procumbens
Prostrate toadflax	*Linaria supina*
Purple broomrape	
	Orobanche purpurea
Purple corydalis	*Corydalis bulbosa*
Purple crocus	*Crocus purpureus*
	Crocus vernus
Purple dead-nettle	
	Lamium purpureum
Purple heather	*Erica cinera*
Purple loosestrife	
	Lythrum salicaria
Purple milk-vetch	
	Astragalus danicus
Purple saxifrage	
	Saxifraga oppositifolia
Purple spurge	*Euphorbia peplis*
Purple toadflax	*Linaria purpurea*
Pyramidal orchid	
	Anacamptis pyramidalis
	Orchis pyramidalis
Pyrenean cranesbill	
	Geranium pyrenaicum
Pyrenean lily	*Lilium pyrenaicum*
Pyrenean valerian	
	Valeriana pyrenaica
Radish	*Raphanus sativus*
Ragged robin	*Lychnis flos-cuculi*
Ragwort	*Senecio jacobaea*
Ramsons	*Attium ursinum*
Rape	*Brassica napus*
Raspberry	*Rubus idaeus*
Red bartsia	*Odontites verna*
Red broomrape	*Orobanche alba*
Red campion	*Lychnis dioica*
	Silene dioica
Red catchfly	*Lychnis viscaria*

Red clover	*Trifolium pratense*
Red dead-nettle	
	Lamium purpureum
Reddish pondweed	
	Potamogeton alpinus
	Potamogeton rufescens
Red hemp-nettle	
	Galeopsis angustifolia
Red pondweed	
	Potamogeton alpinus
	Potamogeton rufescens
Red rattle	*Pedicularis palustris*
Redshank	*Polygonum persicaria*
Red valerian	*Centranthus ruber*
Reflexed stonecrop	
	Sedum reflexum
Restharrow	*Ononis repens*
Reversed clover	
	Trifolium resupinatum
Ribbed melitot	
	Melilotus officinalis
Ribwort plantain	
	Plantago lanceolata
Rigid hornwort	
	Ceratophyllum demersum
River crowfoot	
	Ranunculus fluitans
River water crowfoot	
	Ranunculus fluitans
River water dropwort	
	Oenanthe fluviatilis
Rock cinquefoil	*Potentilla rupestris*
Rock samphire	
	Crithmum maritimum
Rock sea spurrey	
	Spergularia rupicola
Rock stonecrop	
	Sedum forsteranum
Rootless duckweed	*Wolffia arrhiza*
Rosebay willow herb	
	Epilobium angustifolium
Rose of sharon	
	Hypericum calycinum
Roseroot	*Rhodiola rosea*
	Sedum rosea
Rough chervil	
	Chaerophyllum temulentum
Rough clover	*Trifolium scabrum*
Rough comfrey	
	Symphytum asperum
Rough hawkbit	
	Leontodon hispidus
Rough hawk's beard	*Crepis biennis*
Round-headed leek	
	Allium sphaerocephalon
Round-headed rampion	
	Phyteuma orbiculare
	Phyteuma tenerum
Round knotweed	
	Reynoutria japonica
Round-leaved cranesbill	
	Geranium rotundifolium

Round-leaved fluellen
Kickxia spuria

Round-leaved mint
Mentha suaveolens

Round-leaved speedwell
Veronica filiformis

Round-leaved sundew
Drosera rotundifolia

Round-leaved wintergreen
Pyrola rotundifolia

Rock sea-spurrey
Spergularia rupicola

Rue-leaved saxifrage
Saxifraga tridactylites

Russian comfrey
Symphytum × uplandicum

Russian spurge
Euphorbia uralensis

Sainfoin *Onobrychis viciifolia*

Saint Patrick's cabbage
Saxifraga spathularis

Salad burnet *Sanguisorba minor*

Salsify *Tragopogon porrifolius*

Saltwort *Glaux maritima*

Samphire *Crithmum maritimum*

Sand crocus *Romulea columnae*
Romulea parviflora

Sand leek *Allium scorodoprasum*

Sand spurrey *Spergularia rubra*

Sand toadflax *Linaria arenaria*

Sandy nettle *Urtica urens*

Sandy orach *Atriplex laciniata*

Saw-wort *Serratula tinctoria*

Scarce autumn felwort
Gentiana germanica
Gentianella germanica

Scarce water figwort
Scrophularia ehrhartii
Scrophularia umbrosa

Scarlet pimpernel
Anagallis arvensis

Scented agrimony
Agrimonia procera

Scented mayweed
Matricaria recutita

Scentless chamomile
Tripleurospermum maritimum

Scentless mayweed
Tripleurospermum inodorum

Scots lovage *Ligusticum scoticum*

Scottish asphodel *Tofieldia pusilla*

Scottish bluebell
Campanula rotundifolia

Scottish thistle
Onopordon acanthium

Sea aster *Aster tripolium*

Sea beet *Beta vulgaris maritima*

Sea bindweed
Calystegia soldanella

Sea-blite *Suaeda maritima*

Sea campion
Silene vulgaris maritima

Sea carrot
Daucus carota gummifer

Sea centaury *Centaurium littorale*

Sea holly *Eryngium maritimum*

Sea kale *Crambe maritima*

Sea lavender *Limonium vulgare*

Sea mayweed
Tripleurospermum maritimum

Sea milkwort *Glaux maritima*

Sea pea *Lathyrus japonicus*
Lathyrus maritimus

Sea pink *Armeria maritima*
Statice armeria

Sea plantain *Plantago maritima*

Sea purslane
Halimione portulacoides

Sea radish *Raphanus maritimus*

Sea rocket *Cakile maritima*

Seaside crowfoot
Ranunculus baudotii

Seaside pansy *Viola curtisii*

Seaside pea *Lathyrus japonicus*
Lathyrus maritimus

Sea stock *Matthiola sinuata*

Sea storksbill *Erodium maritimum*

Sea spurge *Euphorbia paralias*

Sea wormwood
Artemisia maritima

Self heal *Prunella vulgaris*

Serrated wintergreen
Orthilia secunda

Sharp-leaved fluellen
Kickxia elatine

Sheep's bit *Jasione montana*

Sheep's sorrel *Rumex acetosella*

Shepherd's cress
Teesdalia nudicaulis

Shepherd's purse
Capsella pursa-pastoris

Shining cranesbill
Geranium lucidum

Shining pondweed
Potamogeton lucens

Shoreline orache *Atriplex littoralis*

Shore weed *Littorella lacustris*
Littorella uniflora

Short-fruited willow herb
Epilobium obscurum

Short-leaved forget-me-not
Myosotis stolonifera

Short-haired eyebright
Euphrasia brevipila

Short-pedicelled rose *Rosa dumalis*

Shrubby cinquefoil
Potentilla fruticosa

Shrubby speedwell
Veronica fruticans

Sibbaldia *Sibbaldia procumbens*

Sickle-leaved hare's ear

Sickle-leaved hare's ear	
	Bupleurum falcatum
Sickle medick	*Medicago falcata*
Silverweed	*Potentilla anserina*
Single-flowered wintergreen	
	Moneses uniflora
	Pyrola uniflora
Skullcap	*Scutellaria galericulata*
Slender bedstraw	
	Galium pumilum
Slender-flowered thistle	
	Carduus tenuiflorus
Slender hare's ear	
	Bupleurum tenuissimum
Slender mullein	
	Verbascum virgatum
Slender Saint John's wort	
	Hypericum pulchrum
Slender scottish eyebright	
	Euphrasia scottica
Slender tare	*Vicia tenuissima*
Slender trefoil	
	Trifolium micranthum
Small alpine gentian	
	Gentiana nivalis
Small balsam	*Impatiens parviflora*
Small catchfly	*Silene gallica*
Small cranberry	
	Oxycoccus microcarpus
	Vaccinium microcarpum
Small fleabane	*Pulicaria vulgaris*
Small-flowered buttercup	
	Ranunculus parviflorus
Small-flowered cranesbill	
	Geranium pusillum
Small-flowered sticky eyebright	
	Euphrasia hirtella
Small mallow	*Malva pusilla*
Small medick	*Medicago minima*
Small melilot	*Melilotus indica*
Small nettle	*Urtica urens*
Small restharrow	*Ononis reclinata*
Small scabious	
	Scabiosa columbaria
Small teasel	*Dipsacus pilosus*
Small toadflax	
	Chaenorhinum minus
Small white orchid	
	Habenaria albida
	Leucorchis albida
	Pseudorchis albida
Small yellow balsam	
	Impatiens parviflora
Small yellow foxglove	
	Digitalis lutea
Smith's pepperwort	
	Lepidium heterophyllum
Smooth cat's ear	
	Hypochoeris glabra
Smooth hawkbit	
	Leontodon autumnalis

Smooth hawk's beard	
	Crepis capillaris
Smooth sowthistle	
	Sonchus oleraceus
Smooth tare	*Vicia tetrasperma*
Snake's head fritillary	
	Fritillaria meleagris
Snapdragon	*Antirrhinum majus*
Sneezewort	*Achillea ptarmica*
Snowdon eyebright	
	Euphrasia rivularis
Snowdrop	*Galanthus nivalis*
Soapwort	*Saponaria officinalis*
Soft clover	*Trifolium striatum*
Soft cranesbill	*Geranium molle*
Soft hawk's beard	*Crepis mollis*
Soft-leaved rose	*Rosa villosa*
Soldier orchid	*Orchis militaris*
Solomon's seal	
	Polygonatum multiflorum
Sorrel	*Rumex acetosa*
Sowthistle	*Sonchus oleraceus*
Spanish catchfly	*Silene otites*
Spanish daffodil	
	Narcissus hispanicus
Spathulate fleawort	
	Senecio spathulifolius
Spear-leaved orach	
	Atriplex prostrata
Spear-leaved willow herb	
	Epilobium lanceolatum
Spear mint	*Mentha spicata*
Spear thistle	*Cirsium vulgare*
Spiked rampion	
	Phyteuma spicatum
Spiked speedwell	*Veronica spicata*
Spiked star of Bethlehem	
	Ornithogalum pyrenaicum
Spiked water milfoil	
	Myriophyllum spicatum
Spiny saltwort	*Salsola kali*
Spotted cat's ear	
	Hypochoeris maculata
Spotted dead-nettle	
	Lamium maculatum
Spotted medick	*Medicago arabica*
Spotted rockrose	
	Tuberaria guttata
Spreading bellflower	
	Campanula patula
Spreading hedge parsley	
	Torilis arvensis
Spring cinquefoil	
	Potentilla tabernaemontani
Spring crocus	*Crocus purpureus*
	Crocus vernus
Spring gentian	*Gentiana verna*
Spring snowflake	
	Leucojam vernum
Spring speedwell	*Veronica verna*

Upright hedge parsley

Spring squill	*Scilla verna*
Spring vetch	*Vicia lathyroides*
Square-stemmed Saint John's wort	
	Hypericum tetrapterum
Square-stemmed willow herb	
	Epilobium tetragonum
Star-headed clover	
	Trifolium stellata
Star of Bethlehem	
	Ornithogalum umbellatum
Starry saxifrage	*Saxifraga stellaris*
Star thistle	*Centaurea calcitrapa*
Sticky catchfly	*Lychnis viscaria*
Sticky groundsel	*Senecio viscosus*
Stinging nettle	*Urtica dioica*
Stinking bob	
	Geranium robertianum
Stinking chamomile	
	Anthemis cotula
Stinking hawk's beard	
	Crepis foetida
Stinking hellebore	
	Helleborus foetidus
Stinking iris	*Iris foetidissima*
Stinking tutsan	
	Hypericum hircinum
Stone bramble	*Rubus saxatilis*
Strawberry-headed clover	
	Trifolium fragiferum
Streaky cranesbill	
	Geranium versicolor
Striated catchfly	*Silene conica*
Subterranean clover	
	Trifolium subterraneum
Suffocated clover	
	Trifolium suffocatum
Summer lady's tresses	
	Spiranthes aestivalis
Summer snowflake	
	Leucojam aestivum
Sundew	*Drosera rotundifolia*
Sun spurge	*Euphorbia helioscopia*
Swede	*Brassica napus*
Sweet briar	*Rosa rubiginosa*
Sweet chamomile	
	Chamaemelum nobile
Sweet flag	*Acorus calamus*
Sweet lupin	*Lupinus luteus*
Sweet spurge	*Euphorbia dulcis*
Sweet violet	*Viola odorata*
Sweet william	*Dianthus barbatus*
Sweet woodruff	*Galium odoratum*
Tall broomrape	*Orobanche elatior*
	Orobanche major
Tall golden rod	*Solidago altissima*
	Solidago canadensis
Tall melilot	*Melilotus altissima*
Tansy	*Chrysanthemum vulgare*
	Tanacetum vulgare

Teasel	*Dipsacus fullonum*
	Dipsacus sylvestris
Teasel-headed clover	
	Trifolium squamosum
Teesdale violet	*Viola rupestris*
Tenby daffodil	*Narcissus obvallaris*
Thin-runner willow herb	
	Epilobium obscurum
Thistle broomrape	
	Orobanche reticulata
Thorn-apple	*Datura stramonium*
Thorow-wax	
	Bupleurum rotundifolium
Thread-leaved water crowfoot	
	Ranunculus trichophyllus
Threefold lady's tresses	
	Spiranthes romanzoffiana
Thrift	*Armeria maritima*
	Statice armeria
Thyme-leaved speedwell	
	Veronica serpyllifolia
Toothed medick	
	Medicago polymorpha
Toothwort	*Lathraea squamaria*
Tormentil	*Potentilla erecta*
Touch-me-not	
	Impatiens noli-tangere
Trailing rose	*Rosa arvensis*
Traveller's joy	*Clematis vitalba*
Tree lupin	*Lupinus arboreus*
Triangular-stemmed garlic	
	Allium triquetrum
Tricolor pansy	*Viola tricolor*
Trifid bur marigold	
	Bidens tripartita
Tuberous comfrey	
	Symphytum tuberosum
Tuberous meadow thistle	
	Cirsium tuberosum
Tuberous pea	*Lathyrus tuberosus*
Tuberous vetchling	
	Lathyrus tuberosus
Tubular water dropwort	
	Oenanthe fistulosa
Tufted centaury	
	Centaurium capitatum
Tufted loosestrife	
	Lysimachia thyrsiflora
Tufted saxifrage	*Saxifraga cespitosa*
Tufted vetch	*Vicia cracca*
Turk's cap lily	*Lilium martagon*
Turnip	*Brassica rapa*
Tutsan	*Hypericum androsaemum*
Twayblade	*Listera ovata*
Upright cinquefoil	*Potentilla erecta*
Upright clover	*Trifolium strictum*
Upright spurge	*Euphorbia stricta*
Upright hedge parsley	
	Torilis japonica

Upright yellow sorrel	
	Oxalis europaea
Valerian	*Valeriana officinalis*
Various-leaved pondweed	
	Potamogeton gramineus
	Potamogeton heterophyllus
Venus's looking glass	
	Legousia hybrida
	Specularia hybrida
Vervain	*Verbena officinalis*
Violet helleborine	
	Epipactis purpurata
	Epipactis sessilifolia
Violet-horned poppy	
	Roemeria hybrida
Violet iris	*Iris spuria*
Viper's grass	*Scorzonera humilis*
Wall bedstraw	*Galium parisiense*
Wall germander	
	Teucrium chamaedrys
Wall pepper	*Sedum acre*
Wall rocket	*Diplotaxis muralis*
	Diplotaxis tenuifolia
Wall speedwell	*Veronica arvensis*
Wall whitlowgrass	*Draba muralis*
Warren crocus	
	Romulea columnae
	Romulea parviflora
Water avens	*Geum rivale*
Water chickweed	
	Myosoton aquaticum
Watercress	*Nasturtium officinale*
Water crowfoot	
	Ranunculus aquatilis
Water dock	
	Rumex hydrolapathum
Water dropwort	
	Oenanthe fistulosa
Water figwort	
	Scrophularia auriculata
Water forget-me-not	
	Myosotis scorpioides
Water germander	
	Teucrium scordium
Water lobelia	*Lobelia dortmanna*
Water mint	*Mentha aquatica*
Water pepper	
	Polygonum hydropiper
Water plantain	
	Alisma plantago-aquatica
Water purslane	*Lythrum portula*
Water soldier	*Stratiotes aloides*
Water violet	*Hottonia palustris*
Wavy bitter cress	
	Cardamine flexuosa
Wavy Saint John's wort	
	Hypericum undulatum
Weld	*Reseda luteola*
Welsh poppy	*Meconopsis cambrica*
Welted thistle	
	Carduus acanthoides
Western bladderwort	
	Utricularia major
	Utricularia neglecta
Wetland dock	
	Rumex hydrolapathum
White bryony	*Bryonia dioica*
White campion	*Silene alba*
White clover	*Trifolium repens*
White comfrey	
	Symphytum orientale
White dead-nettle	*Lamium album*
White flax	*Linum catharticum*
White helleborine	
	Cephalanthera damasonium
White horehound	
	Marrubium vulgare
White melilot	*Melilotus alba*
White mullein	*Verbascum lychnitis*
White mustard	*Sinapis alba*
White ramping fumitory	
	Fumaria capreolata
White rockrose	
	Helianthemum apenninum
White stonecrop	*Sedum album*
White waterlily	*Nymphaea alba*
Whitlowgrass	*Erophila verna*
Whorled solomon's seal	
	Polygonatum verticillatum
Whorled water milfoil	
	Myriophyllum verticillatum
Whortleberry	
	Vaccinium myrtillus
Wild angelica	*Angelica sylvestris*
Wild balsam	
	Impatiens noli-tangere
Wild basil	*Clinopodium vulgare*
Wild cabbage	*Brassica oleracea*
Wild carnation	
	Dianthus caryophyllus
Wild carrot	*Daucus carota*
Wild chamomile	
	Matricaria recutita
Wild clary	*Salvia verbenaca*
Wild daffodil	
	Narcissus pseudonarcissus
Wild leek	*Allium ampeloprasum*
Wild lentil	*Anthyllis cicer*
Wild liquorice	
	Astragalus glycyphyllos
Wild madder	*Rubia peregrina*
Wild mignonette	*Reseda lutea*
Wild onion	*Allium vineale*
Wild parsnip	*Pastinaca sativa*
Wild pea	*Lathyrus sylvestris*
Wild pink	*Dianthus plumarius*
Wild radish	
	Raphanus raphanistrum
Wild sage	*Salvia horminoides*
Wild strawberry	*Fragaria vesca*
Wild thyme	*Thymus serpyllum*

Zigzag clover

Wild tulip *Tulipa sylvestris*
Winter aconite *Eranthis hyemalis*
Winter heliotrope
Petasites fragrans
Wood anemone
Anemone nemorosa
Wood avens *Geum urbanum*
Wood bitter vetch *Vicia orobus*
Wood cranesbill
Geranium sylvaticum
Wood forget-me-not
Myosotis sylvatica
Wood garlic *Attium ursinum*
Wood groundsel *Senecio sylvaticus*
Woodland violet
Viola reichenbachiana
Wood sage *Teucrium scorodonia*
Wood sorrel *Oxalis acetosella*
Wood speedwell
Veronica montana
Wood spurge
Euphorbia amygdaloides
Wood stitchwort
Stellaria nemorum
Wood vetch *Vicia sylvatica*
Wood woundwort
Stachys sylvatica
Woody nightshade
Solanum dulcamara
Woolly thistle *Cirsium eriophorum*
Wormwood
Artemisia absinthium
Yarrow *Achillea millefolium*
Yellow alpine milk vetch
Anthyllis frigidus
Yellow archangel
Lamiastrum galeobdolon
Yellow bartsia *Parentucellia biscosa*
Yellow bird's nest
Monotropa hypopitys

Yellow bog saxifrage
Saxifraga hirculus
Yellow chamomile
Anthemis tinctoria
Yellow corydalis *Corydalis lutea*
Yellow figwort
Scrophularia vernalis
Yellow flag *Iris pseudacoris*
Yellow forget-me-not
Myosotis discolor
Yellow-horned poppy
Glaucium flavum
Yellow iris *Iris pseudacorus*
Yellow loosestrife
Lysimachia vulgaris
Yellow lupin *Lupinus luteus*
Yellow meadow vetchling
Lathyrus pratensis
Yellow medick *Medicago falcata*
Yellow melilot *Melilotus altissima*
Yellow mountain saxifrage
Saxifraga aizoides
Yellow pimpernel
Lysimachia nemorum
Yellow rattle *Rhinanthus minor*
Yellow sorrel *Oxalis corniculata*
Yellow star of Bethlehem
Gagea lutea
Yellow star thistle
Centaurea solstitialis
Yellow vetch *Vicia lutea*
Yellow vetchling *Lathyrus aphaca*
Yellow waterlily *Nuphar lutea*
Yellow wort
Blackstonia perfoliata
Zigzag clover *Trifolium medium*

BOTANICAL
NAMES

ALPINES AND ROCKERY PLANTS

Alpines are plants whose natural habitat is the mountainous area above the tree line, but in gardening circles the term includes rock-gardening plants. Very little true alpine gardening is attempted in the English-speaking parts of the world because alpine conditions are rarely encountered naturally and it is very difficult to reproduce them artificially.

Drumstick Primrose –
Primula denticulata

Abies balsamea hudsoniana

Abies balsamea hudsoniana
Hudson balsam fir
Abies balsamea nana
Dwarf balsam fir
Abies cephalonica nana
Dwarf Grecian fir
Acaena anserinifolia
Bidgee-widgee
Bidi-bidi
Bididy-bid
Acaena buchananii
New Zealand bur
Acaena inermis
Blue Mountain bidi-bidi
Acaena microphylla
Blue Mountain bidi-bidi
New Zealand bur
Scarlet bidi-bidi
Acaena novae-zealandiae
Red bidi-bidi
Acantholimon glumaceum
Prickly thrift
Acanthus dioscoridis
Dwarf acanthus
Acer japonicum aconitifolium
Japanese maple
Acer japonicum aureum
Golden Japanese maple
Acer palmatum   Japanese maple
Achillea clavennae   Silvery milfoil
Achillea serbica   Milfoil
Yarrow
Achillea tomentosa   Yellow milfoil
Acinos alpinus   Alpine calamint
Aconitum napellus
Blue monkshood
Common monkshood
Monkshood
Actaea pachypoda
White baneberry
Actaea rubra   Red baneberry
Acthionema saxatile
Candy mustard
Adiantum pedatum
Northern maidenhair fern
Adonis vernalis
European spring adonis
Spring adonis
Yellow adonis
Aethionema grandiflorum
Persian stonecress
Aethionema rotundifolium
Candytuft
Aethionema warleyense
Lebanon candytuft
Store cress
Ajuga reptans   Bugle
Ajuga reptans atropurpurea
Purple bugle
Ajuga reptans variegata
Variegated bugle
Alchemilla alpina
Alpine lady's mantle

Alchemilla conjuncta
Alpine lady's mantle
Alchemilla vulgaris   Bear's foot
Lady's mantle
Lion's foot
Allium oreophilum
Flowering onion
Allium moly   Golden garlic
Alyssum argenteum   Italian alyssum
Alyssum montanum
Mountain alyson
Mountain alyssum
Alyssum saxatile   Gold dust
Golden alyssum
Golden tuft
Anacyclus depressus
Mount Atlas daisy
Anacyclus pyrethrum   Pellitory
Spanish pellitory
Anacyclus pyrethrum depressus
Mount Atlas daisy
Androsace carnea
Pink rock jasmine
Androsace chamaejasme
Bastard jasmine
Androsace helvetica
Swiss rock jasmine
Androsace lactea
Milkwhite rock jasmine
Androsace occidentalis
Rock jasmine
Rock jessamine
Anemone apennina   Anemone
Anemone narcissiflora
Appleblossom alpine anemone
Narcissus-flowered anemone
Anemone nemorosa
Wood anemone
Anemone pulsatilla
Common pasque flower
Pasque flower
Anemone rupicola
Rock windflower
Antennaria alpina   Alpine cat's foot
Antennaria dioica   Cat's foot
Mountain cat's ear
Mountain everlasting
Pussy's toes
Anthyllis montana
Mountain kidney vetch
Aquilegia alpina   Alpine columbine
Aquilegia atrata   Purple columbime
Aquilegia caerulea
Rocky Mountain columbine
Aquilegia canadensis   Rock bells
Aquilegia discolor
Dwarf Spanish columbine
Aquilegia einseleana
Einsel's columbine
Arabis albida   Rock cress
Snow-on-the-mountain
Arabis alpina   Alpine rock cress

Chamaecyparis lawsoniana minima

Arabis caucasica	Garden arabis
	White rock
Arabis ferdinandi-coburgii	
	Rock cress
Arabis hirsuta	Hairy rock cress
Arctostaphylos uva-ursi	
	Bearberry
Arenaria balearica	
	Corsican sandwort
Arenaria ciliata	Fringed sandwort
Arenaria grandiflora	
	Large-flowered sandwort
Arenaria leptoclados	
	Slender sandwort
Arenaria montana	
	Alpine sandwort
	Mountain sandwort
Arenaria norvegica	
	Arctic sandwort
Arenaria purpurascens	
	Pink sandwort
Arenaria serpyllifolia	Thyme-
	leaved sandwort
Arenaria stricta	Rock sandwort
Arenaria tetraquetra	Sandwort
Armeria alliacea	Jersey thrift
Armeria caespitosa	Thrift
Armeria juniperifolia	
	Dwarf pink thrift
Armeria maritima	Sea pink
	Thrift
Armeria maritima alba	
	White sea lavender
	White sea pink
Armeria welwitschii	Spanish thrift
Arnebia echioides	Prophet flower
Arnebia pulchra	Prophet flower
Artemisia nitida	
	Narrow-leaved wormwood
Artemisia schmidtiana nana	
	Japanese dwarf wormwood
Asarum europaeum	Asarabacca
Asperula hirta	Pyrenean woodruff
Asperula nitida	Woodruff
Asphodeline lutea	King's spear
	Yellow asphodel
Asplenium ruta-muraria	
	Wall spleenwort
Asplenium scolopendrium	
	Hart's tongue fern
Asplenium tricomanes	
	Maidenhair spleenwort
Aster alpinus	Alpine aster
	Blue alpine daisy
Aster amellus	Italian starwort
	Michaelmas daisy
Aster linosyris	Golden aster
Aubretia deltoidea	Aubretia
Bellendena montana	
	Mountain rocket
Berberis candidula	Barberry

Berteroa incana	Hoary alison
Betula nana	Dwarf birch
Blechnum spicant	Hard fern
Briza media	Quaking grass
Brunnera macrophylla	
	Siberian bugloss
Calamintha alpina	Alpine calamint
Calandrinia menziesii	Red maids
	Rock purslane
Calceolaria biflora	Slipper flower
Calceolaria plantaginea	
	Slipper flower
Calceolaria polyrrhiza	
	Patagonian slipper flower
Caltha palustris	King cup
	Marsh marigold
Campanula carpatica	
	Tussock bellflower
Campanula cochleariifolia	
	Bellflower
	Fairy's thimble
Campanula elatines	
	Starry bellflower
Campanula garganica	
	Gargano bellflower
Campanula latifolia macrantha	
	Giant bellflower
Campanula persicifolia	
	Peach-leaved bellflower
Campanula pulla	Solitary harebell
Campanula pusilla	Fairy's thimble
Campanula tridentata	
	Large bellflower
Cardamine hirsuta	
	Hairy bitter cress
Cardamine pratensis	Bitter cress
	Cuckoo flower
	Lady's smock
	Mayflower
	Meadow cress
Carex firma	Sedge
Carex montana	Mountain sedge
Carlina acanthifolia	
	Acanthus-leaved carline thistle
Carlina acaulis	Alpine thistle
Carlina acaulis simplex	
	Alpine carline thistle
Cedrus libani sargentii	
	Sargent's cedar
Centaurea montana	
	Mountain cornflower
Centaurea montana sulphurea	
	Yellow mountain cornflower
Centaurea rhapontica	
	Giant knapweed
Cerastium alpinum	
	Alpine mouse-eared chickweed
Cerastium tomentosum	
	Mouse-eared chickweed
	Snow-in-summer
Chamaecyparis lawsoniana	
minima	Dwarf lawson's cypress

*Chamaecyparis lawsoniana
minima glauca*

*Chamaecyparis lawsoniana
minima glauca*
    Dwarf blue lawson's cypress
Chamaecyparis obtusa
    Hinoki cypress
Chamaecyparis obtusa nana
    Dwarf hinoki cypress
*Chamaecyparis pisifera
filifera aurea nana*
    Dwarf golden sawara cypress
*Chamaecyparis pisifera
filifera nana*
    Dwarf sarawa cypress
Chelone obliqua   Turtle head
*Chiastophyllum
oppositifolium*   Lamb's tail
Chionodoxa luciliae
    Glory-of-the-snow
Chrysanthemum arcticum
    Rockery daisy
Chrysogonum virginianum
    Golden star
Cimicifuga racemosa
    Black snakeroot
Cirsium acaule   Stemless thistle
Clematis alpina   Alpine clematis
Colchicum bornmuelleri
    Autumn crocus
Convallaria majalis
    Lily-of-the-valley
Convallaria majalis rosea
    Pink lily-of-the-valley
Cornus canadensis   Dogwood
Corrigiola litoralis   Strapwort
Corydalis cava   Hollow fumitory
Corydalis cheilanthifolia
    Fumitory
Corydalis lutea   Yellow fumitory
Corydalis solida   Fingered fumitory
Corylopsis pauciflora
    Buttercup winter hazel
Cotoneaster horizontalis
    Herringbone cotoneaster
Cotula coronopifolia
    Brass buttons
Cyclamen hederifolium
    Ivyleaf cyclamen
Cyclamen neapolitanum
    Ivyleaf cyclamen
Cyclamen purpurascens
    Alpine violet
    Sowbread
Cymbalaria muralis
    Kenilworth ivy
Cypripedium calceolus
    Lady's slipper orchid
Cypripedium reginae
    Queen's slipper orchid
    Showy lady's slipper orchid
Cytisus decumbens   Broom
Cytisus scoparius   Common broom
Daphne cneorum   Garland flower

Daphne mezereum   Mezereon
Dianthus alpinus   Alpine pink
Dianthus caesius   Cheddar pink
Dianthus carthusianorum
    Carthusian pink
Dianthus deltoides   Maiden pink
Dianthus gratianopolitanus
    Cheddar pink
Dianthus pavonius
    veined pink
Dianthus pinifolius
    Macedonian white pink
Dianthus plumarius   Common pink
    Wild pink
Dianthus superbus   Superb pink
Dicentra exima   Bleeding heart
Dictamnus albus
    White burning bush
Dictamnus rubra
    Red burning bush
Digitalis grandiflora
    Large yellow foxglove
Digitalis purpurea
    Common foxglove
    Dead men's bells
    Fairy thimbles
    Lady's gloves
    Witch's gloves
Dodecatheon meadia
    Shooting star
Draba aizoides
    Yellow whitlow grass
Draba sibirica
    Siberian whitlow grass
Dryas octopetala   Mountain avens
Echioides longiflora
    Prophet flower
Epimedium alpinum
    Alpine barrenwort
    Barrenwort
Eranthis hyemalis   Winter aconite
Erigeron unifloris   Fleabane
Eriophyllum lanatum
    Oregon sunshine
Erodium chamaedryoides
    White storksbill
Erodium cicutarium
    Common storksbill
Erodium maritimum   Sea storksbill
Erodium reichardii
    White storksbill
Eryngium bourgatii
    Pyrenean eryngo
Erysimum alpinum
    Alpine wallflower
Erythronium dens-canis
    Dog's tooth violet
Euphorbia capitulata
    Alpine spurge
Euphorbia myrsinites   Blue spurge
    Milkwort
    Spurge

Iberis sempervirens

Euphorbia palustris	Marsh spurge
Festuca cinerea	Grey fescue
Filipendula vulgaris	Dropwort
Fragaria vesca	Alpine strawberry
	Wild strawberry
Francoa ramosa	Maiden's wreath
Francoa sonchifolia	Bridal wreath
Frankenia laevis	Sea heath
Fritillaria meleagris	
	Snake's head fritillary
Galium odoratum	Sweet woodruff
Genista lydia	Broom
Gentiana acaulis	Stemless gentian
	Trumpet gentian
Gentiana alpina	Alpine gentian
Gentiana andrewsii	Bottle gentian
	Closed gentian
Gentiana asclepiadea	
	Willow gentian
Gentiana autumnalis	
	Pine-barren gentian
Gentiana brachyphylla	
	Short-leaved gentian
Gentiana catesbaei	
	Catesby's gentian
	Sampson's snakeroot
Gentiana clausa	Blind gentian
	Bottle gentian
	Closed gentian
Gentiana cruciata	Cross gentian
Gentiana dinarica	Stemless gentian
Gentiana linearis	Closed gentian
Gentiana lutea	Yellow gentian
Gentiana newberryi	Alpine gentian
Gentiana nipponica	
	Japanese gentian
Gentiana pneumonanthe	
	Calathian violet
	Marsh gentian
Gentiana punctata	Spotted gentian
Gentiana scabrae	Japanese gentian
Gentiana septemfida	
	Crested gentian
	Summer gentian
Gentiana setigera	
	Mendocino gentian
Gentiana sino-ornata	
	Autumn-flowering gentian
Gentiana tergestina	Karst gentian
Gentiana terglouensis	
	Triglav gentian
Gentiana utriculosa	
	Bladder gentian
Gentiana verna	Spring gentian
	Vernal gentian
Gentiana villosa	
	Sampson's snakeroot
Gentianopsis holopetala	
	Sierra gentian
Geranium endressii	
	Western cranesbill
Geranium macrorrhizum	
	Rock cranesbill
Geranium sanguineum	
	Bloody cranesbill
Geranium sanguineum album	
	White bloody cranesbill
Geum montanum	Alpine avens
Geum rivale	Water avens
Gillenia trifoliata	Indian physic
Globularia cordifolia	
	Matted globularia
Globularia trichosantha	
	Globe daisy
Gypsophila repens	
	Alpine gypsophila
	Baby's breath
	Chalk plant
	Creeping gypsophila
Hacquetia epipactis	Hacquetia
Hedysarum obscurum	
	Alpine sainfoin
Helianthemum chamaecistus	
	Common rockrose
Helianthemum nummularium	
	Common rockrose
	Rockrose
Helianthemum oelandicum	
	Alpine rockrose
Helleborus foetidus	
	Stinking hellebore
Helleborus niger	Christmas rose
Hemerocallis minor	
	Grass-leaved day lily
Hepatica nobilis	Liverleaf
Herniaria glabra	
	Glabrous rupturewort
Heuchera sanguinea alba	
	Coral bells
Hieracium aurantiacum	
	Orange hawkweed
Hieracium pilosella	
	Mouse-ear hawkweed
Hieracium villosum	
	Shaggy hawkweed
Horminum pyrenaicum	
	Dragon's mouth
	Pyrenean dragonmouth
Hosta fortunei aurea	
	Gold plantain lily
	Spring plantain lily
Hosta plantaginea	
	Large white plantain lily
Hutchinsia alpina	Chamois cream
Iberis amara	Candytuft
	Rocket candytuft
	Wild candytuft
Iberis gibraltarica	
	Gibraltar candytuft
Iberis saxatilis	Candytuft
Iberis sempervirens	Candytuft
	Edging candytuft
	Evergreen candytuft

Iberis umbellata

Iberis umbellata	Globe candytuft
Iris graminea	Grass-leaved iris
Iris lacustris	American dwarf iris
Iris orientalis	Yellow iris
Iris pseudacorus	Yellow flag
Iris pumila	Dwarf iris
Iris sibirica	Siberian iris
Iris tectorum	Roof iris
Jasminum nudiflorum	Winter jasmine
Jovibarba sobolifera	Hen-and-chickens houseleek
Juniperus chinensis blaauw	Blue Chinese juniper Dwarf Chinese juniper
Juniperus communis compressa	Dwarf common juniper
Juniperus communis depressa	Prostrate juniper
Juniperus communis hornibrookii	Creeping common juniper
Juniperus horizontalis glauca	Blue creeping juniper
Juniperus squamata glauca	Blue flaky juniper
Juniperus virginiana nana compacta	Dwarf pencil cedar
Lathyrus vernus	Spring vetchling
Leontopodium alpinum	Edelweiss Lion's foot
Leuzea rhapontica	Giant knapweed
Lewisia pygmaea	Pygmy lewisia
Lewisia rediviva	Bitter root Bitterwort
Liatris elegans	Snakeroot
Liatris spicata	Blue snakeroot
Linum flavum	Golden flax
Linum flavum compactum	Yellow flax
Linum perenne	Perennial flax
Linum suffruticosum	Shrubby white flax
Lithospermum canescens	Indian paint Indian warpaint Puccoon
Lithospermum distichum	Mexican puccoon
Lithospermum officinale	Common gromwell Gromwell Puccoon
Lobularia maritima	Sea alyssum Sweet alyssum
Lotus corniculatus pleniforus	Bacon and eggs Bird's foot trefoil
Lychnis alpina	Alpine catchfly Alpine lychnis
Lychnis viscaria	German catchfly
Lysimachia nummularia	Creeping jenny
Lysimachia punctata	Large yellow loosestrife
Lythrum salicaria	Purple loosestrife
Lythrum virgatum	Slender loosestrife
Maianthemum bifolium	May lily
Malva moschata	Musk mallow
Malva moschata alba	White musk mallow
Meconopsis integrifolia	Lampshade poppy
Melittis melissophyllum	Bastard balm
Mertensia virginica	Virginian cowslip
Meum athamanticum	Baldmoney Spignel
Minuartia graminifolia	Apennine sandwort
Minuartia hybrida	Fine-leaved sandwort
Minuartia stricta	Rock sandwort
Minuartia verna	Spring sandwort Vernal sandwort
Minuartia viscosa	Sticky sandwort
Moehringia muscosa	Mossy sandwort
Moehringia trinervia	Three-veined sandwort
Muscari botryoides	Grape hyacinth
Myosotis alpestris	Alpine forget-me-not
Myosotis arvensis	Field forget-me-not
Myosotis discolor	Yellow forget-me-not
Myosotis sylvatica	Wood forget-me-not
Narcissus bulbocodium	Hoop-petticoat daffodil
Nepeta × faassenii	Catmint
Oenothera missouriensis	Creeping evening primrose
Omphalodes verna	Blue-eyed mary
Opuntia engelmannii	Prickly-pear cactus
Origanum vulgare compactum	Wild marjoram
Osmunda regalis	Royal fern
Oxalis acetosella	Wood sorrel
Papaver alpinum	Alpine poppy
Papaver burseri	Alpine poppy
Papaver kerneri	Gold alpine poppy
Papaver nudicaule	Iceland poppy
Papaver rhaeticum	Rhaetian poppy

Paradisea liliastrum
Saint Bruno's lily

Parietaria diffusa Wall pellitory

Parietaria judaica Wall pellitory

Petasites fragrans Sweet coltsfoot
Winter heliotrope

Petrocallis pyrenaica Rock beauty

Petrophila sessilis
Prickly conesticks

Petrorhagia saxifraga Tunic flower

Phlox bifida Sand phlox

Phlox borealis Alaskan phlox

Phlox canadensis
Wild sweet william

Phlox divaricata Blue phlox
Wild sweet william

Phlox douglasii Alpine phlox

Phlox maculata Wild sweet william

Phlox nivalis Trailing phlox

Phlox sibirica Siberian phlox

Phlox subulata Moss phlox
Moss pink
Mountain phlox

Phyllitis scolopendrium
Hart's tongue fern

Phyteuma orbiculare
Round-headed rampion

Phyteuma scheuchzeri
Horned rampion

Picea abies compressa
Dwarf Norway spruce

Picea abies echiniformis
Dwarf spruce

Picea abies pumila glauca
Blue dwarf spruce

Picea abies pygmaea
Pygmy spruce

Picea glauca Blue spruce

Pimpinella saxifraga
Burnet saxifrage

Pinguicula caudata
Mexican butterwort

Pinguicula grandiflora
Great butterwort
Large-flowered butterwort

Pinguicula montana
Mountain butterwort

Pinguicula vulgaris Butterwort
Common butterwort

Pinus aristata Bristle-cone pine

Pinus montana Mountain pine

Pinus mugo Mountain pine

Pinus mugo mugo
Swiss mountain pine

Pinus mugo pumila Dwarf pine

Pinus sibirica pumila
Dwarf Siberian pine

Pinus sibirica pumila glauca
Dwarf blue Siberian pine

Pinus strobus nana
Dwarf white pine

Pinus strobus umbraculifera
Dwarf weymouth pine

Pinus sylvestris beuvronensis
Dwarf Scots pine

Pinus sylvestris nana
Dwarf Scots pine

Platycodon grandiflorus
Balloon flower
White bells

Podophyllum hexandrum
Himalayan may apple

Polygonatum commutatum
Giant solomon's seal

Polygonatum hookeri
Dwarf solomon's seal

Polygonatum × hybridum
Solomon's seal

Polygonum bistorta superbum
Bistort
Snakeroot

Polypodium vulgare
Common polypody

Potentilla anserina Goose grass
Goose tansy
Silverweed

Potentilla arenaria Grey cinquefoil

Potentilla argentea
Hoary cinquefoil
Silvery cinquefoil

Potentilla argentea calabra
Hoary cinquefoil

Potentilla aurea Golden cinquefoil

Potentilla cinerea Grey cinquefoil

Potentilla palustris
Marsh cinquefoil
Marsh five-finger

Potentilla pyrenaica
Pyrenean cinquefoil

Potentilla recta warrenii
Sulphur cinquefoil

Potentilla rupestris Prairie tea
Rock cinquefoil

Potentilla simplex
Oldfield cinquefoil

Potentilla tridentata
Three-toothed cinquefoil

Potentilla verna Spring cinquefoil

Primula auricula Auricula
Bear's ear
Dusty miller

Primula denticulata
Drumstick primrose

Primula denticulata alba
White ball primrose

Primula elatior Oxlip

Primula farinosa
Bird's eye primrose

Primula floribunda
Buttercup primrose

Primula florindae Tibetan cowslip

Primula forbesii Baby primrose

Primula laurentiana

Primula laurentiana
　　Bird's eye primrose
Primula malacoides
　　Baby primrose
　　Fairy primrose
Primula mistassinica
　　Bird's eye primrose
Primula obconica
　　German primrose
　　Poison primrose
Primula officinalis　　Cowslip
Primula × polyantha　　Polyanthus
Primula × pubescens　　Auricula
Primula sinensis　　Chinese primrose
Primula veris　　Cowslip
　　Fairy cups
　　Herb peter
　　Keyflower
　　Palsywort
　　Saint Peter's wort
Primula vialii
　　Red-hot poker primrose
Primula vulgaris
　　Common primrose
　　English primrose
　　Primrose
Prunella grandiflora
　　Large self-heal
Prunus tenella
　　Dwarf Russian almond
Pulmonaria angustifolia
　　Blue cowslip
　　Lungwort
*Pulmonaria angustifolia
azurea*
　　Narrow-leaved lungwort
Pulmonaria picta　　Bethlehem sage
Pulmonaria rubra　　Lungwort
Pulmonaria saccharata
　　Bethlehem sage
Pulsatilla alba
　　White pasque flower
Pulsatilla pratensis nigricans
　　Field anemone
Pulsatilla vernalis　　Spring anemone
Pulsatilla vulgaris
　　Common pasque flower
　　Meadow anemone
　　Pasque flower
Pulsatilla vulgaris alba
　　White pasque flower
Pulsatilla vulgaris rubra
　　Red pasque flower
Ramonda myconi
　　Pyrenean primrose
　　Pyrenean ramonda
Ramonda pyrenaica
　　Pyrenean primrose
Ranunculus aconitifolius
　　Fair maids of Kent
　　White buttercup
Ranunculus acris multiplex
　　Meadow buttercup

Ranunculus alpestris
　　Alpine buttercup
　　Mountain buttercup
Ranunculus asiaticus
　　Persian buttercup
　　Persian ranunculus
Ranunculus bulbosus
　　Bulbous buttercup
　　Bulbous crowfoot
Ranunculus ficaria
　　Lesser celandine
　　Pilewort
　　Small celandine
Ranunculus glacialis
　　Glacier crowfoot
Ranunculus gramineus
　　Grass-leaved buttercup
Ranunculus montanus
　　Mountain buttercup
Ranunculus repens　　Butter daisy
　　Creeping buttercup
　　Creeping crowfoot
　　Yellow gowan
Raoulia eximia　　Vegetable sheep
Rhamnus saxitilis　　Rock buckthorn
Rhododendron aperantum
　　Burmese dwarf rhododendron
Rhododendron chrysanthum
　　Prostrate rhododendron
Rhododendron ferrugineum
　　Alpenrose
　　Alpine rose
Rhododendron hirsutum
　　Hairy alpine rose
Rhododendron leucaspis
　　Tibetan rhododendron
Sagina nodosa　　Knotted pearlwort
Sagina pilifera　　Pearlwort
Sagina procumbens
　　Common pearlwort
Sagina subulata
　　Awl-leaved pearlwort
　　Heath pearlwort
Salix alpina　　Mountain willow
Salix hastata wehrhahnii
　　Dwarf willow
Salix reticulata　　Reticulate willow
Salvia argentea　　Silver sage
Santolina chamaecyparissus
　　Lavender cotton
Saponaria caespitosa
　　Tufted soapwort
Saponaria lutea　　Yellow soapwort
Saponaria ocymoides
　　Rock soapwort
　　Tumbling ted
Saponaria officinalis　　Bouncing bet
　　Lather root
Sarothamnus scoparius
　　Common broom
Satureja montana alba
　　Winter savory

ALPINES

Silene caroliniana

Saxifraga aizoides	
	Yellow mountain saxifrage
Saxifraga aizoon major	
	Live-long saxifrage
Saxifraga altissima	
	Host's saxifrage
Saxifraga × arendsii	
	Mossy saxifrage
Saxifraga burserana	
	One-flowered cushion saxifrage
Saxifraga caesia	Blue saxifrage
Saxifraga cespitosa	Tufted saxifrage
Saxifraga cotyledon	
	Pyramidal saxifrage
Saxifraga granulata	
	Fair maids of France
	Meadow saxifrage
Saxifraga hirsuta	Kidney saxifrage
Saxifraga hostii	Host's saxifrage
Saxifraga hypnoides	
	Dovedale moss
	Mossy rockfoil
	Mossy saxifrage
Saxifraga longifolia	
	Pyrenean saxifrage
Saxifraga moschata	Musk saxifrage
Saxifraga muscoides	
	Musky saxifrage
Saxifraga oppositifolia	
	Purple mountain saxifrage
	Purple saxifrage
Saxifraga paniculata major	
	Live-long saxifrage
Saxifraga rivularis	
	Alpine brook saxifrage
Saxifraga rosacea	Irish saxifrage
Saxifraga stellata	Starry saxifrage
Saxifraga stolonifera	
	Beefsteak geranium
	Creeping sailor
	Mother-of-thousands
	Strawberry begonia
	Strawberry geranium
Saxifraga tennesseensis	
	Golden-eye saxifrage
Saxifraga umbrosa	London pride
	Nancy pretty
Saxifraga × urbium	
	Garden London pride
	None-so-pretty
	Saint Patrick's cabbage
Scilla bifolia	Two-leaved scilla
Scilla hispanica	Spanish bluebell
Scutellaria alpina	Alpine scullcap
Scutellaria orientalis pinnatifida	
	Yellow scullcap
Sedum acre	Biting stonecrop
	Golden carpet
	Golden moss
	Wallpepper
	Yellow stonecrop
Sedum adolphi	Golden sedum

Sedum alba	Wallpepper
	Whitecrop
	White stonecrop
Sedum album	White stonecrop
Sedum anacampseros	
	Reddish stonecrop
Sedum anglicum	English stonecrop
Sedum dasyphyllum suendermannii	
	Thick-leaved stonecrop
Sedum forsteranum	
	Rock stonecrop
Sedum morganianum	Beaver's tail
	Burro's tail
	Donkey's tail
	Horse's tail
	Lamb's tail
Sedum multiceps	Baby joshua tree
	Dwarf joshua tree
	Joshua tree
	Little joshua tree
	Miniature joshua tree
Sedum pachyphyllum	
	Jellybean plant
	Jellybeans
	Many fingers
Sedum reflexum	
	Reflexed stonecrop
	Rock stonecrop
Sedum roseum	Roseroot sedum
Sedum × rubrotinctum	
	Christmas cheer
	Pork and beans
Sedum sieboldii	October daphne
	October plant
	Siebold's stonecrop
Sedum spectabile	Ice plant
Sedum telephium	Livelong
	Midsummer men
	Orpine
	Sedum live-forever
Sedum treleasii	Silver stonecrop
Sedum villosum	Hairy stonecrop
Sempervivum arachnoideum	
	Cobweb houseleek
	Spider's web houseleek
Sempervivum calcareum	
	Limestone houseleek
Sempervivum montanum	
	Mountain houseleek
Sempervivum soboliferum	
	Hen-and-chickens houseleek
Sempervivum tectorum	
	Common houseleek
	Hen-and-chickens houseleek
	Houseleek
	Old-man-and-woman
Sempervivum tectorum calcareum	Limestone houseleek
Silene acaulis	Cushion pink
	Moss campion
Silene alpestris	Alpine catchfly
Silene caroliniana	Wild pink

Sea campion

Silene maritima	Sea campion
Silene pendula	Nodding catchfly
Silene uniflora	Sea campion
Sisyrinchium angustifolium	
	Blue-eyed grass
	Satin flower
Soldanella alpina	Alpine snowbell
Soldanella montana	
	Mountain soldanella
	Mountain tassel
Solidago virgaurea	Golden rod
Stachys byzantina	Lamb's ears
Stipa capillata	Feather grass
Tanacetum densum amani	
	Prince of Wales feathers
Taxus baccata	Common yew
	Yew
Taxus baccata compacta	
	Dwarf yew
Taxus baccata nana	Dwarf yew
Taxus cuspidata minima	
	Dwarf Japanese yew
Thalictrum aquilegifolium	
	Great meadow rue
Thlaspi alliaceum	
	Garlic pennycress
Thlaspi alpestre	Alpine pennycress
Thlaspi arvense	Fanweed
	Field pennycress
	French weed
	Mithridate mustard
	Pennycress
	Stinkweed
Thlaspi perfoliatum	
	Perfoliate pennycress
Thymus articus coccineus	
	Wild thyme
Thymus citriodorus	Lemon thyme
Thymus coccineus	Wild thyme
Thymus herba-barona	
	Caraway thyme
Thymus praecox pseudolan	
guinosus	Hairy thyme
Thymus serpyllum	
	Breckland thyme
	Creeping thyme
	Lemon thyme
	Wild thyme
Thymus vulgaris	Common thyme
	Garden thyme
Tiarella cordifolia	Foam flower
Tricyrtis hirta	Japanese toadlily
Trollius pumilus	Globe flower
Tsuga canadensis jeddeloh	
	Dwarf eastern hemlock
Tsuga canadensis nana	
	Dwarf hemlock
Valeriana montana	
	Mountain valerian
Valeriana supina	Valerian
Veratrum album	
	White false helleborine

Veratrum nigrum	
	Black false helleborine
Veronica alpina	Alpine speedwell
	Alpine veronica
Veronica americana	
	American brooklime
Veronica austriaca teucrium	
	Large speedwell
Veronica beccabunga	
	European brooklime
Veronica chamaedrys	Angel's eye
	Bird's eye
	Germander speedwell
Veronica filiformis	Speedwell
Veronica fruticulosa	
	Shrubby speedwell
	Shrubby-stalked speedwell
Veronica longifolia	
	Long-leaved speedwell
Veronica officinalis	
	Common speedwell
	Gypsyweed
Veronica serpyllifolia	
	Thyme-leaved speedwell
Veronica spicata incana	
	Silver speedwell
	Silver spiked speedwell
Vinca minor	Common periwinkle
	Lesser periwinkle
	Myrtle
	Running myrtle
Viola adunca	Hookspur violet
	Western dog violet
Viola beckwithii	Great basin violet
Viola biflora	Yellow wood violet
Viola blanda	Sweet white violet
Viola brittoniana	Coast violet
Viola canadensis	Canadian violet
	Tall white violet
Viola canina	Dog violet
Viola conspersa	
	American dog violet
Viola cornuta	Horned pansy
	Horned violet
Viola × emarginata	
	Triangle-leaved violet
Viola fimbriatula	
	Northern downy violet
Viola flettii	Olympic violet
	Rock violet
Viola glabella	Stream violet
Viola gracilis	Olympian violet
Viola hastata	Halberd-leaved violet
Viola hederacea	Australian violet
	Ivy-leaved violet
	Trailing violet
Viola labradorica	Labrador violet
Viola lanceolata	
	Eastern water violet
	Lance-leaved violet
Viola langsdorfii	Alaska violet
Viola lobata	Yellow wood violet

Viola macloskeyi
    Western sweet white violet

Viola missouriensis   Missouri violet

Viola nephrophylla
    Northern bog violet

Viola nuttallii   Yellow prairie violet

Viola ocellata   Two-eyed violet

Viola odorata   English violet
    Florist's violet
    Garden violet
    Sweet violet

Viola orbiculata
    Western round-leaved violet

Viola palmata   Early blue violet
    Wild okra

Viola palustris   Alpine marsh violet
    Marsh violet

Viola pedata   Bird's foot violet
    Crowfoot violet
    Pansy violet

Viola pedatifida   Larkspur violet
    Purple prairie violet

Viola pendunculata
    California golden violet
    Californian wild pansy
    Johnny-jump-up

Viola priceana   Confederate violet

Viola primulifolia
    Primrose-leaved violet

Viola pubescens
    Downy yellow violet

Viola rafinesquii   Field pansy

Viola renifolia   Kidney-leaved violet
    Northern white violet

Viola riviniana   Dog violet
    Wood violet

Viola rostrata   Long-spurred violet

Viola rotundifolia
    Early yellow violet
    Round-leaved yellow violet

Viola sagittata   Arrow-leaved violet

Viola selkirkii   Great spurred violet

Viola sempervirens
    Evergreen violet
    Redwood violet

Viola septemloba
    Southern coast violet

Viola septentrionalis
    Northern blue violet

Viola sororia   Woolly blue violet

Viola striata   Cream violet
    Pale violet
    Striped violet

Viola tricolor   European wild pansy
    Field pansy
    Johnny-jump-up
    Miniature pansy

Viola trinervata   Sagebrush violet

Viola viarum   Plains violet

Viola × wittrockiana
    Garden pansy
    Heartsease
    Ladies' delight
    Pansy
    Stepmother's flower

Waldsteinia geoides
    Barren strawberry

Yucca filamentosa   Thread agave

Yucca gloriosa   Palm lily
    Spanish dagger

Zauschneria californica
    Californian fuchsia
    Humming-bird's trumpet

AQUATICS

This section applies to plants living usually in fresh water, either rooted in soil or free-floating, also to plants living in bogs, swamps, and around the edges of ponds and lakes.

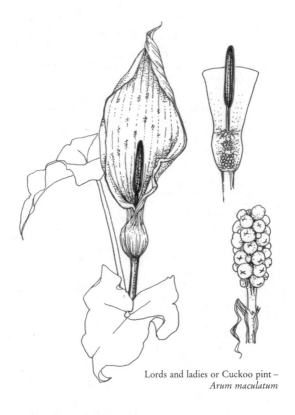

Lords and ladies or Cuckoo pint –
Arum maculatum

Acorus calamus

Acorus calamus	Sweet sedge
Aglaonema simplex	Malayan sword
Alisma plantago-aquatica	
	Mad-dog weed
	Water plantain
Anacharis canadensis	Ditch moss
	Water thyme
	Waterweed
Anacharis densa	
	Brazilian waterweed
Apium inundatum	
	Floating marshwort
Apium nodiflorum	
	Fool's watercress
Aponogeton distachyus	
	Cape asparagus
	Cape pondweed
	Water hawthorn
Aponogeton fenestralis	Laceleaf
	Latticeleaf
	Madagascar lace plant
	Water yam
Aponogeton madagascariensis	
	Laceleaf
	Latticeleaf
	Madagascar lace plant
	Water yam
Azolla caroliniana	Fairy moss
	Mosquito fern
	Mosquito plant
	Water fern
Bacopa monnieri	Baby tears
Baldellia ranunculoides	
	Lesser water plantain
Bidens cernua	
	Nodding bur marigold
Brasenia schreberi	Water shield
Cabomba caroliniana	Fanwort
	Fish grass
	Washington grass
Calla palustris	Bog arum
	Water arum
Callitriche hermaphroditum	
	Autumnal water starwort
Callitriche obtusangula	
	Blunt-fruited water starwort
Callitriche stagnalis	
	Water starwort
Caltha palustris	Kingcup
	Marsh marigold
	Molly blobs
	Water blobs
	Water cowslip
Castalia flava	Yellow water lily
Castalia odorata	Fragrant water lily
	Pond lily
	White water lily
Ceratophyllum demersum	
	American hornwort
	Hornwort
	Rigid hornwort
Ceratophyllum submersum	
	Soft hornwort

Ceratopteris pteridoides	
	Floating fern
Ceratopteris richardii	
	Triangular water fern
Ceratopteris thalictroides	
	Water fern
	Water sprite
Cicuta virosa	Water hemlock
Cryptocoryne affinis	
	Water trumpet
Cyperus alternifolius	
	Umbrella palm
	Umbrella plant
	Umbrella sedge
Cyperus esculentus	
	Nut grass
	Nut sedge
	Yellow nut grass
	Yellow nut sedge
Cyperus isocladus	Dwarf papyrus
	Miniature papyrus
Cyperus papyrus	Biblical bulrush
	Bulrush
	Paper plant
	Papyrus
Echinodorus cordifolius	
	Texas mud-baby
Echinodorus intermedius	
	Pigmy chainsword plant
Echinodorus magdalenensis	
	Dwarf amazon sword plant
Echinodorus martii	
	Pigmy chainsword plant
Echinodorus paniculatus	
	Amazon sword plant
Echinodorus radicans	
	Texas mud-baby
Echinodorus tenellus	
	Amazon sword plant
Eichhornia azurea	Peacock hyacinth
Eichhornia crassipes	
	Water hyacinth
Elatine hexandra	Waterwort
Eleocharis acicularis	Hair grass
	Least spike rush
	Slender spike rush
Eleocharis dulcis	
	Chinese water chestnut
	Ma-tai
Elodea callitrichoides	
	Greater water thyme
Elodea canadensis	
	Canadian pondweed
	Ditch moss
	Water thyme
	Waterweed
Elodea densa	Brazilian waterweed
Elodea nuttallii	
	Nuttall's water thyme
Eriocaulon aquaticum	Pipewort

Nelumbo nucifera

Eupatorium maculatum	
	Joe-pye weed
	Smokeweed
Eupatorium perfoliatum	
	Common boneset
	Thoroughwort
Euryale ferox	Gorgon
	Prickly water lily
Fontinalis antipyretica	
	Fountain moss
	Spring moss
	Water moss
	Willow moss
Geum rivale	Nodding avens
	Water avens
Groenlandia densa	
	Opposite-leaved pondweed
Heteranthera dubia	
	Water star grass
Heteranthera reniformis	
	Kidney mud plantain
Hottonia palustris	Water violet
Hydrocera angustifolia	
	Water balsam
Hydrocharis morsus ranae	
	Frog-bit
Hydrocleys nymphoides	
	Water poppy
Hygrophila difformis	
	Water wistaria
Hypericum elodes	
	Marsh St. John's wort
Ipomoea aquatica	
	Water convolvulus
Iris pseudacorus	Flag iris
	Water flag
	Yellow flag iris
Isnardia palustris	Water purslane
Isoetes echinospora	
	Spiny-spored quillwort
Isoetes engelmannii	
	Engelmann's quillwort
Juncus effusus	Japanese mat rush
	Soft rush
Juncus effusus spiralis	
	Corkscrew rush
	Spiral rush
Juncus lesueurii	Salt rush
Justicia americana	
	American water willow
	Water willow
Lagerosiphon major	
	Curly water thyme
Lemna gibba	Fat duckweed
Lemna miniscula	Lesser duckweed
Lemna minor	Common duckweed
	Duckweed
	Lesser duckweed
Lemna trisulca	Ivy duckweed
	Star duckweed
Lobelia dortmanna	Water lobelia
Ludwigia alternifolia	Rattlebox
	Seedbox

Ludwigia palustris	Water purslane
Luronium natans	
	Floating water plantain
Lysichiton americanum	Bog arum
	Skunk cabbage
Lysimachia clethroides	
	Gooseneck lysimachia
Lysimachia nummularia	
	Creeping charlie
	Creeping jenny
	Moneywort
Lysimachia punctata	
	Garden lysimachia
Lysimachia thyrsiflora	
	Tufted lysimachia
Lysimachia vulgaris	
	Garden lysimachia
Lythrum salicaria	
	Purple loosestrife
	Spiked lythrum
Marsilea quadrifolia	
	European water clover
Mentha aquatica	Water mint
Menyanthes trifoliata	Bog bean
	Brook bean
	Marsh clover
	Marsh trefoil
	Water shamrock
Montia fontana	Water blinks
Myosotis aquaticum	
	Water chickweed
Myosotis scorpioides	
	Water forget-me-not
Myosotis secunda	
	Marsh forget-me-not
Myriophyllum aquaticum	
	Parrot's feather
	Water feather
Myriophyllum hippuroides	
	Red water milfoil
	Western milfoil
Myriophyllum spicatum	
	Spiked water milfoil
Myriophyllum verticillatum	
	Myriad leaf
	Whorled water milfoil
Najas flexilis	Flexible naiad
Najas marina	Greater naiad
Narcissus jonquilla	Wild jonquil
Nelumbo lutea	American lotus
	Pond nuts
	Water chinquapin
	Wonkapin
	Yanquapin
	Yellow nelumbo
Nelumbo nucifera	East Indian lotus
	Sacred lotus

Nelumbo pentapetala

Nelumbo pentapetala
American lotus
Pond nuts
Water chinquapin
Wonkapin
Yanquapin
Yellow nelumbo
Nelumbo speciosa   East Indian lotus
Sacred lotus
Nuphar advena
American spatterdock
Common spatterdock
Spatterdock
Nuphar luteum   Brandy bottle
Yellow water lily
Nymphaea alba
European white water lily
Platterdock
White water lily
Nymphaea caerulea
Blue Egyptian lotus
Blue lotus
Egyptian lotus
Nymphaea capensis   Blue water lily
Cape blue water lily
Nymphaea flava   Yellow water lily
Nymphaea gigantea
Australian water lily
Nymphaea lotus   Egyptian lotus
Egyptian water lily
Lotus
White Egyptian lotus
Nymphaea mexicana
Yellow water lily
Nymphaea odorata
Fragrant water lily
Pond lily
Sweet water lily
White water lily
Nymphaea pygmaea
Pygmy water lily
Nymphaea rubra
Indian red water lily
Nymphaea stellata   Blue lotus
Indian blue lotus
Nile blue lotus
Nymphaea tetragona
Pygmy water lily
Nymphaea tuberosa
Magnolia water lily
Tuberous water lily
Nymphaea venusta
European white water lily
Platterdock
Oenanthe fistulosa
Common water dropwort
Water dropwort
Water lovage
Oenanthe fluviatilis
River water dropwort
Oenanthe lachenalis
Parsley water dropwort
Pistia stratiotes   Shellflower
Water lettuce

Polygonium hydropiper
Smartweed
Water pepper
Pontederia cordata   Pickerel weed
Potamogeton alpinus
Red pondweed
Rusty pondweed
Potamogeton coloratus
Fen pondweed
Potamogeton compressus
Grass-wrack pondweed
Potamogeton crispus
Curled pondweed
Potamogeton densus   Frog's lettuce
Potamogeton epihydrus
American pondweed
Potamogeton filiformis
Slender-leaved pondweed
Potamogeton friesii
Flat-stalked pondweed
Potamogeton gramineus
Various-leaved pondweed
Potamogeton lucens
Shining pondweed
Potamogeton nathans
Floating pondweed
Potamogeton nodosus
Loddon pondweed
Potamogeton obtusifolius
Grassy pondweed
Potamogeton pectinatus
Fennel pondweed
Potamogeton perfoliatus
Perfoliate pondweed
Potamogeton polygonifolius
Bog pondweed
Potamogeton pusillus
Small pondweed
Potamogeton trichoides
Hair-like pondweed
Ranunculus aquatalis
Common water crowfoot
Water buttercup
Water crowfoot
Ranunculus fluitans
River crowfoot
River water crowfoot
Riccia fluitans   Crystalwort
Rumex hydrolapathan
Great water dock
Water dock
Rumex maritimus
Golden water dock
Rumex palustris   Marsh dock
Ruppia cirrhosa
Spider tassel pondweed
Ruppia maritima
Beaked tassel pondweed
Sagittaria cuneata   Wapato
Sagittaria latifolia   Duck potato
Wapato

Zostera noltii

Sagittaria montevidensis
Giant arrowhead

Sagittaria sagittifolia
Common arrowhead
Old world arrowhead
Swamp potato
Swan potato
Water archer

Sagittaria sagittifolia
leucopetala Japanese arrowhead

Sagittaria subulata
Awl-leaf arrowhead

Salvinia auriculata Floating fern

Salvinia rotundifolia
Floating moss

Saururus cernuus
American swamp lily
Swamp lily
Water dragon

Sparganium angustifolium
Floating bur reed

Sparganium emersum
Unbranched bur reed

Sparganium erectum
Branched bur reed

Sparganium minimum
Small bur reed

Sparganium ramosum Bur reed
Common bur reed

Spirodela polyrhiza Duckweed
Great duckweed
Greater duckweed
Water flaxseed

Stratiotes aloides Water aloe
Water soldier

Subularia aquatica Awlwort

Trapa natans Jesuit's nut
Water chestnut

Typha angustifolia Lesser bulrush

Typha latifolia Bulrush reedmace

Typha minima Least bulrush

Vallisneria americana
Water celery
Wild celery

Vallisneria spiralis Eel grass
Italian-type eel grass
Tape grass

Veronica anagallis
Water speedwell

Veronica anagallis-aquatica
Blue water speedwell
Water speedwell

Victoria amazonica
Amazon water lily
Amazon water platter
Queen Victoria water lily
Royal water lily
Victoria water lily
Water maize

Victoria cruziana
Santa Cruz water lily
Santa Cruz water platter

Victoria regia Amazon water lily
Amazon water platter
Queen Victoria water lily
Royal water lily
Victoria water lily
Water maize

Victoria trickeri
Santa Cruz water lily
Santa Crus water platter

Wolffia columbiana
Common wolffia
Wolffia

Wolffiella floridana Bogmat
Mud-midget

Zannichellia palustris
Horned pondweed

Zizania aquatica
Canadian wild rice
Water oats
Wild rice

Zostera angustifolia Eel grass
Narrow-leaved ee -grass
Narrow-leaved grass wrack
Tape grass

Zostera marina Common eel grass
Common grass wrack

Zostera noltii Dwarf eel grass
Dwarf grass wrack

BULBS

Horticulturally the term 'bulb' includes bulbs, corms, tubers, and rhizomes, but here a few stoloniferous subjects have been included (some lilies, for instance) where confusion exists in the minds of some amateur gardeners. Orchids are listed separately elsewhere in the book.

Sieber's crocus –
Crocus sieberi

Achimenes longiflora

Achimenes longiflora
Hot-water plant

Agapanthus africanus   African lily
Blue agapanthus
Blue lily
Lily-of-the-Nile

Agapanthus umbellatus
African lily
Blue agapanthus
Lily-of-the-Nile

Allium amabile
Chinese ornamental onion

Allium ampeloprasum   Wild leek

Allium angulosum   Mouse garlic

Allium bakeri   Rakkyo

Allium canadense   Meadow leek
Rose leek
Wild garlic
Wild onion

Allium carinatum   Keeled garlic

Allium cepa   Ever-ready onion
Multiplier onion
Potato onion
Shallot

Allium cernuum
American ornamental onion
Lady's leek
Nodding onion
Wild onion

Allium chinense   Rakkyo

Allium christophii   Stars-of-Persia

Allium cuthbertii   Striped garlic

Allium cyaneum   Blue onion
Purple onion

Allium fistulosum   Ciboule
Japanese bunching onion
Spanish onion
Two-bladed onion
Welsh onion

Allium flavum   Yellow onion

Allium giganteum
Giant ornamental onion

Allium haematochiton
Red-skinned onion

Allium moly   Golden garlic
Lily leek
Yellow-star ornamental onion

Allium neapolitanum
Daffodil garlic
Florist's allium
Flowering onion

Allium oleraceum   Field garlic

Allium paradoxum
Few-flowered leek

Allium ramosum
Fragrant-flowered garlic

Allium roseum   Rosey garlic

Allium sativum   Garlic
Rocambole
Serpent garlic

Allium schoenoprasum   Chive
Cive
Schnittlauch

Allium scorodoprasum   Giant garlic
Rocambole
Sand leek
Spanish garlic

Allium senescens   German garlic

Allium sikkimense
Pendant ornamental onion

Allium sphaerocephalum
Round-headed garlic
Round-headed leek

Allium stellatum   Prairie onion

Allium tanguticum
Lavender globe lily

Allium tricoccum   Ramp
Wild leek

Allium triquetum
Triangular-stemmed garlic
Triquetros leek

Allium tuberosum   Chinese chive
Garlic chive
Oriental garlic

Allium ursinum   Bear's garlic
Buckrams
Gipsy onion
Hog's garlic
Ramsons
Ramsons wood garlic
Wild garlic

Allium validum   Swamp onion

Allium vineale   Crow garlic
Field garlic
Stag's garlic

Alstroemeria aurantiaca
Lily-of-the-Incas
Peruvian lily

Alstroemeria haemantha
Herb lily

Alstroemeria leontochir
ovallei   Lion's paw

Amaryllis aulica   Lily-of-the-palace

Amaryllis aurea   Golden African lily
Golden hurricane lily
Golden spider lily

Amaryllis belladonna
Belladonna lily
Cape belladonna
Jersey lily
March lily

Amaryllis hallii   Magic lily
Resurrection lily

Amaryllis × johnsonii
Saint Joseph's lily

Amaryllis phaedranassa   Queen lily

Amaryllis radiata   Red spider lily
Spider lily

Amaryllis reticulata   Lacework lily
Meshed lily

Amphisiphon stylosa   Cape hyacinth

Anemone appennina   Blue anemone
Windflower

Anemone blanda
Greek windflower
Windflower

Calochortus amoenus

Anemone coronaria
 De caen anemone
 Poppy anemone
 Saint Brigid anemone
 Windflower
Anemone cylindrica
 Long-headed anemone
 Thimbleweed
Anemone fulgens Scarlet anemone
 Scarlet windflower
Anemone hupehensis
 Japanese anemone
Anemone nemorosa
 European wood anemone
 Wood anemone
Anemone nuttalliana
 Hartshorn plant
 Lion's beard
 Pasque flower
 Prairie smoke
 Wild crocus
Anemone patens Pasque flower
Anemone pavonina Windflower
Anemone pulsatilla Pasque flower
Anemone riparia Thimbleweed
Anemone sylvestris
 Snowdrop windflower
Anemone virginiana Thimbleweed
Aponogeton distachyos
 Cape pondweed
 Water hawthorn
Arisaema dracontium Dragonroot
 Green dragon
Arisaema speciosum Cobra lily
Arisaema triphyllum Dragonroot
 Indian turnip
 Jack-in-the-pulpit
Arisarum proboscideum
 Mouse plant
Arum italicum Italian arum
Arum maculatum Adam-and-eve
 Cuckoo pint
 Lords and ladies
Arum palaestinum Black calla
 Solomon's lily
Arum picton Black calla
Babiana rubrocyanea
 Baboon flower
 Blue-and-red baboon root
 Winecups
Begonia albo-coccinea
 Elephant's ear begonia
Begonia boweri Eyelash begonia
 Miniature begonia
Begonia × crestabruchii
 Lettuce-leaf begonia
Begonia davisii Davis begonia
Begonia discolor Hardy begonia
Begonia dregei Grape-leaf begonia
 Maple-leaf begonia
Begonia × erythrophylla
 Beefsteak begonia
 Kidney begonia

Begonia francisii
 Nasturtium-leaf begonia
Begonia goegoensis
 Fire-king begonia
Begonia gracilis Hollyhock begonia
Begonia grandis Hardy begonia
Begonia heracleifolia Star begonia
 Star-leaf begonia
Begonia hydrocotylifolia
 Miniature pond-lily begonia
 Pennywort begonia
Begonia limmingheiana
 Shrimp begonia
Begonia martiana
 Hollyhock begonia
Begonia masoniana
 Iron-cross begonia
Begonia nelumbiifolia
 Lily-pad begonia
 Pond-lily begonia
Begonia parvifolia
 Grape-leaf begonia
 Maple-leaf begonia
Begonia rex King begonia
 Painted-leaf begonia
Begonia × speculata
 Grape-leaf begonia
Begonia × tuberhybrida
 Hybrid tuberous begonia
 Pendulous begonia
Begonia versicolor
 Fairy-carpet begonia
Begonia weltoniensis
 Grapevine begonia
 Maple-leaf begonia
Belamcanda chinensis
 Blackberry lily
Bellevalia romana Roman hyacinth
Brimeura ida-maia
 Californian firecracker
Brodiaea coronaria
 Harvest brodiaea
Brodiaea elegans
 Harvest brodiaea
Brodiaea uniflora Spring starflower
Brunsvigia josephinae
 Candelabra flower
 Josephine's lily
Brunsvigia radulosa
 Pink candelabra
Caladium bicolor Heart-of-Jesus
Caladium × hortulanum
 Fancy-leaved caladium
Calochortus albus Fairy lantern
 Globe lily
Calochortus amabilis
 Golden fairy lantern
 Golden globe tulip
Calochortus amoenus
 Fairy lantern
 Purple globe tulip

BOTANICAL NAMES

Calochortus coeruleus

Calochortus coeruleus
    Beavertail grass
    Cat's ear
Calochortus concolor
    Golden-bowl mariposa
Calochortus kennedyi
    Desert mariposa
Calochortus luteus
    Yellow mariposa
Calochortus macrocarpus
    Green-banded mariposa
Calochortus maweanus Pussy ears
Calochortus nudus Sierra star tulip
Calochortus nuttallii
    Mariposa lily
    Sego lily
Calochortus pulchellus
    Yellow fairy lantern
Calochortus splendens
    Lilac mariposa
Calochortus tolmiei Pussy ears
Calochortus uniflorus
    Pink star tulip
Calochortus venustus
    White mariposa
Calochortus vestae
    Goddess mariposa
Calochortus weedii
    Weed's mariposa
Camassia quamash Camass
    Camosh
    Common camass
    Quamash
Camassia scilloides Eastern camass
    Indigo squill
    Meadow hyacinth
    Wild hyacinth
Canna edulis Achira
    Edible canna
    Queensland arrowroot
    Tous-les-mois
Canna × generalis
    Common garden canna
Canna indica Indian shot
Canna × orchiodes
    Orchid-flowered canna
Cardiocrinum cathayanum
    Spotted Chinese lily
Cardiocrinum cordatum
    Japanese lily
Cardiocrinum giganteum
    Easter lily
    Giant Himalayan lily
    Giant lily
Chamaescilla corymbosa
    Blue star
Chionodoxa gigantea
    Glory-of-the-snow
Chlidanthus fragrans
    Perfumed fairy lily
    Peruvian daffodil
Cleome spinosa Giant spider plant

Clivia mineata Bush lily
    Fire lily
Colchicum × agrippinum
    Meadow saffron
Colchicum autumnale
    Autumn crocus
    Fall crocus
    Meadow saffron
    Mysteria
    Naked ladies
    Wonder bulb
Colocasia esculenta Dasheen
    Taro
Convallaria majalis
    Lily-of-the-valley
Corydalis aurea Spring fumitory
Corydalis bulbosa Fumewort
Corydalis sempervirens
    Rock harlequin
    Roman wormwood
Crinum americanum
    Southern swamp crinum
    Swamp lily
Crinum asiaticum
    Asiatic poison bulb
Crinum lugardiae Veldt lily
Crinum × powellii Cape lily
    Elephant lily
    Indian elephant flower
Crocosmia × crocosmiiflora
    Falling stars
    Montbretia
Crocus angustifolius Cloth-of-gold
Crocus biflorus Scotch crocus
Crocus byzantinus
    Iris-flowered crocus
Crocus flavus Dutch yellow crocus
Crocus iridiflorus
    Iris-flowered crocus
Crocus korolkowii
    Celandine crocus
Crocus sativus Saffron crocus
Crocus susianus Cloth-of-gold
Crocus vernus Dutch crocus
Curtonus paniculatus Natal iris
Cyclamen hederifolium
    Alpine violet
    Baby cyclamen
Cyclamen persicum
    Florist's cyclamen
    Persian violet
Cyrtanthus mackennii Fire lily
    Ifafa lily
Cyrtanthus obliquus Kynassa lily
Cyrtanthus purpureus George lily
    Scarborough lily
Dahlia imperialis Bell tree dahlia
    Candelabra dahlia
    Tree dahlia
Dahlia merckii Bedding dahlia
Dichelostemma ida-maia
    Californian firecracker

Goodyera pubescens

Dierama pendulum
Angel's fishing rod
Grassy bell

Dierama pulcherrimum
Wandflower

Doryanthes excelsa
Australian giant lily

Dracunculus vulgaris Black arum
Dragon arum

Endymion campanulata
Spanish bluebell

Endymion hispanicus
Bell-flowered squill
Spanish bluebell
Spanish jacinth

Endymion italicus Italian squill

Endymion non-scriptus
English bluebell
Harebell

Endymion nutans English bluebell

Eranthis hyemalis Winter aconite

Eremurus robustus Desert candle
Foxtail lily

Erythronium albidum
Blonde lilian
White dog's tooth violet

Erythronium americanum
Adder's tongue
Amberbell
Serpent's tongue
Trout lily
Yellow adder's tongue
Yellow snowdrop

Erythronium californicum
Fawn lily

Erythronium dens-canis
Dog's tooth violet

Erythronium giganteum
Avalanche lily

Erythronium grandiflorum
Avalanche lily

Erythronium montatum
Avalanche lily

Erythronium obtusatum
Avalanche lily

Erythronium revolutum
American trout lily

Eucharis amazonica Amazon lily

Eucharis grandiflora Amazon lily
Eucharist lily
Lily-of-the-amazon
Madonna lily

Eucomis autumnalis
Pineapple flower

Eucomis bicolor Pineapple flower

Eucomis comosa Pineapple flower

Eucomis pallidiflora
Giant pineapple flower
Pineapple flower

Eucomis undulata
Pineapple flower

Eucomis zambesiaca
Pineapple flower

Freesia hybrida Outdoor freesia

Fritillaria biflora Mission bells

Fritillaria camschatcensis
Black fritillary
Black lily
Black sarana
Kamchatka lily

Fritillaria camtschatcensis
Black fritillary
Black lily
Black sarana
Kamchatka lily

Fritillaria imperialis Cr
imperial fritillary

Fritillaria lanceolata Checker lily
Narrow-leaved fritillary

Fritillaria liliacea White fritillary

Fritillaria meleagris
Checkered daffodil
Checkered lily
Guinea-hen tulip
Snake's head fritillary

Fritillaria pluriflora Adobe lily
Pink fritillary

Fritillaria pudica Yellow fritillary

Fritillaria pyrenaica
Pyrenean fritillary

Fritillaria recurva Scarlet fritillary

Fritillaria uva-vulpis Fox's grape

Galanthus elwesii Giant snowdrop
Turkish snowdrop

Galanthus nivalis
Common snowdrop
Fair maids of February

Galanthus reginae-olgae
Autumn snowdrop

Galtonia candicans Berg lily
Galtonia
Giant summer hyacinth
Spire lily
Summer hyacinth

Gladiolus callianthus Acidanthera

Gladiolus dalenii
Rhodesian gladiolus

Gladiolus × hortulanus
Garden gladiolus
Sword lily

Gladiolus primulinus
Maid of the mist

Gladiolus segetum Corn flag

Gladiolus tristis
Yellow marsh afrikander

Gloriosa rothschildiana Flame lily
Gloriosa lily
Glory lily

Gloriosa superba Flame lily
Gloriosa lily
Glory lily

Gloxinia perennis Canterbury bells

Gloxinia sylvatica
Peruvian redbird

Goodyera pubescens
Rattlesnake plantain

Haemanthus katharinae
Catherine-wheel

Haemanthus magnificus
Giant stove brush

Haemanthus multiflorus
African blood lily
Blood lily
Paintbrush

Haemanthus natalensis Blood lily
Natal paintbrush
Paintbrush

Haemanthus puniceus
Royal paintbrush

Hedychium coccineum
Red ginger lily
Scarlet ginger lily

Hedychium coronarium
Butterfly ginger lily
Butterfly lily
Cinnamon jasmine
Garland flower
Ginger lily
White ginger lily

Hedychium flavescens
Yellow ginger lily

Hedychium gardneranum
Gingerwort
Kahili ginger lily
Kahli ginger

Hermodactylus tuberosus
Widow iris

Hippeastrum aulicum
Lily-of-the-palace

Hippeastrum edule Barbados lily

Hippeastrum equestre
Barbados lily

Hippeastrum × johnsonii
Saint Joseph's lily

Hippeastrum reginae Mexican lily

Hippeastrum reticulatum
Lacework lily
Meshed lily

Hippeastrum robustum
Lily-of-the-palace

Homoglossum merianella Flames

Hyacinthoides campanulata
Spanish bluebell

Hyacinthoides hispanica
Giant bluebell
Spanish bluebell

Hyacinthoides non-scriptus
Culverkeys
English bluebell
Ring-of-bells
Wild hyacinth
Wood bells
Wood hyacinth

Hyacinthoides nutans
English bluebell

Hyacinthus candicans
Summer hyacinth

Hyacinthus orientalis
Common hyacinth
Dutch hyacinth
Garden hyacinth
Hyacinth
Roman hyacinth

Hyacinthus romanus
Roman hyacinth

Hymenocallis amancaes
Peruvian daffodil

Hymenocallis × festalis
Basket flower
Spider lily

Hymenocallis narcissiflora
Basket flower
Peruvian daffodil
Spider lily

Hymenocallis palmeri Alligator lily

Ionoxalis violacea
Violet wood sorrel

Ipheion uniflorum
Spring starflower

Iris anglica English iris

Iris basaltica Mourning iris
Palestine iris

Iris brevicaulis Lamance iris

Iris cristata Crested iris
Dwarf crested iris

Iris ensata Sword-leaved iris

Iris foetidissima Gladwin iris
Scarlet-seeded iris
Stinking gladwin iris
Stinking iris

Iris fulva Copper iris
Red iris

Iris × germanica Flag
Fleur-de-lis

Iris graeberiana Blue fall iris
Yellow fall iris

Iris hartwegii Sierra iris

Iris kaemferi Japanese iris

Iris latifolia English iris

Iris missouriensis Western blue flag

Iris ochroleuca Butterfly iris

Iris odoratissima Orris

Iris orchidoides Orchid iris

Iris pallida Orris

Iris persica Persian iris

Iris prismatica Slender blue flag

Iris pseudacorus Water flag
Yellow flag
Yellow iris

Iris reticulata Netted iris

Iris sibirica Siberian iris

Iris spuria Butterfly iris
Spuria iris

Iris susiana Mourning iris
Palestine iris

Iris tectorum Roof iris
Wall iris

Iris tingitana Morocco iris

Lilium parryi

Iris tuberosa	Widow iris
Iris verna	Dwarf iris
	Violet iris
Iris versicolor	Blue flag
	Poison flag
	Wild iris
Iris virginica	Blue flag
	Southern blue flag
Iris xiphioides	English iris
Iris xiphium	
	Portuguese iris
	Spanish iris
Ismene calathina	Spider lily
Ixia incarnata	Clanwilliam bluebell
Ixia maculata	African corn lily
	Corn lily
	Wand flower
Ixia viridiflora	African corn lily
	Corn lily
	Green ixia
	Wand flower
Kniphofia triangularis	
	Red hot poker
	Torch lily
Lachenalia aloides	Cape cowslip
Lachenalia bulbifera	
	Cape cowslip
Lachenalia contaminata	
	Cape cowslip
	Wild hyacinth
Lachenalia glaucina	Cape cowslip
Lachenalia mutabilis	Cape cowslip
Lachenalia ribida	Cape cowslip
Lachenalia tricolor	Cape cowslip
Leucocoryne ixioides	
	Glory-of-the-snow
	Glory-of-the-sun
Leucocoryne odorata	
	Glory-of-the-sun
Leucocoryne uniflora	
	Spring starflower
Leucojum aestivum	
	Giant snowflake
	Loddon lily
	Summer snowflake
Leucojum vernum	
	Spring snowflake
Liatris aspera	Blazing star
	Gayfeather
	Prickly blazing star
Liatris spicata	Spiky gayfeather
Lilium auratum	Gold-banded lily
	Golden-banded lily
	Golden-rayed lily
	Mountain lily
Lilium bolanderi	Thimble lily
Lilium bulbiferum	Fire lily
	Orange lily
Lilium canadense	Canada lily
	Meadow lily
	Wild yellow lily
	Yellow-bell lily
	Yellow lily

Lilium candidum	Madonna lily
Lilium carolinianum	Leopard lily
	Pine lily
	Southern red lily
Lilium catesbaei	Leopard lily
	Pine lily
	Southern red lily
Lilium chalcedonicum	
	Scarlet martagon lily
	Scarlet turk's cap lily
Lilium columbianum	
	Columbia lily
	Oregon lily
Lilium concolor	Star lily
Lilium davidii	
	Orange turk's cap lily
Lilium duchartrei	
	Marble martagon lily
Lilium excelsum	Nankeen lily
Lilium formosum	
	Chinese white lily
Lilium grayi	Bell lily
	Gray's lily
	Orange-bell lily
	Roan lily
Lilium hansonii	
	Japanese turk's cap lily
Lilium × hollandicum	
	Candlestick lily
Lilium humboldtii	Humbold lily
Lilium iridollae	Pot-of-gold lily
Lilium japonicum	Japanese lily
Lilium krameri	Japanese lily
Lilium lancifolium	Japanese lily
	Showy Japanese lily
	Showy lily
	Tiger lily
Lilium leucanthum	
	Chinese white lily
Lilium longiflorum	Bermuda lily
	Easter lily
	Trumpet lily
	White trumpet lily
Lilium mackliniae	Manipur lily
Lilium makinoi	Japanese lily
Lilium maritimum	Coastal lily
Lilium martagon	Martagon lily
	Turban lily
	Turk's cap lily
Lilium medeoloides	Wheel lily
Lilium michauxii	Carolina lily
	Turk's cap lily
Lilium michiganense	Michigan lily
Lilium monadelphum	
	Caucasian lily
Lilium occidentale	Eureka lily
	Western lily
Lilium pardalinum	Leopard lily
	Panther lily
	Sunset lily
Lilium parryi	Lemon lily

Lilium parvum

Lilium parvum	Alpine lily
	Sierra lily
	Small tiger lily
Lilium philadelphicum	
	Orange-cup lily
	Wild orange-red lily
	Wood lily
Lilium pomponium	
	Lesser turk's cap lily
	Little turk's cap lily
	Minor turk's cap lily
	Turban lily
Lilium pumilum	Coral lily
Lilium pyrenaicum	
	Yellow turk's cap lily
Lilium regale	Regal lily
	Royal lily
Lilium rubescens	Chamise lily
	Chaparral lily
	Redwood lily
Lilium speciosum	Japanese lily
	Showy lily
	Showy japanese lily
Lilium superbum	
	American turk's cap lily
	Lily-royal
	Swamp lily
	Turk's cap lily
Lilium tenuifolium	Coral lily
Lilium × testaceum	Nankeen lily
Lilium tigrinum	Tiger lily
Lilium washingtonianum	
	Washington lily
Lilium × umbellatum	
	Candlestick lily
Littonia modesta	Climbing lily
Lloydia serotina	
	Mountain spiderwort
Lycoris africana	
	Golden African lily
	Golden hurricane lily
	Golden spider lily
	Spider lily
Lycoris aurea	Golden spider lily
	Spider lily
Lycoris radiata	Red spider lily
	Spider lily
Lycoris squamigera	Magic lily
	Resurrection lily
Milla biflora	Mexican star
Milla uniflora	Spring starflower
Muscari azureum	
	Turkish grape hyacinth
Muscari botryoides	
	Common grape hyacinth
	Small grape hyacinth
Muscari comosum	Cipollino
	Feather hyacinth
	Tassel hyacinth
Muscari macrocarpum	
	Yellow grape hyacinth

Muscari moschatum	
	Grape hyacinth
	Musk hyacinth
	Nutmeg hyacinth
Muscari muscarini	Musk hyacinth
Muscari neglectum	
	Grape hyacinth
Muscari racemosum	
	Grape hyacinth
	Musk hyacinth
	Nutmeg hyacinth
Muscari tubergenianum	
	Cambridge grape hyacinth
	Oxford grape hyacinth
Narcissus assoanus	Dwarf jonquil
Narcissus bulbocodium	
	Hoop-petticoat daffodil
	Petticoat daffodil
Narcissus calathinus	
	Campernelle jonquil
Narcissus canaliculatus	
	Chinese sacred lily
	Polyanthus narcissus
Narcissus hispanicus	
	Spanish daffodil
Narcissus jonquilla	Jonquil
Narcissus × medioluteus	
	Poetaz narcissus
	Primrose peerless narcissus
Narcissus obvallaris	Tenby daffodil
Narcissus × odorus	
	Campernelle jonquil
Narcissus papyraceus	
	Paperwhite narcissus
Narcissus poeticus	
	Pheasant's eye narcissus
	Poet's narcissus
Narcissus poeticus physaloides	
	Chinese lantern
Narcissus pseudonarcissus	Daffodil
	Lent lily
	Trumpet narcissus
	Wild daffodil
Narcissus requienii	Dwarf jonquil
Narcissus tazetta	
	Bunch-flowered narcissus
	Chinese sacred lily
	Polyanthus narcissus
Narcissus triandrus	Angel's tears
Narcissus watieri	
	North African narcissus
Neomarica caerulea	
	Twelve apostles
Neomarica northiana	Walking iris
Nerine bowdenii	Guernsey lily
Nerine sarniensis	Guernsey lily
Nerine undulata	Nodding nerine
Nomocharis pardanthina	
	Blotched panther-lily
Ornithogalum arabicum	
	Star-of-Bethlehem

Ornithogalum caudatum
False sea onion
German onion
Sea onion

Ornithogalum narbonense
Star-of-Bethlehem

Ornithogalum nutans
Drooping star-of-Bethlehem
Nodding star-of-Bethlehem

Ornithogalum pyrenaicum
Bath asparagus
French asparagus
Prussian asparagus
Star-of-Bethlehem

Ornithogalum saundersiae
Giant chincherinchee

Ornithogalum thyrsoides
African wonder flower
Chincherinchee
Wonder flower

Ornithogalum umbellatum
Dove's dung
Nap-at-noon
Star-of-bethlehem
Summer snowflake

Ostrowskia magnifica
Giant bellflower

Oxalis acetosella
European wood sorrel
Irish shamrock

Oxalis cernua   Bermuda buttercup

Oxalis deppei   Good-luck leaf
Good-luck plant
Lucky clover

Oxalis enneaphylla
Falklands scurvy grass
Scurvy grass

Oxalis oregana   Redwood sorrel

Oxalis pes-caprae
Bermuda buttercup

Oxalis violacea   Violet wood sorrel

Pancratium maritimum
Mediterranean lily
Sea daffodil
Sea lily

Paradisea liliastrum   Paradise lily
Saint Bruno's lily

Paradisea lusitanica
Portuguese paradise lily

Polianthes geminiflora
Orange pearl
Orange tuberose

Polianthes tuberosa   The pearl
Tuberose

Pulsatilla patens   Pasque flower

Puschkinia libanotica
Lebanon squill
Striped squill

Puschkinia scilloides
Lebanon squill

Pyrolirion tubiflorum
Peruvian mountain daffodil

Ranunculus acris
Bachelor's buttons

Ranunculus asiaticus   Crowfoot
Persian buttercup
Persian ranunculus
Turban buttercup

Ranunculus bulbosus
Bulbous buttercup
Bulbous crowfoot

Ranunculus ficaria
Lesser celandine
Pilewort
Small celandine

Rhodohypoxis baurii
Drakensberg star

Rhodophiala bifida   Hurricane lily

Richardia rehmannii   Pink calla lily
Red calla lily

Romulea columnae   Sand crocus

Sandersonia aurantiaca
Chinese lantern lily
Christmas bells

Sauromatum guttatum
Lizard flower

Sauromatum venosum   Voodoo lily

Saururu cernuus
American swamp lily
Swamp lily
Water dragon

Scadoxus multiflorus   Blood lily
Fireball lily

Scadoxus puniceus
Royal paintbrush

Schizostylis coccinea   Crimson flag
Kaffir lily
River lily

Scilla amoena   Star hyacinth

Scilla autumnalis   Autumn squill
Starry hyacinth

Scilla bifolia   Alpine squill
Nodding squill

Scilla chinensis   Chinese squill
Japanese jacinth

Scilla hyacinthoides
Hyacinth squill

Scilla monophyllos   Dwarf squill

Scilla natalensis   South African squill

Scilla peruviana   Cuban lily
Hyacinth-of-Peru
Peruvian jacinth

Scilla scilloides   Chinese squill
Japanese jacinth

Scilla siberica   Siberian squill

Scilla verna   Sea onion
Spring squill

Simethis planifolia   Kerry lily

Sinningia cardinalis
Cardinal flower
Helmet flower

Sinningia leucotricha
Brazilian edelweiss

Sinningia regina   Cinderella slippers
Violet slipper gloxinia

Sinningia speciosa

Sinningia speciosa	Brazilian gloxinia
	Florist's gloxinia
	Gloxinia
	Violet slipper gloxinia
Sparaxis elegans	Harlequin flower
Sparaxis grandiflora	
	Harlequin flower
Sparaxis pendula	
	Angel's fishing rod
	Grassy bell
Sparaxis tricolor	Harlequin flower
Spigelia marilandica	Indian pink
	Maryland pink root
Sprekelia formosissima	Aztec lily
	Jacobean lily
	Maltese cross
	Orchid amaryllis
	Saint James's lily
Sternbergia lutea	Autumn daffodil
	Lily-of-the-field
	Winter daffodil
	Yellow starflower
Streptanthera elegans	
	Harlequin flower
Tecophilaea cyanocrocus	
	Chilean crocus
Tigridia pavonia	
	Mexican shell flower
	Peacock tiger flower
	Tiger flower
	Tiger lily
Trillium cernuum	
	Nodding trillium
Trillium cuneatum	
	Whippoorwill flower
Trillium erectum	Brown beth
	Purple trillium
	Squawroot
	Stinking benjamin
Trillium erectum album	
	Wax trillium
Trillium flavum	Brown beth
	Purple trillium
	Squawroot
	Stinking benjamin
Trillium grandiflorum	
	Wake robin
	White wake robin
Trillium nivale	
	Dwarf white trillium
	Snow trillium
Trillium ovatum	Coast trillium
Trillium recurvatum	
	Bloody butcher
	Purple toadshade
	Purple trillium
	Purple wake robin
Trillium sessile	Toadshade
	Wake robin
Trillium undulatum	
	Painted trillium
Trillium vaseyi	Sweet beth
Trillium viride	Wood trillium

Tulbaghia simmleri	
	Pink agapanthus
	Sweet garlic
	Wild garlic
Tulipa acuminata	Horned tulip
	Turkish tulip
Tulipa clusiana	Lady tulip
Tulipa cornuta	Turkish tulip
Tulipa kaufmanniana	
	Waterlily tulip
Tulipa occulus-solis	
	Persian sun's eye
Tulipa sylvestris	Wild tulip
Tulipa turkistanica	Turkish tulip
Urginea maritima	
	Crusader's spears
	Red squill
	Sea onion
	Sea squill
	White squill
Vallota speciosa	George lily
	Scarborough lily
Veltheimia brachteata	Forest lily
Veltheimia viridiflora	Forest lily
Zantedeschia aethiopica	Arum lily
	Calla lily
	Florist's calla lily
	Garden calla lily
	Pig lily
	Trumpet lily
Zantedeschia africana	Arum lily
	Calla lily
	Florist's calla lily
	Garden calla lily
	Pig lily
	Trumpet lily
Zantedeschia albomaculata	
	Black-throated calla lily
	Spotted calla lily
Zantedeschia elliottiana	
	Golden calla lily
	Yellow calla lily
Zantedeschia melanoleuca	
	Black-throated calla lily
	Spotted calla lily
Zantedeschia rehmannii	
	Pink arum lily
	Pink calla lily
	Red calla lily
Zephyranthes atamasco	
	Atamasco lily
	Easter lily
Zephyranthes candida	Fairy lily
	Flower of the western wind
	Flower of the wind
	Rain lily
	Storm lily
	West wind lily
	Windflower
	Zephyr lily
Zephyranthes grandiflora	
	Windflower
Zephyranthes rosea	Windflower

CACTI AND SUCCULENTS

The majority of cacti are succulent, arid- or desert-area plants with thickened stems that serve the plant both for water-storage and as photosynthetic organs, replacing the leaves which are usually miniscule or totally absent. Most are armed with vicious spines and should be handled with care. They should be kept well away from children and domestic pets.

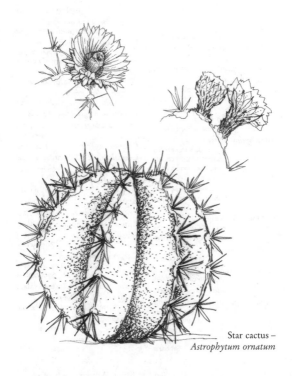

Star cactus –
Astrophytum ornatum

Adenium obesum

Adenium obesum	Desert rose
Agave americana	American aloe
	Century plant
Agave cantala	Cantala
Agave fourcroydes	Henequen
Agave lophantha	Lechuguilla
Agave palmeri	Blue century plant
Aichryson × domesticum	
	Cloud grass
	Youth-and-old-age
Aloe africana	Spiny aloe
Aloe barbadensis	Barbados aloe
	Medicinal aloe
	Unguentine cactus
Aloe ferox	Cape aloe
	Red aloe
Aloe humilis	Crocodile jaws
	Hedgehog aloe
	Spider aloe
Aloe perryi	Socotrine aloe
	Zanzibar aloe
Aloe variegata	Falcon feather
	Kanniedood
	Partridge-breasted aloe
	Pheasant's wing
	Tiger aloe
Aloe victoriae reginae	
	Queen Victoria's aloe
Ancistrocactus scheeri	
	Fish-hook cactus
Aporocactus flagelliformis	
	Rat's tail cactus
Ariocarpus fissuratus	Living rock
	Star cactus
Ariocarpus retusus	Seven stars
Astrophytum asterias	
	Sand dollar cactus
	Sea urchin cactus
	Silver dollar cactus
Astrophytum capricorne	
	Goat's horn cactus
Astrophytum myriostigma	
	Bishop's cap cactus
	Bishop's hood cactus
	Monkshood cactus
Astrophytum ornatum	
	Bishop's cap cactus
	Ornamental monkshood
	Star cactus
Borzicactus celsianus	
	Old man of the mountains
	South American old man
Borzicactus fossulatus	
	Mountain cereus
Borzicactus trollii	
	Old man of the Andes
Caralluma woodii	String of hearts
Carnegiea gigantea	Arizona giant
	Giant cactus
	Giant saguaro
	Pitahaya
	Saguaro
	Sahuaro

Cephalocereus chrysacanthus	
	Golden old man
	Golden spines
Cephalocereus fulviceps	
	Mexican giant
Cephalocereus palmeri	
	Bald old man
	Yellow old man
	Woolly torch cactus
Cephalocereus senilis	
	Old man cactus
	Old man's head
Cephalophyllum alstonii	
	Red spike
Cereus baumannii	Scarlet bugler
Cereus peruvianus	Apple cactus
	Hedge cactus
	Peruvian apple
	Peruvian apple cactus
Chamaecereus silvestrii	
	Peanut cactus
Chiastophyllum oppositifolium	
	Lamb's tail
Cleistocactus baumannii	
	Scarlet bugler
Cleistocactus smaragdiflorus	
	Firecracker cactus
Cleistocactus strausii	Silver torch
Coryphantha echinus	
	Hedgehog-cory cactus
Coryphantha missouriensis	
	Missouri pincushion cactus
Coryphantha runyonii	
	Big nipple cactus
	Dumpling cactus
Coryphantha vivipara	
	Pincushion cactus
Coryphantha vivipara arizonica	Beehive cactus
Cotyledon simplicifolia	Lamb's tail
Crassula arborescens	Jade plant
	Money tree
Crassula argentea	Jade plant
Crassula barklyi	Rattlesnake tail
Crassula deltoides	Silver beads
Crassula lycopodioides	
	Rat-tail plant
Crassula ovata	Jade tree
Crassula portulacea	Jade plant
Crassula rhomboidea	Silver beads
Crassula rupestris	Bead vine
	Rosary vine
Dasylirion leiophyllum	
	Desert candle
	Spoon plant
Deamia testudo	Tortoise cactus
Dorotheanthus bellidiformis	
	Annual mesembryanthemum
Doryanthes excelsa	
	Globe spear-lily
Doryanthes palmeri	
	Palmer spear-lily

Gymnocalycium denudatum

Dracaena draco	Dragon tree
Echeveria affinis	Black echeveria
Echeveria derenbergii	Painted lady
Echeveria glauca	Blue echeveria
Echeveria leucotricha	
	Chenille plant
Echeveria multicaulis	
	Copper roses
Echeveria pulvinata	Plush plant
Echeveria rosea	Desert rose
Echeveria secunda glauca	
	Blue echeveria
Echeveria setosa	Firecracker plant
Echinocactus dasyacanthus	
	Texas rainbow cactus
Echinocactus grusonii	
	Barrel cactus
	Golden ball
	Golden barrel cactus
	Mother-in-law's armchair
Echinocactus horizonthalonius	
	Eagle-claws cactus
	Mule-crippler cactus
Echinocactus ingens	
	Blue barrel cactus
	Giant barrel cactus
	Large barrel cactus
	Mexican giant barrel
Echinocereus delaetii	
	Lesser old man cactus
Echinocereus dubius	Purple pitaya
Echinocereus enneacanthus	
	Strawberry cactus
Echinocereus horizonthalonius	
	Blue barrel cactus
Echinocereus pectinatus	
	Hedgehog cactus
Echinocereus pectinatus rigidissimus	
	Rainbow cactus
Echinocereus reichenbachii	
	Lace cactus
Echinocereus salm-dyckianus	
	Strawberry cactus
Echinocereus sarissophorus	
	Purple hedgehog cereus
Echinocereus viridiflorus	
	Green-flowered pitaya
	Green-flowered torch cactus
Echinofossulocactus zacatecasensis	Brain cactus
Echinopsis aurea	Golden lily cactus
Echinopsis eyriesii	
	Sea urchin cactus
Echinopsis multiplex	Barrel cactus
	Easter-lily cactus
	Pink easter-lily cactus
Epiphyllum ackermannii	
	Orchid cactus
Epiphyllum anguliger	
	Fishbone cactus
Epiphyllum × hybridum	
	Orchid cactus
Epiphyllum oxypetalum	
	Night-blooming cereus
	Queen of the night
Epiphyllum truncatum	
	Christmas cactus
	Crab's claw cactus
Epithelantha bokei	Button cactus
Epithelantha micromeris	
	Button cactus
Escobaria tuberculosa	Cob cactus
Espostoa lanata	Cotton-ball cactus
	Snowball cactus
Euphorbia antisyphilitica	
	Candelilla
	Wax plant
Euphorbia caput-medusae	
	Medusa's head spurge
Euphorbia gorgonis	Dragon's head
	Gorgon's head
Euphorbia mammillaris	
	Corncob cactus
Euphorbia milii	Crown of thorns
Euphorbia milii splendens	
	Crown of thorns
Euphorbia obesa	Gingham golf ball
Euphorbia pseudocactus	
	Cactus spurge
Euphorbia tirucalli	Finger tree
	Milk bush
	Rubber spurge
	Stick cactus
Faucaria tigrina	Tiger jaws
Fenestraria rhopalophylla	
	Window plant
Ferocactus corniger	Devil's tongue
Ferocactus glaucescens	
	Blue barrel cactus
Ferocactus hamatacanthus	
	Turk's head
Ferocactus histrix	Electrode cactus
Ferocactus latispinus	
	Devil's tongue
	Fish-hook cactus
Ferocactus rectispinus	
	Hatpin cactus
Ferocactus setispinus	
	Strawberry cactus
Ferocactus viridescens	
	Small barrel cactus
Ferocactus wislizenii	
	Fish-hook cactus
Fritha pulchra	Fairy-elephant's feet
Furcraea foetida	Green aloe
	Mauritius hemp
Furcraea hexapetala	Cuban hemp
Graptopetalum paraguayense	
	Ghost plant
	Mother-of-pearl plant
Graptophyllum pictum	
	Caricature plant
Gymnocalycium denudatum	
	Spider cactus

Gymnocalycium gibbosum

Gymnocalycium gibbosum
Chin cactus
Gymnocalycium mihanovichii
Plaid cactus
Plain cactus
Gymnocalycium schickendantzii
White chin cactus
Gymnocalycium quehlianum
Rose-plaid cactus
Hamatocactus uncinatus
Cat claw cactus
Hariota salicornioides   Bottle plant
Dancing bones
Drunkard's dream
Spice cactus
Harrisia jusbertii   Moon cactus
Harrisia martinii   Moon cactus
Hatiora salicornioides
Bottle plant
Dancing bones
Drunkard's dream
Spice cactus
Haworthia chalwinii
Aristocrat plant
Column-of-pearls
Haworthia cymbiformis
Window aloe
Window cushion
Window plant
Haworthia fasciata
Zebra haworthia
Haworthia limifolia
Fairy washboard
Haworthia margaritifera
Pearl plant
Haworthia papillosa   Pearly-dots
Haworthia pumila   Pearl plant
Haworthia setata   Lace haworthia
Haworthia tessellata
Star window plant
Hechtia scariosa   Fairy agave
Heliocereus speciosus   Sun cactus
Hylocereus undatus
Honolulu queen
Queen of the night
Night-blooming cereus
Kalanchoe tomentosa   Panda plant
Panda-bear plant
Plush plant
Pussy-ears
Lemaireocereus eruca
Creeping-devil cactus
Lemaireocereus gummosus
Dagger cactus
Lemaireocereus marginatus
Organ-pipe cactus
Lemaireocereus pruinosus
Powder-blue cereus
Lemaireocereus thurberi
Organ-pipe cactus
Lemaireocereus weberi
Candelabra cactus

Leuchtenbergia principis
Agave cactus
Prism cactus
Lithops dorotheae   Stone plant
Lithops julii   Living stones
Stone plant
Lithops lesliei   Pebble cactus
Lobivia aurea   Golden lily cactus
Lobivia bruchii
South American golden-barrel
Lobivia hertrichiana   Cob cactus
Lophocereus schottii   Senita
Totem-pole cactus
Whisker cactus
Lophophora williamsii
Devil's root cactus
Dry whisky
Dumpling cactus
Mescal button
Peyote cactus
Lothophora williamsii
Devil's root cactus
Dry whisky
Dumpling cactus
Mescal button
Peyote cactus
Machaerocereus eruca
Creeping devil cactus
Mammillaria bocasana
Fish-hook cactus
Powder-puff
Snowball cactus
Mammillaria camptotricha
Bird's nest cactus
Golden bird's nest cactus
Mammillaria candida
Snowball pincushion
Mammillaria elongata
Golden star
Gold lace
Lace cactus
Lady finger
Mammillaria fragilis
Thimble cactus
Thimble mammillaria
Mammillaria gracilis   Powder puff
Mammillaria hahniana
Old lady cactus
Old lady of Mexico
Old woman cactus
Mammillaria heyderi
Coral cactus
Cream cactus
Mammillaria karwinskiana
Royal-cross cactus
Mammillaria lanata
Old lady cactus
Mammillaria plumosa
Feather cactus
Mammillaria pringlei
Lemon-ball cactus
Mammillaria prolifera
Little candles
Silver cluster cactus

CACTI

Oreocereus trollii

Mammillaria tetracantha
Ruby dumpling

Mammillaria wildii
Fish-hook pincushion cactus

Mammillaria zeilmanniana
Rose pincushion

Melocactus communis
Melon cactus
Turk's cap cactus
Turk's head cactus

Melocactus intortus
Turk's cap cactus

Melocactus maxonii
Turk's head cactus

Morangaya pensilis
Sprawling cactus
Wandering cactus

Myrtillocactus geometrizans
Blue candle
Blue flame
Blue myrtle cactus

Neobessya similis   Nipple cactus

Nopalea cochenillifera
Cochineal plant

Nopalxochia ackermannii
Orchid cactus
Red orchid cactus

Nopalxochia phyllanthoides
Empress of Germany
Pond-lily cactus

Notocactus haselbergii
Scarlet ball cactus

Notocactus leninghausii
Golden ball

Notocactus scopa   Silver ball cactus

Notocactus submammulosus
Lemon ball cactus

Nyctocereus serpentinus
Night-blooming cereus
Queen of the night
Serpent cactus
Snake cactus

Obregonia denegrii
Artichoke cactus

Opuntia arbuscula   Pencil cholla

Opuntia articulata   Paper cactus
Spruce cones

Opuntia basilaris   Beavertail cactus
Rose tuna

Opuntia bigelovii
Teddy-bear cactus
Teddy-bear cholla

Opuntia chlorotica   Flapjack cactus

Opuntia clavarioides   Black fingers
Crested opuntia
Fairy castles
Gnome's throne
Sea-coral

Opuntia cylindrica   Cane cactus
Emerald-idol

Opuntia elata   Orange tuna

Opuntia erectoclada   Dominoes

Opuntia erinacea ursina
Grizzly bear cactus

Opuntia ficus-indica
Burbank's spineless cactus
Indian fig
Prickly pear
Spineless cactus

Opuntia floccosa   Cushion cactus
Woolly sheep

Opuntia fulgida   Jumping cactus

Opuntia mamillata   Boxing glove
Club cactus

Opuntia imbricata
Chain-link cactus

Opuntia leptocaulis
Desert christmas cactus
Tasajillo

Opuntia leucotricha   Aaron's beard

Opuntia megacantha   Nopal

Opuntia microdasys   Bunny ears
Golden opuntia
Goldplush
Rabbit ears
Yellow bunny ears
Yellow rabbit ears

*Opuntia microdasys
albispina*   Honey-bunny
Polka dots
Prickly pear

Opuntia microdasys rufida
Blind pear
Cinnamon cactus
Red bunny ears

Opuntia prolifera   Jumping cholla

Opuntia ramosissima   Pencil cactus

Opuntia rufida   Cinnamon cactus
Red bunny-ears

Opuntia santa-rita
Purple prickly pear

Opuntia schickendantzii
Lion's tongue
Mule's ears

Opuntia schottii   Devil cactus
Dog cholla

Opuntia soehrensii   Fairy needles

Opuntia sphaerica   Thimble tuna

Opuntia subulata   Eve's pin cactus

Opuntia velutina   Velvet opuntia

Opuntia versicolor   Staghorn cholla

Opuntia vestita
Cotton-pole cactus
Old man opuntia

Opuntia vilis   Little tree opuntia
Mexican dwarf tree cactus

Opuntia violacea santa-rita
Blue-blade
Dollar cactus

Opuntia vulgaris   Barbary fig
Irish mittens
Joseph's coat cactus
Prickly pear

Oreocereus trollii
Old man of the Andes

*Pachycereus pecten-
aboriginum*

Pachycereus pecten-	
aboriginum	Comb cactus
	Hairbrush cactus
	Indian comb
	Native's comb
Pachycereus pringlei	
	Giant Mexican cereus
	Mexican giant cactus
Pachyphytum glutinicaule	
	Sticky moonstones
Pachyphytum oviferum	
	Moonstones
	Sugar-almond plant
Pachypodium lamerei	
	Madagascan palm
Parodia aureispina	
	Golden tom thumb
	Tom thumb cactus
Pedilanthus tithymaloides	
	Devil's backbone
	Japanese poinsettia
	Jewbush
	Redbird cactus
	Redbird flower
	Ribbon cactus
	Slipper flower
Pediocactus simpsonii	
	Snowball cactus
Pelargonium carnosum	
	Fleshy-stalked pelargonium
Pelargonium ceratophyllum	
	Horned-leaf pelargonium
Pelargonium cotyledonis	
	Hollyhock-leaved pelargonium
	Old-father-live-forever
Pelargonium crithmifolium	
	Samphire-leaved pelargonium
Pelargonium gibbosum	
	Gouty pelargonium
Pelargonium hystrix	
	Porcupine pelargonium
Pelargonium tetragonum	
	Square-stemmed pelargonium
Peniocereus greggii	
	Night-blooming cereus
	Reina-de-la-noche
Pereskia aculeata	
	Barbados gooseberry
	Leafy cactus
	Lemon vine
Pereskia bleo	Wax rose
Pereskia grandiflora	Rose cactus
Pilocereus celsianus	
	Old man of the mountains
Pleiospilos bolusii	
	African living rock
	Living rock cactus
	Mimicry plant
Pleiospilos nelii	Cleftstone
	Mimicry plant
	Splitrock
Plumiera acuminata	
	Frangipani tree
Portulaca grandiflora	Sun plant

Pulque agave	Agave salmiana
Rebutia grandiflora	
	Scarlet crown cactus
Rebutia kupperana	
	Red crown cactus
Rebutia minuscula	
	Mexican sunball
	Red crown cactus
Rebutia pseudodeminuta	
	Wallflower crown
Rebutia senilis	Fire crown cactus
Rhipsalidopsis gaertneri	
	Easter cactus
Rhipsalis baccifera	
	Mistletoe cactus
Rhipsalis cassutha	
	Mistletoe cactus
Rhipsalis cereuscula	Coral cactus
	Popcorn cactus
	Rice cactus
Rhipsalis houlletiana	
	Snowdrop cactus
Rhipsalis paradoxa	Chain cactus
	Chainlink cactus
	Link plant
Rhipsalis salicornioides	
	Dancing bones
	Drunkard's dream
	Spice cactus
Rhipsalis warmingiana	
	Popcorn cactus
Rochea coccinea	
	Hyacinth-scented rochea
Sansevieria hahnii	
	Bird's nest sansevieria
Sansevieria trifasciata laurentii	
	Mother-in-law's tongue
Schlumbergera bridgesii	
	Christmas cactus
	Crab's claw cactus
Schlumbergera gaertneri	
	Whitsun cactus
Schlumbergera gaertneri	
makoyana	Cat's whiskers
Schlumbergera truncata	
	Christmas cactus
	Claw cactus
	Crab cactus
	Thanksgiving cactus
	Zygocactus
Sedum anglicum	English stonecrop
	Stonecrop
Sedum pachyphyllum	
	Jellybean plant
Sedum spectabile	Ice plant
Selenicereus grandiflorus	
	Night-blooming cereus
	Queen of the night
Selenicereus macdonaldiae	
	Queen of the night
Selenicereus pteranthus	
	Princess of the night

CACTI

Zygocactus truncatus

Sempervivum arachnoides	*Titanopsis schwantesii* White jewel
Cobweb houseleek	*Trichocereus spachianus*
Spider houseleek	Golden column
Sempervivum arachnoideum	White torch cactus
Cobweb houseleek	*Trichodiadema barbatum*
Sempervivum tectorum	Pickle plant
Hen-and-chickens	*Trichodiadema densum*
Roof houseleek	Desert rose
Senecio articulatus Candle plant	*Wigginsia vorwerkiana*
Senecio rowleyanus String of beads	Colombian ball cactus
String of pearls	*Wilcoxia poselgeri* Dahlia cactus
Stenocereus thurberi	Scented cactus
Organ pipe cactus	*Wilcoxia schmollii*
Stetsonia coryne Toothpick cactus	Lamb's tail cactus
Thelocactus bicolor Glory of Texas	*Yucca brevifolia* Joshua tree
Texas pride	*Zygocactus truncatus*
Thelocactus setispinus	Christmas cactus
Strawberry cactus	Crab's claw cactus
Titanopsis calcarea Jewel plant	

CARNIVOROUS PLANTS

Otherwise known as insectivorous plants, these are plants that have developed special mechanisms for trapping and digesting mainly, but not exclusively, small insects. There are several types of carnivorous plants including the pitcher plants, the sticky-leaved sundews and butterworts, and the spring-trap leaves of Venus's flytrap. The strange nature of these plants makes them popular subjects for exhibiting at horticultural shows and school study groups. They are also extensively grown indoors and in greenhouses as a natural control for flying insects.

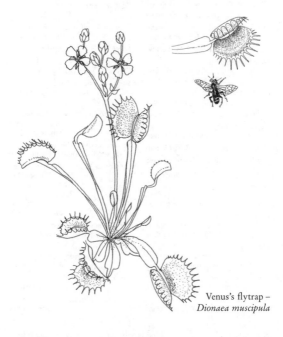

Venus's flytrap –
Dionaea muscipula

Aldrovanda vesiculosa

Aldrovanda vesiculosa
Floating pondtrap
Spikey pondtrap
Waterbug trap
Waterwheel plant

Byblis gigantea
Erect Australian sundew

Byblis liniflora
Sprawling Australian sundew

Cephalotus follicularis
Australian pitcher plant
Swamp pitcher plant

Chrysamphora californica
California pitcher plant
Cobra lily
Cobra orchid
Cobra plant
Pitcher plant

Darlingtonia californica
California pitcher plant
Cobra lily
Cobra orchid
Cobra plant
Pitcher plant

Dionaea muscipula
Flytrap sensitive
Sensitive flytrap
Tipitiwitchet
Venus's flytrap

Drosera anglica   Great sundew

Drosera auriculata   Eared sundew

Drosera brevifolia   Dwarf sundew

Drosera capensis   Sundew

Drosera capillaris   Pink sundew

Drosera filiformis   Dew thread
Thread-leaved sundew

Drosera menziesii   Pink rainbow

Drosera rotundifolia
Common sundew
Round-leaved sundew

Drosophyllum lusitanicum
Portuguese sundew

Genlisea aurea   Golden flytrap

Genlisea filiformis
Spiral-leaved flytrap

Genlisea hispidula   Bristly flytrap

Genlisea pygmaea   Dwarf genlisea

Genlisea repens   Creeping genlisea

Heliamphora minor
Small marsh pitcher
Small sun pitcher

Heliamphora nutans
Guayan sun pitcher plant
Marsh pitcher plant

Heliamphora tatei
Giant marsh pitcher plant
Giant sun pitcher plant

Nepenthes × atrosanguinea
Mottled purple pitcher plant

Nepenthes hookerana
Funnelform pitcher plant

Nepenthes merrilliana
Monkey pitcher
Monkey's larder
Monkey's rice pot
Mouse pitcher

Nepenthes sanguinea
Deep red pitcher plant

Nepenthes ventricosa
Pale green pitcher plant

Pinguicula alpina
Alpine butterwort

Pinguicula caerulea
Blue butterwort

Pinguicula crenatiloba
Toothed butterwort

Pinguicula grandiflora
Greater butterwort

Pinguicula ionantha
Violet butterwort

Pinguicula lilacina
Pale violet butterwort
Violet butterwort

Pinguicula lusitanica
Portuguese butterwort

Pinguicula lutea   Yellow butterwort

Pinguicula pumila
Dwarf butterwort

Pinguicula ramosa
Branched butterwort

Pinguicula variegata
Variegated butterwort

Pinguicula villosa
Shaggy butterwort

Pinguicula vulgaris   Butterwort
Common butterwort

Polypompholyx multifida
Pink petticoat

Polypompholyx tenella   Pink fans

Sarracenia alabamensis
Alabama canebrake pitcher
plant

Sarracenia alata   Pale pitcher
Winged trumpets
Yellow trumpets

Sarracenia drummondii
White trumpet pitcher plant

Sarracenia flava   Huntsman's horn
Trumpet leaf
Trumpets
Umbrella trumpets
Watches
Yellow pitcher plant

Sarracenia leucophylla
White trumpet pitcher plant

Sarracenia minor
Hooded pitcher plant
Rainhat trumpet

Sarracenia oreophila
Green pitcher plant

Sarracenia psittacina
Parrot pitcher plant

CARNIVOROUS PLANTS

Utricularia spiralis

Sarracenia purpurea
Common pitcher plant
Huntsman's cup
Indian cup
Northern pitcher plant
Pitcher plant
Purple pitcher plant
Side-saddle flower
Southern pitcher plant
Sweet pitcher plant

Sarracenia rubra
Sweet pitcher plant
Sweet trumpet

Sarracenia sledgei   Pale pitcher

Utricularia amethystina
Amethyst bladderwort

Utricularia cornuta
Horned bladderwort

Utricularia gibba
Humped bladderwort

Utricularia inflata
Floated bladderwort

Utricularia macrorhiza
Great bladderwort

Utricularia minor
Lesser bladderwort

Utricularia nova-zealandiae
New Zealand bladderwort

Utricularia pubescens
Downy bladderwort

Utricularia resupinata
Inverted bladderwort

Utricularia spiralis
Twining bladderwort

FERNS AND FERN ALLIES

These are flowerless plants bearing leaves (fronds) and reproducing by spores on the lower surface of the mature foliage. Ferns and fern allies share a similar life-cycle and are treated horticulturally in a similar manner. For example, the so-called asparagus fern *Asparagus setaceus* is not a fern but one of the *liliaceae*.

Toothed Davallia –
Davallia denticulata

Acrostichum aureum

Acrostichum aureum Leather fern

Adiantum aethiopicum
Common maidenhair

Adiantum capillus-veneris
Maidenhair
Southern maidenhair
True maidenhair
Venus' hair

*Adiantum capillus-veneris
incisum* Cleft maidenhair
Cleft true maidenhair

Adiantum cuneatum
Delta maidenhair

Adiantum decorum
Delta maidenhair

Adiantum diaphanum
Filmy maidenhair

Adiantum formosum
Australian maidenhair
Blackstem maidenhair
Giant maidenhair

Adiantum hispidulum
Five-fingered jack
Rosy maidenhair
Rough maidenhair

Adiantum pedatum
American maidenhair
Maidenhair
Northern maidenhair

Adiantum peruvianum
Silver dollar fern

Adiantum raddianum
Delta maidenhair

Adiantum reniforme
Kidney maidenhair

Adiantum tenerum
Brittle maidenhair

Adiantum trapeziforme
Diamond maidenhair

*Aglaomorpha goeringianum
pictum* Japanese painted fern

Aglaomorpha pycnocarpon
Glade fern

Aglaomorpha thelypteroides
Silver glade fern

Allosorus crispus Parsley fern

Aneimia adiantifolia Pine fern

Angiopteris evecta Giant fern
King fern

Anogramma leptophylla
Jersey fern

Arachniodes aristata
Prickly shield fern

Asplenium adiantum-nigrum
Black maidenhair spleenwort
Black spleenwort

*Asplenium adiantum-
nigrum ramosum*
Branched black maidenhair
spleenwort

*Asplenium adiantum-
nigrum variegatum*
Variegated black maidenhair
spleenwort
Variegated black spleenwort

Asplenium billottii
Lanceolate spleenwort

Asplenium bulbiferum
Hen-and-chickens fern
Mother fern
Mother spleenwort

Asplenium ceterach
Rusty back fern
Scaly spleenwort

Asplenium ceterach kalon
Wide-fronded scaly spleenwort

Asplenium daucifolium
Mauritius spleenwort

Asplenium flabellifolium
Necklace fern

Asplenium flaccidum
Hanging spleenwort
Weeping spleenwort

Asplenium fontanum
Rock spleenwort

*Asplenium fontanum
multifidum*
Cloven rock spleenwort

*Asplenium fontanum
refractum*
Refracted rock spleenwort

Asplenium germanicum
Alternate spleenwort

*Asplenium germanicum
acutidentatum*
Sharp-toothed alternate
spleenwort

Asplenium lanceolatum
Lanceolate spleenwort

Asplenium lanceolatum kalon
Twin-fronded lanceolate
spleenwort

Asplenium majus
Hanging spleenwort
Weeping spleenwort

Asplenium marinum
Sea spleenwort

Asplenium marinum ramosum
Branched sea spleenwort

Asplenium mayi
Hanging spleenwort
Weeping spleenwort

Asplenium nidus Bird's nest fern
Crow's nest fern

Asplenium platyneuron
Ebony spleenwort

Asplenium ruta-muraria
Rue-leaved spleenwort
Wall rue

*Asplenium ruta-muraria
unilaterale*
One-sided rue-leaved
spleenwort

FERNS

Cystopteris fragilis furcans

Asplenium septentrionale
Forked spleenwort

Asplenium trichomanes
Maidenhair fern
Maidenhair spleenwort

Asplenium trichomanes confluens
Confluent maidenhair fern
Confluent maidenhair spleenwort

Asplenium trichomanes multifidum
Clove maidenhair fern
Cloven maidenhair spleenwort

Asplenium trichomanes ramosum
Branched maidenhair fern
Branched maidenhair spleenwort

Asplenium viride
Green spleenwort

Asplenium viride multifidum
Cloven green spleenwort

Asplenium viviparum
Mauritius spleenwort

Athyrium australe
Australian lady fern

Athyrium filix-femina   Lady fern

Athyrium filix-femina corymbiferum
Tasselled lady fern

Athyrium filix-femina multifidum   Cloven lady fern

Athyrium filix-femina multifidum nanum
Dwarf cloven lady fern

Athyrium filix-femina multifurcatum
Multi-forked lady fern

Athyrium filix-femina victoriae
Queen lady fern

Athyrium thelypteroides
Silvery glade fern
Silvery spleenwort

Azolla caroliniana   Fairy moss
Mosquito fern
Mosquito plant
Water fern

Blechnum capense   Palm-leaf fern

Blechnum cartilagineum
Australian water fern
Bristle fern

Blechnum discolor   Crown fern

Blechnum nudum
Fishbone rib fern
Fishbone water fern

Blechnum occidentale
Hammock fern

Blechnum patersonii   Strap rib fern
Strap water fern

Blechnum spicant   Deer fern
Hard fern
Ladder fern

Blechnum spicant contractum-ramosum
Narrow-branched hard fern

Blechnum spicant duplex
Double-fronded hard fern

Blechnum spicant flabellata
Fan-like hard fern

Blechnum spicant furcans
Forked hard fern

Blechnum spicant ramosum
Branched hard fern

Blechnum watsii   Hard water fern

Botrychium lunaria   Moonwort

Botrychium lunaria incisum
Cleft moonwort

Botrychium multifidum
Leathery moonwort

Botrychium virginianum
Rattlesnake fern
Virginian moonwort

Camptosorus rhizophyllus
Walking fern

Ceratopteris pteridoides
Floating fern

Ceratopteris richardii
Triangular water fern

Ceratopteris thalictroides
Water fern

Ceterach officinarum
Rusty back fern
Scaly spleenwort

Cheilanthes lamosa   Hairy-lip fern

Cibotium glaucum   Hapu tree fern
Hawaiian tree fern

Cibotium schiedei
Mexican tree fern

Cryptogramma crispa   Parsley fern

Ctenitis sloanei
American tree fern
Florida tree fern

Culcita dubia
Common ground fern
False bracken
Rainbow fern

Cyathea arborea   Tree fern
West Indian tree fern

Cyathea australis   Rough tree fern

Cyathea baileyana   Wig tree fern

Cyathea cooperi
Australian tree fern

Cyathea dealbata   Ponga
Silver king fern

Cyathea medullaris   Black tree fern
Korau
Mamaku

Cyrtomium falcatum
House holly fern
Japanese holly fern

Cystopteris bulbifera
Berry bladder fern
Bulbil bladder fern

Cystopteris fragilis
Brittle bladder fern
Fragile bladder fern

Cystopteris fragilis furcans
Forked brittle bladder fern

Cystopteris montana

Cystopteris montana
   Mountain bladder fern

Cystopteris regia
   Alpine bladder fern

Davallia bullata Ball fern
   Squirrel's foot fern

Davallia canariensis
   Deer's foot fern
   Rabbit's foot fern

Davallia denticulata
   Toothed davallia

Davallia fejeensis Hare's foot fern
   Rabbit's foot fern

Davallia mariesii Ball fern
   Hare's foot fern
   Squirrel's foot fern

Davallia trichomanoides
   Squirrel's foot fern

Dennstaedia davallioides
   Lacy ground fern

Dennstaedia punctilobula
   Boulder fern
   Hay-scented fern

Dicksonia antarctica
   Soft tree fern
   Tasmanian tree fern
   Woolly tree fern

Dicksonia fibrosa
   Golden tree fern
   Wheki-ponga
   Woolly tree fern

Dicksonia punctilobula
   Boulder fern
   Hay-scented fern

Dicksonia squarrosa
   New Zealand tree fern
   Rough dicksonia
   Wheki

Dicksonia youngiae
   Bristly tree fern

Diplazium acrostichoides
   Silvery glade fern
   Silvery spleenwort

Doodia aspera Hacksaw fern
   Prickly rasp fern
   Rasp fern

Doodia caudata Small rasp fern

Doodia media Common rasp fern
   Hacksaw fern

Doryopteris pedata Hand fern

Drynaria quercifolia Oak-leaf fern

Dryopteris aemula
   Hay-scented buckler fern

Dryopteris affinis Scaly male fern

Dryopteris austriaca
   Broad buckler fern

Dryopteris austriaca spinulosa
   Toothed wood fern

Dryopteris borreri
   Golden-scaled male fern

Dryopteris carthusiana
   Toothed wood fern

Dryopteris cristata
   Crested buckler fern
   Crested wood fern

Dryopteris dilatata
   Broad buckler fern
   Florist's fern

Dryopteris erythrosora
   Autumn fern
   Japanese buckler fern
   Japanese shield fern

Dryopteris filix-mas
   Common buckler fern
   Male fern

Dryopteris goldiana
   Giant wood fern
   Goldie's fern

Dryopteris marginalis
   Leatherwood fern
   Marginal buckler fern
   Marginal shield fern

Dryopteris oreades
   Mountain male fern

Dryopteris spinulosa
   Toothed wood fern

Dryopteris submontana
   Rigid buckler fern

Dryopteris thelypteris Marsh fern

Equisetum hyemale
   Common horsetail
   Common scouring brush

Equisetum scirpoides
   Dwarf horsetail
   Dwarf scouring brush

Equisetum variegatum
   Variegated horsetail
   Variegated scouring brush

Gleichnia dicarpa
   Pouched coral fern
   Tangle fern

Gleichnia dichotoma Savannah fern

Gleichnia linearis Savannah fern

Gleichnia microphylla
   Parasol fern
   Scrambling coral fern
   Umbrella fern

Grammitis australis Finger fern

Grammitis billardieri Finger fern

Gymnocarpium dryopteris
   Common oak fern
   Oak fern

Gymnocarpium robertianum
   Limestone oak fern
   Northern oak fern

Gymnogramma leptophylla
   Annual maidenhair

Helminthostachys zeylanica
   Malayan flowering fern

Histiopteris incisa Bat's wing fern
   Oak fern

Humata tyermannii
   Bear's foot fern

Hymenophyllum tunbridgense
   Tunbridge filmy fern

FERNS

Osmunda regalis cristata

Hymenophyllum unilaterale
One-sided filmy fern

Hypolepis punctata   Bramble fern

Hypolepis millefolia
Thousand-leaved fern

Isoetes echinospora
Spiny-spored quillwort

Isoetes engelmannii
Engelmann's quillwort

Lastrea cristata
Crested buckler fern

Lastrea dilatata
Broad buckler fern

Lastrea dilatata cristata
Crested broad buckler fern

Lastrea dilatata lepidota
Scaly broad buckler fern

Lastrea filix-mas
Common buckler fern
Male fern

Lastrea filix-mas cristata
Crested male fern

Lastrea filix-mas furcans
Forked male fern

Lastrea filix-mas multi-cristata
Multi-crested male fern

Lastrea filix-mas ramosa
Branched male fern

Lastrea montana
Mountain buckler fern

Lastrea montana cristata
Crested mountain buckler fern

Lastrea montana furcans
Forked mountain buckler fern

Lastrea recurva
Hay-scented buckler fern

Lastrea rigida   Rigid buckler fern

Lastrea spinulosa
Prickly buckler fern

Lastrea thelypteris
Marsh buckler fern

Lastreopsis hispida
Bristly shield fern

Lastreopsis microsora
Creeping shield fern

Leptopteris superba
Pince of Wales' feather

Lycopodium clavatum
Ground pine
Running pine

Lycopodium complanatum
Ground cedar
Ground pine

Lycopodium lucidulum
Shining club moss

Lycopodium obscurum
Ground pine
Princess pine

Lycopodium tristachyum
Ground cedar

Lygodium circinatum
Malay climbing fern

Lygodium japonicum
Japanese climbing fern

Lygodium palmatum
Climbing fern
Hartford fern

Marattia douglasii   Pala

Marattia fraxinea   King fern
Para fern
Potato fern

Marattia salicina   King fern
Para fern
Potato fern

Marsilea crenata
Floating pepperwort

Marsilea pubescens
Stemless pepperwort

Marsilea quadrifolia
European water clover

Marsilea strigosa
Stemless pepperwort

Matteuccia struthiopteris
Ostrich-feather fern
Shuttlecock fern

Microsorium diversifolium
Kangaroo fern

Microsorium pustulatum
Fragrant fern

Microsorium scandens
Fragrant fern

Nephrolepis cordifolia
Common fishbone fern
Erect sword fern
Fishbone fern
Herringbone fern
Sword fern

Nephrolepis exaltata   Boston fern
Fishbone fern
Sword fern

Onoclea sensibilis
American oak fern
Bead fern
Sensitive fern

Ophioglossum californicum
Californian adder's tongue

Ophioglossum engelmannii
Engelmann's adder's tongue

Ophioglossum lusitanicum
Little adder's tongue

Ophioglossum pendulum
Drooping adder's tongue

Ophioglossum vulgatum
Adder's tongue

Oreopteris limbosperma
Lemon-scented fern

Osmunda cinnamomea   Buckhorn
Cinnamon fern
Fiddleheads

Osmunda claytoniana
Interrupted fern

Osmunda regalis   Flowering fern
Royal fern

Osmunda regalis cristata
Crested royal fern

Pellaea atropurpurea

Pellaea atropurpurea
Purple cliff brake

Pellaea falcata
Australian cliff brake
Sickle fern

Pellaea rotundifolia   Button fern
New Zealand cliff brake

Pellaea viridis   Green cliff brake

Phegopteris connectilis   Beech fern
Narrow beech fern
Northern beech fern

Phlebodium aureum
Golden polypody
Hare's foot fern
Rabbit's foot fern

Phyllitis scolopendrium
Hart's tongue fern

Phymatodes quercifolia
Oak-leaf fern

Phymatodes scolopendrium
Wart fern

Pityrogramma austroamericana
Gold fern

Pityrogramma calomelanos
Silver fern

Pityrogramma chrysophylla
Gold fern

Platycerium bifurcatum   Elkhorn
Staghorn

Platycerium hillii
Northern elkhorn

Platycerium superbum   Elkhorn
Moosehorn
Staghorn

Platycerium veitchii   Silver elkhorn

Polypodium alpestre
Alpine polypody

Polypodium alpestre flexile
Flexible alpine polypody

Polypodium alpestre laciniatum
Fringed alpine polypody

Polypodium angustifolium
Narrow-leaved strap fern

Polypodium aureum
Golden polypody
Hare's foot fern
Rabbit's foot fern

Polypodium calcareum
Limestone polypody

Polypodium californicum
California polypody

Polypodium dryopteris
Common oak fern
Oak fern
Triple-branched polypody

Polypodium fraxinifolium
Ash-leaf polypody

Polypodium glycyrrhiza
Licorice fern

Polypodium hesperium
Western polypody

Polypodium incanum
Resurrection fern

Polypodium integrifolium
Climbing bird's nest fern

Polypodium integrifolium cristatum
Crested climbing bird's nest fern

Polypodium interjectum
Western polypody

Polypodium irioides
Climbing bird's nest fern

Polypodium phegopteris
Beech fern
Mountain polypody

Polypodium phegopteris multifidum
Cloven mountain polypody

Polypodium phyllitidis
Ribbon fern
Strap fern

Polypodium phymatodes
Wart fern

Polypodium polypodioides
Resurrection fern

Polypodium punctatum
Climbing bird's nest fern

Polypodium pustulatum
Fragrant fern

Polypodium scandens
Fragrant fern

Polypodium scolopendria
Wart fern

Polypodium scouleri
Leathery polypody

Polypodium subauriculatum
Jointed pine

Polypodium virginianum
American wall fern
Rock polypody

Polypodium vulgare   Adder's fern
Common polypody
European polypody
Wall fern
Wall polypody
Wart fern

Polypodium vulgare acutum
Tapering polypody

Polypodium vulgare auritum
Ear-lobed polypody

Polypodium vulgare bifidum
Double-pinnuled polypody

Polypodium vulgare cambricum
Welsh polypody

Polypodium vulgare crenatum
Notched polypody

Polypodium vulgare cristatum
Crested polypody

Polypodium vulgare semilacerum
Irish polypody

Polystichum acrostichoides
Christmas fern
Dagger fern

FERNS

*Scolopendrium vulgare
multiforme*

Polystichum aculeatum
Hard shield fern
Soft shield fern

*Polystichum aculeatum
furcatum*
Forked hard shield fern

*Polystichum aculeatum
pulchrum*
Beautiful hard shield fern

Polystichum angulare
Soft prickly shield fern

*Polystichum angulare
cristatum*
Crested soft prickly shield fern

*Polystichum angulare
depauperatum*
Skeleton soft-prickly shield fern

*Polystichum angulare
grandiceps*
Grand-tasselled shield fern

Polystichum angulare lineare
Narrow-lined shield fern

*Polystichum angulare
ramulosum*
Branch-crested shield fern

*Polystichum angulare
semitripinnatum*
Divided soft prickly shield fern

*Polystichum angulare
tripinnatum*
Tri-pinnate soft prickly
shield fern

*Polystichum angulare
truncatum*
Truncate shield fern

Polystichum braunii
Braun's holly fern

Polystichum falcatum
Japanese holly fern

Polystichum lonchitis Holly fern
Mountain holly fern

Polystichum lonchitis confertum
Dense holly fern
Irish holly fern

Polystichum munitum
Giant holly fern
Sword fern

Polystichum proliferum
Mother shield fern

Polystichum setiferum Hedge fern
Soft shield fern

Polystichum tsus-simense
Dwarf leather fern
Tsusima holly fern

Polystichum vestitum
Prickly shield fern

Psilotum nudum Whisk fern

Pteridium aquilinum Bracken
Common bracken

Pteridium esculentum
Australian bracken

Pteridium latiusculum
Eastern bracken

Pteridium pubescens
Western bracken

Pteris aquilinum Bracken
Common bracken

Pteris aquilina bisulca
Split-fronded bracken

Pteris aquilina cristata
Crested bracken

Pteris cretica Cretan brake
Cretan fern
Ribbon brake
Ribbon fern

Pteris dentata Fine-toothed brake

Pteris ensiformis
Australian slender brake
Slender brake
Snow brake
Sword brake

Pteris flabellata
Fine-toothed brake

Pteris flaccida Fine-toothed brake

Pteris longifolia Chinese brake
Ladder brake
Rusty brake

Pteris multifida Chinese fern
Huguenot fern
Spider fern

Pteris serrulata Chinese fern
Huguenot fern
Spider fern

Pteris tremula Australian brake
Toothed brake
Trembling brake
Trembling fern

Pteris tripartita Giant brake
Trisect brake

Pteris vittata Chinese brake
Ladder brake
Rusty brake

Pyrossia lingua Japanese felt fern
Tongue fern

Pyrossia rupestris Rock felt fern

Rumohra adiantiformis
Leather fern
Leather shield fern
Shield hare's foot

Salvinia rotundifolia
Floating moss

Schizaea pusilla Curly-grass fern

*Scolopendrium vulgare
crispum-latum*
Broad curly hartstongue

*Scolopendrium vulgare
cristatum*
Crested hartstongue

Scolopendrium vulgare duplex
Double-fronded hartstongue

*Scolopendrium vulgare
laceratum*
Jagged-edge hartstongue

*Scolopendrium vulgare
multiforme*
Many-shaped hartstongue

Scolopendrium vulgare
ramo-cristatum

Scolopendrium vulgare ramo-cristatum	
Branch-crested hartstongue	
Scolopendrium vulgare ramo-palmatum	
Twin-fronded hartstongue	
Scolopendrium vulgare reniforme	
Kidney-shaped hartstongue	
Scolopendrium vulgare truncatum	
Short-fronded hartstongue	
Scolopendrium vulgare undulato-ramosum	
Wavy hartstongue	
Selaginella apoda	
Basket selaginella	
Selaginella braunii	
Treelet spike moss	
Selaginella cuspidata	Moss fern
	Sweat plant
Selaginella densa	Basket selaginella
Selaginella douglasii	
Douglas's spike moss	
Selaginella kraussiana	
Mat spike moss	
Trailing selaginella	
Trailing spike moss	
Selaginella lepidophylla	
Resurrection plant	
Rose of Jericho	
Selaginella pallescens	Moss fern
	Sweat plant
Selaginella rupestris	
Dwarf lycopod	
Rock selaginella	
Selaginella uncinata	
Blue selaginella	
Peacock moss	
Rainbow fern	
Trailing selaginella	
Selaginella willdenovii	
Peacock fern	
Willdenow's selaginella	

Stenochlaena palustris	
Climbing swamp fern	
Sticherus flabellatus	Shiny fan fern
	Umbrella fern
Sticherus tener	Silky fan fern
Struthiopteris germanica	
Ostrich-feather fern	
Shuttlecock fern	
Tectaria cicutaria	Button fern
	Snail fern
Tectaria gemmifera	Button fern
	Snail fern
Thelypteris hexagonoptera	
Broad beech fern	
Southern beech fern	
Thelypteris noveboracensis	
New York fern	
Thelypteris oreopteris	
Mountain buckler fern	
Thelypteris palustris	Marsh fern
Thelypteris phegopteris	
Beech fern	
Broad beech fern	
Northern beech fern	
Trichomanes radicans	
European bristle fern	
Trichomanes radicans furcans	
Forked bristle fern	
Tricholomopsis rutilans	
Plums and custard fern	
Woodsia alpina	Alpine woodsia
Woodsia ilvensis	Oblong woodsia
	Rusty woodsia
Woodsia obtusa	
Blunt-lobed woodsia	
Common woodsia	
Woodwardia chamissoi	
Giant chain fern	
Woodwardia fimbriata	
Giant chain fern	
Woodwardia radicans	
European chain fern	
Woodwardia virginica	
Virginia chain fern	

FUNGI

A large group of simple plants lacking chlorophyll is covered by the word fungi. This section is concerned only with the larger edible, inedible and poisonous fungi which are visible to the naked eye. Many edible fungi are matched in appearance with inedible, or even poisonous types so it is reckless to gather wild fungi unless you are experienced and familiar with the subtle difference. In the following list, entries have been classified by the use of bracketed initions thus:

(E) – Edible, but some may be allergic to them.

(I) – Inedible for a variety of reasons.

(P) – Poisonous, but not usually fatal.

(F) – Can be fatal if eaten, sometimes within minutes.

Field mushroom –
Agaricus campestris

Agaricus arvensis

Agaricus arvensis   Horse agaric (E)
Horse mushroom (E)

Agaricus bisporus
Cultivated mushroom (E)

Agaricus bitorquis
Pavement mushroom (E)
Urban mushroom (E)

Agaricus campestris
Field mushroom (E)
Pink bottom mushroom (E)

Agaricus sylvaticus
Brown wood mushroom (E)

Agaricus silvicola
Wood mushroom (E)

Agaricus xanthodermus
Yellow stainer (P)

Aleuria aurantia
Orange-peel fungus (E)

Amanita caesarea
Caesar's mushroom (I)

Amanita citrina   False death cap (I)

Amanita fulva   Sheathed agaric (E)
Tawny grisette (E)

Amanita mappa   False death cap (I)

Amanita muscaria   Fly agaric (P)

Amanita pantherina
False blusher (P)
Panther cap (P)

Amanita phalloides   Death cap (F)
Mortician cap (F)

Amanita regalis   Royal amanita (P)

Amanita rubescens
The blusher (E)
Woodland pink mushroom (E)

Amanita vaginata
Common grisette (E)

Amanita verna
Fool's mushroom (F)
Spring mushroom (F)

Amanita virosa   Coffin filler (F)
Destroying angel (F)

Armillaria mellea
Bootlace fungus (E)
Honey fungus (E)

Auricularia auricula-judae
Judas's ear (E)

Auricularia mesenterica
Tripe fungus (I)

Auriscalpium vulgare
Ear-pick fungus (I)

Boleta speciosus
Showy mushroom (E)

Boletus aereus   Bronze boletus (E)

Boletus aestivalis
Summer boletus (E)

Boletus badius   Bay boletus (E)

Boletus calopus   Olive boletus (I)

Boletus chrysenteron
Red cracked boletus (I)
Crazed boletus (I)

Boletus cyanescens
Indigo boletus (E)

Boletus edulis   Cep (E)
Penny bun fungus (E)

Boletus erythropus
Red-leg boletus (E)

Boletus leucophaeus
Brown birch boletus (E)
Cow fungus (E)

Boletus luridus   Lurid boletus (E)

Boletus regius   Royal boletus (E)

Boletus reticulatus
Summer boletus (E)

Boletus rhodoxanthus
Purple boletus (P)

Boletus satanas   Devil's boletus (P)
Satan's mushroom (P)

Boletus subtomentosus
Downy boletus (E)
Goat's lip mushroom (E)

Bulgaria inquinans
Bachelor's buttons (E)
Black bulgar (E)

Calocybe gambosa
Saint George's mushroom (E)

Calvatia gigantea
Giant puffball (E)

Calyptella capula
Stinging-nettle fungus (I)

Cantharellus cibarius
Chanterelle (E)

Cantharellus cornucopioides
Horn of plenty (E)

Chlorociboria aeruginacens
Greenstain fungus (I)
Green wood cup (I)

Chlorosplenium aeruginosum
Greenstain fungus (I)
Green wood cup (I)

Chondrostereum purpureum
Silverleaf fungus (I)

Clathrus cancellatus
Basket fungus (E)

Clathrus ruber   Basket fungus (E)

Clavaria argillacea   Moor club (I)

Clavariadelphus pistillaris
Giant club (E)

Claviceps purpurea   Ergot (P)
Spurred rye (P)

Clavulina cinerea
Grey coral fungus (P)

Clavulina coralloides
Crested coral fungus (P)

Clavulina cristata
Crested coral fungus (P)

Clavulina rugosa
Wrinkled club (P)

Clitocybe clavipes   Club foot (I)

Clitocybe geotropa
Trumpet agaric (E)

Clitocybe illudens
Copper trumpet (P)

Helvella elastica

Clitocybe mellea
Bootlace fungus (E)
Honey fungus (E)

Clitocybe nebularis
Clouded agaric (E)
Clouded clitocybe (E)

Clitocybe odora
Aniseed toadstool (E)

Clitocybe olearia
Copper trumpet (P)

Clitopilus prunulus
Miller mushroom (E)
Plum agaric (E)
Sweetbread mushroom (E)

Collybia butyracea
Buttery collybia (E)
Butter cap (E)

Collybia confluens
Clustered tough-shank (I)

Collybia dryophila
Oaktree collybia (E)
Wood agaric (E)

Collybia fusipes   Spindle shank (E)

Collybia maculata   Foxy spot (I)
Spotted tough-shank (I)

Collybia peronata
Wood woollyfoot (I)

Collybia velutipes
Velvet shank (E)
Winter mushroom (E)

Conocybe lactea
Milky conocybe (I)

Coprinus atramentarius
Grey inkcap (E)
Inkcap (E)

Coprinus comatus   Inkcap (E)
Lawyer's wig (E)
Shaggy cap (E)
Shaggy inkcap (E)

Coprinus disseminatus
Fairy bonnets (P)
Trooping crumble cap (P)

Coprinus micaceus
Glistening inkcap (I)
Mica inkcap (I)
Shining inkcap (I)

Coprinus niveus
Horse-dung fungus (I)

Coprinus picaceus
Magpie fungus (P)

Coprinus plicatilis
Japanese sunshade (P)
Little Japanese umbrella (P)

Cordyceps militaris
Scarlet caterpillar fungus (P)

Coricolus versicolor
Multi-zoned bracket fungus (P)

Cortinarius traganus
Goaty smell cortinarius (I)

Craterellus cornucopioides
Horn of plenty (E)

Crucibulum crucibuliforme
Bird's nest fungus (I)

Crucibulum laeve
Bird's nest fungus (I)

Cyathus striatus   Splash cup (I)

Daedalea quercina   Maze gill (P)

Daedaleopsis confragosa
Blushing bracket fungus (I)

Daldinia concentrica
King Alfred's cakes (I)
King Alfred's balls (I)
King Alfred's cramp balls (I)

Dentinum repandum
Wood hedgehog (E)

Echinodontium tinctorium
Indian paint fungus (I)

Entoloma clypeatum
Buckler agaric (I)

Entoloma lividum
Livid entoloma (P)

Entoloma sinuatum
Livid entoloma (P)

Exidia glandulosa
Witch's butter (I)

Fistulina hepatica
Beefsteak fungus (E)
Rusty oak fungus (E)

Flammulina velutipes
Velvet shank (E)
Winter mushroom (E)

Fomes fomentarius   Hoof fungus (I)
Tinder fungus (I)

Geastrum fimbriatum   Earthstar (I)

Geastrum fornicatum   Earthstar (I)

Geastrum pectinatum   Earthstar (I)

Geastrum sessile   Earthstar (I)

Geastrum triplex   Earthstar (I)

Gomphidius glutinosus
Woodland black spot (E)

Gomphidius rutilis
Pine forest mushroom (E)

Grifola frondosa
Hen-of-the-woods (E)

Gymnosporangium
clavariaeforme
Knotted fungus (I)

Gyromitra esculenta
False morel (P)

Gyromitra infula   Turban fungus (E)

Gyroporus castaneus
Bitter boletus (E)
Chestnut boletus (E)

Gyroporus cyanescens
Indigo boletus (E)

Hebeloma crustuliniforme
Fairy cake fungus (F)
Poison pie (F)

Hebeloma sacchariolens
Bitter-sweet fungus (P)

Helvella crispa
Common white helvella (P)

Helvella elastica
Distorted helvella (P)

Helvella fusca

Helvella fusca
            Black-saddle helvella (P)
Helvella infula   Turban fungus (E)
Helvella lacunosa
            Black helvella (P)
Helvella villosa   Cupped villosa (P)
Hericium erinaceum
            Hedgehog fungus (E)
Heterobasidion annosum
            Root fomes (I)
Hirneola auricula-judae
            Jew's ear (P)
Hydnum imbricatum
            Scaly hydnum (E)
Hydnum repandum
            Hedgehog fungus (E)
            Wood hedgehog (E)
Hydnum scabrosum
            Bitter hydnum (I)
Hygrocybe conica
            Conical wax cap (I)
Hygrocybe pratensis
            Meadow wax cap (I)
Hygrocybe psittacina
            Parrot fungus (E)
            Parrot wax cap (E)
Hygrocybe virginea
            Snowy wax cap (I)
Hygrophoropsis aurantiaca
            False chanterelle (I)
Hygrophorus eburneus
            Ivory wax cap (I)
Hygrophorus marzuolus
            March mushroom (E)
Hygrophorus psittacinus
            Parrot fungus (E)
Hypholoma capnoides
            Smokey-gilled woodlover (I)
Hypholoma fasciculare
            Sulphur tuft (P)
Hypholoma sublateritium
            Brick-red agaric (I)
Inocybe fastigiata
            Conical inocybe (P)
Inocybe geophylla
            Common white inocybe (P)
Inocybe napipes
            Skullcap inocybe (P)
Inocybe patouillardii
            Red-staining inocybe (F)
Kuehneromyces mutabilis
            Changeable mutabilis (E)
            Rusty agaric (E)
Laccaria amethystea
            Amethyst deceiver (E)
Laccaria laccata   Deceiver (E)
            The deceiver (E)
Lacrymaria lacrymabunda
            Weeping widow (I)
Lacrymaria velutina   Velvet cap (I)
            Weeping widow (I)

Lactarius deliciosus   Milk agaric (E)
            Milk cap (E)
            Saffron milk cap (E)
Lactarius glyciosmus
            Coconut-scented milk cap (I)
Lactarius necator
            Ugly mushroom (I)
Lactarius plumbeus
            Ugly mushroom (I)
Lactarius quietus   Oak milk cap (I)
Lactarius rufus   Rusty milk cap (I)
Lactarius torminosus
            Shaggy milk cap (I)
            Woolly milk cap (I)
Lactarius turpis
            Ugly mushroom (I)
Lactarius vellereus
            Fleecy milk cap (I)
Laetiporus sulphureus
            Sulfur polypore (E)
            Sulphur polypore (E)
Langermannia gigantea
            Giant puffball (E)
Lapista nuda   Wood blewits (E)
Leccinum aurantiacum
            Orange cap boletus (E)
Leccinum carpini
            Hornbeam boletus (E)
Leccinum duriusculum
            White poplar mushroom (E)
Leccinum griseum
            Hornbeam boletus (E)
Leccinum holopus
            White birch boletus (E)
Leccinum melaneum
            Black birch boletus (E)
Leccinum quercinum
            Oaktree boletus (E)
Leccinum scabrum
            Brown birch boletus (E)
            Cow fungus (E)
Leccinum testaceoscabrum
            Orange birch boletus (E)
Leccinum versipelle
            Orange birch boletus (E)
Lentinula edodes
            Shiitake fungus (E)
Leota lubrica   Jellybaby fungus (E)
            Jellybean fungus (E)
Lepiota castanea
            Tasselated toadstool (P)
Lepiota procera   Fairy sunshade (E)
            Parasol mushroom (E)
Lepiota rhacodes
            Shaggy parasol (E)
Lepista irina   Bigelow's blewit (E)
Lepista nebularis
            Clouded agaric (E)
            Clouded clitocybe (E)
Lepista nuda   Wood blewitts (E)
Lepista saeva   Blewits (E)

Russula claroflava

Lycoperdon echinatum
Hedgehog fungus (E)

Lycoperdon gemmatum
Common puffball (E)
Poor man's sweetbread (E)

Lycoperdon maximum
Giant puffball (E)

Lycoperdon perlatum
Common puffball (E)
Poor man's sweetbread (E)

Macrolepiota procera
Fairy sunshade (E)
Parasol mushroom (E)

Macrolepiota rhacodes
Shaggy parasol (E)

Marasmius androsaceus
Horsehair fungus (I)

Marasmius oreades
Clover windling (E)
Fairy ring mushroom (E)

Masseola crispa
Cauliflower fungus (E)

Meripilus giganteus
Giant polypore (I)

Microglossum viride
Green earth tongue (I)

Morchella conica
Conical morel (E)

Morchella elata   Morel (E)

Morchella esculenta
Common morel (E)

Mutinus caninus   Dog stinkhorn (I)

Mycena galericulata
Bonnet mycena (I)

Mycena ribula   Little nail fungus (P)

Mycena vitilis   Pixie's cap fungus (I)

Namatoloma fasciculare
Sulphur tuft (P)

Nectria cinnabarina
Coral spot fungus (I)

Neogyromitra gigas
Giant giromitra (E)

Omphalotus olearius
Copper trumpet (P)

Otidea onotica
Donkey's ear fungus (E)
Hare's ear fungus (I)

Oudemansiella mucida
Beech tuft (I)
Poached egg fungus (I)

Oudemansiella radicata
Long root mushroom (E)

Paxillus involutus
Brown roll-rim (P)
Roll-rim fungus (P)

Peziza ammophila
Brown star fungus (P)

Peziza aurantia
Orange-peel fungus (E)

Phallus hadriani
Devil's egg fungus (I)

Phallus impudicus   Stinkhorn (E)

Pholiota caperata
Gipsy mushroom (E)

Pholiota squarrosa
Shaggy pholiota (I)

Piptoporus betulinus
Birch bracket fungus (E)
Birch polypore (E)
Razor-strop fungus (E)

Pleurotus cornucopiae
Branched oyster fungus (E)

Pleurotus ostreatus
Oyster mushroom (E)

Pluteus atricapillus
Deer mushroom (E)
Fawn agaric (E)

Pluteus cervinus
Deer mushroom (E)
Fawn agaric (E)

Polyporus betulinus
Birch bracket fungus (E)
Razor-strop fungus (E)

Polyporus frondosus
Hen-of-the-woods (E)

Polyporus squamosus
Dryad's saddle (E)
Scaly polypore (E)

Polyporus versicolor
Multi-zoned bracket fungus (P)

Psalliota arvensis   Horse agaric (E)
Horse mushroom (E)

Psalliota campestris
Field mushroom (E)
Pink bottom mushroom (E)

Psalliota silvicola
Wood mushroom (E)

Psalliota sylvatica
Brown wood mushroom (E)

Pseudocoprinus disseminatus
Trooping crumble cap (P)

Pseudohydnum gelatinosum
Jello tongue (P)
Jelly tongue (P)

Psilocybe semilanceata
Liberty cap (P)

Pycnoporus cinnabarinus
Cinnabar polypore (I)

Ramaria formosa
Multi-branched fungus (P)

Rhytisma acerinum
Sycamore tarspot (I)

Rhyzina undulata
Pine fire fungus (P)

Rozites caperata
Gipsy mushroom (E)

Russula aeruginea
Grass-green russula (E)

Russula alutacea
Leathery russula (E)

Russula aurata   Golden russula (E)

Russula claroflava
Yellow swamp russula (E)

Russula cyanoxantha

Russula cyanoxantha
    Blue and yellow russula (E)
    Green agaric (E)

Russula decolorans
    Faded russula (E)

Russula emetica
    Sickener mushroom (P)
    Emetic russula (P)

Russula flava
    Yellow swamp russula (E)

Russula fragilis   Fragile russula (P)

Russula mairei
    Beechwood sickener (I)

Russula nigricans
    Blackening russula (I)

Russula ochroleuca
    Yellow russula (I)

Russula olivacea
    Olive-green russula (E)

Russula undulata
    Purple-black russula (P)

Russula vesca
    Bare-tooth russula (E)

Russula virescens   Green agaric (E)
    Green cracking russula (E)

Russula xerampelina
    Dark red russula (E)

Sarcodon imbricatum
    Scaly hydnum (E)

Sarcoscypha coccinea   Elf cup (E)
    Pixie cup (E)
    Scarlet cup (E)
    Scarlet elf cap (E)

Schizophyllum commune
    Split gill fungus (P)

Scleroderma citrinum
    Common earthball (P)

Scutellinia scutellata
    Eyelash cup fungus (I)
    Eyelash fungus (I)

Sparassis crispa
    Cauliflower fungus (E)
    Brain fungus (E)

Sparassis ramosa
    Cauliflower fungus (E)

Sphaerotheca pannosa
    Rose mildew (I)

Stereum hirsutum
    Hairy stereum (I)

Stropharia aeruginosa
    Verdigris fungus (E)

Suillus bovinus   Cow boletus (E)
    Jersey cow boletus (E)
    Rainfall boletus (E)

Suillus grevillei   Larch boletus (E)

Suillus luteus   Ringed boletus (E)
    Slippery jack (E)

Suillus variegatus
    Variegated boletus (E)

Taphrina betulina
    Witch's broom (P)

Thelephora terrestris
    Carpet fungus (I)
    Earth fan (I)

Trametes versicolor
    Multi-zoned bracket fungus (P)
    Multi-zoned polypore (P)

Tremella mesenterica
    Jellybrain fungus (I)
    Yellow brain fungus (I)

Tricholoma equestre
    Firwood agaric (E)
    Man on horseback (E)

Tricholoma flavovirens
    Firwood agaric (E)
    Man on horseback (E)

Tricholoma gambosa
    Saint George's mushroom (E)

Tricholoma georgii
    Saint George's mushroom (E)

Tricholoma nudum
    Wood blewits (E)

Tricholoma pardalotum
    Tiger tricholoma (P)

Tricholoma pardinum
    Tiger tricholoma (P)

Tricholoma personatum
    Blewits (E)

Tricholoma portentosum
    Dingy agaric (E)

Tricholoma sulphureum
    Yellow agaric (P)
    Gas tar fungus (P)

Tricholoma terreum
    Grey agaric (E)

Tricholomopsis rutilans
    Plums and custard (E)

Tuber aestivum   Summer truffle (E)
    Truffle (E)

Tuber magnatum   White truffle (E)
    Piedmont truffle (E)

Tuber melanosporum
    French truffle (E)
    Perigord truffle (E)

Tylopilus felleus   Bitter bolete (I)
    Bitter cep (I)

Ustilago maydis   Corn smut (P)

Volvaria speciosa
    Rose-gilled grisette (E)

Volvaria volvacea
    Padi-straw fungus (E)

Volvariella speciosa
    Rose-gilled grisette (E)

Volvariella volvacea
    Padi-straw fungus (E)

Xerocomus badius   Bay boletus (E)

Xerocomus subtomentosus
    Downy boletus (E)
    Goat's lip mushroom (E)

Xylaria hypoxylon
    Candlesnuff fungus (I)

Xylaria polymorpha
    Dead man's fingers (I)

GRASSES, REEDS, SEDGES, BAMBOOS, VETCHES, ETC.

True grasses are members of the family *Gramineae* and are found in almost every corner of the planet in one form or another, but in this section we are concerned with the ornamentals, meadow grasses, timber frasses (bamboos), soil-holding or sand-binding grasses, as well as wetland subjects such as sedges and reeds. Some grasses may also be listed under 'Aquatics'.

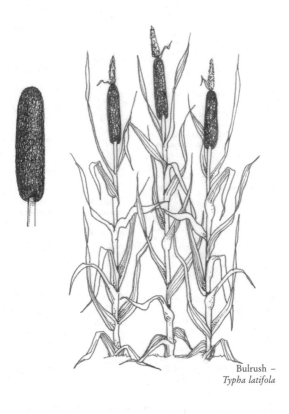

Bulrush –
Typha latifola

Agropogon littoralis

Agropogon littoralis
Perennial beard grass
Agropyron caninum
Bearded twitch
Bearded couch
Dog grass
Agropyron cristatum
Crested wheatgrass
Fairy crested wheatgrass
Agropyron donianum Don's twitch
Agropyron elongatum
Tall wheatgrass
Agropyron intermedium
Intermediate wheatgrass
Agropyron junceiforme
Sand couch
Sand twitch
Agropyron pungens Sea couch
Sea twitch
Agropyron repens
Creeping twitch grass
Couch grass
Quack grass
Quick grass
Quitch grass
Scutch grass
Twitch grass
Witch grass
Agropyron sibiricum
Desert wheatgrass
Siberian wheatgrass
Standard crested wheatgrass
Agropyron smithii
Western wheatgrass
Agropyron spicatum
Bluebunch wheatgrass
Agropyron trachycaulum
Slender wheatgrass
Agropyron trichophorum
Pubescent wheatgrass
Stiff hair wheatgrass
Agrostis canina Brown bent
Velvet bent
Agrostis curtisii Bristle bent
Agrostis gigantea Black bent
Red top
Agrostis nebulosa Cloud bent
Cloud grass
Agrostis perennans Autumn bent
Brown bent grass
Upland bent
Upland bent grass
Agrostis semiverticillata
Water bent
Agrostis setacea Bristly bent
Agrostis stolonifera Creeping bent
Agrostis tenuis Brown top
Colonial bent
Common bent
Fine bent
New Zealand bent
Rhode Island bent
Aira caryophyllea Silvery hair grass
Aira praecox Early hair grass

Alopecurus aequalis
Orange foxtail
Short-awn(ed) foxtail
Alopecurus alpinus Alpine foxtail
Alopecurus arundinaceus
Creeping foxtail
Reed foxtail
Alopecurus bulbosus
Bulbous foxtail
Tuberous foxtail
Alopecurus geniculatus
Floating foxtail
Marsh foxtail
Alopecurus myosuroides
Black grass
Black twitch
Slender foxtail
Alopecurus pratensis
Common foxtail
Meadow foxtail
Alopecurus pratensis 'aureus'
Golden foxtail
Ammocalamagrostis baltica
Hybrid marram grass
Ammophila arenaria
European beach grass
Marram grass
Ammophila breviligulata
American beach grass
Ampelodesmos mauritanicus
Mauritania vine reed
Andropogon gerardii
Big bluestem grass
Andropogon hallii
Sand bluestem grass
Anthoxanthum odoratum
Scented vernal grass
Sweet vernal grass
Anthoxanthum puelii
Annual vernal grass
Anthyllis montana Mountain vetch
Spanish vetch
Anthyllis vulneraria Kidney vetch
Lady's fingers
Woundwort
Apera interrupta Dense silky bent
Apera spica-venti Loose silky bent
Wind grass
Arenula pratensis
Meadow oat grass
Arrhenatherum bulbosum
Variegated oat grass
Arrhenatherum elatius
False oat grass
Oat grass
Tall oat grass
Variegated oat grass
Arundinaria amabilis
Tonkin bamboo
Tonkin cane
Tsingli cane
Arundinaria anceps
Himalayan bamboo
Ringal

GRASSES

Bromus interruptus

Arundinaria disticha
Dwarf fern leaf bamboo

Arundinaria gigantea
Canebrake bamboo
Cane reed
Giant cane
Southern cane

Arundinaria japonica
Arrow bamboo
Metake

Arundinaria pumila
Dwarf bamboo

Arundinaria pygmaea
Pigmy bamboo

Arundinaria quadrangularis
Square-stemmed bamboo

Arundinaria simonii
Simon bamboo

Arundinaria tecta Small cane
Switch cane

Arundinaria variegata
Dwarf white-stripe bamboo

Arundo donax Carrizo
Cana brava
Giant reed

Avena barbata Slender wild oat

Avena fatua Common wild oat
Drake
Flaver
Potato oat
Spring wild oat
Tartarian oat
Wild oat

Avena ludoviciana
Winter wild oat grass
Winter wild oat

Avena pratensis
Perennial oat grass

Avena sativa Oats

Avena sterilis Animated oat grass
Animated oats
Winter wild oat

Avena strigosa Bristle oat
Small oat

Axonopus affinis Carpet grass
Common carpet grass

Bambusa arundinacea
Giant thorny bamboo

Bambusa beecheyana
Beechey bamboo

Bambusa glaucescens
Hedge bamboo
Oriental hedge bamboo

Bambusa multiplex
Hedge bamboo

Bambusa oldhamii
Oldham bamboo

Bambusa tuldoides
Punting-pole bamboo

Bambusa ventricosa
Buddha bamboo
Buddha's belly bamboo

Bambusa vulgaris
Common bamboo

Blysmus compressus
Sedge-like club rush
Flat sedge

Blysmus rufus Red blysmus

Boehmeria nivea China grass
Ramie fibre

Bothriochloa caucasica
Caucasian bluestem

Bothriochloa intermedia
Australian bluestem

Bothriochloa ischaemum
Turkestan bluestem
Yellow bluestem

Bothriochloa saccharoides
Silver beard grass

Bouteloua curtipendula
Sideoats grama

Bouteloua eriopoda Black grama

Bouteloua gracilis Blue grama
Mosquito grass

Bouteloua hirsuta Hairy grama

Bouteloua repens Slender grama

Brachiaria subquadripara
Creeping signal grass

Brachypodium pinnatum
Chalk false brome
Heath false brome
Tor grass

Brachypodium sylvaticum
False brome
Tor grass
Wood false brome

Briza maxima Large quaking grass
Pearl grass

Briza media
Common quaking grass
Doddering dillies
Quaking grass
Totter grass

Briza minor Lesser quaking grass
Little quaking grass
Small quaking grass

Bromus arvensis Field brome

Bromus briziformis
Rattlesnake brome
Rattlesnake chess

Bromus canadensis Fringed brome

Bromus carinatus
Californian brome

Bromus commutatus
Meadow brome

Bromus diandrus Great brome

Bromus erectus Erect brome
Upright brome

Bromus hordeaceus
Soft brome grass

Bromus inermis Awnless brome
Hungarian brome
Smooth brome

Bromus interruptus
Interrupted brome

Bromus japonicus

Bromus japonicus	Japanese brome
	Japanese chess
Bromus lanceolatus	
	Mediterranean brome
Bromus lepidus	Slender brome
Bromus madritensis	
	Compact brome
	Madrid brome
Bromus marginatus	
	Mountain brome
Bromus mollis	Soft brome grass
	Soft chess
Bromus recemosus	Smooth brome
Bromus ramosus	Hairy brome
	Woodland brome
Bromus rigidus	Stiff brome
Bromus rubens	Foxtail chess
Bromus secalinus	Rye brome
Bromus sterilis	Barren brome
Bromus tectorum	
	Drooping brome
Bromus unioloides	Prairie brome
	Rescue brome
	Rescue grass
Bromus wildenowii	Rescue brome
Buchloe dactyloides	Buffalo grass
Butomus umbellatus	
	Flowering rush
	Grassy rush
	Water gladiolus
Cabomba caroliniana	Fanwort
	Fish grass
	Washington grass
Calamagrostis canescens	
	Purple small reed
Calamagrostis epigejos	Bush grass
	Wood small reed
Calamagrostis nana	Small reed
Calamagrostis scotia	
	Scottish small reed
Calamagrostis stricta	
	Narrow small reed
Calamus rotang	Rattan
	Rattan cane
Calamus scipionum	Malacca cane
Carex acuta	Narrow-spiked sedge
	Slender-spiked sedge
Carex acutiformis	
	Lesser pond sedge
Carex appropinquata	
	Lesser tussock sedge
Carex aquatilis	
	Mountain water sedge
	Water sedge
Carex arenaria	Sand sedge
Carex atrata	Black sedge
Carex bigelowii	Stiff sedge
Carex binervis	
	Green-ribbed sedge
Carex buxbaumii	Club sedge
	Dark sedge

Carex capillaris	Hair sedge
Carex caryophyllea	Spring sedge
Carex curta	Pale sedge
	White sedge
Carex demissa	Low sedge
Carex depauperata	
	Starved wood sedge
Carex diandra	
	Double-stemmed sedge
	Lesser tussock sedge
Carex digitata	Fingered sedge
Carex dioica	Dioecious sedge
Carex distans	Distant sedge
Carex disticha	Brown sedge
	Creeping brown sedge
Carex divisa	Salt meadow sedge
Carex divulsa	Grey sedge
Carex echinata	Star sedge
Carex elata	Tufted sedge
Carex elongata	Elongated sedge
Carex ericetorum	Heath sedge
Carex extensa	Long-bracted sedge
Carex filiformis	
	Downy-fruited sedge
Carex flacca	Glaucous sedge
Carex flava	Yellow sedge
Carex grayi	Mace sedge
Carex hirta	Hairy sedge
Carex hostiana	Tawny sedge
Carex humilis	Dwarf sedge
Carex lachenalii	Hare's foot sedge
Carex laevigata	Smooth sedge
Carex lasiocarpa	
	Downy-fruited sedge
Carex limosa	Bog sedge
	Mud sedge
Carex loliacea	Darnel sedge
Carex maritima	Curved sedge
Carex montana	Mountain sedge
Carex muricata	Prickly sedge
Carex nigra	Common sedge
Carex norvegica	Alpine sedge
Carex ornithopoda	
	Bird's foot sedge
Carex otrubae	False fox sedge
Carex ovalis	Oval sedge
Carex pallescens	Pale sedge
Carex panicea	Carnation grass
	Carnation sedge
Carex paniculata	
	Greater tussock sedge
	Great tussock sedge
Carex pauciflora	
	Few-flowered sedge
Carex paupercula	
	Broad-leaved mud sedge
Carex pendula	Drooping sedge
	Pendulous sedge
	Sedge grass

Dendrocalamus gigantea

Carex pilulifera	Pill sedge
Carex polyphylla	Chalk sedge
Carex pseudocyperus	
	Cyperus sedge
Carex punctata	
	Dotted sedge
Carex puticaris	Flea sedge
Carex rariflora	
	Few-flowered sedge
Carex recta	Caithness sedge
Carex remota	
	Distant-flowered sedge
Carex riparia	Great pond sedge
	Pond sedge
Carex rostrata	Beaked sedge
	Bottle sedge
Carex rupestris	Rock sedge
Carex saxatilis	Russet sedge
Carex serotina	
	Late-flowering sedge
Carex spicata	Spiked sedge
Carex stenolepis	
	Thin-glumed sedge
Carex strigosa	
	Loose-spiked wood sedge
	Thin-spiked wood sedge
Carex sylvatica	Wood sedge
Carex vaginata	
	Wide-sheathed sedge
Carex vesicaria	Bladder sedge
Carex vulpina	Fox sedge
Catabrosa aquatica	
	Water whorl grass
	Whorl grass
Catapodium marinum	
	Stiff sand grass
Catapodium rigidum	Fern grass
Chasmanthium latifolium	
	American wild oats
	Wild oats
Chimonobambusa falcata	
	Sickle bamboo
Chimonobambusa quadrangularis	
	Square bamboo
	Square-stem bamboo
Chionochloa conspicua	
	Hunangemoho
Chloris berroi	Giant finger grass
	Uruguay finger grass
Chloris gayana	Rhodes grass
Chloris truncata	
	Creeping windmill grass
	Star grass
Chloris ventricosa	
	Australian windmill grass
Chusquea culeou	Culeu
Cladium mariscus	Fenland sedge
	Great fenland sedge
Coix lacryma-jobi	Job's tears
Cortaderia argentea	Pampas grass

Cortaderia richardii	Toe-toe
Cortaderia selloana	Pampas grass
Corynephorus canescens	
	Grey hair grass
Cymbopogon citratus	
	Fevergrass
	Lemongrass
	West Indian lemongrass
Cymbopogon nardus	
	Citronella grass
	Nard grass
Cynodon dactylon	
	Bermuda grass
	Creeping dog's tooth grass
	Creeping finger grass
	Doob grass
	Kweek grass
Cynodon transvaalensis	
	African Bermuda grass
	Transvaal dogtooth grass
Cynosurus cristatus	
	Crested dogstail
Cynosurus echinatus	
	Rough dogstail
Cyperus alternifolius	
	Umbrella grass
	Umbrella palm
	Umbrella plant
	Umbrella sedge
Cyperus articulatus	Adrue
	Guinea rush
Cyperus esculentus	Chufa
	Earth almond
	Nut grass
	Nut sedge
	Rush nut
	Tiger nut
	Yellow nut grass
	Yellow nut sedge
	Zulu nut
Cyperus fuscus	Brown cyperus
Cyperus isocladus	Dwarf papyrus
	Miniature papyrus
Cyperus longus	Galingale
	Sweet galingale
Cyperus papyrus	Bulrush (biblical)
	Egyptian paper rush
	Paper plant
	Paper reed
	Papyrus
Cyperus tagetiformis	
	Chinese mat grass
Dactylis glomerata	Cocksfoot
	Orchard grass
Danthonia decumbens	
	Heath grass
Danthonia semiannularis	
	Australian danthonia
	Australian oat grass
Danthonia setacea	Wallaby grass
Dendrocalamus gigantea	
	Giant bamboo

Dendrocalamus strictus
Calcutta bamboo
Male bamboo

Deschampsia alpina
Alpine hair grass

Deschampsia caespitosa
Tufted hair grass

Deschampsia flexuosa
Crinkled hair grass
Wavy hair grass

Deschampsia setacea
Bog hair grass
Dog-hair grass

Dezmazeria rigida   Fern grass

Dezmazeria sicula   Spike grass

Dichanthium annulatum
Brahman grass
Diaz bluestem
Kleberg grass
Ringed beard grass

Dichanthium aristatum
Angleton bluestem
Angleton grass

Dichanthium ischaemum
Dogstooth grass

Digitaria decumbens   Pangola grass

Digitaria didactyla
Blue couch grass
Blue finger grass

Digitaria ischaemum
Smooth finger grass
Red millet

Digitaria pentzii
Pentz finger grass

Digitaria sanguinalis   Crab grass
Hairy crab grass
Hairy finger grass

Digitaria serotina
Creeping finger grass

Echinochloa crus-galli   Barn grass
Barnyard grass
Barnyard millet
Cockspur grass
Japanese millet

Ehrharta calycina
Perennial veldt grass

Eleocharis acicularis   Hair grass
Least spike rush
Needle spike rush
Slender spike rush

Eleocharis dulcis
Chinese water chestnut
Ma-tai
Water chestnut

Eleocharis effusum   Saw grass

Eleocharis multicaulis
Many-stemmed spike rush

Eleocharis palustris
Common spike rush

Eleocharis parvula
Dwarf spike rush

Eleocharis tuberosa
Chinese water chestnut

Eleocharis uniglumis
Single-glumed spike rush
Slender spike rush

Eleusine coracana   African millet
Finger millet
Korakan
Ragi

Eleusine indica   Goose grass
Wire grass

Elymus angustus   Altai wild rye

Elymus aralensis   Aral wild rye

Elymus arenarius   Dune grass
European dune grass
Lyme grass
Rancheria grass
Sea lyme grass

Elymus canadensis
Canadian wild rye

Elymus caninus   Bearded couch

Elymus chinensis   Chinese wild rye
False wheatgrass

Elymus condensatus
Giant wild rye

Elymus farctus   Sand couch

Elymus glaucous   Blue wild rye

Elymus junceus   Russian wild rye

Elymus pycnanthus   Sea couch

Elymus racemosus   Volga wild rye

Elymus repens   Common couch
Couch grass
Creeping twitch grass
Twitch grass
Witchgrass

Elymus sibiricus   Siberian wild rye

Elymus virginicus
Virginian wild rye

Equisetum hyemale
Common horsetail
Common scouring brush
Dutch rush
Horsetail
Scouring brush

Eragrostis amabilis
Feather love grass
Japanese love grass

Eragrostis capillaris   Lace grass

Eragrostis chloromelas
Blue love grass
Boer love grass

Eragrostis curvula
Drooping love grass
Weeping love grass

Eragrostis elegans   Love grass

Eragrostis lehmanniana
Lehmann's love grass

Eragrostis trichodes
Sand love grass

Eremochloa ophiuroides
Centipede grass
Lazy man's grass

Erianthus ravennae   Plume grass
Ravenna grass
Woolly beard grass

Eriochloa aristata
Branched cup grass
Mexican everlasting grass

Eriochloa polystachya Carib grass
Malogilla
Malojilla
Malojillo

Eriochloa villosa Hairy cup grass

Eriophorum angustifolium
Common cotton grass

Eriophorum gracile
Slender cotton grass

Eriophorum latifolium
Broad-leaved cotton grass

Eriophorum vaginatum
Hare's tail cotton grass
Hare's tail grass

Erophila versa Whitlow grass

Festuca altissima Reed fescue
Wild fescue
Wood fescue

Festuca amethystina Tufted fescue

Festuca arundinacea Tall fescue
Tall festuca

Festuca elatior Alta fescue
Reed fescue
Tall fescue

Festuca gigantea Giant fescue

Festuca glauca Blue fescue
Grey fescue

Festuca heterophylla
Various-leaved fescue

Festuca juncifolia
Rush-leaved fescue

Festuca longifolia Hard fescue

Festuca ovina Sheep's fescue

Festuca ovina glauca Blue fescue

Festuca pratensis English bluegrass
Meadow fescue

Festuca rubra Chewing's fescue
Creeping fescue
Red fescue
Shade fescue

Festuca tenuifolia
Awnless sheep's fescue
Fine-leaved sheep's fescue
Hair fescue

Festulolium loliaceum
Hybrid fescue

Gastridium ventricosum
Nit grass

Glyceria declinata
Glaucous sweet grass

Glyceria fluitans
Floating sweet grass

Glyceria maxima
Reed sweet grass

Glyceria pedicellata
Hybrid sweet grass

Glyceria plicata
Plicate sweet grass

Gynerium sagittatum Arrow cane
Moa grass
Uva grass
Wild cane

Helictotrichon pratense
Meadow oat grass

Helictotrichon pubescens
Downy oat grass
Hairy oat grass

Helictotrichon sempervirens
Oat grass

Heteranthera dubia
Water star grass

Heteranthera graminae
Water star grass

Hierochloe odorata Holy grass
Vanilla grass

Hilaria belangeri Curley mesquite

Hilaria jamesii Galleta

Hilaria mutica Tobosa grass

Hilaria rigida Big galleta

Holcus lanatus Meadow soft grass
Tufted soft grass
Velvet grass
Yorkshire fog

Holcus mollis Creeping soft grass
Soft grass

Holcus mollis variegatus
Variegated creeping soft grass

Holcus sorghum Sorghum

Holcus virgatus Tunis grass

Hordelymus europaeus
Wood barley

Hordeum brevisubulatum
Short-awned barley

Hordeum bulbosum
Bulbous barley

Hordeum distichon Pearl barley

Hordeum jubatum Foxtail barley
Squirrel-tail barley
Squirrel-tail grass

Hordeum marinum Sea barley
Squirrel-tail grass

Hordeum murinum Wall barley
Wild barley

Hordeum secalinum
Meadow barley

Hordeum vulgare Barley
Common barley
Six-rowed barley
Nepal barley

Hutchinsea alpina Chamois grass

Juncus acutiflorus
Sharp-flowered rush

Juncus acutus Sharp rush
Sharp sea rush

Juncus alpinoarticulatus
Alpine rush

Juncus articulatus Jointed rush

Juncus ballicus Baltic rush

Juncus biglumis Two-flowered rush

Juncus bufonius Toad rush

Juncus bulbosus

Juncus bulbosus	Bulbous rush
Juncus capitatus	Capitate rush
	Dwarf rush
Juncus castaneus	Chestnut rush
Juncus communis	Common rush
Juncus compressus	
	Round-fruited rush
Juncus effusus	Japanese-mat rush
	Soft rush
Juncus filiformis	Thread rush
Juncus gerardii	Salt mud rush
Juncus inflexus	Hard rush
Juncus lesuerii	Salt rush
Juncus maritimus	Sea rush
Juncus mutabilis	Pigmy rush
Juncus squarrosus	Heath rush
Juncus subnodulosus	
	Blunt-flowered rush
Juncus subuliflorus	Common rush
Juncus tenuis	Slender rush
Juncus trifidus	Three-leaved rush
Juncus triglumis	
	Three-flowered rush
Kobresia simpliciuscula	
	False sedge
Koeleria cristata	Crested hair grass
Koeleria gracilis	Crested hair grass
Koeleria macrantha	
	Crested hair grass
Koeleria vallesiana	Somerset grass
Lagurus ovatus	Hare's tai
	Hare's tail grass
	Rabbit's tail grass
Lamarckia aurea	Goldentop
Lathyrus grandiflorus	
	Everlasting pea
	Two-flowered pea
Lathyrus hirsutus	Caley pea
	Rough pea
	Singletary pea
	Wild winter pea
	Winter pea
Lathyrus japonicus	Beach pea
	Heath pea
	Seaside pea
Lathyrus latifolius	Everlasting pea
	Perennial pea
Lathyrus littoralis	Beach pea
Lathyrus pratensis	
	Meadow vetchling
	Yellow vetchling
Lathyrus splendens	
	Pride of California
Lathyrus sylvestris	Everlasting pea
	Flat pea
	Perennial pea
Lathyrus tuberosus	Dutch mice
	Earth nut pea
	Tuberous vetch
Lathyrus vernus	Spring vetch

Leersia oryzoides	Cat grass
	Cut grass
	Rice
Lolium multiflorum	
	Australian rye grass
	Italian rye grass
Lolium perenne	English rye grass
	Lyme grass
	Perennial rye grass
	Rye grass
	Strand wheat
	Terrell grass
Lolium temulentum	Darnel
Lotus corniculatus	
	Bird's foot trefoil
Luzula arcuata	Curved wood rush
Luzula campestris	
	Common wood rush
	Field wood rush
	Wood rush
Luzula forsteri	
	Forster's wood rush
Luzula multiflora	
	Heath wood rush
	Many-flowered wood rush
Luzula nivea	Snowy wood rush
Luzula pallescens	
	Fenland wood rush
Luzula pilosa	Hairy wood rush
Luzula spicata	Spiked wood rush
Luzula sylvatica	Great wood rush
Medicago hispida	Bur clover
	Toothed bur clover
Medicago lupulina	Black medick
	Hop clover
	Nonesuch
	Trefoil
	Yellow trefoil
Medicago sativa	Alfalfa
	Lucerne
Melica altissima	
	Siberian melic grass
Melica ciliata	Eyelash pearl grass
	Silky melic grass
	Silky spike melica
Melica nutans	Mountain melick
	Nodding melick
Melica uniflora	Wood melick
Melinis minutiflora	Molasses grass
Mibora minima	Early sand grass
	Sand bent
Milium effusum	Wood millet
Milium effusum aureum	
	Bowles' golden grass
Miscanthus nepalensis	
	Himalaya fairy grass
	Nepal silver grass
Miscanthus sacchariflorus	
	Amur silver grass
Miscanthus sinensis	
	Chinese silver grass
	Eulalia

Phylloastachys flexuosa

Miscanthus sinensis zebrinus
Tiger grass
Zebra grass

Molinia caerulea — Blue bent
Flying bent
Indian grass
Moor grass
Purple moor grass

Nardus stricta — Mat grass
Moor mat grass

Neyraudia reynaudiana
Burma reed

Onobrychis sativa — Sainfoin

Onobrychis viciifolia — Esparcet
Holy clover
Sainfoin

Oplismenus hirtellus — Basket grass

Ornithopus perpusillus
Common bird's foot

Oryza sativa — Rice

Oryzopsis hymenoides
Indian millet
Indian rice
Silk grass

Oryzopsis miliacea — Smilo grass

Panicum antidotale
Blue panic grass
Giant panic grass

Panicum bulbosum
Bulbous panic grass

Panicum capillare — Old-witch grass
Witchgrass

Panicum maximum — Guinea grass

Panicum miliaceum — Broomcorn
Broomcorn millet
Common millet
Hog millet
Millet
Proso

Panicum obtusum — Vine mesquite

Panicum purpurascens — Para grass

Panicum ramosum
Browntop millet

Panicum texanum — Colorado grass

Panicum virgatum — Switch grass

Parapholis incurva
Curved sea hard grass
Sickle grass

Parapholis strigosa — Hard grass
Sea hard grass

Paspalum dilitatum — Dallis grass

Paspalum malacophyllum
Ribbed paspalum

Paspalum notatum — Bahia grass

Paspalum racemosum
Peruvian paspalum

Paspalum urvillei — Vasey grass

Pennisetum alopecuroides
Chinese fountain grass
Chinese pennisetum

Pennisetum americanum
African millet
Indian millet
Pearl millet

Pennisetum ciliare — Buffel grass

Pennisetum glaucum — Cuscus
Pearl millet

Pennisetum latifolium
Uruguay pennisetum

Pennisetum purpureum
Elephant grass
Napier grass

Pennisetum setaceum
African fountain grass
Fountain grass

Pennisetum villosum
Abyssinian feathertop
Feathertop

Phalaris arundinacea
Reed canary grass

Phalaris arundinacea picta
Gardener's garters
Ribbon grass

Phalaris canariensis — Birdseed grass
Canary grass

Phleum alpinum — Alpine cat's tail
Alpine timothy

Phleum arenarium — Sand cat's tail

Phleum bertolonii — Lesser cat's tail

Phleum commutatum
Alpine cat's tail
Alpine timothy

Phleum nodosum — Small cat's tail

Phleum phleoides
Purple-stemmed cat's tail

Phleum pratense — Cat's tail grass
Common timothy
Herd's grass
Meadow cat's tail
Mountain timothy
Timothy grass

Phragmites australis — Carrizo
Common reed

Phragmites communis
Common reed

Phyllostachys aurea
Fishpole bamboo
Golden bamboo

Phyllostachys aureosulcata
Forage bamboo
Golden groove bamboo
Stake bamboo
Yellow groove bamboo

Phyllostachys bambusoides
Giant timber bamboo
Hardy timber bamboo
Japanese timber bamboo
Madake
Timber bamboo

Phyllostachys dulcis
Sweetshoot bamboo

Phylloastachys flexuosa
Zigzag bamboo

BOTANICAL NAMES

Phyllostachys heterocycla

Phyllostachys heterocycla
    Tortoiseshell bamboo
Phyllostachys meyeri
    Meyer's bamboo
Phyllostachys nigra   Black bamboo
Phyllostachys nigra henon
    Henon bamboo
Phyllostachys pubescens
    Moso bamboo
Poa alpina   Alpine meadow grass
Poa ampla   Big bluegrass
Poa angustifolia
    Narrow-leaved meadow grass
Poa annua   Annual bluegrass
    Annual meadow grass
    Dwarf meadow grass
    Low spear grass
    Six weeks grass
Poa arachnifera   Texas bluegrass
Poa balfouri
    Balfour's meadow grass
Poa bulbosa   Bulbous bluegrass
    Bulbous meadow grass
Poa chaixii
    Broad-leaved meadow grass
Poa compressa   Canadian bluegrass
    Flat-stalked meadow grass
    Wire grass
Poa flexuosa   Wavy meadow grass
Poa glauca   Glaucous meadow grass
Poa infirma   Early meadow grass
    Scilly Isles meadow grass
Poa nemoralis   Wood bluegrass
    Woodland meadow grass
    Wood meadow grass
Poa palustris   Marsh meadow grass
    Swamp meadow grass
Poa pratensis
    Common meadow grass
    June grass
    Kentucky bluegrass
    Meadow grass
    Smooth meadow grass
    Smooth-stalked meadow grass
    Spear grass
Poa sandbergii
    Sandberg's bluegrass
Poa subcaerulea
    Spreading meadow grass
Poa trivialis   Rough bluegrass
    Rough meadow grass
    Rough-stalked bluegrass
    Rough-stalked meadow grass
Polypogon monspeliensis
    Annual beard grass
    Rabbit-foot grass
    Rabbit's foot
Polypogon viridis   Water bent
Poterium sanguisorba
    Lesser burnet

Pseudosasa japonica
    Arrow bamboo
    Hardy bamboo
    Metake
Puccinellia distans
    Reflexed salt-marsh grass
Puccinellia fasciculata
    Borrer's salt-marsh grass
    Tufted salt-marsh grass
Puccinellia maritima
    Common salt-marsh grass
Puccinellia rupestris
    Stiff salt-marsh grass
Rhynchelytrum repens   Natal grass
    Ruby grass
Rhynchospora alba
    White beak sedge
Rhynchospora fusca
    Brown beak sedge
Saccharum officinarum   Sugarcane
Saccharum sinense
    Chinese sweet cane
Sasa veitchii   Kuma bamboo grass
Schizachyrium scoparium
    Bluestem
    Broom
    Broom beard grass
    Bunch grass
    Little bluestem
    Prairie beard grass
    Wire grass
Schoenus ferrugineus
    Brown bog rush
Schoenus nigricans   Black bog rush
Scirpus americanus
    Jersey club rush
Scirpus cespitosus   Deer grass
Scirpus fluitans   Floating mud rush
Scirpus holoschoenus
    Round-headed club rush
Scirpus lacustris   Club rush
    Common bulrush
Scirpus maritimus   Sea club rush
Scirpus setaceus   Bristle club rush
    Bristle scirpus
Scirpus sylvaticus   Wood club rush
Scirpus tabernaemontani
    Glaucous club rush
Scirpus triquetrus
    Triangular club rush
Scolochloa festucacea   Swamp grass
Secale cereale   Common rye
Semiarundinaria fastuosa
    Narihira bamboo
Sesleria albicans   Blue moor grass
Sesleria caerulea   Blue moor grass
Sesleria heufleriana
    Balkan bluegrass
Setaria glauca
    Glaucous bristle grass
    Yellow bristle grass

Trifolium pratense praecox

Setaria italica	Bengal grass
	Foxtail bristle grass
	Foxtail grass
	Foxtail millet
	Hungarian grass
	Italian millet
	Japanese millet
Setaria macrostachya	
	Plains bristle grass
Setaria palmifolia	Palm grass
Setaria poiretiana	
	Poiret's bristle grass
Setaria viridis	Bristle grass
	Green bristle grass
Sieglingia decumbens	
	Heath grass
	Mountain heath grass
Sorghastrum avenaceum	
	Indian grass
	Wood grass
Sorghum album	Johnson grass
Sorghum bicolor	Sorghum
Sorghum halepense	Aleppo grass
	Grass sorghum
	Great millet
	Egyptian millet
	Johnson grass
	Means grass
Sorghum saccharatum	
	Chinese sugar maple
Sorghum sudanese	Sudan grass
Sorghum vulgare	Broomcorn
	Sorghum
Sorghum vulgare caffrorum	
	Hegari
	Kaffir corn
Sorghum vulgare caudatum	
	Feterita
Sorghum vulgare cernuum	
	White durra
Sorghum vulgare durra	
	Brown durra
	Durra
Sorghum vulgare roxburghii	
	Shallu
Sorghum vulgare saccharatum	
	Sorgo
	Sugar sorghum
	Sweet sorghum
Sparganium angustifolium	
	Floating bur reed
Sparganium emersum	
	Unbranched bur reed
Sparganium erectum	
	Branched bur reed
Sparganium minimum	
	Small bur reed
Spartina alterniflora	
	American cord grass
	Multi-spiked cord grass
	Smooth cord grass
Spartina anglica	
	Common cord grass

Spartina maritima	Marsh grass
	Shore-line cord grass
Spartina pectinata	Cord grass
	Freshwater cord grass
	Prairie cord grass
	Slough grass
Spartina pectinata aureo-marginata	
	Variegated cord grass
Spartina townsendii	Rice grass
	Townsend's cord grass
Sporobolus cryptandrus	
	Sand dropseed
Stenotaphrum secundatum	
	Buffalo grass
	Saint Augustine's grass
	Shore grass
Stipa arundinacea	Pheasant grass
	Pheasant-tail grass
Stipa comata	Needle-and-thread
Stipa elegantissima	
	Australian feather grass
Stipa gigantea	Golden oats
Stipa leucotricha	Texas needlegrass
	Texas winter grass
Stipa pennata	
	Common feather grass
	European feather grass
	Feather grass
Stipa pulcherrima	
	Golden feather grass
Stipa splendens	Chee grass
Stipa tenacissima	Esparto grass
Stipa viridula	Feather bunchgrass
	Green needlegrass
Thamnocalamus spathaceus	
	Muriel bamboo
Tricholaena rosea	Natal grass
	Ruby grass
Trifolium agrarium	Hop clover
	Yellow clover
Trifolium alpinum	Alpine clover
Trifolium ambiguum	Kura clover
Trifolium campestre	
	Large hop clover
	Low clover
Trifolium dubium	
	Yellow suckling clover
Trifolium fragiferum	
	Strawberry clover
	Strawberry-headed clover
Trifolium hybridum	Alsike clover
Trifolium incarnatum	
	Crimson clover
	Italian clover
	Trifolium
Trifolium pannonicum	
	Hungarian clover
Trifolium pratense	Red clover
Trifolium pratense praecox	
	Broad red clover

Trifolium pratense serotinum
Late-flowering red clover
Single-cut cow grass

Trifolium pratense spontaneum
Wild red clover

Trifolium procumbens
Cow hop clover
Irish shamrock
Shamrock
Small hop clover
Yellow clover
Yellow suckling clover

Trifolium repens
Dutch white clover
Shamrock
White clover
White Dutch clover

Trifolium resupinatum
Persian clover
Reversed clover

Trifolium subterraneum   Subclover
Subterranean clover

Triglochia palustris Sea arrow grass

Trisetum flavescens
Golden oat grass
Yellow oat grass

Triticum aestivum   Bread wheat
Common wheat

Triticum compactum   Club wheat
Dwarf wheat
Hedgehog wheat

Triticum dicoccon
Double grain spelt
Double grain wheat
Emmer
German wheat
Rice wheat
Starch wheat

Triticum durum   Durum wheat
Hard wheat

Triticum monococcum   Einkorn
One-grained wheat
Single grain wheat

Triticum polonicum   Giant rye
Polish wheat

Triticum turgidum   Alaska wheat
English wheat
Mediterranean wheat
Poulard wheat
River wheat

Typha angustifolia
Narrow-leaved cat-tail
Narrow-leaved reedmace
Small bulrush
Soft flag

Typha elephanta   Elephant grass

Typha latifolia   Bulrush
Common cat-tail
Cossack asparagus
Nail-rod
Reedmace

Uniola paniculata   Sea oats
Spike grass

Vallisneria americana
Water celery
Wild celery

Vallisneria spiralis   Eel grass

Vetiveria zizanioides   Khas-khas
Khus-khus
Vetiver

Vicia angustifolia   Common vetch

Vicia benghalensis   Purple vetch

Vicia cracca   Bird vetch
Canada pea
Cow vetch
Tufted vetch

Vicia dasycarpa   Woolly-pod vetch
Woolly vetch

Vicia ervilia   Bitter vetch
Ervil

Vicia faba   Broad bean
English bean
European bean
Field bean
Horse bean
Tick bean
Windsor bean

Vicia gigantea   Sitka vetch

Vicia sativa   Common vetch
Spring vetch
Tare

Vicia villosa   Hairy vetch
Large Russian vetch
Winter vetch

Vulpia ambigua   Bearded fescue

Vulpia bromoides
Squirrel-tail fescue

Vulpia fasciculata   Dune fescue

Vulpia membranacea   Dune fescue

Vulpia myuros   Rat's tail fescue

Vulpia unilateralis Mat grass fescue

Xerophyllum tenax   Bear grass
Elk grass
Fire lily
Indian basket grass
Squaw grass

Zea mays   Corn silk
Corn
Indian corn
Maize

Zea mexicana   Teosinte

Zizania aquatica   Annual wild rice
Indian rice
Water oats

Zoysia japonica Japanese lawn grass
Korean grass
Korean lawn grass
Korean velvet grass

Zoysia matrella   Flawn
Japanese carpet grass
Manila grass
Zoysia grass

Zoysia tenuifolia   Korean grass
Korean lawn grass
Korean velvet grass
Mascarene grass

HERBS

Horticulturally, not botanically, the word 'herb' covers a range of plants used in cooking for flavouring and seasoning, as garnishes, and as domestic remedies. They are also widely used in orthodox and homeopathic medicines. The following list includes many subjects that are extremely poisonous, and drastic reactions, even occasional deaths are not unknown from the ignorant use of them. Experimentation by the general public cannot be too strongly condemned. In addition to the common and medicinal herbs, a few herbal trees have been included.

English lavender –
Lavandula angustifolia

Achillea decolorans

Achillea decolorans	English mace
Achillea millefolium	Devil's nettle
	Milfoil
	Nosebleed
	Old man's pepper
	Sneezewort
	Staunchweed
	Toothache weed
	Yarrow
Acinos arvensis	Basil thyme
Aconitum napellus	Aconite
	Blue rocket
	Friar's cap
	Helmet flower
	Monkshood
Acorus calamus	Calamus
	Sweet flag
Actaea spicata	Baneberry
	Bugbane
	Herb christopher
	Toadroot
Adonis vernalis	Adonis
Aesculus hippocastanum	
	Buckeye
	Horse chestnut
Aethusa cynapium	Dog poison
	Fool's parsley
	Lesser hemlock
Agastache anethiodora	
	Anise hyssop
	Fennel hyssop
Agastache foeniculum	
	Anise hyssop
	Fennel hyssop
Agastache rugosa	Korean mint
Agrimonia eupatoria	Agrimony
	Church steeples
	Cocklebur
	Liverwort
	Sticklewort
	Tall agrimony
Ajuga reptans	Bugle
	Bugleweed
Alchemilla vulgaris	Bear's foot
	Lady's mantle
	Lion's foot
	Nine hooks
Alchemilla xanthochlora	
	Bear's foot
	Lady's mantle
	Lion's foot
	Nine hooks
Alliaria petiolata	Garlic mustard
	Jack-by-the-hedge
Allium ampeloprasum	
	Elephant garlic
	Great-headed garlic
Allium ascalonicum	Shallot
Allium cepa aggregatum	
	Egyptian onion
Allium cepa proliferum	
	Tree onion
Allium cepa viviparum	
	Egyptian onion

Allium fistulosum	Welsh onion
Allium perutile	Everlasting onion
Allium porrum	Leek
Allium sativum	Garlic
	Poor man's treacle
Allium schoenoprasum	Chive
Allium schoenoprasum sibiricum	
	Giant chive
Allium tuberosum	Garlic chive
	Oriental chive
Allium ursinum	London lily
	Ramsons
	Stink bombs
	Stinking lily
	Stinking nanny
	Wild garlic
	Wood garlic
Aloe barbadensis	Aloe
	Barbados aloe
	Bitter aloe
	True aloe
Aloe vera	Aloe
	Barbados aloe
	True aloe
Aloe vulgaris	Aloe
	Barbados aloe
	True aloe
Aloysia citriodora	Lemon verbena
Aloysia triphylla	Herb louisa
	Lemon verbena
Althaea officinalis	Althea
	Marshmallow
Amaracus dictamnus	
	Crete dittany
	Hop marjoram
Anchusa officinalis	Alkanet
	Bugloss
Anchusa sempervirens	Anchusa
Anethum graveolens	Dill
Angelica archangelica	Angelica
	Garden angelica
Anisum vulgare	Anise
	Aniseed
Anthemis nobilis	Bowman
	Chamomile
	Common chamomile
	Manzanilla
	Maythen
	Roman chamomile
Anthriscus cerefolium	Chervil
Apium graveolens	Smallage
	Wild celery
Aquilegia vulgaris	Columbine
	Culverwort
	European crowfoot
	Garden columbine
	Gran's bonnet
Arctium lappa	Beggar's buttons
	Burdock
	Clot-bur
	Cockle-bur
	Fox's clote
	Gypsy's rhubarb
	Love leaves

Capsicum frutescens

Armoracia rusticana	
	Horseradish
	Mountain radish
	Red cole
Arnica montana	Arnica
	Leopard's bane
	Mountain tobacco
Artemisia abrotanum	Artemisia
	Lad's love
	Old man
	Southernwood
Artemisia absinthium	Absinthe
	Artemisia
	Common wormwood
	Green ginger
	Old woman
	Wormwood
Artemisia annua	
	Sweet wormwood
Artemisia arbuscula	
	Low sagebrush
Artemisia californica	
	California sagebrush
Artemisia dranunculoides	
	False tarragon
	Russian tarragon
Artemisia dracunculus	Artemisia
	Estragon
	False tarragon
	French tarragon
	Little dragon
	Russian tarragon
	Tarragon
Artemisia filifolia	Sand sage
Artemisia lactiflora	
	White mugwort
Artemisia laxa	Alpine wormwood
Artemisia ludoviciana	
	Cudweed
	Western mugwort
	White sage
Artemisia maritima	Wormseed
Artemisia pontica	Old warrior
Artemisia stellerana	
	Beach wormwood
	Dusty miller
	Old woman
Artemisia tridentata	
	Basin sagebrush
	Common sagebrush
	Sagebrush
Artemisia vulgaris	Felon herb
	Mugwort
	Saint John's plant
Arthrocnemum perenne	
	Marsh samphire
Asclepias curassavica	
	Blood flower
Asperula odorata	Sweet woodruff
	Woodrova
	Woodruff
	Wuderove

Atriplex hortensis	French spinach
	Orach
	Orache
	Sea purslane
Atriplex hortensis rubra	
	Red orache
Atropa belladonna	Belladonna
	Deadly nightshade
	Devil's cherries
Ballota nigra	Black horehound
Balsamita major	Alecost
	Mint geranium
Balsamita major tomentosum	Camphor plant
Bellis perennis	Bruisewort
	Daisy
	English daisy
Betula pendula	Silver birch
Betula verrucosa	Silver birch
Borago officinalis	Bee bread
	Borage
	Burrage
	Cool tankard
	Herb of gladness
	Star flower
	Tailwort
Brassica alba	White mustard
	Yellow mustard
Brassica hirta	White mustard
Brassica juncea	Brown mustard
Brassica nigra	Black mustard
Bryonia dioica	English mandrake
	Ladies' seal
	Tetterbury
	White bryony
	Wild vine
Buddleia salviifolia	
	South African wood sage
Calamintha acinos	Basil thyme
	Calamint
	Mother of thyme
	Mountain balm
Calamintha grandiflora	Calamint
Calendula officinalis	Calendula
	Golds
	Marigold
	Marygold
	Mary gowles
	Pot marigold
	Ruddes
Calluna vulgaris	Heather
Caltha palustris	Kingcup
	Marsh marigold
Calycanthus floridus	Allspice
	Carolina allspice
	Jamaica pepper
	Pineapple shrub
	Strawberry shrub
Cannabis sativa	Bhang
	Marihuana
Capsicum annuum	Pepper
Capsicum frutescens	Pimento

BOTANICAL NAMES

Cardamine pratensis

Cardamine pratensis   Bitter cress
      Cuckoo flower
      Lady's smock

Carthamus tinctorus
      Bastard saffron
      False saffron
      Safflower
      Saffron thistle

Carum carvi   Caraway

Carum petroselinum tuberosum   Hamburg parsley

Cedronella canariensis
      Balm of Gilead

Cedronella triphylla
      Balm of Gilead

Centaurium minus   Centaury
      Christ's ladder
      Red centaury

Centaurium umbellatum
      Centaury

Ceratonia siligua   Carob

Cetraria islandica   Iceland moss

Chaerophyllum temulentum
      Chervil

Chamaemelum nobile   Chamomile

Chamomilla recutira
      German chamomile
      Scented mayweed

Chenopodium album   Fat hen

Chenopodium bonus-henricus
      Good King Henry
      Smearwort

Chenopodium botrys   Ambrosia

Chenopodium vulvaria
      Stinking goosefoot
      Stinking motherwort
      Wild arrach

Chrysanthemum balsamita
      Alecost
      Alespice
      Balsam herb
      Balsamita
      Bible leaf
      Costmary
      Mint geranium

Chrysanthemum parthenium
      Featherfew
      Featherfoil
      Feverfew
      Flirtwort

Chrysanthemum vulgare   Alecost
      Bitter buttons
      Tansy

Cichorium intybus   Chicory
      Succory

Cinnamomum camphora
      Camphor tree

Cinnamomum zelanicum
      Cinnamon

Claytonia perfoliata
      Miner's lettuce
      Winter purslane

Cnicus benedictus   Blessed thistle

Cochlearia armoracia
      Horseradish
      Mountain radish
      Red cole

Colchicum autumnale
      Autumn crocus

Commiphora opobalsamum
      Balm of Gilead

Conium maculatum
      Beaver poison
      Hemlock
      Kecksies
      Kex
      Poison hemlock
      Poison parsley
      Spotted hemlock

Convalleria majalis
      Ladder to heaven
      Lily-of-the-valley
      May lily
      Our Lady's tears

Coriandrum sativum
      Chinese parsley
      Coriander

Costus speciosus   Cape ginger
      Crape ginger
      Grape ginger

Crataegus oxyacantha
      Bread-and-cheese tree
      Hawthorn
      Ladies' meat
      May
      Quickthorn

Crithmum maritimum   Samphire
      Sea fennel

Crocus sativus   Saffron

Cuminum cyminum   Cumin

Cuminum odorum   Cumin

Curcuma longa   Turmeric

Cynara cardunculus   Cardoon
      Wild artichoke

Cytisus scoparius   Broom
      Cat's peas
      Golden chair
      Lady's slipper
      Pixie's slipper

Datura stramonium   Jimson weed
      Thornapple

Delphinium ajacis   Knight's spur
      Lark's heel
      Larkspur

Dianthus caryophyllus   Gillyflower
      July flower
      Pink

Dictamnus albus   Burning bush
      False dittany
      Fraxinella
      Gas plant
      White dittany

Dictamnus fraxinella   Burning bush
      False dittany
      Fraxinella
      Gas plant
      White dittany

Digitalis lutea	Straw foxglove
Digitalis purpurea	Bloody fingers
	Dead man's bells
	Fairy thimbles
	Foxglove
	Witches' gloves
Digitalis thrapsi	
	Dwarf purple foxglove
Dipsacus fullonum	Teasel
Dipsacus sylvestris	Teasel
Dipterix odorata	Tonka bean
	Tonquin bean
Drosera rotundifolia	Sundew
Echallium elaterium	
	Squirting cucumber
Echinacea angustifolia	
	Coneflower
	Echinacea
Echium vulgare	Viper's bugloss
Eonymus europaeus	Burning bush
	Fusoria
	Indian arrowroot
	Skewerwood
	Spindle tree
	Wahoo
Equisetum arvense	Field horsetail
Eruca vesicaria	Arugula
	Salad rocket
Eryngium maritimum	Sea holly
Erythraea centaurium	Centaury
Eucalyptus globulus	Blue gum
	Tasmanian blue gum
Eucalyptus gunnii	Gum tree
Eugenia aromatica	Cloves
Eupatorium cannabinum	
	Hemp agrimony
Eupatorium purpureum	
	Gravelroot
	Trumpet weed
Euphrasia officinalis	Eyebright
Ferula communis	Giant fennel
Ferula foetida	Asafetida
Ferula gabaniflua	Galbanum
Ferula suaveolens	Sambal
Ferula sumbul	Sumbul
Filipendula hexapetala	Dropwort
	Meadowsweet
Filipendula ulmaria	Dropwort
	Meadowsweet
Filipendula vulgaris	Dropwort
	Meadowsweet
Foeniculum dulce	Finocchio
	Florence fennel
Foeniculum officinale	Fenkel
	Fennel
Foeniculum vulgare	Fenkel
	Fennel
Foeniculum vulgare azoricum	
	Anise
	Finocchio
	Florence fennel

Fragaria vesca	Alpine strawberry
	Wild strawberry
Frangula alnus	Alder buckthorn
	Black dogwood
Galega officinalis	Goat's rue
	Italian fitch
Galium odoratum	Sweet woodruff
	Woodrova
	Woodruff
	Wuderove
Galium verum	Cheese rennet
	Curdwort
	Lady's bedstraw
	Maid's hair
	Our Lady's bedstraw
	Peasant's mattress
	Yellow bedstraw
Gaultheria procumbens	
	Checkerberry
	Mountain tea
	Teaberry
	Wintergreen
Gentiana lutea	Gentian
	Yellow gentian
Glechoma hederacea	Alehoof
	Ground ivy
Glycyrrhiza glabra	Black sugar
	Licorice
	Spanish juice
Gratiola officinalis	Hedge hyssop
Hamamelis virginiana	
	Spotted alder
	Winterbloom
	Witch-hazel
Helianthus annuus	Sunflower
Helichrysum angustifolium	
	Curry plant
Helichrysum italicum	Curry plant
Heliotropium arborescens	
	Cherry pie
	Common heliotrope
Heliotropium corymbosum	
	Cherry pie
	Common heliotrope
Heliotropium peruvianum	
	Cherry pie
	Common heliotrope
Helleborus niger	Black hellebore
	Christmas rose
Herniaria glabra	
	Glabrous rupture wort
Hesperis matronalis	Dame's violet
	Rocket
	Roquette
	Sweet rocket
	Vesper flower
Heuchera richardsonii	Alum root
	Coral bells
Humulus lupulus	Hop
Humulus lupulus aureus	
	Golden hop

Hydrastis canadensis	Golden seal
	Orange root
	Yellow puccoon
	Yellow root
Hyoscyamus niger	Cassilata
	Henbane
	Hog's bean
	Stinking nightshade
Hypericum perforatum	
	Saint John's wort
Hyssopus aristatus	Rock hyssop
Hyssopus officinalis	Hyssop
Ilex aquifolium	Holly
Inula helenium	Elecampane
	Horseheal
	Scabwort
	Velvet dock
	Yellow starwort
Ipomoea pes-caprae	
	Beach morning glory
Iris florentina	Orris
	Orris root
Iris germanica	Orris
	Orris root
Iris germanica florentina	Flag iris
	Florentine iris
	Orris
	Orris root
Isatis tinctoria	Dyer's weed
	Woad
Jasminum officinale	
	Common jasmine
	Jasmine
	Tea jasmine
	White jasmine
Juniperus communis	Hack matack
	Horse savin
	Juniper
Laburnum anagyroides	
	Golden chain
	Laburnum
Lamium maculatum	
	Cobbler's bench
	Dead nettle
	Spotted dead nettle
Laurus nobilis	
(Do not confuse with Kalmia latifolia (poisonous))	Bay
	Bay laurel
	Bay leaf
	Roman laurel
	Sweet bay
	True laurel
Lavandula alba	White lavender
Lavandula angustifolia	
	English lavender
	True lavender
Lavandula dentata	
	French lavender
	Fringed lavender
Lavandula officinalis	
	English lavender
	True lavender

Lavandula spica	English lavender
	True lavender
Lavandula stoechas	
	French lavender
	Spanish lavender
Lavandula vera	English lavender
	True lavender
Leonotus leonurus	Wild dagga
Leontodon taraxacum	Blowball
	Common dandelion
	Dandelion
	Peasant's clock
	Priest's crown
	Swine's snout
Leonurus cardiaca	Motherwort
Lepidium sativum	Garden cress
	Pepper cress
Levisticum officinale	Italian lovage
	Lovage
Levisticum scoticum	Scots lovage
Ligusticum paludapifolium	
	Italian lovage
	Lovage
Lilium candidum	Madonna lily
Linaria vulgaris	Buttered haycocks
	Churnstaff
	Eggs and bacon
	Flaxweed
	Fluellin
	Pattens and clogs
	Toadflax
Linum usitatissimum	Flax
Lippia citriodora	Herb louisa
	Lemon verbena
Lippia triphylla	Herb louisa
	Lemon verbena
Lonicera caprifolium	
	Perfoliate honeysuckle
Lonicera periclymenum	
	Goat's leaf
	Honeysuckle
	Woodbine
Loranthus europaeus	
	False mistletoe
Lupinus polyphyllus	Lupin
Majorana hortensis	
	Knotted marjoram
	Sweet marjoram
Malva sylvestris	Common mallow
Mandragora officinarum	
	Devil's apple
	Mandrake
	Satan's apple
Marrubium incanum	
	White horehound
Marrubium vulgare	
	Common horehound
	Hoarhound
	Horehound
	White horehound
Matricaria chamomilla	
	German chamomile
	Scented mayweed
	Wild chamomile

*Ocimum basilicum
lactucafolium*

Matricaria eximia	Featherfew
	Featherfoil
	Feverfew
	Flirtwort
Matricaria recutita	
	German chamomile
Medicago sativa	Buffalo herb
Melilotus officinalis	Melilot
Melissa officinalis	Balm
	Bee balm
	Common balm
	Lemon balm
	Sweet balm
	Sweet bay
Melissa officinalis aurea	
	Golden balm
Melissa officinalis variegata	
	Variegated lemon balm
Melittis melissophyllum	
	Bastard balm
Mentha aquatica	Water mint
Mentha arvensis piperascens	
	Field mint
	Japanese mint
Mentha citrata	Bergamot mint
	Eau-de-cologne Mint
	Orange mint
Mentha crispa	Curly mint
	Spearmint
Mentha × gentilis	Bushy mint
	Ginger mint
	Red mint
	Scotch mint
Mentha longifolia	Horse mint
Mentha × piperita	Brandy mint
	Double mint
	Peppermint
Mentha × piperita citrata	
	Eau-de-cologne mint
Mentha pulegium	
	English pennyroyal
	Lurk-in-the-ditch
	Pennyroyal
	Pudding grass
	Run-by-the-ground
Mentha requienii	Corsican mint
	Crème de menthe plant
	Menthella
Mentha rotundifolia	Apple mint
	Egyptian mint
	Round-leaved mint
	Woolly mint
Mentha spicata	Common mint
	Garden mint
	Lamb mint
	Mackerel mint
	Sage of Bethlehem
	Spearmint
	Spire mint
Mentha suaveolens	Applemint
Mentha suaveolens variegata	
	Pineapple mint
	Variegated applemint

Mentha viridis	Garden mint
	Lamb mint
	Mackerel mint
	Sage of Bethlehem
	Spearmint
	Spire mint
Monarda citriodora	
	Lemon bergamot
Monarda didyma	Balm
	Bee balm
	Bergamot
	Indian plume
	Oswega tea
	Scarlet monarda
Monarda fistulosa	Wild bergamot
Monarda punctata	Horse mint
Morus alba	White mulberry
Morus nigra	Black mulberry
Muscari botryoides	Grape hyacinth
	Starch hyacinth
Myosotis arvensis	Forget-me-not
Myosotis sylvatica	Forget-me-not
Myristica fragrans	Mace
	Nutmeg
Myrrhis odorata	Anise
	Anise fern
	British myrrh
	Great chervil
	Shepherd's needle
	Smooth cicely
	Sweet bracken
	Sweet chervil
	Sweet cicely
Myrtus communis	Myrtle
Myrtus communis tarentina	
	Myrtle
	Tarentum myrtle
Nasturtium officinale	Watercress
Nepeta cataria	Catmint
	Catnep
	Catnip
Nepeta cataria citriodorum	
	Lemon catmint
Nepeta hederacea	Alehoof
	Creeping charlie
	Field balm
	Gill-go-over-the-ground
	Ground ivy
	Haymaids
	Lizzy-run-up-the-hedge
Nepeta mussini	Border catmint
Ocimum americanum	
	Lemon balm
	Lemon basil
Ocimum basilicum	Basil
	Sweet basil
Ocimum basilicum aurauascens	
	Purple basil
Ocimum basilicum lactucafolium	
	Lettuce-leaved basil
	Monster-leaved basil

Ocimum basilicum
purpurescens

Ocimum basilicum	
purpurescens	Dark opal basil
Ocimum citriodorum	Lemon basil
Ocimum minimum	Bush basil
	Fine-leaved basil
	Greek basil
Ocimum sanctum	Holy basil
Oenothera biennis	
	Evening primrose
	King's cure-all
	Moonflower
Onobrychis viciifolia	Sainfoin
Ononis spinosa	Restharrow
Onopordon acanthium	
	Cotton thistle
Origanum dictamnus	
	Crete dittany
	Hop marjoram
Origanum heracleoticum	
	Pot marjoram
	Winter marjoram
	Wintersweet marjoram
Origanum majorana	
	Annual marjoram
	Knotted marjoram
	Sweet marjoram
Origanum onites	Pot marjoram
Origanum vulgare	Marjoram
	Organy
	Origano
	Pot marjoram
	Wild marjoram
Osmanthus fragrans	Sweet olive
Panax ginseng	Ginseng
Papaver somniferum	
	Opium poppy
Pastinaca sativa	Wild parsnip
Pelargonium × fragrans	
	Nutmeg-scented geranium
Pelargonium graveolens	
	Scented geranium
Pelargonium tomentosum	
	Peppermint-scented geranium
Pentaglottis sempervirens	
	Anchusa
Perilla frutescens crispa	
	Beefsteak plant
	Purple perilla
	Summer coleus
Petroselinum crispum	
	Curly parsley
	Parsley
	Petersylinge
Petroselinum crispum fusiformis	
	Hamburg parsley
	Turnip-rooted parsley
Petroselinum neapolitanum	
	French parsley
	Italian parsley
Petroselinum tuberosum	
	Hamburg parsley
Peucedanum graveolens	Dill

Philadelphus coronarius	
	Mock orange
Phytolacca americana	Pokeweed
Pimpinella anisum	Anise
	Aniseed
Pimpinella major	
	Greater burnet saxifrage
Pimpinella saxifraga	
	Burnet saxifrage
Polemoneum coeruleum	Charity
	Greek valerian
	Jacob's ladder
Polygonum bistorta	Bistort
Polygonum hydropiper	
	Water pepper
Populus balsamifera	
	Balsam poplar
Portulaca oleracea	Purslane
	Summer purslane
Potentilla anserina	Goose tansy
	Silver weed
Potentilla erecta	
	Common tormentil
Potentilla tormentilla	
	Common tormentil
Poterium sanguisorba	Burnet
	Salad burnet
Primula officinalis	Cowslip
Primula parthenium	Featherfew
	Featherfoil
	Feverfew
	Flirtwort
Primula veris	Cowslip
Primula vulgaris	Primrose
Prunus dulcis	Almond
Pulmonaria officinalis	
	Jerusalem cowslip
	Lungwort
Quercus petraea	Durmast oak
Quercus rober	Common oak
	Oak
Quercus sessilis	Durmast oak
Ranunculus acris	
	Bachelor's buttons
	Buttercup
	Gold knots
Reseda alba	White mignonette
Reseda lutea	Wild mignonette
Reseda odorata	Mignonette
Rhamnus catharticus	Buckthorn
	Common buckthorn
	Hartshorn
	Highway thorn
	Ram's horn
	Ram's thorn
Rheum palmatum	Rhubarb
Rheum rhabarbarum	
	Garden rhubarb
Rorippa nasturtium-aquaticum	
	Watercress
Rosa centifolia muscosa	Moss rose
Rosa eglanteria	Sweetbriar

Sesamum indicum

Rosa gallica officinalis
    Apothecary's rose

Rosmarinus lavandulaceus
    Compass weed
    Polar plant
    Rosemary

Rosmarinus officinalis
    Compass weed
    Polar plant
    Rosemary

Rubia tinctoria   Madder root

Rubus australis   Bush lawyer

Rubus fruticosus   Bramble

Rumex acetosa   Cuckoo's meat
    Cuckoo sorrow
    Garden sorrel
    Green sauce
    Sorrel
    Sour sabs

Rumex crispus   Curled dock

Rumex obtusifolius
    Broad-leaved dock
    Butter dock
    Dock

Rumex scutatus   Buckler-leaf sorrel
    French sorrel

Ruta graveolens   Herb of grace
    Rue

Sabatia angularis
    American centaury
    Bitter broom
    Bitter clover
    Wild succory

Salix alba   European willow
    White willow

Salix nigra   Black American willow
    Pussy willow
    Willow

Salvia apiana   Greasewood
    White sage

Salvia argentea   Silver sage

Salvia aurea   Golden sage

Salvia azurea   Blue sage

Salvia barrelieri   Spanish sage

Salvia elegans   Pineapple sage

Salvia officinalis   Garden sage
    Sage

Salvia officinalis icterina
    Golden sage

Salvia officinalis purpurescens
    Red sage

Salvia officinalis tricolor
    Tricolor sage
    Variegated sage

Salvia purpurescens   Purple sage

Salvia rutilans   Pineapple sage

Salvia sclarea   Clary
    Clary sage
    Clear eye

Sambucus nigra   Bore tree
    Elder
    Pipe tree
    Sweet elder

Sanguisorba minor   Burnet
    Salad burnet

Santolina chamaecyparissus
    Cotton lavender
    French lavender
    Grey santolina
    Lavender cotton
    Santolina

Santolina incana   Cotton lavender
    French lavender
    Grey santolina
    Lavender cotton
    Santolina

Santolina tomentosa
    Cotton lavender
    French lavender
    Grey santolina
    Lavender cotton
    Santolina

Saponaria officinalis   Bouncing bet
    Bruisewort
    Fuller's herb
    Soapwort
    Wild sweet william

Sarothamnus scoparius   Broom
    Cat's peas
    Golden chair
    Lady's slipper
    Pixie's slipper

Satureia (Satureja) hortensis
    Bean herb
    Savory
    Summer savory

Satureia (Satureja) montana
    Bean herb
    Savory
    Winter savory

Satureia (Satureja) repanda
    Creeping savory

Scrophularia marilandica
    Carpenter's square

Scutellaria galericulata
    Helmet flower
    Madweed
    Scullcap
    Skullcap
    Virginian scullcap

Scutellaria laterifolia
    Helmet flower
    Madweed
    Scullcap
    Skullcap
    Virginian scullcap

Selinum tenuifolium
    Himalayan parsley

Sempervivum tectorum
    Hen-and-chickens
    Houseleek
    St. Patrick's cabbage
    Thunder plant

Sesamum indicum   Semsem
    Sesame

Sesamum orientale

Sesamum orientale	Semsem
	Sesame
Sinapis alba	White mustard
	Yellow mustard
Sinapis juncea	Brown mustard
Sinapis nigra	Black mustard
Sium sisarum	Skirret
Smyrnium olusatrum	Alexanders
Solanum dulcamara	Bittersweet
	Bittersweet nightshack
	Felonwood
	Violet bloom
	Woody nightshade
Solanum laciniatum	
	Kangaroo apple
Spiraea filipendula	Dropwort
	Meadowsweet
Stachys lanata	Donkey's ears
	Lamb's ears
	Woolly betony
Stachys officinalis	Betony
	Bishopswort
	Wood betony
	Woundwort
Stachys olympica	Donkey's ears
	Lamb's ears
	Woolly betony
Stellaria media	Chickweed
Succisa pratensis	
	Devil's bit scabious
Swertia chirata	Chiretta
Symphytum grandiflorum	
	Creeping comfrey
Symphytum officinale	Ass's ear
	Blackwort
	Boneset
	Comfrey
	Common comfrey
	Consound
	Healing herb
	Knitbone
Symphytum perigrinum	
	Russian comfrey
Symphytum uplandicum	
	Russian comfrey
Tamus communis	Black bryony
	Blackeye root
Tanacetum balsamita	Alecost
	Alespice
	Costmary
Tanacetum parthenium	
	Bachelor's buttons
	Feverfew
Tanacetum vulgare	Alecost
	Bitter buttons
	Buttons
	Golden buttons
	Tansy
Taraxacum officinale	Blowball
	Common dandelion
	Dandelion
	Peasant's clock
	Priest's crown
	Swine's snout

Taxus baccata	English yew
	Yew
Taxus brevifolia	Western yew
	Yew
Teucrium chamaedrys	Germander
	Wall germander
Teucrium scorodonia	Wood sage
Thymus alba	White thyme
Thymus × citriodorus	
	Lemon thyme
	Silver queen
Thymus × citriodorus aureus	
	Golden-edged thyme
	Golden thyme
Thymus herba-barona	
	Caraway thyme
Thymus mastichina	
	Herb masticke
Thymus mastichina × didi	Didi
Thymus pulegioides	
	Broad-leaved thyme
Thymus serpyllum	
	Creeping thyme
	Lemon thyme
	Wild thyme
Thymus vulgaris	Common thyme
	English thyme
	French thyme
	Garden thyme
	Silver posy
	Thyme
Thymus vulgaris aureus	
	Golden thyme
Tilia cordata	Small-leaved lime
Trifolium incarnatum	
	Crimson clover
	Red clover
	Trefoil
Trigonella foenum-graecum	
	Bird's foot
	Fenugreek
	Greek hayseed
Tropaeolum majus	Indian cress
	Nasturtium
Tussilago farfara	Bullsfoot
	Coltsfoot
	Coughwort
	Foalfoot
	Horsehoof
Ulmaria filipendula	Dropwort
	Meadowsweet
Ulmus fulva	Indian elm
	Moose elm
	Red elm
	Slippery elm
Urtica dioica	Nettle
	Stinging nettle
Vaccinium myrtillus	Bilberry

Valeriana officinalis	All-heal
	Cat's valerian
	Garden heliotrope
	Heal-all
	Phew plant
	Phu
	Setwall
	Valerian
Valerianella locusta	Corn salad
	Lamb's lettuce
Valerianella olitoria	Corn salad
	Lamb's lettuce
Vanilla fragrans	Vanilla
Veratrum album	
	European white hellebore
	White hellebore
Veratrum viride	False hellebore
	Green hellebore
	White hellebore
Verbascum thapsus	Aaron's rod
	Adam's flannel
	Blanket leaf
	Great mullein
	Hag taper
	Jacob's staff
	Mullein
	Our Lady's candle
	Shepherd's club
	Torches
Verbena officinalis	
	European vervain
	Herb of grace
	Herb of the cross
	Pigeon's grass

	Simpler's joy
	Vervain
	Verbena triphylla
	Herb louisa
	Lemon verbena
Veronica officinalis	Speedwell
Vinca major	Greater periwinkle
	Periwinkle
Vinca major alba	
	White periwinkle
Viola odorata	Florist's violet
	Sweet violet
	Violet
Viola odorata alba	White violet
Viola pallida plena	Parma violet
Viola riviana	Dog violet
Viola selkirkii	
	Great spurred violet
Viola tricolor	Call-me-to-you
	Heartsease
	Herb constancy
	Kiss-her-in-the-buttery
	Kit runabout
	Love-in-idleness
	Love-lies-bleeding
	Pink-of-my-john
	Wild pansy
Viscum alba	Birdlime mistletoe
	Lignum crucis
	Mistletoe
Vitex agnus-castus	Monk's pepper
Zingiber cassumunar	Bengal root
Zingiber officinale	Ginger

HOUSE PLANTS

This section deals with those plants normally associated with hotel foyers, civic halls, restaurants, and public functions in addition to those found in private houses, sun rooms, and conservatories. Many subjects have been duplicated elsewhere in this book which, in their natural environment, would grow too large for indoor use but by regular pruning and root restriction can be controlled within acceptable limits. With a few exceptions bulbs and cacti are listed elsewhere under those headings.

Flaming katy –
Kalanchoe blossfeldiana

BOTANICAL NAMES

Acacia armata

Acacia armata	Kangaroo thorn
Acacia baileyana	
	Cootamunda wattle
Acacia cultriformis	Knife acacia
Acacia cyanophylla	
	Blue-leaved wattle
Acacia dealbata	Florist's mimosa
	Silver wattle
Acacia longifolia	
	Sydney golden wattle
Acacia lophantha	Plume albizia
Acacia melanoxylon	
	Blackwood acacia
Acacia podalyriifolia	
	Mount Morgan wattle
	Queensland silver wattle
Acacia pravissima	Oven's wattle
Acacia prominens	Golden rain
	Gosford wattle
Acacia retinodes	Wirilda
Acacia verticillata	Prickly moses
Acalypha godseffiana	
	Lance copperleaf
Acalypha hispida	Chenille plant
	Red-hot cat's tail
Acalypha wilkesiana	Copperleaf
Acca sellowiana	Pineapple guava
Acokanthera spectabilis	
	Wintersweet
Acorus gramineus	Sweet flag
Acrostichum aureum	Leather fern
Acrostichum crinitum	
	Elephant's ear fern
Adenium obesum	Desert rose
Adiantum capillus-veneris	
	Common maidenhair fern
	Maidenhair fern
Adiantum formosum	
	Australian maidenhair fern
Adiantum hispidulum	
	Rose maidenhair fern
Adiantum pubescens	
	Rose maidenhair fern
Adromischus maculatus	
	Calico hearts
Aechmea bromeliifolia	Wax torch
Aechmea fasciata	Silver vase
	Urn plant
Aechmea fulgens	Coralberry
Aerides fieldingii	Fox brush orchid
Aeschynanthus lobbianus	
	Lipstick vine
Aeschynanthus pulcher	
	Royal red bugler
Agathaea coelestis	
	Blue marguerite
Agave americana	American aloe
	Century plant
	Maguey
Agave filifera	Thread agave

Aglaonema modestum	
	Chinese evergreen
Albizia distachya	Plume albizia
Albizia julibrissin	Silk tree
Albizia lophantha	Plume albizia
Allamanda cathartica	
	Golden trumpet
Allium schoenoprasum	Chives
Allophyton mexicanum	
	Mexican foxglove
	Mexican violet
Alocasia macrorrhiza	
	Giant elephant's ear
Alocasia sanderiana	Kris plant
Aloe barbadensis	Burn plant
	Magic plant
	Medicine plant
Aloe ciliaris	Climbing aloe
Aloe striata	Coral aloe
Alonsoa warscewiczii	Mask flower
Aloysia citriodora	Lemon verbena
Alpinia calcarata	Indian ginger
Alpinia magnifica	
	Philippine wax flower
	Torch ginger
Alpinia nutans	Pink porcelain lily
	Shell ginger
Alpinia purpurata	Red ginger
Alpinia sanderae	Variegated ginger
Alpinia speciosa	Pink porcelain lily
	Shell ginger
Alpinia zerumbet	
	Pink porcelain lily
	Shell ginger
Alsophila smithii	Soft tree fern
Alsophila tricolor	Ponga
	Silver tree fern
Alstroemeria pelegrina alba	
	Lily of the incas
Alternanthera bettzickiana	
	Calico plant
Alternanthera ficoidea	Parrot leaf
Amaranthus caudatus	
	Love-lies-bleeding
Amaranthus hybridus	
	Prince's feather
Amaranthus hypochondriacus	
	Prince's feather
Amaryllis belladonna	
	Belladonna lily
Amaryllis formosissima	Aztec lily
	Jacobean lily
Amorphophallus bulbifer	
	Devil's tail
Amorphophallus rivieri	
	Devil's tongue
Ampelopsis brevipendunculata	
	Porcelain berry
Ampelopsis heterophylla	
	Porcelain berry

Babiana stricta

Ananas bracteatus	Red pineapple
	Wild pineapple
Anastatica hierochuntica	
	Resurrection plant
	Rose of Jericho
Anemia adiantifolia	Pine fern
Anemone coronaria	
	Poppy anemone
Anguloa clowesii	Tulip orchid
Anguloa uniflora	Tulip orchid
Anigozanthos flavidus	
	Tall kangaroo paw
Anigozanthos humilis	Catspaw
Anigozanthos manglesii	
Common green kangaroo paw	
	Mangle's kangaroo paw
Anigozanthos preissii	
	Albany catspaw
Anigozanthos rufus	
	Red kangaroo paw
Anigozanthos viridis	
	Green kangaroo paw
Annona cherimolia	Cherimoya
Annona reticulata	Bullock's heart
	Custard apple
Annona squamosa	Sugar apple
	Sweet sop
Anoectochilus regalis	King plant
Ansellia africana	Leopard orchid
Anthericum comosum	
	Spider plant
Anthericum elatum	Spider plant
Anthurium andreanum	Tailflower
Anthurium crystallinum	
	Crystal anthurium
Anthurium scherzerianum	
	Flamingo flower
Anthurium veitchii	
	King anthurium
Anthurium warocqueanum	
	Queen anthurium
Antigonon leptopus	Corallita
	Coral vine
Antirrhinum majus	Snapdragon
Aphelandra squarrosa	Zebra plant
Arachis hypogaea	Groundnut
	Peanut
Arachnoides aristata	
	East Indian holly fern
Aralia japonica	
	False castor-oil plant
Aralia papyrifera	Rice-paper plant
Aralia sieboldii	
	False castor-oil plant
Araucaria angustifolia	
	Candelabra tree
	Parana pine
Araucaria cunninghamii	
	Hoop pine
	Moreton Bay pine

Araucaria excelsa	
	Norfolk Island pine
Araucaria heterophylla	
	Norfolk Island pine
Araujia sericofera	Cruel plant
Arbutus unedo	Strawberry tree
Archontophoenix alexandrae	
	Alexandra palm
	Northern bungalow palm
Archontophoenix cunningh	
amiana	Illawarra palm
	Piccabeen bungalow palm
	Piccabeen palm
Ardisia crispa	Coralberry
	Marlberry
	Spiceberry
Aregelia marmorata	Marble plant
Arisarum proboscideum	
	Mouse plant
Aristolochia elegans	Birth wort
	Calico flower
Arum dracunculus	Dragon plant
Arundinaria amabilis	
	Tonkin bamboo
Arundo donax	Giant reed
Asclepias curassavica	Blood flower
Asparagus asparagoides	Smilax
Asparagus asparagoides	
myrtifolius	Baby smilax
Asparagus falcatus	Sicklethorn
Asparagus medeoloides	Smilax
Asparagus plumosus	
	Asparagus fern
Asparagus setaceus	Asparagus fern
Aspidistra elatior	Cast iron plant
Aspidistra lurida	Cast iron plant
Aspidium falcatum	Holly fern
Asplenium bulbiferum	
	Mother spleenwort
Asplenium daucifolium	
	Mother fern
Asplenium nidus	Bird's nest fern
Asplenium nidus-avis	
	Bird's nest fern
Asplenium platyneuron	
	Ebony spleenwort
Asplenium scolopendrium	
	Hart's tongue fern
Asplenium viviparum	
	Mother fern
Aster bergeriana	Kingfisher daisy
Athyrium esculentum	Paco
	Vegetable fern
Averrhoa bilimbi	Bilimbi
Averrhoa carambola	
	Carambola tree
Azalea obtusa	Japanese azalea
Azolla caroliniana	Mosquito plant
	Water fern
Babiana stricta	Baboon flower

Beaucarnea recurvata
Elephant foot
Pony tail

Begonia bowerii
Miniature eyelash begonia

Begonia × cheimantha
Christmas begonia
Lorraine begonia

Begonia coccinea
Angel's wing begonia

Begonia cubensis
Holly-leaved begonia

Begonia dregei
Miniature maple-leaf begonia

Begonia × erythrophylla
Beefsteak begonia

Begonia × feastii
Beefsteak begonia

Begonia foliosa   Fernleaf begonia

Begonia fuchsioides
Fuchsia begonia

Begonia haageana
Elephant's ear begonia

Begonia heracleifolia   Star begonia

Begonia leptotricha   Woolly bear

Begonia luxurians
Palm-leaf begonia

Begonia masoniana
Iron cross begonia

Begonia metallica
Metal leaf begonia

Begonia rex   Fan begonia
Painted leaf begonia

Begonia scharffii
Elephant's ear begonia

Begonia semperflorens
Wax begonia

Beloperone guttata   Shrimp plant

Berberidopsis corallina   Coral plant

Bertolonia hirsuta   Jewel plant

Bessera elegans   Coral drops

Bignonia callistegioides
Trumpet vine

Bigninia capensis
Cape honeysuckle

Bignonia purpurea   Love charm

Bignonia speciosa   Trumpet vine

Bignonia stans   Yellow bells
Yellow elder

Bignonia venusta   Flame flower
Flame vine
Flaming trumpets

Billbergia nutans   Friendship plant
Queen's tears

Biophytum sensitivum
Sensitive plant

Blechnum patersonii
Strap water fern

Bloomeria crocea   Golden stars

Boehmeria nivea   China grass
Ramie

Bougainvillea glabra   Paper flower

Bouvardia ternifolia
Scarlet trompetilla

Bouvardia triphylla
Scarlet trompetilla

Bowiea volubilis   Climbing onion

Brassaia actinophylla
Octopus tree
Queensland umbrella tree

Breynia nivosa   Snow bush

Brodiaea coccinea   Fire cracker

Brodiaea ida-maia   Fire cracker

Brodiaea laxa   Grass nut
Ithuriel's spear

Brodiaea volubilis   Snake lily

Browallia speciosa   Bush violet

Brunfelsia americana
Lady of the night

Brunfelsia calycina
Yesterday, today and tomorrow

Brunfelsia paucifolia calycina
Yesterday, today and tomorrow

Brunfelsia undulata
White rain tree

Brunfelsia violacea
Lady of the night

Brunsvigia josephinae
Josephine's lily

Caesalpinia gilliesii
Bird of paradise flower

Caesalpinia pulcherrima
Barbados pride

Caladium × hortulanum
Angel's wings

Calathea insignis   Rattlesnake plant

Calathea lancifolia
Rattlesnake plant

Calathea makoyana   Peacock plant

Calathea zebrina   Zebra plant

Calceolaria × herbeohybrida
Pocket book plant
Slipper flower

Calendula officinalis
Common marigold
Pot marigold

Callisia elegans   Wandering jew

Callistemon citrinus
Lemon bottlebrush

Callistemon lanceolatus
Lemon bottlebrush

Callistemon speciosus   Bottlebrush

Calonyction bona-nox
Moonflower

Campanula isophylla
Italian bellflower

Campanula pyramidalis
Chimney bellflower

Campyloneurum phyllitidis
Ribbon fern
Strap fern

Canna × generalis   Indian shot

Citrus limetta

Capsicum annuum	Bell pepper
	Chillies
	Green pepper
	Paprika
	Red pepper
	Sweet pepper
Cardiospermum halicacabum	
	Balloon vine
	Heart-pea
	Heart-seed
Carex morrowii	Japanese sedge
Carica papaya	Papaya
	Pawpaw
Carissa grandiflora	Natal plum
Carpobrotus acinaciformis	
	Hottentot fig
Carpobrotus chilensis	Sea fig
Carpobrotus edulis	Hottentot fig
Caryota urens	Sago palm
	Toddy palm
	Wine palm
Cassia artemisioides	
	Wormwood cassia
Catharanthus roseus	
	Madagascar periwinkle
Cattleya bowringiana	
	Cluster cattleya
Cattleya citrina	Tulip cattleya
Cattleya intermedia	
	Cocktail orchid
Cattleya labiata	Autumn cattleya
Cattleya labiata dowiana	
	Queen cattleya
Cattleya labiata mossiae	
	Easter orchid
	Spring cattleya
Cattleya labiata trianaei	
	Christmas orchid
	Winter cattleya
Cattleya mossiae	Easter orchid
	Spring cattleya
Cattleya trianaei	Christmas orchid
	Winter cattleya
Celosia argentea cristata	
	Cockscomb
Celosia argentea pyramidalis	
	Prince of Wales feathers
Celosia cristata	Cockscomb
Celosia plumosa	
	Prince of Wales feathers
Celosia pyramidalis	
	Prince of Wales feathers
Centaurea cineraria	Dusty miller
Centaurea cyanus	Cornflower
Centaurea gymnocarpa	
	Dusty miller
Centaurea moschata	Sweet sultan
Ceratonia siliqua	Algaroba
	Carob
	Locust
	Saint John's bread
Ceratopteris cornuta	Water fern

Ceratopteris pteridoides	
	Floating fern
	Water fern
Ceratopteris siliquosa	Water fern
Ceratopteris thalictroides	
	Water fern
Ceropegia caffrorum	Lamp flower
Ceropegia sandersonii	
	Fountain flower
	Parachute plant
Ceropegia woodii	
	Chinese lantern plant
	Rosary vine
	String of hearts
Chamaedorea elegans	
	Dwarf mountain palm
	Mexican dwarf palm
	Parlour palm
Chamaedorea erumpens	
	Bamboo palm
Chamaedorea metallica	
	Miniature fishtail palm
Chamaedorea seifrizii	Reed palm
Chamaelaucium uncinatum	
	Geraldton wax flower
Chamaerops humilis	
	Dwarf fan palm
Cheiranthus × allionii	
	Siberian wallflower
Cheiranthus cheiri	
	Common wallflower
	Wallflower
Chlorophytum capense	
	Spider plant
Chlorophytum comosum	
	Spider plant
Chlorophytum elatum	Spider plant
Choisya ternata	
	Mexican orangeblossom
Chrysalidocarpus lutescens	
	Areca palm
	Golden feather palm
Cineraria maritima	Dusty miller
Cissus antarctica	Kangaroo vine
Cissus capensis	Cape grape
Cissus discolor	Rex begonia vine
Cissus rhombifolia	Grape ivy
Cissus sicyoides	Princess vine
Cissus voinierianum	Chestnut vine
× Citrofortunella mitis	
	Calamondin
Citrus aurantifolia	Lime
Citrus aurantium	Seville orange
Citrus aurantium sinense	
	Sweet orange
Citrus bigarardia	Seville orange
Citrus decumanus	Pummelo
	Shaddock
Citrus grandis	Pummelo
	Shaddock
Citrus limetta	Lime

Citrus maxima

Citrus maxima	Pummelo
	Shaddock
Citrus medica	Citron
Citrus mitis	Calamondin
Citrus nobilis	Mandarin
	Satsuma
	Tangerine
Citrus paradisi	Grapefruit
Citrus reticulata	Mandarin
	Satsuma
	Tangerine
Citrus reticulata × fortunella	
	Calamondin
Citrus sinensis	Sweet orange
Clerodendrum fallax	
	Java glorybean
Clerodendrum paniculatum	
	Pagoda flower
Clerodendrum speciosissimum	
	Java glorybean
Clerodendrum thomsonae	
	Bleeding heart vine
	Glory bower
Clerodendrum ugandense	
	Blue glory bower
Clethra arborea	
	Lily-of-the-valley tree
Clianthus dampieri	Damper's pea
	Glory pea
Clianthus formosus	Damper's pea
	Glory pea
Clianthus puniceus	Kaka beak
	Lobster claw
	Parrot's bill
Clivia miniata	Kaffir lily
Clusia grandiflora	Scotch attorney
Clusia rosea	Autograph tree
	Balsam apple
	Fat pork tree
Clytostoma binatum	Love charm
Clytostoma callistegioides	
	Trumpet vine
Clytostoma purpureum	
	Love charm
Cobaea scandens	Cathedral bells
	Cup-and-saucer creeper
Coccoloba uvifera	Seaside grape
Cocos nucifera	Coconut palm
Codiaeum variegatum pictum	
	Croton
	Joseph's coat
Coffea arabica	Arabian coffee
Colchicum autumnale	
	Autumn crocus
	Meadow saffron
Coleus blumei	Flame nettle
	Painted nettle
Colocasia esculenta	Cocoyam
	Dasheen
	Elephant's ear
	Taro
Columnea gloriosa	Goldfish plant

Columnea microphylla	
	Goldfish plant
Commelina benghalensis	
	Indian day flower
Commelina coelestis	Day flower
Convallaria majalis	
	Lily-of-the-valley
Cordyline australis	Cabbage palm
	Cabbage tree
Cordyline terminalis	Ti tree
Coronilla emerus	Scorpion senna
Corynocarpus laevigatus	
	Karaka
	New Zealand laurel
Corypha elata	Gebang
	Philippines sugar plum
Corypha gembanga	Gebang
	Philippines sugar plum
Corypha umbraculifera	
	Talipot palm
Costus igneus	Spiral ginger
Costus speciosus	Crepe ginger
	Malay ginger
Cotyledon maculata	Calico-hearts
Crassula argentea	Jade plant
Crassula barklyi	Rattlesnake tail
Crassula deltoidea	Silver beads
Crassula lycopodioides	
	Rat tail plant
Crassula maculata	Calico hearts
Crassula portulacea	Jade plant
Crassula rhomboidea	Silver beads
Crassula rupestris	Bead vine
	Rosary vine
Crassula teres	Rattlesnake tail
Crinum × powellii	Swamp lily
Crossandra infundibuliformis	
	Firecracker plant
Crossandra undulifolia	
	Firecracker plant
Crotalaria agatiflora	
	Canary bird bush
Crotalaria juncea	Sunn hemp
Cryptanthus acaulis	
	Green earth star
Cryptanthus bromelioides tricolor	Rainbow star
Cunonia capensis	Red alder
Cuphea hyssopifolia	False heather
Cuphea ignea	Cigar flower
Cuphea platycentra	Cigar flower
Curculigo capitulata	Palm grass
Curculigo recurvata	Palm grass
Curcuma domestica	Turmeric
Curcuma longa	Turmeric
Cyanotis kewensis	Teddy bear vine
Cyanotis somaliensis	Pussy ears
Cyathea arborea	
	West Indian tree fern

Cyathea dealbata — Ponga
Silver tree fern

Cyathea medullaris —
Black tree fern
Sago fern

Cyathea smithii — Soft tree fern
Sago fern palm

Cycas circinalis — Fern palm
Sago fern palm

Cycas revoluta — Japanese sago palm

Cycnoches ventricosum —
Swan orchid

Cymbalaria muralis —
Ivy-leaved toadflax
Kenilworth ivy

Cypella plumbea — Blue tiger lily

Cyperus alternifolius —
Umbrella grass
Umbrella plant

Cyperus diffusus — Umbrella plant

Cyperus esculentus — Chufa nut
Tiger nut

Cyperus papyrus —
Egyptian paper reed
Papyrus

Cyphomandra betacea —
Tree tomato

Cyrtanthus mackenii — Ifafa lily

Cyrtomium falcatum — Holly fern

Cyrtostachys lakka —
Sealing wax palm

Cytisus canariensis — Florist's genista

Daedalacanthus nervosum —
Blue sage

Darlingtonia californica —
Californian pitcher plant
Cobra plant

Datura × candida — Angel's trumpet

Davallia canariensis —
Hare's foot fern

Davallia dissecta —
Squirrel's foot fern

Davallia fijiensis — Rabbit's foot fern

Davallia mariesii — Ball fern
Squirrel's foot fern

Davallia pyxidata —
Squirrel's foot fern

Davallia solida fijiensis —
Rabbit's foot fern

Davallia trichomanoides —
Squirrel's foot fern

Dennstaedtia davallioides —
Lacy ground fern

Dianthus caryophyllus — Carnation

Dianthus chinensis — Indian pink

Dicentra spectabilis —
Bleeding heart
Dutchman's breeches
Lady-in-the-bath

Dicksonia antarctica — Soft tree fern

Dieffenbachia amoena —
Giant dumb cane

Dieffenbachia maculata —
Common dumb cane

Dionaea muscipula — Venus flytrap

Dioon edule — Chestnut dioon
Mexican fern palm

Dioscorea batatus — Chinese yam
Cinnamon yam

Dioscorea bulbifera — Aerial yam
Air potato

Dioscorea discolor —
Ornamental yam

Dioscorea opposita — Chinese yam
Cinnamon yam

Diplazium esculentum — Paco
Vegetable fern

Dischidia rafflesiana —
Malayan urn vine

Dodonaea viscosa — Akeake
Hop bush

Dolichos lablab — Hyacinth bean
Indian bean
Lablab bean

Dolichos lablab lignosus —
Australian pea

Dolichos lignosus — Australian pea

Dorotheanthus bellidiformis —
Livingstone daisy

Doxantha unguis-cati —
Cat's claw vine

Dracaena draco — Dragon's tree

Dracaena fragrans — Corn plant

Dracaena godseffiana —
Gold dust dracaena

Dracaena marginata —
Madagascar dragon tree

Dracaena sanderiana — Ribbon plant

Dracaena surculosa —
Gold dust dracaena

Dracunculus vulgaris —
Dragon plant

Drimys aromatica —
Mountain pepper

Drimys colorata — Pepper tree

Drimys lanceolata —
Mountain pepper

Drimys winteri — Winter's bark

Drosanthemum speciosum —
Dewflower

Drosera anglica — Great sundew

Drosera menziesii — Pink rainbow

Drosera rotundifolia —
Common sundew

Drynaria quercifolia — Oak fern

Dryopteris borreri —
Golden male fern

Dryopteris erythrosora —
Japanese buckler fern
Japanese shield fern

Dryopteris filix-mas — Male fern

Dryopteris pseudomas —
Golden male fern

Duranta plumieri

Duranta plumieri	
	Golden dewdrop
	Pigeon berry
	Skyflower
Duranta repens	Golden dewdrop
	Pigeon berry
	Skyflower
Ecballium elaterium	
	Squirting cucumber
Eccremocarpus scaber	
	Chilean glory flower
Echeveria affinis	Black echeveria
Echeveria derenbergii	
	Painted lady
Echeveria glauca	Blue echeveria
Echeveria leucotricha	
	Chenille plant
Echeveria multicaulis	
	Copper roses
Echeveria paraguayense	
	Ghost plant
	Mother-of-pearl plant
Echeveria pulvinata	Plush plant
Echeveria rosea	Desert rose
Echeveria secunda glauca	
	Blue echeveria
Echeveria setosa	Firecracker plant
Echium lycopsis	
	Purple viper's bugloss
Echium plantagineum	
	Purple viper's bugloss
Eichhornia crassipes	
	Water hyacinth
Eichhornia speciosa	
	Water hyacinth
Elaphoglossum crinitum	
	Elephant's ear fern
Encephalartos altensteinii	
	Bread tree cycad
Epidendrum cochleatum	
	Cockle-shell orchid
Epipremnum aureum	Devil's ivy
	Golden pothos
	Taro vine
Episcia cupreata	Flame violet
Episcia dianthiflora	Lace flower
Eranthemum nervosum	Blue sage
Eranthemum pulchellum	
	Blue sage
Erica carnea	Winter heath
Erica herbacea	Winter heath
Eriobotrya japonica	Loquat
Ervatamia coronaria	
	Cape jasmine
	East Indian rosebay
Erythrina crista-galli	
	Cockspur coral tree
Eschscholzia californica	
	Californian poppy
Eucalyptus citriodora	
	Lemon-scented gum
Eucalyptus ficifolia	
	Red-flowering gum
Eucalyptus globulus	Blue gum
	Tasmanian blue gum
Eucalyptus macrocarpa	Mottlecah
Eucalyptus nitens	Shining gum
Eucharis amazonica	Amazon lily
Eucharis grandiflora	Amazon lily
Eucomis comosa	Pineapple flower
Eucomis pole-evansii	
	Giant pineapple flower
Eugenia australis	Brush cherry
Eugenia jambos	Malabar plum
	Rose apple
Eugenia myrtifolia	Brush cherry
Eugenia paniculata	Brush cherry
Eugenia uniflora	Pitanga
	Surinam cherry
Euphorbia caput-medusae	
	Medusa's head
Euphorbia fulgens	Scarlet plume
Euphorbia grandicornis	
	Cow's horn
Euphorbia milii	Crown of thorns
Euphorbia pseudocactus	
	Cactus spurge
Euphorbia pulcherrima	Poinsettia
Euphorbia tirucalli	Finger tree
	Milk bush
	Rubber spurge
Eustoma grandiflorum	
	Prairie gentian
Eustoma russellianum	
	Prairie gentian
Fatsia japonica	
	False castor-oil plant
Fatsia papyrifera	Rice-paper plant
Faucaria felina	Cat's jaws
Faucaria tigrina	Tiger jaws
Feijoa sellowiana	Pineapple guava
Felicia amelloides	Blue marguerite
Felicia bergeriana	Kingfisher daisy
Ficus benghalensis	Banyan
Ficus benjamina	Weeping fig
Ficus deltoidea	Mistletoe fig
Ficus diversifolia	Indoor fig
	Mistletoe fig
Ficus elastica	Rubber plant
Ficus lyrata	Fiddle-leaf fig
Ficus macrocarpa	Laurel fig
Ficus macrophylla	
	Australian banyan
	Moreton bay fig
Ficus nitida	Laurel fig
Ficus pandurata	Fiddle-leaf fig
Ficus pumila	Climbing fig
	Creeping fig
Ficus religiosa	Bo tree
	Pepul
	Sacred fig tree

Helichrysum petiolatum

Ficus repens	Climbing fig
	Creeping fig
Ficus retusa	Laurel fig
Ficus rubiginosa	Port Jackson fig
Ficus sycamorus	Mulberry fig
	Sycamore fig
Fittonia argyroneura	Silver net leaf
Fittonia verschaffeltii	Nerve plant
	Painted net leaf
Fortunella crassifolia	
	Meiwa kumquat
Fortunella hindsii	
	Hong Kong kumquat
Fortunella margarita	
	Nagami kumquat
	Oval kumquat
Francoa appendiculata	
	Bridal wreath
Fritillaria camtschatcensis	
	Black sarana
Fritillaria imperialis	
	Crown imperial
Fritillaria meleagris	
	Snake's head fritillary
Furcraea foetida	Mauritius hemp
Furcraea gigantea	Mauritius hemp
Galanthus elwesii	Giant snowdrop
Galanthus nivalis	
	Common snowdrop
	Snowdrop
Galtonia candicans	
	Summer hyacinth
Gardenia citriodora	Wild coffee
Gardenia florida	Cape jasmin
	Common gardenia
	Gardenia
Gardenia grandiflora	Cape jasmin
	Common gardenia
	Gardenia
Gardenia jasminoides	
	Cape jasmine
	Common gardenia
	Gardenia
Gasteria verrucosa	Ox tongue
Gelsemium sempervirens	
	Carolina yellow jasmine
	False yellow jasmine
Genista canariensis	
	Florist's genista
Geogenanthus undatus	
	Seersucker plant
Gerbera jamesonii	
	Barberton daisy
	Transvaal daisy
Gesneria cardinalis	
	Cardinal flower
Gilia capitata	Blue thimble flower
Gilia coronipifolia	Skyrocket
	Standing cypress
Gilia rubra	Skyrocket
	Standing cypress
Gilia tricolor	Bird's eyes

Gladiolus blandus	Painted lady
Gladiolus carneus	Painted lady
Gladiolus primulinus	
	Maid of the mist
Glechoma hederacea	Ground ivy
Gloriosa rothschildiana	Glory lily
Gloxinia speciosa	Gloxinia
Gomphrena globosa	
	Globe amaranth
Gossypium arboreum	Tree cotton
Gossypium herbaceum	
	Levant cotton
	Turkish cotton
Gossypium sturtianum	
	Sturt's desert rose
Gossypium sturtii	
	Sturt's desert rose
Graptopetalum paraguayense	
	Ghost plant
	Mother-of-pearl plant
Graptophyllum pictum	
	Caricature plant
Grevillea robusta	Silky oak
Guzmania berteroana	
	Flaming torch
Guzmania lingulata	Scarlet star
Guzmania picta	
	Blushing bromeliad
Gynura aurantiaca	Velvet plant
Haemanthus albiflos	Paintbrush
Haemanthus katherinae	
	Blood flower
Haemanthus multiflorus	
	Blood flower
Hardenbergia comptoniana	
	Wild sarsparilla
Haworthia fasciata	
	Zebra haworthia
Haworthia margaritifera	
	Pearl plant
Haworthia pumila	Pearl plant
Haworthia tessellata	
	Star window plant
Hedera canariensis	
	Canary Island ivy
Hedera colchica	Persian ivy
Hedera helix	Common ivy
	English ivy
Hedychium coccineum	
	Red ginger lily
	Scarlet ginger lily
Hedychium coronarium	
	Butterfly ginger lily
	Garland flower
Hedychium gardnerianum	
	Kahili ginger
Helichrysum angustifolium	
	Curry plant
Helichrysum italicum	Curry plant
Helichrysum petiolatum	
	Liquorice plant

Helichrysum serotinum
Curry plant

Heliotropium arborescens
Common heliotrope

Heliotropium peruvianum
Common heliotrope

Helxine soleirolii Baby's tears
Mind your own business

Hemigraphis alternata Flame ivy
Red ivy

Hemigraphis colorata Flame ivy
Red ivy

Hemitelia smithii Soft tree fern

Hesperoyucca whipplei
Our Lord's candle

Hibbertia scandens Guinea flower
Snake vine

Hibbertia volubilis Guinea flower
Snake vine

Hibiscus rosa-sinensis
Rose of china

Hibiscus schizopetalus
Japanese lantern

Hippeastrum × ackermannii
Amaryllis

Hippeastrum × acramanii
Amaryllis

Hippeastrum procerum
Blue amaryllis

Homeria breyniana Cape tulip

Homeria collina Cape tulip

Howeia belmoreana Curly palm
Sentry palm

Howeia forsterana Kentia palm
Paradise palm
Sentry palm
Thatchleaf palm

Hoya bella Miniature wax plant

Hoya carnosa Wax plant

Huernia zebrina Owl eyes

Humulus japonicus Japanese hop

Humulus lupulus Common hop

Hydrangea hortensis
Common hydrangea
French hydrangea

Hydrangea macrophylla
Common hydrangea
French hydrangea

Hylocereus undatus
Queen of the night

Hymenocallis calathina
Basket flower
Peruvian daffodil

Hymenocallis narcissiflora
Basket flower
Peruvian daffodil

Hypoestes phyllostachya
Freckle face
Polka dot plant

Imantophyllum miniatum
Kaffir lily

Impatiens balsamina Rose balsam
Touch-me-not

Impatiens holstii Busy lizzie
Patient lucy

Impatiens sultanii Busy lizzie
Patient lucy

Impatiens walleriana Busy lizzie
Patient lucy

Ipheion uniflorum
Spring star flower

Ipomoea batatas Sweet potato

Ipomoea bona-nox Moonflower

Ipomoea coccinea
Red morning glory
Star ipomoea

Ipomoea noctiflora Moonflower

Ipomoea purpurea
Common morning glory
Morning glory

Ipomoea quamoclit Cypress vine

Ipomopsis rubra Skyrocket
Standing cypress

Iresine herbstii Beefsteak plant

Iresine lindenii Blood leaf

Ixora coccinea Flame-of-the-woods

Ixora javanica Jungle geranium

Jacobinia carnea King's crown

Jacobinia velutina King's crown

Jasminum mesneyi
Primrose jasmine

Jasminum nitidum Star jasmine
Windmill jasmine

Jasminum officinale
Common white jasmine
Poet's jessamine
Summer jasmine

Jasminum polyanthum
Chinese jasmine
Pink jasmine

Jasminum primulinum
Primrose jasmine

Jasminum sambac Arabian jasmine

Jatropha multifida Coral plant
Physic nut

Jatropha podagrica Gout plant
Tartogo

Jubaea chilensis Chilean wine palm
Coquito
Honey palm

Jubaea spectabilis
Chilean wine palm
Coquito
Honey palm

Justicia carnea King's crown

Justicia coccinea Cardinal's guard

Kaempferia roscoana
Dwarf ginger lily
Peacock plant

Kaempferia rotunda
Tropical crocus

Kalanchoe beharensis Velvet leaf

HOUSE PLANTS

Maranta leuconeura kerchoveana

Kalanchoe blossfeldiana
Flaming katy

Kalanchoe diagremontiana
Devil's backbone

Kalanchoe pilosa — Panda plant
Pussy ears

Kalanchoe tomentosa — Panda plant
Pussy ears

Kalanchoe tubiflora
Chandelier plant

Kennedia nigricans — Black bean

Kennedia prostrata
Running postman

Kennedia rubicunda
Dusky coral pea

Kentia belmoreana — Curly palm
Sentry palm

Kentia forsterana — Kentia palm
Paradise palm
Sentry palm
Thatchleaf palm

Kleinia articulatus — Candle plant

Kochia scoparia — Burning bush
Summer cypress

Lachenalia aloides — Cape cowslip

Lachenalia glaucina
Opal lachenalia

Lachenalia mutabilis
Fairy lachenalia

Lachenalia tricolor — Cape cowslip

Lagenaria siceraria — Bottle gourd
Calabash gourd

Lagenaria vulgaris — Bottle gourd
Calabash gourd

Lagerstroemia indica — Cape myrtle

Lamium maculatum
Spotted dead nettle

Lantana camara — Yellow sage

Lapageria rosea
Chilean bell flower

Latania aurea — Yellow latan palm

Latania borbonica — Red latan palm

Latania commersonii
Red latan palm

Latania loddigesii — Silver latan palm

Latania lontaroides — Red latan palm

Latania verschaffeltii
Yellow latan palm

Laurus nobilis — Bay
Sweet bay

Lavandula angustifolia
Common lavender
English lavender

Lavandula stoechas
French lavender

Leptospermum nitidum
Shiny tea tree

Leptospermum scoparium — Manuka
New Zealand tea tree

Leucadendron argenteum
Silver tree

Leucocoryne ixioides
Glory-of-the-snow

Leucospermum cordifolium
Nodding pincushion

Leucospermum nutans
Nodding pincushion

Limonia aurantifolia — Lime

Limonium sinuatum
Notch-leaf statice
Winged statice

Limonium suworowii
Rat's tail statice
Russian statice

Linaria cymbalaria
Ivy-leaved toadflax
Kenilworth ivy

Linaria maroccana — Toadflax

Lippia citriodora — Lemon verbena

Lisianthus russellianus
Prairie gentian

Livistona australis
Australian fountain palm
Gippsland fountain palm

Livistona chinensis
Chinese fountain palm

Lotus berthelotii — Coral gem
Parrot's beak
Winged pea

Luffa aegyptica — Loofah
Sponge gourd

Luffa cylindrica — Loofah
Sponge gourd

Lycopersicon esculentum — Tomato

Lycopersicon lycopersicum — Tomato

Lycopersicon lycopersicum cerasiforme — Cherry tomato
Salad tomato

Lycopersicon pimpinellifolium
Currant tomato

Lycopodium phlegmaria
Queensland tassel fern

Malpighia coccigera
Miniature holly

Malpighia glabra — Barbados cherry

Mandevilla laxa — Chilean jasmine

Mandevilla suaveolens
Chilean jasmine

Manettia bicolor — Firecracker vine

Manettia inflata — Firecracker vine

Manettia luteo-rubra
Firecracker vine

Manihot esculenta — Cassava
Manioc
Tapioca

Manihot utilissima — Cassava
Manioc
Tapioca

Maranta arundinacea — Arrowroot

Maranta leuconeura — Prayer plant

Maranta leuconeura kerchoveana — Rabbit's foot
Rabbit tracks

Marsilea drummondi	Water clover
Marsilea quadrifolia	Water clover
Matthiola incana	Stock
Medinilla magnifica	Rose grape
Melaleuca leucodendron	Cajeput
	River tea tree
Melia azederach	Bead tree
	Persian lilac
Melianthus major	Honeybush
Mesembryanthemum criniflorum	Livingstone daisy
Metrosideros collina	Lehua
Metrosideros excelsa	New Zealand christmas tree
Metrosideros tomentosa	New Zealand christmas tree
Mimosa pudica	Humble plant
	Sensitive plant
Mimulus guttatus	Monkey flower
	Monkey musk
Mimulus luteus	Monkey flower
	Monkey musk
Mimulus moschatus	Musk
Mitriostigma axillare	Wild coffee
Momordica balsamina	Balsam apple
Momordica charantia	Balsam pear
Mondo jaburan	White lily turf
Monstera deliciosa	Cerimen
	Mexican bread fruit
	Swiss cheese plant
Monstera latevaginata	Shingle plant
Monstera pertusa	Ceriman
	Mexican bread fruit
	Swiss cheese plant
Muehlenbeckia complexa	Maidenhair vine
	Wire vine
Murraya exotica	Orange jasmine
Murraya koenigii	Curry leaf
Murraya paniculata	Orange jasmine
Musa basjoo	Japanese banana
Musa coccinea	Flowering banana
	Scarlet banana
Musa × paradisiaca	Common banana
Musa × sapientum	Common banana
Myosotis sylvatica	Wood forget-me-not
Myriophyllum aquaticum	Water feather
Myriophyllum brasiliense	Water feather
Myriophyllum proserpinacoides	Water feather
Myrtus communis	Common myrtle

Nandina domestica	Heavenly bamboo
	Sacred bamboo
Neanthe elegans	Dwarf mountain palm
	Parlour palm
Nelumbium lutea	American lotus
Nelumbo lutea	American lotus
Nelumbo pentapetala	American lotus
Nemophila maculata	Five spot
Nemophila menziesii	Baby blue eyes
Neoregelia marmorata	Marble plant
Neoregelia spectabilis	Fingernail plant
Nephrolepis cordifolia	Ladder fern
	Sword fern
Nephrolepis exaltata	Sword fern
Nerine sarniensis	Guernsey lily
Nerium oleander	Oleander
	Rose bay
Nertera depressa	Coral moss
Nertera granadensis	Bead plant
Nicolaia elatior	Philippine wax flower
	Torch ginger
Nidularium fulgens	Blushing bromeliad
Nidularium innocentii	Bird's nest bromeliad
Nidularium pictum	Blushing bromeliad
Nolina recurvata	Elephant foot
	Pony tail
Nolina tuberculata	Elephant foot
	Pony tail
Nymphaea caerula	Egyptian blue lotus
Nymphaea capensis	Blue water lily
Nymphaea pygmaea	Pigmy water lily
Nymphaea stellata	Blue lotus
Nymphaea tetragona	Pigmy water lily
Ochna serrulata	Mickey mouse plant
Odontoglossum crispum	Lace orchid
Odontoglossum pulchellum	Lily-of-the-valley orchid
Olea europaea	Olive
Oncidium flexuosum	Dancing doll orchid
Oncidium ornithorhynchum	Dove orchid
Oncidium papilio	Butterfly orchid
Oncidium tigrinum	Tiger orchid
Oncidium varicosum	Golden butterfly orchid

Peperomia glabella

Ophiopogon jaburan
White lily turf

Oplismenus hirtellus   Basket grass

Ornithogalum arabicum
Star of bethlehem

Ornithogalum nutans
Drooping star of bethlehem

Ornithogalum saundersiae
Giant chincherinchee

Ornithogalum thyrsoides
Chincherinchee

Ornithogalum umbellatum
Star of bethlehem

Oxalis cernua   Bermuda buttercup

Oxalis deppei   Lucky clover

Oxalis pes-caprae
Bermuda buttercup

Pachystachys cardinalis
Cardinal's guard

Pachystachys coccinea
Cardinal's guard

Pachystachys lutea   Lollipop plant

Pancratium maritimum
Sea daffodil
Sea lily

Pandanus baptisii   Blue screw pine

Pandanus utilis
Common screw pine

Pandorea jasminoides
Bower plant

Pandorea pandorana
Wonga-wonga vine

Parochetus communis
Blue shamrock pea

Passiflora caerulea
Blue passion flower
Common passion flower

Passiflora coccinea   Red granadilla
Red passion flower

Passiflora edulis   Passion fruit
Purple granadilla

Passiflora ligularis   Sweet granadilla

Passiflora mollissima
Banana passion fruit
Curuba

Passiflora quadrangularis
Giant granadilla

Passiflora racemosa
Red passion flower

Pedilanthes tithymaloides
Redbird cactus
Slipper cactus

*Pedilanthes tithymaloides
smallii*   Jacob's ladder

Pelargonium capitatum
Rose-scented geranium
Rose-scented pelargonium

Pelargonium crispum
Lemon geranium
Lemon pelargonium
Prince Rupert geranium
Prince Rupert pelargonium

Pelargonium × domesticum
Fancy geranium
Fancy pelargonium
Lady Washington geraniums
Lady Washington pelargonium
Martha Washington geranium
Martha Washington pelargonium
Regal geranium
Regal pelargonium
Show geranium
Show pelargonium

Pelargonium echinatum
Cactus geranium
Cactus pelargonium

Pelargonium × fragrans
Nutmeg geranium
Nutmeg pelargonium

Pelargonium graveolens
Rose geranium
Rose pelargonium

Pelargonium × hortorum
Zonal geranium
Zonal pelargonium

Pelargonium odoratissimum
Apple-scented geranium
Apple-scented pelargonium

Pelargonium peltatum
Ivy-leaved geranium
Ivy-leaved pelargonium

Pelargonium quercifolium
Oak-leaved geranium
Oak-leaved pelargonium

Pelargonium tomentosum
Peppermint-scented geranium
Peppermint-scented pelargonium

Pelargonium zonale
Horse-shoe geranium
Horse-shoe pelargonium
Zonal geranium
Zonal pelargonium

Pellaea adiantoides
Green brake fern

Pellaea atropurpurea
Purple-stemmed cliff brake

Pellaea hastata   Green brake fern

Pellaea rotundifolia   Button fern

Pellaea viridis   Green brake fern

Pellionia daveauana
Trailing water-melon begonia

Pentas carnea   Egyptian star cluster
Star cluster

Pentas lanceolata
Egyptian star cluster
Star cluster

Peperomia argyreia
Water-melon begonia

Peperomia caperata
Emerald ripple

Peperomia clusiifolia
Baby rubber plant

Peperomia fraseri
Flowering mignonette

Peperomia glabella   Wax privet

Peperomia griseoargentea
Ivyleaf peperomia
Silverleaf peperomia

Peperomia hederifolia
Ivyleaf peperomia
Silverleaf peperomia

Peperomia maculosa
Radiator plant

Peperomia magnoliifolia
Desert privet

Peperomia marmorata Silver heart

Peperomia polybotrya Coin leaf

Peperomia prostrata
Creeping peperomia

Peperomia pulchella
Whorled peperomia

Peperomia rotundifolia pilosior
Creeping peperomia

Peperomia sandersii
Water-melon begonia

Peperomia scandens
Cupid peperomia

Peperomia verticillata
Whorled peperomia

Pereskia aculeata
Barbados gooseberry

Peristeria elata Dove orchid
Holy ghost plant

Persea americana Avocado

Persea gratissima Avocado

Petrea volubilis Purple wreath
Queen's wreath

Phaeomeria magnifica
Philippine wax flower
Torch ginger

Phaius bicolor Nun's hood orchid

Phaius blumei Nun's hood orchid

Phaius grandiflorus
Nun's hood orchid

Phaius gravesii Nun's hood orchid

Phaius tankervillae
Nun's hood orchid

Phaius wallichii Nun's hood orchid

Pharbitis purpurea
Common morning glory
Morning glory

Philodendron andreanum
Black-gold philodendron
Velour philodendron

Philodendron auritum Five fingers

Philodendron bipinnatifidum
Tree philodendron

Philodendron cordatum Heart leaf

Philodendron domesticum
Elephant's ear
Spade leaf

*Philodendron domesticum ×
erubescens*
Red-leaf philodendron

Philodendron eichleri
Tree philodendron

Philodendron erubescens
Blushing philodendron

Philodendron gloriosum Satin leaf

Philodendron hastatum
Elephant's ear
Spade leaf

Philodendron × mandaianum
Red-leaf philodendron

Philodendron melanochrysum
Black-gold philodendron
Velour philodendron

Philodendron pertusum Ceriman
Mexican bread fruit
Swiss cheese plant

Philodendron scandens Heart leaf
Sweetheart plant

Philodendron selloum
Lacy tree philodendron

Phlebodium aureum
Hare's foot fern

Phoenix canariensis
Canary Islands date palm

Phoenix dactylifera Date palm

Phoenix roebelenii
Pigmy date palm

Phormium colensoi Mountain flax

Phormium cookianum
Mountain flax

Phormium tenax
New Zealand flax

Phyllanthus nivosa Snow bush

Phyllitis scolopendrium
Hart's tongue fern

Phyllostachys aurea
Fishpole bamboo
Golden bamboo

Pilea cadierei Aluminium plant

Pilea involucrata Friendship plant
Panamiga

Pilea microphylla Artillery plant

Pilea muscosa Artillery plant

Pilea nummulariifolia
Creeping charlie

Pilea repens Black leaf panamiga

Pinguicula grandiflora
Greater butterwort

Pinguicula vulgaris Butterwort

Piper crocatum
Ornamental pepper

Piper nigrum Black pepper

Piper ornatum crocatum
Ornamental pepper

Pisonia umbellifera
Bird-catching tree
Parapara

Pistia stratiotes Water lettuce

Pittosporum crassifolium Karo

Pittosporum tenuifolium Kohuhu

Pittosporum tobira
Japanese pittosporum
Tobira

Pittosporum undulatum
Victorian box

Platycerium alcicorne
Common stag's horn fern

Platycerium angolense
Elephant's ear fern

Platycerium biforme
Crown stag's horn

Platycerium bifurcatum
Common stag's horn fern

Platycerium coronarium
Crown stag's horn

Platycerium grande
Regal elkhorn fern

Platycerium hillii   Elkhorn fern

Plectranthus coleoides   Candle plant

Plectranthus oertendahlii
Brazilian coleus
Swedish ivy

Plumbago auriculata
Blue cape leadwort
Blue cape plumbago

Plumbago capensis
Blue cape leadwort
Blue cape plumbago

Plumbago indica   Scarlet leadwort

Plumbago rosea   Scarlet leadwort

Plumeria acuminata   Frangipani

Plumeria acutifolia   Pagoda tree

Plumeria rubra   Frangipani

Plumeria rubra acutifolia
Pagoda tree

Plumeria rubra lutea
Yellow pagoda tree

Poinciana pulcherrima
Barbados pride

Polianthes tuberosa   Tuberose

Polypodium angustifolium
Narrow-leaved ribbon fern
Narrow-leaved strap fern

Polypodium aureum
Hare's foot fern

Polypodium phyllitidis
Ribbon fern
Strap fern

Polypodium subauriculatum
Lacy pine fern

Polypodium vulgare
Common polypody

Polyscias balfouriana
Dinner-plate aralia

Polyscias filicifolia   Fern-leaf aralia

Polyscias guilfoylei   Wild coffee

Polystichum acrostichoides
Christmas fern

Polystichum aculeatum
Hard shield fern
Prickly shield fern

Polystichum aristatum
East Indian holly fern

Polystichum falcatum   Holly fern

Polystichum munitum
Giant holly fern
Western sword fern

Polystichum setiferum
Soft shield fern

Polystichum tsus-simense
Tsusina holly fern

Portulaca grandiflora   Sun plant

Pothos aureus
Devil's ivy
Golden pothos
Taro vine

Primula malacoides
Fairy primrose

Primula sinensis   Chinese primrose

Primula × tommasinii   Polyanthus

Primula vulgaris   Primrose

Pritchardia filifera
Desert fan palm
Petticoat palm

Prostanthera melissifolia
Balm mint bush

Prostanthera nivea   Snowy mint bush

Protea mellifera   Honey flower
Sugarbush

Protea repens   Honey flower
Sugarbush

Pseudowintera colorata
Pepper tree

Pteris cretica   Cretan brake
Ribbon fern
Table fern

Pteris ensiformis   Sword brake

Pteris multifida   Spider fern

Pteris serrulata   Spider fern

Pteris vittata   Ladder fern

Ptychosperma elegans
Alexander palm
Solitaire palm

Punica granatum   Pomegranate

Pyrostegia ignea   Flame flower
Flame vine
Flaming trumpets

Pyrostegia venusta   Flame flower
Flame vine
Flaming trumpets

Quisqualis indica
Rangoon creeper

Raphidophora aurea   Devil's ivy
Golden pothos
Taro vine

Ravenala madagascariensis
Traveller's tree

Reinwardtia indica   Yellow flax

Reinwardtia trigyna   Yellow flax

Rhaphidophora celatocaulis
Shingle plant

Rhaphidophora decursiva
Shingle plant

Rhapis excelsa

Rhapis excelsa	Bamboo palm
	Ground rattan
	Little lady palm
	Miniature fan palm
Rhodochiton atrosanguineum	
	Purple bell vine
Rhodochiton volubile	
	Purple bell vine
Rhoeo discolor	Boat lily
	Moses-in-a-boat
Rhoeo spathacea	Boat lily
	Moses-in-a-boat
Rhoicissus capensis	Cape grape
Ricinus communis	Castor oil plant
Rivina humilis	Rouge plant
Rivina laevis	Rouge plant
Rosmarinus lavandulaceum	
	Prostrate rosemary
Rosmarinus officinalis	Rosemary
Rosmarinus officinalis prostratus	
	Prostrate rosemary
Roystonea regia	Cuban royal palm
Ruellia mackoyana	Monkey plant
Rumohra adiantiformis	
	Leather fern
Ruscus hypoglossum	
	Butcher's broom
Russelia equisetiformis	Coral plant
	Fountain bush
Russelia juncea	Coral plant
	Fountain bush
Salpiglossis sinuata	Painted tongue
Salvia elegans	Pineapple sage
Salvia officinalis	Common sage
Salvia rutilans	Pineapple sage
Salvia splendens	Scarlet sage
Salvinia auriculata	Floating moss
Sansevieria grandicuspis	
	Star sansevieria
Sansevieria trifasciata	
	Mother-in-law's tongue
	Snake plant
Sansevieria zeylanica	
	Devil's tongue
Sarracenia flava	Huntsman's horn
	Yellow pitcher plant
Sarracenia purpurea	
	Common pitcher plant
Sauromatum guttatum	
	Monarch of the east
	Voodoo lily
Saxifraga sarmentosa	
	Mother of thousands
	Strawberry geranium
Saxifraga stolonifera	
	Mother of thousands
	Strawberry geranium
Saxifraga × urbium	London pride
Schefflera actinophylla	
	Octopus tree
	Queensland umbrella tree

Schinus molle	Peruvian mastic tree
	Peruvian pepper tree
Schizanthus pinnatus	
	Poor man's orchid
Scilla sibirica	Siberian squill
Scindapsus aureus	Devil's ivy
	Golden pothos
	Taro vine
Seaforthia elegans	Alexander palm
	Solitaire palm
Sedum morganianum	Ass's tail
	Burro's tail
	Donkey's tail
Sedum weinbergii	Ghost plant
	Mother-of-pearl plant
Selaginella lepidophylla	
	Resurrection plant
	Rose of Jericho
Selaginella uncinata	Peacock moss
	Rainbow fern
Senecio articulatus	Candle plant
Senecio bicolor cineraria	
	Dusty miller
Senecio cineraria	Dusty miller
Senecio confusus	
	Mexican flame vine
Senecio elegans	Purple ragwort
Senecio macroglossus	Cape ivy
	Natal ivy
	Wax vine
Senecio maritimus	Dusty miller
Senecio mikanioides	German ivy
Senecio rowleyanus	String of beads
	String of pearls
Setcreasea striata	Wandering jew
Sinningia cardinalis	
	Cardinal flower
Sinningia speciosa	Gloxinia
Smilax asparagoides	Smilax
Smithiana cinnabarina	
	Temple bells
Solandra guttata	Cup of gold
Solandra hartwegii	Cup of gold
Solandra maxima	Cup of gold
Solandra nitida	Cup of gold
Solanum aviculare	Kangaroo apple
Solanum capsicastrum	
	Winter cherry
Solanum crispum	
	Chilean potato tree
Solanum jasminoides	
	Jasmine nightshade
	Potato vine
Solanum macranthum	
	Brazilian potato tree
Solanum melongena	Aubergine
	Egg-plant
Solanum pseudocapsicum	
	Jerusalem cherry
Solanum rantonnetii	
	Blue potato bush

Trachelium caeruleum

Solanum wendlandii
Giant potato vine

Soleirolia soleirolii   Baby's tears
Mind your own business

Sollya fusiformis
Australian bluebell
Australian bluebell creeper

Sollya heterophylla
Australian bluebell
Australian bluebell creeper

Sparaxis tricolor   Harlequin flower
Wand flower

Sparmannia africana
African hemp

Spathiphyllum wallisii   Peace lily
White sails

Sprekelia formosissima   Aztec lily
Jacobean lily

Stapelia hirsuta   Hairy toad plant

Stenolobium stans   Yellow bells
Yellow elder

Stenotaphrum secundatum
Saint Augustine grass

Stephanotis floribunda
Bridal flower
Madagascar jasmine
Waxy jasmine

Stigmaphyllon ciliatum
Butterfly vine
Golden creeper
Golden vine

Strelitzia reginae
Bird of paradise flower

Streptanthera cuprea
Orange kaleidoscope flower

Streptanthera elegans
White kaleidoscope flower

Streptocarpus dunnii
Red nodding bells

Streptocarpus rexii   Cape primrose

Streptocarpus saxorum
False African violet

Streptosolen jamesonii
Marmalade bush

Strobilanthes dyerianus
Persian shield

Stropholirion californicum
Snake lily

Stylidium graminifolium
Common trigger plant
Grass trigger plant
Trigger plant

Swainsona greyana
Darling river pea

Syngonium auritum   Five fingers

Syngonium podophyllum
Arrowhead vine
Goosefoot plant

Syzygium jambos   Malabar plum
Rose apple

Syzygium paniculatum
Brush cherry

Tabernaemontana coronaria
Cape jasmine
East Indian rosebay

Tabernaemontana divaricata
Cape jasmine
East Indian rosebay

Tacca chantrieri   Bat flower
Cat's whiskers

Tacca leontopetaloides
East Indian arrowroot
South sea arrowroot

Tacca pinnatifida
East Indian arrowroot
South sea arrowroot

Tacsonia mollissima
Banana passion fruit
Curuba

Tagetes erecta   African marigold
American marigold

Tagetes patula   French marigold

Tagetes signata   Signet marigold
Tagetes

Tagetes tenuifolia   Signet marigold
Tagetes

Tasmannia latifolia
Mountain pepper

Tecoma capensis   Cape honeysuckle

Tecoma stans   Yellow bells
Yellow elder

Tecomaria capensis
Cape honeysuckle

Teline canariensis   Florist's genista

Telopea speciosissima   Waratah

Tetranema mexicanum
Mexican foxglove
Mexican violet

Tetranema roseum
Mexican foxglove
Mexican violet

Tetrapanax papyriferus
Rice-paper plant

Tetrastigma voinierianum
Chestnut vine

Thevetia peruviana
Yellow oleander

Thrinax microcarpa   Key palm

Thrinax morrisii   Key palm

Thrinax parviflora
Palmetto thatch palm

Thunbergia alata
Black-eyed susan

Thunbergia grandiflora
Blue trumpet vine

Tibouchina semidecandra
Glory bush

Tibouchina urvilleana   Glory bush

Tillandsia usneioides   Spanish moss

Tolmiea menziesii   Piggyback plant

Torenia fournieri
Wishbone flower

Trachelium caeruleum
Common throatwort

Trachycarpus fortunei

Trachycarpus fortunei	Chusan palm
	Windmill palm
Tradescantia albiflora	Wandering jew
Tradescantia fluminensis	Wandering jew
Trevesia palmata	Snowflake aralia
Trichosanthes anguina	Snake gourd
Triteleia laxa	Grass nut
	Ithuriel's spear
Tropaeolum canariensis	Canary creeper
Tropaeolum majus	Nasturtium
Tropaeolum peregrinum	Canary creeper
Tulbaghia fragrans	Sweet garlic
Turnera trioniflora	Sage rose
Turnera ulmifolia	Sage rose
Vallota purpurea	Guernsey lily
	Scarborough lily
Vallota speciosa	Guernsey lily
	Scarborough lily
Veitchia merrillii	Christmas palm
Verbena triphylla	Lemon verbena
Viburnum tinus	Laurustinus
Vinca rosea	Madagascar periwinkle
Viola hederacea	Ivy-leaved violet
Viola × hortensis	Garden pansy
	Pansy
Viola × wittrockiana	Garden pansy
	Pansy
Vitis capensis	Cape grape
Vitis voinierianum	Chestnut vine
Vittaria lineata	Florida ribbon fern
Vriesea carinata	Lobster claws
Vriesea hieroglyphica	King of the bromeliads
Vriesea psittacina	Painted feathers
Vriesea splendens	Flaming sword

Washingtonia filifera	Desert fan palm
	Petticoat palm
Westringia fruticosa	Australian rosemary
Westringia rosmariniformis	Australian rosemary
Wintera aromatica	Winter's bark
Winterania latifolia	Mountain pepper
Worsleya rayneri	Blue amaryllis
Xanthorrhoea arborea	Botany bay gum
Xanthorrhoea preissii	Blackboy
	Common blackboy
Xanthosma lindenii	Indian kale
Xanthosma violaceum	Blue taro
	Yautia
Yucca brevifolia	Joshua tree
Yucca elata	Palmella
	Soap tree
Yucca elephantipes	Spineless yucca
Yucca gigantea	Spineless yucca
Yucca gloriosa	Spanish dagger
Yucca guatemalensis	Spineless yucca
Yucca whipplei	Our Lord's candle
Zamia floridana	Coontie
	Seminole bread
Zamia furfuracea	Florida arrowroot
Zamia pumila	Florida arrowroot
Zebrina pendula	Silvery inch plant
	Wandering jew
Zebrina purpusii	Bronze inch plant
Zephyranthes candida	Flower of the west wind
Zingiber officinale	Common ginger
	Ginger

ORCHIDS

Orchids are members of the very large family *Orchidaceae*. Various genera are grown by commercial specialists for the cut flower market, and by hobbyists as garden plants, in greenhouses or the home. Many artificial hybrids have been produced and are often to be found in private collections. The following list includes those most commonly offered for sale in nurseries and garden centres.

Bee orchid –
Ophrys apifera

Aceras anthropophorum
    Man orchid

Aceras anthropophora
    Green man orchid

Aerides odorata   Foxtail orchid

Amesia gigantea   Chatterbox
    Giant helleborine
    Giant orchid
    Stream orchid

Anacamptis pyramidalis
    Pyramidal orchid

Anagraecum sesquipedale
    Star-of-Bethlehem orchid

Anguloa clowesii   Cradle orchid

Anguloa uniflora
    Cradle orchid

Anoectochilus setaceus
    Jewel orchid
    King-of-the-forest

Aplectrum hymale   Adam-and-Eve
    Puttyroot

Aplectrum spicatum
    Adam-and-Eve
    Puttyroot

Arachnis cathcartii
    Scorpion orchid

Arachnis clarkei   Esmaralda

Arachnis flos-aeris   Spider orchid

Arachnis moschifera   Spider orchid

Arachnis × maingayi
    Pink scorpion orchid
    Spider orchid

Arethusa bulbosa   Bog orchid
    Bog rose orchid
    Dragon's mouth orchid
    Swamp pink orchid
    Wild pink orchid

Arpophyllum giganteum
    Hyacinth orchid

Arpophyllum spicatum
    Hyacinth orchid

Arundina bambusifolia
    Bamboo orchid

Arundina graminifolia
    Bamboo orchid

Bletia alta   Pine pink orchid

Bletia purpurea   Pine pink orchid

Brassavola nodosa
    Lady-of-the-night

Bulbophyllum refractum
    Windmill orchid

Calopogon tuberosus   Grass pink
    Swamp pink

Calypso borealis   Calypso
    Cytherea
    Fairy-slipper orchid
    Pink-slipper orchid

Calypso bulbosa   Bog orchid
    Calypso
    Cytherea
    Fairy-slipper orchid
    Pink-slipper orchid

Catasetum macrocarpum
    Jumping orchid

Catasetum tridentatum
    Jumping orchid

Cattleya citrina   Tulip cattleya

Cattleya dowiana   Queen cattleya

Cattleya gaskelliana
    Summer cattleya

Cattleya labiata   Autumn cattleya
    Christmas cattleya
    Queen cattleya

Cattleya mossiae   Easter cattleya
    Spring cattleya

Cattleya percivaliana
    Christmas cattleya

Cattleya trianaei
    Christmas orchid
    Winter cattleya

Caularthron bicornutum
    Virgin Mary orchid
    Virgin orchid

Caularthron bilamellatum
    Little virgin orchid

Cephalanthera damasonium
    White helleborine

Cephalanthera rubra
    Red helleborine

Chamorchis alpina
    False musk orchid

Cleistes divaricata
    Funnel-crest orchid
    Lily-leaved pogonia
    Rosebud orchid
    Spreading pogonia

Coeloglossum viride   Frog orchid

Coelogyne pandurata
    Black orchid

Corallorhiza maculata
    Large coralroot
    Spotted coralroot

Corallorhiza multiflora
    Large coralroot
    Spotted coralroot

Corallorhiza mertensiana
    Western coralroot

Corallorhiza odontorhiza
    Autumn coralroot
    Chicken toes
    Crawley root
    Dragon's claw
    Late coralroot
    Small coralroot

Corallorhiza trifida
    Coralroot orchid

Coryanthes macrantha
    Bucket orchid
    Monkey orchid

Coryanthes speciosa   Bat orchid
    Bucket orchid

Cychnoches pentadactylon
    Swan orchid

ORCHIDS

Habenaria bifolia

Cypripedium acaule
Moccasin flower
Moccasin orchid
Nerveroot
Pink lady's slipper
Two-leaved lady's slipper

Cypripedium calceolus
Lady's slipper orchid

Cypripedium pubescens
Yellow lady's slipper

Cyrtopodium punctatum
Bee-swarm orchid
Cigar orchid
Cow-horn orchid

Dactylorhiza fuchsii
Common spotted orchid

Dactylorhiza incarnata
Early marsh orchid
Meadow orchid

Dactylorhiza latifolia
Marsh orchid

Dactylorhiza maculata
Heath spotted orchid

Dactylorhiza majalis
Broad-leaved marsh orchid

Dactylorhiza praetermissa
Southern marsh orchid

Dactylorhiza purpurella
Northern marsh orchid

Dendrobium bigibbum
Cooktown orchid

Dendrobium chrysotoxum
Fried-egg orchid

Dendrobium crumenatum
Pigeon orchid

Dendrobium speciosum   Rock lily

Dendrochilum filiforme
Golden chain orchid

Dendrochilum glumaceum
Hay-scented orchid
Silver chain

Diacrium bicornutum
Virgin Mary orchid
Virgin orchid

Diacrium bilamellatum
Little virgin orchid

Epidendrum atropurpureum
Spice orchid

Epidendrum boothianum
Booth's epidendrum
Dollar orchid

Epidendrum cochleatum
Clam-shell orchid
Cockle-shell orchid

Epidendrum conopseum
Green-fly orchid

Epidendrum erythronioides
Booth's epidendrum
Dollar orchid

Epidendrum macrochilum
Spice orchid

Epidendrum × obrienianum
Baby orchid
Butterfly orchid
Scarlet orchid

Epidendrum phoeniceum
Chocolate orchid

Epidendrum prismatocarpum
Rainbow orchid

Epidendrum tampense
Butterfly orchid
Florida butterfly orchid

Epipactis atrorubans
Dark red helleborine

Epipactis dunensis
Dune helleborine

Epipactis gigantea   Chatterbox
Giant helleborine
Giant orchid
Stream orchid

Epipactis helleborine
Bastard helleborine
Broadleaved helleborine

Epipactis latifolia
Bastard helleborine
Broadleaved helleborine

Epipactis leptochila
Narrow-lipped orchid

Epipactis palustris
Marsh helleborine

Epipactis phyllanthes
Green-flowered helleborine

Epipactis purpurata
Violet helleborine

Epipogium aphyllum
Ghost orchid

Eulophia alta   Ground coco
Wild coco

Galeandra lacustris
Helmet orchid

Goodyera oblongifolia
Giant rattlesnake plantain
Menzies' rattlesnake plantain
Rattlesnake plantain

Goodyera pubescens
Downy rattlesnake orchid
Downy rattlesnake plantain
Scrofula weed

Goodyera repens
Creeping lady's tresses
Dwarf rattlesnake plantain
Lesser rattlesnake plantain
Northern rattlesnake plantain

Goodyera tesselata
Checkered rattlesnake plantain
Smooth rattlesnake plantain

Grammatophyllum sanderanum   Queen of orchids

Grammatophyllum speciosum
Queen of orchids

Gymnadenia conopsea
Fragrant orchid
Sweet-scented orchid

Habenaria bifolia
Lesser butterfly orchid

Habenaria blephariglottis
  Snowy orchid
  White fringed orchid
Habenaria ciliaris Fringed orchis
  Orange fringe
  Orange plume
  Yellow fringed orchid
Habenaria clavellata Frog spike
  Green rein orchid
  Green woodland orchid
  Little club-spur orchid
  Southern rein orchid
Habenaria cristata
  Crested fringed orchid
  Crested rein orchid
  Crested yellow orchid
  Golden fringed orchid
  Orange crest
Habenaria dilitata Bog candle
  Bog orchid
  Leafy white orchid
  Scent bottle
  Tall white bog orchid
Habenaria gracilis
  Slender bog orchid
Habenaria hookeri
  Hooker's orchid
Habenaria hyperborea
  Leafy northern green orchid
  Northern green orchid
Habenaria lacera
  Green fringed orchid
  Ragged fringed orchid
  Ragged orchid
Habenaria nivea Bog torch
  Frog spear
  Savannah orchid
  Snowy orchid
  Southern small white orchid
  White frog-arrow
  White rein orchid
Habenaria obusata Blunt-
  leaf orchid
  Northern small bog orchid
  One-leaf rein orchid
  Single-leaf rein orchid
  Small bog orchid
Habenaria orbiculata Heal-all
  Moon-set
  Round-leaved orchid
Habenaria peramoena
  Pride-of-the-peak
  Purple fret-lip
  Purple fringeless orchid
  Purple spire orchid
Habenaria psycodes
  Butterfly orchid
  Fairy fringe
  Lesser purple-fringed orchid
  Small purple-fringed orchid
  Soldier's plume
Habenaria saccata
  Slender bog orchid
Habenaria unalascensis
  Alaskan orchid
  Alaska piperia

Haemaria discolor
  Gold-lace orchid
Hammarbya paludosa
  Little bog orchid
Herminium monorchis
  Musk orchid
  Musk orchis
Himantoglossum hircinum
  Lizard orchid
Ipsea speciosa Daffodil orchid
Isotria verticillata
  Green adderling
  Purple five-leaved orchid
  Whorled pogonia
Leucorchis albida
  Small white orchid
Limnorchis hyperborea
  Leafy northern green orchid
  Northern green orchid
Limodorum abortiva
  Violet bird's nest orchid
Liparis liliifolia Large twayblade
  Mauve sleekwort
  Purple scutcheon
Liparis loeselii Bog twayblade
  Fen orchid
  Loesel's twayblade
  Olive scutcheon
  Russet witch
  Yellow twayblade
Listeria convallarioides
  Broad-leaved twayblade
  Broad-lipped twayblade
Listeria cordata
  Heart-leaf twayblade
  Lesser twayblade
Listeria ovata Common twayblade
Lycaste skinneri Nun orchid
  White nun orchid
Lycaste virginalis Nun orchid
  White nun orchid
Lysias orbiculata Heal-all
  Moon-set
  Round-leaved orchid
Malaxis unifolia Adder's tongue
  Green adder's mouth
  Green malaxis
  Tenderwort
  Wide adder's mouth
Miltonia candida Pansy orchid
Miltonia flavescens Pansy orchid
Miltonia spectabilis Pansy orchid
Miltonia vexillaria Pansy orchid
Neotinea maculata
  Dense-flowered orchid
Neottia nidus-avis
  Bird's nest orchid
Odontoglossum crispum
  Lace orchid
Odontoglossum grande
  Tiger orchid
Odontoglossum pulchellum
  Lily-of-the-valley orchid

Spiranthes spiralis

Oncidium cheirophorum	*Pholidota imbricata*
Colombia buttercup	Rattlesnake orchid
Oncidium flexuosum	*Phragmipedium caudatum*
Dancing-doll orchid	Lady's slipper orchid
Dancing-lady orchid	*Physosiphon tubatus* Bottle orchid
Oncidium krameranum	*Platanthera bifolia*
Butterfly orchid	Lesser butterfly orchid
Oncidium papilio Butterfly orchid	*Platanthera chlorantha*
Ophrys apifera Bee orchid	Butterfly orchid
Ophrys fuciflora	Greater butterfly orchid
Late spider orchid	*Pleione maculata* Indian crocus
Spider orchid	*Pleurothallis macrophylla*
Ophrys holoserica	Widow orchid
Late spider orchid	*Pogonia divaricata*
Ophrys insectifera Fly orchid	Funnel-crest orchid
Ophrys speculum Mirror-of-venus	Lily-leaved pogonia
Mirror orchid	Rosebud orchid
Ophrys sphegodes	Spreading pogonia
Early spider orchid	*Pogonia ophioglossoides*
Ophrys tenthredinifera	Adder's mouth orchid
Sawfly orchid	Adder's tongue-leaved pogonia
Orchis coriophora Bug orchid	Beardflower
Orchis latifolia Marsh orchid	Crested ettercap
Orchis laxiflora Jersey orchid	Ettercap
Orchis maderensis Madeira orchid	Rose crest-lip
Orchis mascula Blue butcher	Rose pogonia
Dead man's fingers	Snake's mouth orchid
Early purple orchid	*Pogonia verticillata*
Orchis militaris Soldier orchid	Green adderling
Orchis morio Gandergoose	Purple five-leaved orchid
Green-winged orchid	Whorled pogonia
Salep orchid	*Polyradicion lindenii*
Orchis purpurea Lady orchid	Palm-polly
Orchis rotundifolia Small round-	White butterfly orchid
leaved orchid	*Polyrrhiza lindenii* Palm-polly
Spotted kirtle-pink orchid	White butterfly orchid
Orchis simia Monkey orchid	*Rodriguezia secunda* Coral orchid
Orchis spectabilis	*Spiranthes aestivalis*
Kirtle-pink orchid	Summer lady's tresses
Purple-hooded orchid	*Spiranthes cernua*
Showy orchid	Common lady's tresses
Woodland orchid	Lady's tresses
Orchis tridentata Toothed orchid	Nodding lady's tresses
Orchis ustulata Burnt orchid	Screw-augur
Burnt-tip orchid	*Spiranthes gracilis*
Dwarf orchid	Green pearl-twist
Paphiopedilum concolor	Long tresses
Cypripedium	Slender lady's tresses
Lady's slipper orchid	Southern lady's tresses
Slipper orchid	*Spiranthes grayi*
Paphiopedilum insigne	Little lady's tresses
Cypripedium	Little pearl-twist
Lady's slipper	*Spiranthes praecox*
Slipper orchid	Giant lady's tresses
Peristeria elata Dove flower	Grass-leaved lady's tresses
Dove orchid	Water tresses
Holy Ghost flower	*Spiranthes romanzoffiana*
Phaius tankervilliae	Hooded lady's tresses
Nun's hood orchid	Irish lady's tresses
Nun's orchid	Romanzoff's lady's tresses
Phalaenopsis amabilis	*Spiranthes spiralis*
Moth orchid	Autumn lady's tresses

Tipularia discolor	Cranefly orchid	*Vanilla articulata*	Link vine
	Crippled cranefly		Worm vine
	Elfin-spur		Wormwood
	Mottled cranefly orchid	*Vanilla barbellata*	Link vine
Tipularia unifolia	Cranefly orchid		Worm vine
	Crippled cranefly		Wormwood
	Elfin-spur	*Vanilla grandiflora*	
	Mottled cranefly orchid		Pompona vanilla
			West Indian vanilla
Traunsteinera globosa		*Vanilla planifolia*	Vanilla
	Round-headed orchid	*Vanilla pompona*	Pompona vanilla
			West Indian vanilla
Vanda coerulea	Blue orchid		

POPULAR GARDEN PLANTS

Annuals, biennials, and perennials that are not usually listed under specific sections elsewhere. Popular flowering plants grown in many gardens as bedding plants, container plants, or climbers are covered.

Marguerite –
Argyranthemum frutescens

Abronia latifolia

Abronia latifolia	Sand verbena
Abronia umbellata	
	Pink sand verbena
Acaena anserinifolia	
	Bidgee-widgee
Acaena inermis	
	Blue mountains bidi-bidi
Acaena microphylla	
	Blue mountains bidi-bidi
	New Zealand burr
	Scarlet bidi-bidi
Achillea ageratum	Sweet nancy
Achillea eupatorium	
	Fernleaf yarrow
Achillea filipendulina	
	Fernleaf yarrow
Achillea millefolium	
	Common yarrow
	Milfoil
	Nosebleed
	Thousand weed
	Yarrow
Achillea ptarmica	Goosewort
	Sneezeweed
	Sneezewort
Achillea tomentosa	Alpine yarrow
Achlys triphylla	Vanilla leaf
Aciphylla aurea	Golden spaniard
Aciphylla colensoi	Spaniard
	Wild spaniard
Aciphylla scott-thomsonii	
	Giant spaniard
Aconitum anthora	
	Yellow monkshood
Aconitum lycoctonum	Wolfsbane
Aconitum lycoctonum lycotonum	Northern wolfsbane
Aconitum lycoctonum vulgaria	Wolfsbane
Aconitum lycoctonum vulparia	Foxbane
Aconitum napellus	
	Common monkshood
	Friar's cap
	Monkshood
Aconitum napellus lycoctonum	
	Yellow aconite
Aconitum orientale	
	Russian aconite
Aconitum vulparia	Wolfsbane
Actaea alba	White baneberry
	White cohosh
Actaea erythocarpa	Red baneberry
Actaea pachypoda	
	White baneberry
Actaea rubra	Red baneberry
Actaea spicata	Black baneberry
Actaea spicata alba	
	White baneberry
Actaea spicata rubra	
	Red baneberry
Actinea grandiflora	
	Pigmy sunflower
Actinotus helianthi	Flannel flower
Adenophora lilifolia	
	Common ladybell
	Grand bellflower
	Ladybell
Adonis aestivalis	Summer adonis
	Summer pheasant's eye
Adonis annua	Pheasant's eye
Adonis vernalis	Spring adonis
	Yellow pheasant's eye
Adoxa moschatellina	
	Five-faced bishop
	Moschatel
	Town hall clock
Aethusa cynapium	Dog poison
	Fool's parsley
Agathaea coelestis	Blue daisy
	Blue marguerite
Ageratum houstonianum	
	Floss flower
Ageratum mexicanum	
	Floss flower
Agrimonia eupatoria	Agrimony
	Church steeples
	Common agrimony
	Sticklewort
Agrostemma coronaria	
	Mullein pink
	Rose campion
Agrostemma flos-jovis	
	Flower of Jove
Agrostemma githago	
	Corn campion
	Corn cockle
	Corn pink
Ajuga chamaepitys	Ground pine
	Yellow bugle
Ajuga genevensis	Blue bugle
Ajuga pyramidalis	Pyramidal bugle
Ajuga reptans	Bugle
	Carpenter's herb
	Common bugle
Alcea chinensis	Garden hollyhock
	Hollyhock
Alcea ficifolia	Antwerp hollyhock
	Figleaf hollyhock
Alcea rosea	Garden hollyhock
	Hollyhock
Alchemilla alpina	
	Alpine lady's mantle
Alchemilla mollis	
	Common lady's mantle
	Lady's mantle
Alchemilla vulgaris	Bear's foot
	Common lady's mantle
	Lady's mantle
	Lion's foot
Aletris farinosa	Ague root
	Blazing star
	Colic root
	Starwort

POPULAR GARDEN PLANTS

Arnebia echioides

Alliaria petiolata
Jack-by-the-hedge
Alonsoa warscewiczii Mask flower
Aloysia triphylla Lemon verbena
Althaea chinensis
Garden hollyhock
Hollyhock
Althaea ficifolia
Antwerp hollyhock
Figleaf hollyhock
Althaea rosea Garden hollyhock
Hollyhock
Alyssum argenteum Yellow tuft
Alyssum maritimum
Sweet alyssum
Alyssum murale Yellow tuft
Alyssum saxatile
Common yellow alyssum
Gold dust
Alyssum spinosum Spiny alyssum
Amaracus dictamnus
Cretan dittany
Amaranthus albus White amaranth
Amaranthus caudatus
Love-lies-bleeding
Tassel flower
Amaranthus graecizans
Tumbleweed
Amaranthus hybridus
Green amaranthus
Amaranthus hypochondriacus
Prince's feather
Amaranthus retroflexus
Amaranth
Common amaranth
Amaranthus tricolor Joseph's coat
Ammi majus Bishop's flower
False bishop's weed
Ammobium alatum Sand flower
Winged everlasting
Amsonia roseus
Madagascar periwinkle
Rose periwinkle
Anacyclus depressus
Mount Atlas daisy
Anacyclus pyrethrum Pellitory
Pellitory of Spain
Spanish pellitory
Anagallis arvensis
Poor man's weather glass
Scarlet pimpernel
Shepherd's barometer
Anagallis foemina Blue pimpernel
Anagallis linifolia Blue pimpernel
Anagallis minima Chaff weed
Anagallis monelli linifolia
Blue pimpernel
Anagallis tenella Bog pimpernel
Anchusa arvensis Bugloss
Anchusa azurea Italian bugloss
Large blue alkanet
Anchusa caespitosa Tufted alkanet

Anchusa capensis Annual anchusa
Cape forget-me-not
Summer forget-me-not
Anchusa italica Italian bugloss
Anchusa officinalis Alkanet
Androsace carnea
Pink rock jasmine
Androsace chamaejasme
Bastard jasmine
Androsace lactea
Milkwhite rock jasmine
Milky rock jasmine
Anemone vernalis Spring anemone
Anemonopsis californica
Apache beads
Yerba mansa
Anemonopsis macrophylla
False anemone
Anthemis arabicus
Palm Springs daisy
Anthemis nobilis
Common chamomile
Garden chamomile
Green chamomile
Roman chamomile
Russian chamomile
Anthemis sancti-johannis
Saint John's chamomile
Anthemis tinctoria
Golden marguerite
Ox-eye chamomile
Antirrhinum majus Snapdragon
Weasel's snout
Aquilegia canadensis
American wild columbine
Aquilegia vulgaris
European wild columbine
Granny's bonnet
True columbine
Arctotis breviscapa African daisy
Arctotis grandis African daisy
Argemone grandiflora
Prickly poppy
Argemone mexicana Devil's fig
Mexican poppy
Argemone platyceras
Crested poppy
Argyranthemum frutescens
Marguerite
Paris daisy
Aristolochia durior
Dutchman's pipe
Aristolochia elegans Calico plant
Aristolochia macrophylla
Dutchman's pipe
Aristolochia sipho
Dutchman's pipe
Armeria maritima Common thrift
Sea pink
Armeria plantaginea Border thrift
Armeria pseudarmeria
Border thrift
Arnebia echioides Prophet flower

Artemisia abrotanum	Lad's love
	Old man
	Southernwood
Artemisia absinthium	Absinthe
	Wormwood
Artemisia gnaphalodes	Cudweed
	Western mugwort
	White sage
Artemisia lactiflora	
	White mugwort
Artemisia ludoviciana	Cudweed
	Western mugwort
	White sage
Artemisia purshiana	Cudweed
	Western mugwort
	White mugwort
	White sage
Artemisia stellerana	
	Beach wormwood
	Dusty miller
	Old woman
Artemisia tridentata	Sage brush
Aruncus dioicus	Goat's beard
Aruncus sylvester	Goat's beard
Aruncus vulgaris	Goat's beard
Aster amellus	Italian aster
Aster ericoides	Heath aster
Aster linosyris	Goldilocks
Aster novae-angliae	
	Michaelmas daisy
	New England aster
Aster novi-belgii	New York aster
	True michaelmas daisy
Aster tenacetifolius	Tahoka daisy
Atriplex hortensis rubra	
	Red mountain spinach
	Red orach
Atriplex lociniata	Frosted orach
Atriplex patula	Common orach
Aubretia deltoides	
	Common aubretia
Aurinia saxatilis	
	Common yellow alyssum
	Gold dust
Baptisia tinctoria	Horsefly
	Wild indigo
Bartonia aurea	Blazing star
Begonia semperflorens	
	Bedding begonia
	Fibrous-rooted begonia
Bellis perennis	Common daisy
	English daisy
	Meadow daisy
Betonica grandiflora	Betony
	Woundwort
Betonica macrantha	Betony
	Woundwort
Betonica officinalis	Bishop's wort
	Wood betony
Bilderdykia aubertii	
	China fleece flower
	Silver lace vine
Bilderdykia baldschuanicum	
	Bokhara fleece flower
	Russian vine
Bocconia cordata	Plume poppy
Bocconia microcarpa	
	Lesser plume poppy
Brachycome iberidifolia	
	Swan river daisy
Brassica oleracea acephala	
	Ornamental cabbage
Brunnera macrophylla	
	Siberian bugloss
Buphthalum salicifolium	
	Willow-leaf ox-eye
	Yellow ox-eye
Bupleurum fruticosum	Bupleurum
Calamintha grandiflora	Calamint
Calamintha nepeta nepeta	
	Calamint
Calandrinia menziesii	Red maids
Calandrinia umbellata	
	Rock purslane
Calendula officinalis	Golds
	Pot marigold
	Scotch marigold
Calendula officinalis prolifera	
	Hen and chickens marigold
Callistephus chinensis	
	Annual aster
	China aster
Calluna vulgaris	Heather
	Ling
Caltha palustris	Kingcup
	Marsh marigold
	Water cowslip
Caltha polypetala	Giant kingcup
Campanula alliarifolia	
	Spurred bellflower
Campanula allionii	
	Alpine campanula
Campanula alpestris	
	Alpine campanula
Campanula barbata	
	Bearded bellflower
Campanula carpatica	
	Carpathian bellflower
	Carpathian harebell
	Tussock bellflower
Campanula cochleariifolia	
	Fairy's thimbles
Campanula elatines garganica	
	Adriatic bellflower
Campanula garganica	
	Adriatic bellflower
Campanula glomerata	
	Clustered bellflower
Campanula grandiflora	
	Canterbury bell
Campanula grandis	
	Peach-leaved bellflower
	Willow bellflower

POPULAR GARDEN PLANTS

Chrysanthemum carinatum

Campanula isophylla
Italian bellflower
Star of Bethlehem

Campanula lactiflora
Milky bellflower

Campanula latifolia
Giant bellflower

Campanula latiloba
Peach-leaved bellflower
Willow bellflower

Campanula medium
Canterbury bell

*Campanula medium
calycanthema*
Cup-and-saucer Canterbury bell

Campanula patula
Spreading bellflower

Campanula persicifolia
Peach-leaved bellflower
Willow bellflower

Campanula portenschlagiana
Wall harebell

Campanula poscharskyana
Serbian bellflower

Campanula pusilla
Fairy's thimbles

Campanula pyramidalis
Chimney bellflower
Steeple bellflower

Campanula rapunculus   Rampion
Rampion bellflower

Campanula rotundifolia   Bluebell
English harebell
Harebell
Harebell bellflower
Scottish bluebell

Campanula thyrsoides
Yellow bellflower

Campanula trachelium
Bats-in-the-belfry
Coventry bells
Nettle-leaved bellflower

Campanula urticifolia
Bats-in-the-belfry
Coventry bells
Nettle-leaved bellflower

Campsis chinensis
Chinese trumpet creeper

Campsis grandiflora
Chinese trumpet creeper

Campsis radicans
American trumpet creeper
Trumpet vine

Capsicum annuum   Red pepper

Catananche coerulea
Blue cupidone
Blue succory
Cupid's dart

Catharanthus roseus
Madagascar periwinkle
Rose periwinkle

Celosia argentea cristata
Cockscomb

Celosia cristata childsii
Cockscomb

Celosia cristata plumosa
Prince of Wales' feathers

Celosia plumosa   Plume flower

Celsia arcturus   Cretan bear's tail
Cretan mullein

Cenia barbata   Pincushion plant

Centaurea americana
Basket flower

Centaurea calcitrapa
Red star thistle
Star thistle

Centaurea candidissima
Dusty miller

Centaurea cineraria   Dusty miller

Centaurea cyanus
Blue cornflower
Bluebottle
Cornflower

Centaurea dealbata
Perennial cornflower

Centaurea gymnocarpa
Dusty miller

Centaurea imperialis   Sweet sultan

Centaurea macrocephala
Globe centaurea
Great golden knapweed

Centaurea montana
Mountain bluet
Mountain knapweed
Perennial cornflower

Centaurea moschata   Sweet sultan

Centaurea nigra   Black knapweed
Hardheads
Knapweed
Lesser knapweed

Centaurea phrygia   Wig knapweed

Centaurea ragusina   Dusty miller

Centaurea rhaponticum
Giant knapweed

Centaurea scabiosa
Great knapweed

Centaurea solstitalis
Saint Barnaby's thistle
Yellow star thistle

Centranthus ruber   Bouncing bess
Drunken sailor
Jupiter's beard
Pretty betsy
Red valerian

Cephalaria gigantea
Giant scabious

Cephalaria tatarica   Giant scabious

Cheiranthus × allionii
Siberian wallflower

Cheiranthus cheiri
Common wallflower
English wallflower

Cheiranthus maritimus
Virginian stock

Chrysanthemum carinatum
Painted daisy

Chrysanthemum coccineum
    Painted daisy
    Pyrethrum

Chrysanthemum coronarium   Crown daisy

Chrysanthemum leucanthemum   Moon daisy
    Ox-eye daisy

Chrysanthemum maximum
    Shasta daisy

Chrysanthemum parthenium
    Feverfew

Chrysanthemum rubellum
    Korean chrysanthemum

Chrysanthemum segetum
    Corn marigold

Chrysanthemum serotinum
    Giant daisy
    Hungarian daisy

Chrysanthemum × superbum
    Shasta daisy

Chrysanthemum uliginosum
    Giant daisy
    Hungarian daisy

Cicendia filiformis   Yellow centaury

Cimicifuga americanum
    American bugbane

Cimicifuga racemosa
    Black snakeroot

Cissus antarctica   Kangaroo vine

Clarkia amoena
    Farewell-to-spring
    Satin flower

Clarkia concinna   Red ribbons

Clarkia elegans   Flore pleno

Clarkia unguiculata   Clarkia
    Mountain garland

Cleome spinosa   Spider flower

Cleretum bellidiforme
    Livingstone daisy

Clianthus dampieri   Glory pea

Clianthus formosus   Glory pea

Clianthus puniceus   Lobster claw
    Parrot's bill

Clianthus speciosus   Glory pea
    Sturt's desert pea

Cnicus benedictus   Blessed thistle
    Holy thistle
    Saint Benedict's thistle

Cobaea scandens   Cathedral bells
    Cup-and-saucer vine
    Mexican ivy
    Monastery bells
    Steeple bells

Collinsia bicolor   Chinese houses
    Innocence

Collinsia grandiflora
    Blue-eyed mary
    Blue lips

Collinsia heterophylla
    Chinese houses

Consolida ambigua   Larkspur
    Rocket larkspur

Consolida regalis
    Branched larkspur
    Forking larkspur

Convolvulus cneorum
    Bush morning glory

Convolvulus minor
    Dwarf morning glory

Convulvulus tricolor
    Dwarf morning glory

Coreopsis grandiflora   Tickseed

Cortaderia argentea   Pampas grass

Cortaderia selloana   Pampas grass

Cosmos atrosanguineus
    Black cosmos
    Chocolate cosmos

Cosmos bipinnatus   Mexican aster

Cotula barbata   Pincushion plant

Cotula cenia   Pincushion plant

Cotula coronopifolia
    Brass buttons

Crepis biennis
    Greater hawk's beard
    Rough hawk's beard

Crepis capillaris
    Smooth hawk's beard

Crepis foetida
    Stinking hawk's beard

Crepis incana   Pink dandelion

Crepis mollis
    Northern hawk's beard

Crepis nicaeensis
    French hawk's beard

Crepis paludosa
    Marsh hawk's beard

Crepis pygmaea
    Pigmy hawk's beard

Crepis rubra   Hawk's beard

Crepis vesicaria
    Beaked hawk's beard

Cryophytum crystallinum
    Ice plant
    Sea fig

Cuphea ignea   Cigar flower
    Firecracker plant

Cyananthus lobatus
    Trailing bellflower

Cynoglossum amabile
    Chinese forget-me-not
    Hound's tongue

Dahlia juarezii   Cactus dahlia

Dahlia merckii   Bedding dahlia

Daubentonia tripetii
    Scarlet wisteria

Delphinium ajacis   Rocket larkspur

Delphinium cardinale
    Scarlet larkspur

Delphinium consolida   Larkspur

Delphinium nudicaule
    Christmas horns

Delphinium staphisagria
    Stavesacre

POPULAR GARDEN PLANTS

Echinops ritro

Desmodium gyrans	
	Telegraph plant
Dianthus × allwoodii	Garden pinks
Dianthus alpinus	Alpine pink
Dianthus arenarius	Sand pink
Dianthus armeria	Deptford pink
Dianthus × arvernensis	
	Auvergne pink
Dianthus barbatus	Sweet william
Dianthus caesius	Cheddar pink
Dianthus caryophyllus	
	Border carnation
	Carnation
	Clove pink
	Cottage pink
	Gilliflower
	Picotee
	Wild carnation
Dianthus chinensis	Annual pink
	Chinese pink
	Indian pink
	Rainbow pink
Dianthus deltoides	Maiden pink
Dianthus glacialis	Glacier pink
Dianthus gratianopolitanus	
	Cheddar pink
Dianthus × heddewigii	
	Japanese pink
Dianthus inodorus	Wood pink
Dianthus × latifolius	Button pink
Dianthus neglectus	Glacier pink
Dianthus pavonius	Glacier pink
Dianthus plumarius	Border pink
	Common pink
	Cottage pink
	Grass pink
	Wild pink
Dianthus sinensis	Chinese pink
Dianthus superbus	Fringed pink
	Large pink
	Lilac pink
Dianthus sylvestris	Wood pink
Diascia barberae	Twinspur
Dicentra canadensis	Squirrel corn
	Staggerweed
	Turkey corn
	Turkey pea
Dicentra chrysantha	
	Golden eardrops
Dicentra eximia	
	Fringed bleeding heart
Dicentra formosa	
	Western bleeding heart
Dicentra spectabilis	
	Bleeding heart
	Chinaman's breeches
	Common bleeding heart
	Dutchman's breeches
	Lady's locket
	Lyre flower
Dictamnus albus	Burning bush
	Dittany
	Gas plant

Dictamnus fraxinella	
	Burning bush
	Dittany
	Gas plant
Didiscus caerulea	Blue lace flower
Digitalis ferruginea	Rusty foxglove
Digitalis grandiflora	
	Large yellow foxglove
Digitalis lanata	Woolly foxglove
Digitalis lutea	
	Small yellow foxglove
Digitalis purpurea	
	Common foxglove
	Dead men's bells
	Fairy thimbles
	Foxglove
	Lady's gloves
	Witch's gloves
Dimorphotheca annua	
	Annual cape marigold
	Cape marigold
Dimorphotheca aurantiaca	
	Star of the veldt
Dimorphotheca barberae	
	Dwarf cape marigold
Dimorphotheca osteospermum	African daisy
Dimorphotheca pluvialis	
	Namaqualand daisy
	Rain daisy
Dimorphotheca sinuata	
	Namaqualand daisy
	Star of the veldt
Diplacus glutinosus	
	Bush monkey flower
Diplarrhena moraea	Butterfly flag
Diplotaxis erucoides	White rocket
Diplotaxis muralis	
	Annual wall rocket
Diplotaxis tenuifolia	
	Perennial wall rocket
	Wall mustard
	Wall rocket
Dodecatheon alpinum	
	Shooting star
Dodecatheon jeffreyi	
	Sierra shooting star
Dodecatheon meadia	
	Common shooting star
	Shooting star
Doronicum caucasicum	
	Leopard's bane
Doronicum pardalianches	
	Great leopard's bane
Dorotheanthus bellidiflorus	
	Livingstone daisy
Dregea sinensis	
	Chinese wax flower
Echinacea angustiflora	
	Kansas niggerhead
Echinacea purpurea	Black samson
	Purple coneflower
Echinops ritro	Globe thistle

Echites andrewsii

Echites andrewsii	Savannah flower
Echium lycopsis	Viper's bugloss
Echium plantagineum	
	Purple bugloss
	Viper's bugloss
Echium vulgare	Blueweed
	Viper's bugloss
Emmenanthe penduliflora	
	Californian golden bells
	Californian whispering bells
Eomecon chionanthe	Snow poppy
Epilobium alsinifolium	
	Chickweed willowherb
Epilobium angustifolium	
	Fireweed
	Rosebay willowherb
Epilobium brunnescens	
	New Zealand willowherb
Epilobium ciliatum	
	American willowherb
Epilobium hirsutum	
	Codlins and cream
	Great willowherb
Epilobium lanceolatum	
	Spear-leaved willowherb
Epilobium montanum	
	Broad-leaved willowherb
Epilobium obcordatum	
	Rock fringe willowherb
Epilobium palustra	
	March willowherb
Epilobium parviflorum	
	Hairy willowherb
	Hoary willowherb
Epilobium roseum	
	Pale willowherb
	Pink willowherb
Epilobium tetragonum	
	Square-stalked willowherb
Episcia cupreata	Flame violet
Episcia dianthiflora	Lace flower
Erechtites hieracifolia	Fireweed
	Pilewort
Erica vulgaris	Heather
	Ling
Erigeron acre	Blue fleabane
Erigeron annuus	Sweet scabious
Erigeron aurantiacus	
	Double orange daisy
	Orange daisy
Erigeron borealis	Alpine fleabane
Erigeron karkinskianus	
	Australian fleabane
	Mexican fleabane
	Wall daisy
Erinus alpinus	Fairy foxglove
	Summer starwort
Eritrichium nanum	
	Herald of heaven
	King of the alps
Eryngium maritimum	Sea holly
Erysimum alpinum	
	Alpine wallflower
Erysimum asperum	
	Siberian wallflower
Erysimum cheiranthoides	
	Treacle mustard
Erysimum perofskianum	
	Siberian wallflower
Eschscholzia californica	
	Californian poppy
Eucharidium concinna	
	Red ribbons
Eupatorium ageratoides	
	White sanicle
	White snakeroot
Eupatorium cannabinum	
	Dutch agrimony
	Hemp agrimony
	Holy rope
Eupatorium coelestinum	
	Blue boneset
	Hardy ageratum
	Mistflower
Eupatorium maculatum	
	Joe-pye weed
Eupatorium perfoliatum	
	Thoroughwort
Eupatorium purpureum	
	Gravel root
	Gravel weed
	Joe-pye weed
	Purple boneset
	Queen of the meadow
Eupatorium rugosum	Hardy age
	Mist flower
	White sanicle
	White snakeroot
Eupatorium teucrifolium	
	Wild horehound
	Wild horsehound
Euphorbia angydaloides	
	Wood spurge
Euphorbia arctica	Great eyebright
Euphorbia caput-medusae	
	Medusa's head
Euphorbia corollata	
	Flowering spurge
	White purslane
Euphorbia cyparissias	
	Cypress spurge
Euphorbia dulcis	Sweet spurge
Euphorbia epithymoides	
	Cushion spurge
Euphorbia esula	Leafy spurge
Euphorbia exigua	Dwarf spurge
Euphorbia fulgens	Scarlet plume
Euphorbia helioscopia	Sun spurge
Euphorbia heterophylla	
	Annual poinsettia
	Fire on the mountain
	Mexican fire plant
	Painted spurge

POPULAR GARDEN PLANTS

Geranium phaeum

Euphorbia hirta	Asthma weed
	Pill-bearing spurge
Euphorbia hyberna	Irish spurge
Euphorbia lathyris	Caper spurge
	Mole plant
Euphorbia maculata	Milk purslane
Euphorbia marginata	
	Mountain snow
	Snow on the mountain
Euphorbia myrsinites	Blue spurge
Euphorbia palustris	Bog spurge
	Fen spurge
Euphorbia paralias	Sea spurge
Euphorbia peplis	Petty spurge
	Purple spurge
Euphorbia peplus	Petty spurge
	Purple spurge
Euphorbia pilulifera	Cat's hair
Euphorbia platyphyllos	
	Broad-leaved spurge
Euphorbia polychroma	
	Cushion spurge
Euphorbia pulcherrima	
	Christmas star
	Fire plant
	Lobster plant
	Poinsettia
Euphorbia robbiae	Robb's bonnet
Euphorbia stricta	Upright spurge
Euphorbia villosa	Hairy spurge
Eutoca viscida	Sticky phacelia
Exacum affine	Arabian violet
	German violet
	Persian violet
Felicia amelloides	Blue daisy
	Blue marguerite
Felicia bergerana	Kingfisher daisy
Festuca glauca	Blue fescue
Festuca ovina	Sheep's fescue
Festuca ovina glauca	Blue fescue
Filipendula hexapetala	Dropwort
Filipendula rubra	
	Queen of the prairie
Filipendula ulmaria	
	Common meadowsweet
	Meadowsweet
	Queen of the meadow
Filipendula vulgaris	Dropwort
Francoa ramosa	Maiden's wreath
Francoa sonchifolia	Bridal wreath
Fumaria capreolata	
	Ramping fumitory
	White fumitory
Fumaria indica	American fumitory
Fumaria muralis	
	Common ramping fumitory
	Wall fumitory
Fumaria officinalis	
	Common fumitory
	Earth smoke
	Fumitory

Fumaria vesicaria	
	Bladder fumitory
Gaillardia aristata	Blanket flower
Gaillardia pulchella	
	Annual gaillardia
	Blanket flower
Galega officinalis	French lilac
	Goat's rue
Galeobdolon luteum	
	Golden dead nettle
	Yellow archangel
Gazania × hybrids	
	Treasure flowers
Gazania leucolaena	
	Trailing gazania
Gazania rigens leucolaena	
	Trailing gazania
Gazania uniflora	Trailing gazania
Gentiana acaulis	Stemless gentian
	Trumpet gentian
Gentiana asclepiadea	
	Willow gentian
Gentiana brachyphylla	
	Short-leaved gentian
Gentiana cruciata	Cross gentian
Gentiana kochiana	
	Trumpet gentian
Gentiana lutea	Yellow gentian
Gentiana pneumonanthe	
	Calathian violet
	Marsh gentian
Gentiana punctata	Spotted gentian
Gentiana scabrae	Japanese gentian
Gentiana septemfida	
	Crested gentian
Gentiana utriculosa	
	Bladder gentian
Gentiana verna	Spring gentian
	Vernal gentian
Gentiana verna angulosa	
	Spring gentian
Geranium columbinum	
	Dove's foot cranesbill
Geranium dissectum	
	Cut-leaved cranesbill
Geranium endressii	
	French cranesbill
Geranium ibericum	
	Iberian cranesbill
Geranium lucidum	
	Shining cranesbill
Geranium macrorrhizum	
	Balkan cranesbill
	Rock cranesbill
Geranium maculatum	
	American cranesbill
	Wild cranesbill
	Wild geranium
Geranium molle	Soft cranesbill
Geranium phaeum	
	Dusky cranesbill
	Mourning widow

Geranium pratense

Geranium pratense
Meadow cranesbill
Meadow geranium

Geranium purpurea
Lesser herb robert

Geranium pyrenaicum
Hedgerow cranesbill
Mountain cranesbill

Geranium robertianum
Herb robert
Stinking bob

Geranium sanguineum
Blood-red geranium
Bloody cranesbill

Geranium sylvaticum
Crow flower
Wood cranesbill

Gerbera jamesonii
Barberton daisy
Transvaal daisy

Geum chiloense Chilean avens

Geum coccineum Chilean avens

Geum montanum Alpine avens

Geum quellyon Chilean avens

Geum reptans Creeping avens

Geum rivale Indian chocolate
Nodding avens
Purple avens
Water avens

Geum urbanum Avens
Cloveroot
Herb bennet
Indian chocolate
Water flower
Wood avens

Gilia californica Prickly phlox

Gilia capitata Blue thimble flower
Queen Anne's thimbles

Gilia coronopifolia Skyrocket
Standing cypress

Gilia hybrida Stardust

Gilia lutea Stardust

Gilia rubra Skyrocket
Standing cypress

Gilia tricolor Bird's eyes

Glaucium corniculatum
Red horned poppy
Sea poppy

Glaucium flavum Horned poppy
Yellow horned poppy

Glaucium grandiflorum
Red horned poppy
Sea poppy

Glaucium luteum
Yellow horned poppy

Godetia amoena
Farewell to spring

Godetia grandiflora Godetia
Satin flower

Gomphrena globosa
Globe amaranth

Gypsophila elegans Baby's breath

Gypsophila muralis
Annual gypsophila

Gypsophila paniculata
Baby's breath
Chalk plant

Gypsophila repens
Alpine gypsophila

Hebe hulkeana New Zealand lilac

Hedyotis caerulea Bluets
Innocence

Helenium autumnale Helenium
Sneezeweed

Helianthus annuus
Annual sunflower
Common sunflower
Giant marigold
Marigold of Peru
Peruvian marigold
Sunflower

Helichrysum angustifolium
Curry plant
White-leaved everlasting

Helichrysum bellidiodes
Everlasting daisy

Helichrysum bracteatum
Strawflower

Helichrysum ledifolium
Kerosene bush

Helichrysum petiolatum
Liquorice plant

Helichrysum seotinum
Curry plant

Helichrysum stoechas Goldilocks

Helichrysum thyrsoideum
Snow in summer

Heliophila longifolia Cape stock

Heliopsis helianthoides scabra Heliopsis

Heliopsis scabra Heliopsis

Heliotropium arborescens
Cherry pie

Heliotropium corymbosum
Cherry pie

Heliotropium hybridum
Cherry pie

Heliotropium peruvianum
Cherry pie

Helipterum manglesii
Sunray
Swan river everlasting

Helleborus argutifolius
Corsican hellebore

Helleborus corsicus
Corsican hellebore

Helleborus foetidus Setterwort
Stinking hellebore

Helleborus lividus
Corsican hellebore

Helleborus niger Black hellebore
Christmas rose

Helleborus orientalis Lenten rose

Helleborus viridis
Green hellebore

Ipomoea quamoclit

Hesperis matronalis	Damask violet
	Dame's rocket
	Dame's violet
	Rocket
	Sweet rocket
Heuchera hispida	Satin leaf
Heuchera sanguinea	Coral bells
	Coral flower
Hieracium aurantiacum	
	Devil's paintbrush
	Grim collier
	Orange hawkweed
Hieracium brunneocroceum	
	Orange hawkweed
Hieracium lanatum	
	Woolly hawkweed
Hieracium maculatum	
	Spotted hawkweed
Hieracium umbellatum	
	Leafy hawkweed
Hieracium villosum	
	Shaggy hawkweed
Hieracium vulgatum	
	Common hawkweed
Homogyne alpina	Alpine coltsfoot
Horminum pyranaicum	
	Dragon mouth
Hypericum hirsutum	
	Hairy Saint John's wort
Hypericum humifusum	
	Creeping Saint John's wort
Hypericum lanuginosum	
	Downy Saint John's wort
Hypericum linarifolium	
	Flax-leaved Saint John's wort
Hypericum perforatum	
	Perforate Saint John's wort
Hypericum tetrapterum	
	Square-stemmed Saint John's wort
Hyssopus aristatus	Hyssop
Iberis amara	Candytuft
	Rocket candytuft
	Wild candytuft
Iberis coronaria	Rocket candytuft
Iberis umbellata	Annual candytuft
	Candytuft
	Common candytuft
	Globe candytuft
Iliamna rivularis	
	Mountain hollyhock
Impatiens aurea	Balsam weed
	Jewel weed
Impatiens balsamina	Balsam
	Rose balsam
	Touch-me-not
Impatiens biflora	Jewel weed
	Spotted touch-me-not
Impatiens capensis	Jewel weed
	Orange balsam
	Spotted touch-me-not

Impatiens glandulifera	
	Himalayan balsam
	Himalayan touch-me-not
	Indian balsam
	Jumping jack
	Policeman's helmet
Impatiens noli-tangere	
	Touch-me-not
	Wild balsam
Impatiens parviflora	Small balsam
Impatiens roylei	
	Himalayan touch-me-not
Impatiens sultanii	Busy lizzie
	Patience
	Patient lucy
Impatiens wallerana	Busy lizzie
	Patience
	Patient lucy
Incarvillea delavayi	
	Chinese trumpet flower
Incarvillea grandiflora	
	Chinese trumpet flower
Inula conyza	
	Ploughman's spikenard
Inula crithmoides	Golden samphire
Inula ensifolia	
	Narrow-leaved inula
Inula helenium	Elecampane
	Scabwort
	Wild sunflower
Inula magnifica	Giant inula
Inula royleana	
	Himalayan elecampane
Ionopsidium acaule	Carpet plant
	Diamond flower
	Violet cress
Ipomoea acuminata	
	Blue dawn flower
	Morning glory
	Perennial morning glory
Ipomoea alba	Moonflower
	Moonvine
Ipomoea bona-nox	Moonflower
	Moonvine
Ipomoea cardinalis	
	Cardinal climber
Ipomoea coccinea	
	Red morning glory
	Star ipomoea
Ipomoea lobata	
	Crimson star glory
Ipomoea × multifida	
	Cardinal climber
Ipomoea nil	
	Japanese morning glory
Ipomoea noctiflora	Moonflower
	Moonvine
	Queen of the night
	Moonvine
Ipomoea purpurea	
	Common morning glory
	Morning glory
Ipomoea quamoclit	Cypress vine

Ipomoea roxburghii

Ipomoea roxburghii
Moonflower

Jasminum mesnyi
Japanese jasmine
Primrose jasmine

Jasminum nudiflorum
Winter jasmine

Jasminum officinale
Common white jasmine
Poet's jessamine
Summer jasmine
White jasmine

Jasminum parkeri Dwarf jasmine

Jasminum polyanthum
Chinese jasmine

Jasminum primulinum
Japanese jasmine
Primrose jasmine

Jovibarba soboliferum
Hen-and-chickens houseleek

Kalistroemia grandiflora
Arizona poppy

Kentranthus ruber Red valerian

Kirengeshoma palmata Waxbells

Knautia arvensis Field scabious

Kochia scoparia trichophylla
Belvedere
Broom cypress
Burning bush
Fire bush
Mock cypress
Summer cypress

Kochia trichophylla Burning bush

Lamiastrum galeobdolon
Golden deadnettle
Yellow archangel

Lamium album Archangel
Bee nettle
Blind nettle
Weasel's snout
White deadnettle

Lamium amplexicaule
Henbit
Henbit deadnettle

Lamium galeobdolon
Golden deadnettle
Weasel's snout
Yellow archangel

Lamium maculatum Dead nettle
Spotted deadnettle

Lamium orvala Giant deadnettle

Lamium purpureum
Purple archangel
Purple deadnettle
Red deadnettle

Lathyrus aphaca Yellow vetchling

Lathyrus hirsutus Hairy vetchling

Lathyrus japonicus Sea pea

Lathyrus latifolius Everlasting pea
Perennial pea
Perennial sweet pea

Lathyrus littoralis Beach pea

Lathyrus maritimus
Beach pea
Sea pea

Lathyrus montanus Bitter vetch

Lathyrus nervosus
Lord Anson's pea

Lathyrus nissolia Grass vetchling

Lathyrus odoratus Sweet pea

Lathyrus palustris Marsh pea

Lathyrus rotundifolius
Persian everlasting pea

Lathyrus sativus Chickling pea
Grass pea

Lathyrus splendens
Californian pea
Pride of California

Lathyrus sylvestris Wild pea

Lathyrus tingitanus Tangier pea
Tangier scarlet pea

Lathyrus tuberosus Tuberous pea

Lathyrus vernuus Spring vetch

Lavatera trimestris Annual mallow
Rose mallow

Layia platyglossa Tidy tips

Layia elegans Tidy tips

Legousia speculum-veneris
Venus's looking glass

Leontopodium alpinum
Common edelweiss
Edelweiss
Lion's foot

Leptosiphon hybridus Stardust

Leucanthemella serotina
Hungarian daisy
Moon daisy

Leucanthemopsis alpina
Alpine moon daisy

Leucanthemum vulgare Dog daisy
Dun daisy
Field daisy
Horse daisy
Marguerite
Moon daisy
Ox-eye daisy
White weed

Leucanthemum × superbum
Shasta daisy

Lewisia rediviva Bitter root

Liatris callilepis Kansas gayfeather

Liatris odoratissima Deer's tongue
Vanilla leaf
Wild vanilla

Liatris pycnostachya
Cat-tail gayfeather
Kansas gayfeather
Prairie blazing star

Liatris scariosa Tall gayfeather

Liatris spicata Blazing star
Button snakeroot
Gayfeather
Spike gayfeather

Liatris squarrosa
Rattlesnake master

POPULAR GARDEN PLANTS

Lupinus pubescens

Ligularia japonica	Giant ragwort
Limnanthes douglasii	
	Marsh flower
	Meadow foam
	Poached egg flower
	Poached egg plant
Limonium bellidifolium	
	Matted sea lavender
Limonium binervosum	
	Rock sea lavender
Limonium bonduellii	
	Algerian statice
Limonium carolinianum	
	American sea lavender
Limonium latifolium	
	Border sea lavender
	Broad-leaved sea lavender
	Statice
Limonium sinuatum	
	Notch-leaf statice
	Statice
	Winged statice
Limonium suworowii	
	Pink pokers
	Rat's-tail statice
	Russian statice
Linanthus grandiflorus	
	Mountain phlox
Linaria alpina	Alpine toadflax
Linaria arenaria	Sand toadflax
Linaria maroccana	Bunny rabbits
	Toadflax
Linaria purpurea	Purple toadflax
Linaria repens	Pale toadflax
Linaria vulgaris	Brideweed
	Butter and eggs
	Calve's snout
	Common toadflax
	Eggs and bacon
	Flaxweed
	Pedlar's basket
	Yellow toadflax
Lindheimera texana	Star daisy
Linum bienne	Pale flax
Linum catharticum	Fairy flax
	White flax
Linum flavum	Golden flax
	Yellow flax
Linum grandiflorum	Red flax
	Scarlet flax
Linum lewisii	Prairie flax
Linum monogynum	
	New Zealand flax
Linum perenne	Blue flax
Linum rubrum	Red flax
	Scarlet flax
Linum usitatissimum	
	Annual blue flax
	Common blue flax
	Flax
	Linseed
Lippia citriodora	Lemon verbena

Lobelia cardinalis	Cardinal flower
	Scarlet lobelia
Lobelia erinus	Bedding lobelia
	Edging lobelia
Lobelia erinus pendula	
	Trailing lobelia
Lobelia fulgens	Cardinal flower
Lobelia inflata	Asthma weed
	Indian tobacco
Lobelia siphililea	Great lobelia
Lobelia siphilitica	Blue lobelia
Lobelia urens	Blue lobelia
	Heath lobelia
Lobularia maritima	Sea alyssum
	Sweet alison
	Sweet alyssum
Lonicera × brownii	
	Scarlet trumpet honeysuckle
Lonicera caprifolium	
	Goat-leaf honeysuckle
Lonicera hildebrandiana	
	Giant honeysuckle
Lonicera involucrata	Twinberry
Lonicera japonica	
	Japanese honeysuckle
Lonicera nitida	Box honeysuckle
Lonicera periclymenum	
	Common honeysuckle
	European honeysuckle
	Honeysuckle
	Wild woodbine
	Woodbine
Lonicera pileata	
	Privet honeysuckle
Lonicera sempervirens	
	Trumpet honeysuckle
Lonicera tatarica	
	Tartarian honeysuckle
Lonicera tragophylla	
	Chinese woodbine
Lonicera xylosteum	
	Fly honeysuckle
Lotus berthelotii	Parrot's beak
	Winged pea
Lotus corniculatus	Bacon and eggs
	Bird's foot trefoil
Lotus scoparius	Deer weed
Lunaria annua	Honesty
	Money plant
	Moonwort
	Satin flower
Lunaria biennis	Honesty
	Money plant
	Moonwort
	Satin flower
Lunaria rediviva	Honesty
	Perennial honesty
Lupinus albus	White lupin
Lupinus lepidus	Prairie lupin
Lupinus luteus	Yellow lupin
Lupinus polyphyllus	Garden lupin
Lupinus pubescens	Downy lupin

Lupinus subcarnosus

Lupinus subcarnosus	
	Texas blue bonnet
Lupinus texensis	Texas blue bonnet
Lupinus vallicola	Valley lupin
Lychnis alpina	Alpine campion
Lychnis chalcedonica	
	Jerusalem cross
	Maltese cross
Lychnis coeli-rosa	Rose of heaven
Lychnis coronaria	Mullein pink
	Rose campion
Lychnis dioica	Red campion
Lychnis flos-cuculi	Ragged robin
Lychnis flos-jovis	Flower of Jove
Lychnis silene	Catchfly
Lychnis viscaria	Catchfly
	German catchfly
	Red catchfly
	Sticky catchfly
Lychnis vulgaris	Catchfly
	German catchfly
Lycopsis arvensis	Lesser bugloss
Lycopus europaeus	Gipsywort
Lycopus virginicus	Bugleweed
	Gipsyweed
	Sweet bugle
	Water bugle
Lynosyris vulgaris	Goldilocks
Lysimachia clethroides	
	Chinese loosestrife
	Shepherd's crook
Lysimachia nemorum	
	Wood pimpernell
	Yellow pimpernell
Lysimachia nummularia	
	Creeping jenny
	Moneywort
	String of sovereigns
	Wandering jenny
Lysimachia punctata	
	Circle flower
	Dotted loosestrife
	Garden loosestrife
	Spotted loosestrife
	Yellow loosestrife
Lysimachia vulgaris	
	Wood pimpernel
	Yellow loosestrife
	Yellow willowherb
Lythrum salicaria	
	Purple loosestrife
	Purple willowherb
	Spiked loosestrife
Macleaya cordata	Plume poppy
Macleaya microcarpa	
	Lesser plume poppy
Malcolmia maritima	
	Virginian stock
Malope trifolia	Mallow wort
Malva alcea	Cut-leaved mallow
	Hollyhock mallow
Malva crispa	Curled mallow
	Curly mallow

Malva moschata	Musk mallow
Malva neglecta	Dwarf mallow
Malva pusilla	Small mallow
Malva sylvestris	Common mallow
Maranta leuconeura	Prayer plant
Martynia louisiana	Unicorn plant
Matricaria eximia	Feverfew
Matthiola bicornis	Evening stock
	Night-scented stock
Matthiola incana	Brompton stock
	Gillyflower
	Stock
Matthiola incana annua	
	Ten-week stock
Meconopsis baileyi	
	Himalayan blue poppy
Meconopsis betonicifolia	
	Blue poppy
	Himalayan blue poppy
Meconopsis cambrica	Welsh poppy
Meconopsis integrifolia	
	Lampshade poppy
	Yellow chinese poppy
Meconopsis napaulensis	
	Satin poppy
Meconopsis quintuplinerva	
	Harebell poppy
Medicago arabica	Spotted medick
Medicago echinus	Calvary clover
Medicago falcata	Yellow medick
Medicago lupulina	Black medick
	Nonsuch
	Shamrock
Medicago minima	Bur medick
Melandrium diurnum	
	Red campion
Mentzelia lindleyi	Blazing star
Mertensia ciliata	American bluebell
Mertensia maritima	
	Northern shore wort
	Oyster plant
Mertensia pulmonarioides	
	Virginian cowslip
Mertensia virginica	
	Virginia bluebell
	Virginia cowslip
Mesembryanthemum criniflorum	Livingstone daisy
Mesembryanthemum crystallinum	Ice plant
	Sea fig
Mimulus aurantiacus	
	Bush monkey flower
	Shrubby musk
Mimulus cardinalis	
	Scarlet monkey flower
Mimulus glutinosus	
	Bush monkey flower
Mimulus guttatus	
	American monkey flower
Mimulus luteus	Monkey flower
	Monkey musk

Ononis repens

Mimulus moschatus	Musk
Mimulus ringens	
	Allegheny monkey flower
Mina lobata	Crimson star glory
Minuartia rubella	
	Mountain sandwort
Minuartia sedoides	Cyphel
	Mossy cyphel
Minuartia verna	Spring sandwort
	Vernal sandwort
Minuartia viscosa	Sticky sandwort
Mirabilis jalapa	Four o'clock plant
	Marvel of Peru
Mirabilis longiflora	
	Sweet four o'clock
Moluccella laevis	Bells of Ireland
	Molucca balm
	Shell flower
Monarda didyma	Bee balm
	Bergamot
	Oswego tea
	Sweet bergamot
Monotropa hypophega	
	Yellow bird's nest
Monotropa uniflora	Ghost flower
	Indian pipe
Montia perfoliata	Spring beauty
Montia sibirica	Pink purslane
Morina longifolia	
	Himalayan whorlflower
	Whorlflower
Mutisia decurrens	
	Climbing gazania
Myosotidium hortensia	
	Antarctic forget-me-not
	Chatham Island lily
	Giant forget-me-not
Myosotidium nobile	
	Antarctic forget-me-not
	Chatham island lily
	Giant forget-me-not
Myosotis alpestris	
	Alpine forget-me-not
	Forget-me-not
Myosotis arvensis	
	Field forget-me-not
Myosotis discolor	Yellow forget-
	me-not
Myosotis oblongata	
	Forget-me-not
Myosotis palustris	
	Water forget-me-not
Myosotis sylvatica	Forget-me-not
	Wood forget-me-not
Nemophila maculata	Five-spot
	Five-spot nemophila
Nemophila insignis	
	Baby blue-eyes
Nemophila menziesii	
	Baby blue-eyes
Nemophila phacelia	
	Californian bluebell

Nepeta cataria	Catmint
	Catnep
	Catnip
Nepeta × faassenii	Catmint
Nepeta mussinii	Catmint
Nicandra physalodes	
	Apple of Peru
	Shoo-fly
Nicotiana affinis	Tobacco plant
Nicotiana alata	Tobacco plant
Nicotiana glauca	
	Yellow bush tobacco
Nicotiana sylvestris	
	Flowering tobacco
Nicotiana rustica	Turkish tobacco
Nicotiana tabacum	
	Common tobacco
	Tobacco
Nierembergia caerulea	
	Lavender cup
Nierembergia hippomanica	
	Lavender cup
Nierembergia repens	White cup
Nigella damascena	Love-in-a-mist
Nolana acuminata	
	Chilean bellflower
Nolana rupicola	Chilean bellflower
Oenothera biennis	
	Evening primrose
	Field primrose
Oenothera fruticosa	
	American sundrops
	Sundrops
Oenothera glazioviana	
	Large-leaved evening primrose
Oenothera linearis	
	American sundrops
	Sundrops
Oenothera macrocarpa	
	Ozark sundrops
Oenothera missouriensis	
	Ozark sundrops
	Prairie evening primrose
Oenothera perennis	
	Dwarf sundrops
Oenothera pumila	
	Dwarf sundrops
Oenothera rosea	
	Pink evening primrose
Oenothera stricta	
	Fragrant evening primrose
Omphalodes cappadocica	
	Navelwort
Omphalodes linifolia	
	Venus's navelwort
Omphalodes umbilicus	Navelwort
Omphalodes verna	
	Blue-eyed mary
	Creeping forget-me-not
Ononis natrix	Goat root
	Yellow rest harrow
Ononis repens	Rest harrow

Onopordum acanthium
Cotton thistle
Downy thistle
Scotch thistle
Woolly thistle

Onopordum arabicum
Arabian thistle
Heraldic thistle
Silver thistle

Onopordum nervosum
Arabian thistle
Heraldic thistle
Silver thistle

Onosma pyramidale   Donkey plant
Himalayan comfrey

Onosma tauricum   Golden drop

Origanum dictamnus
Cretan dittany

Origanum vulgare
Common marjoram
Marjoram
Oregano
Wild marjoram

Osteospermum barberae
Dwarf cape marigold

Ourisia macrophylla
Mountain foxglove

Oxalis acetosella   Shamrock
Wood sorrel

Oxalis articulata   Pink oxalis

Oxalis corniculata   Yellow sorrel

Oxalis deppei   Good-luck plant

Oxyria dignya   Mountain sorrel

Ozothamnus ledifolium
Kerosene bush

Ozothamnus thyrsoideum
Snow-in-summer

Pachysandra procumbens
Allegheny spurge

Pachysandra terminalis
Japanese spurge

Paeonia albiflora   Chinese paeony

Paeonia arborea   Moutan paeony
Tree paeony

Paeonia lactiflora   Chinese paeony

Paeonia moutan   Moutan paeony
Tree paeony

Paeonia officinalis
European wild paeony
Wild paeony

Paeonia suffruticosa
Moutan paeony
Tree paeony

Paeonia tenuifolia
Fern-leaved paeony

Papaver alpinum   Alpine poppy

Papaver burseri   Alpine poppy

Papaver commutatum
Ladybird poppy

Papaver dubium
Long-headed poppy

Papaver glaucum   Tulip poppy

Papaver nudicaule   Arctic poppy
Iceland poppy

Papaver orientale   Oriental poppy

Papaver pavonium
Peacock poppy

Papaver radicatum   Arctic poppy

Papaver rhoeas
Common red poppy
Corn poppy
Field poppy
Flanders poppy

Papaver rupifragum
Spanish poppy

Papaver somniferum
Opium poppy
White poppy

Passiflora caerulea
Blue passionflower
Common blue passionflower

Passiflora edulis   Purple granadilla

Passiflora laurifolia
Jamaica honeysuckle

Passiflora mollissima
Banana passionfruit
Curuba

Passiflora racemosa
Red passion flower

Pedicularis palustris   Red rattle

Pedicularis sylvatica   Lousewort

Pelargonium capitatum
Rose-scented geranium
Rose-scented pelargonium

Pelargonium citriodorum
Lemon-scented geranium
Lemon-scented pelargonium

Pelargonium crispum
Lemon-scented geranium
Lemon-scented pelargonium

Pelargonium × domesticum
Lady Washington geraniums
Lady Washington pelargoniums
Martha Washington geraniums
Martha Washington pelargoniums
Regal geraniums
Regal pelargoniums
Show geraniums
Show pelargoniums

Pelargonium × hortorum
Zonal geraniums
Zonal pelargoniums

Pelargonium peltatum
Ivy-leaved geraniums
Ivy-leaved pelargoniums

Peltiphyllum peltatum
Umbrella plant

Penstemon alpinus
Alpine penstemon

Penstemon cordifolius
Shrubby penstemon

Phacelia campanularia
California bluebell

Phacelia tanacetifolia
Tansy phacelia

POPULAR GARDEN PLANTS

Potentilla verna

Phacelia viscida	Sticky phacelia
Phacelia whitlavia	
	Californian bluebell
Phlox bifida	Sand phlox
Phlox canadensis	Blue phlox
Phlox decussata	Garden phlox
Phlox divaricata	Blue phlox
Phlox douglasii	Alpine phlox
Phlox drummondii	Annual phlox
Phlox paniculata	Garden phlox
Phlox reptans	Creeping phlox
Phlox setacea	Moss phlox
	Moss pink
Phlox stolonifera	Creeping phlox
Phlox subulata	Moss phlox
	Moss pink
Physalis alkekengi	Bladder cherry
	Chinese lantern
Physostegia virginiana	
	False dragonhead
	Lion's heart
	Obedient plant
Phyteuma comosum	
	Horned rampion
Phyteuma nigrum	Black rampion
Phyteuma spicatum	
	Spiked rampion
Pinguicula grandiflora	
	Greater butterwort
	Large-flowered butterwort
Pinguicula vulgaris	Butterwort
	Common butterwort
Platycodon grandiflorum	
	Balloon flower
	Chinese bellflower
Platystemon californicus	
	Cream cups
Podophyllum emodii	
	Himalayan mayflower
Podophyllum peltatum	May apple
Polemonium caeruleum	
	Greek valerian
	Jacob's ladder
Polemonium reptans	Abcess root
	American Greek valerian
	Blue bells
	Creeping jacob's ladder
	Sweat root
Polygala calcarea	Chalk milkwort
	Milkwort
Polygala chamaebuxus	
	Bastard box
	Ground box
Polygala senega	Rattlesnake root
	Senega
	Snakeroot
Polygala serpyllifolia	
	Heath milkwort
Polygala vulgaris	
	Common milkwort
Polygonum amplexicaule	
	Mountain fleece

Polygonum ariculare	Armstrong
Polygonum aubertii	
	China fleece flower
	Silver lace vine
Polygonum aviculare	
	Bird's tongue
	Common knotgrass
	Knotgrass
	Nine-joints
Polygonum baldschuanicum	
	Bokhara fleece flower
	Russian vine
Polygonum bistorta	Snakeweed
Polygonum campanulatum	
	Himalayan knotweed
Polygonum capitatum	
	Pink-head knotweed
Polygonum erectum	
	Russian knotgrass
Polygonum lapathifolium	
	Pale persicaria
Polygonum persicaria	
	Common persicaria
	Red legs
	Red shank
Polygonum viviporum	
	Alpine bistort
Portulaca grandiflora	Rose moss
	Sun plant
Potentilla alba	White cinquefoil
Potentilla anglica	
	Procumbent cinquefoil
Potentilla anserina	Silver weed
	Silvery cinquefoil
Potentilla argentea	
	Hoary cinquefoil
Potentilla cinerea	Grey cinquefoil
Potentilla crantzii	
	Alpine cinquefoil
Potentilla erecta	Bloodroot
	Ewe daisy
	Red root
	Tormentil
Potentilla palustris	
	Marsh cinquefoil
Potentilla recta	
	Sulfer cinquefoil
	Sulphur cinquefoil
	Upright cinquefoil
Potentilla reptans	
	Creeping cinquefoil
	Five fingers
	Five-leaved grass
Potentilla rupestris	
	Rock cinquefoil
Potentilla sterilis	
	Barren strawberry
Potentilla tabernaemontani	
	Spring cinquefoil
Potentilla tormentilla	
	Shepherd's knot
Potentilla verna	Spring cinquefoil

Poterium canadensis

Poterium canadensis	
	American burnet
	Canadian burnet
Primula acaulis	English primrose
	Primrose
Primula alpicola	
	Moonlight primula
Primula auricula	Alpine auricula
	Auricula
	Bear's breeches
	Bear's ear
	Dusty miller
Primula denticulata	
	Drumstick primula
Primula elatior	Oxlip
	Paigles
Primula farinosa	
	Bird's eye primrose
Primula florindae	Giant cowslip
	Himalayan cowslip
Primula helodoxa	
	Glory-of-the-marsh
Primula japonica	
	Japanese primrose
Primula malacoides	
	Fairy primrose
Primula microdonta alpicola	
	Moonlight primula
Primula obconica	
	German primrose
Primula officinalis	Cowslip
Primula polyantha	Polyanthus
Primula sikkimensis	
	Himalayan cowslip
Primula veris	Cowslip
	Fairy cups
	Herb peter
	Keyflower
	Palsywort
	Saint Peter's wort
Primula vulgaris	English primrose
	Primrose
Proboscidea jussieui	Unicorn plant
Psylliostachys suworowii	
	Rat's-tail statice
	Russian statice
Pulmonaria angustifolia	
	Blue cowslip
Pulmonaria officinalis	
	Jerusalem cowslip
	Jerusalem sage
	Soldiers and sailors
	Spotted dog
Pulmonaria saccharata	
	Bethlehem sage
Pulsatilla alba	
	White pasque flower
Pulsatilla alpina	Alpine anemone
Pulsatilla vernalis	Spring anemone
Pulsatilla vulgaris	
	Meadow anemone
	Pasque flower
Puya alpestris	Puya

Puya berteroniana	Puya
Pyrethrum hybridum	
	Painted daisy
	Pyrethrum
Pyrethrum roseum	Painted daisy
	Pyrethrum
Quamoclit lobata	
	Crimson star glory
Ranunculus aconitifolius	
	Aconite-leaved buttercup
	Fair maids of Kent
	White bachelor's buttons
Ranunculus acris	
	Bachelor's buttons
	Gold cup
	Meadow bloom
	Meadow buttercup
	Yellow bachelor's buttons
Ranunculus alpestris	
	Alpine buttercup
Ranunculus amplexicaulis	
	White buttercup
Ranunculus arvensis	
	Corn buttercup
	Corn crowfoot
Ranunculus asiaticus	
	Persian buttercup
	Turban buttercup
Ranunculus auricumus	Goldilocks
Ranunculus bulbosus	
	Bulbous buttercup
Ranunculus ficaria	
	Lesser celandine
	Pilewort
Ranunculus flammula	
	Lesser spearwort
Ranunculus glacialis	
	Glacier crowfoot
Ranunculus hederaceus	
	Ivy-leaved crowfoot
Ranunculus lingua	
	Great spearwort
Ranunculus lyalli	Rockwood lily
Ranunculus repens	
	Creeping buttercup
Ranunculus sceleratus	
	Celery-leaved crowfoot
Rehmannia angulata	
	Chinese foxglove
Reseda alba	White mignonette
Reseda lutea	Wild mignonette
Reseda luteola	Dyer's rocket
	Weld
Reseda odorata	Mignonette
Reseda phyteuma	
	Corn mignonette
Rhazya orientalis	
	Oriental periwinkle
Rhodanthe manglesii	Sunray
	Swan river everlasting
Rhodiola rosea	Roseroot
Ricinus communis	Castor bean
	Castor-oil plant

POPULAR GARDEN PLANTS

Sedum anglicum

Rodgersia pinnata	Feathered bronze leaf
Rudbeckia fulgida	Cone flower
Rudbeckia hirta	Annual rudbeckia
	Black-eyed susan
	Yellow daisy
Rudbeckia triloba	Brown-eyed susan
Ruta graveolens	Herb of grace
	Rue
Sagina maritima	Sea pearlwort
Sagina nodosa	Knotted pearlwort
Sagina procumbens	Common pearlwort
Sagina subulata	Heath pearlwort
Saintpaulia ionantha	African violet
Salpiglossis sinuata	Painted tongue
	Salpiglossis
	Velvet flower
Salvia argentea	Silver sage
Salvia azurea	Blue sage
Salvia farinacea	Mealy-cup sage
Salvia fulgens	Cardinal sage
	Cardinal salvia
	Mexican red sage
Salvia glutinosa	Jupiter's distaff
	Sticky sage
Salvia horminoides	Wild sage
Salvia horminum	Annual clary
	Bluebeard
Salvia nemorosa	Wild sage
Salvia officinalis	Common sage
	Sage
Salvia officinalis purpurascens	Purple sage
Salvia patens	Blue sage
	Gentian sage
Salvia pratensis	Meadow clary
Salvia rutilans	Pineapple sage
Salvia sclarea	Clary
	Clear eye
	True clary
Salvia splendens	Scarlet sage
Salvia superba	Perennial sage
Salvia uliginosa	Bog sage
Salvia verbenacea	Wild clary
Salvia viridis	False clary
Sanguinaria canadensis	Bloodroot
Sanguisorba canadensis	American burnet
	Canadian burnet
Sanvitalia procumbens	Creeping zinnia
Saponaria ocymoides	Rock soapwort
	Tumbling ted
Saponaria officinalis	Bouncing bet
	Soapwort
Saponaria vaccaria	Cow herb
	Dairy pink
Satureia montana	Winter savory
Saxifraga aizoides	Yellow mountain saxifrage
Saxifraga caesia	Blue saxifrage
Saxifraga cespitosa	Tufted saxifrage
Saxifraga cotyledon	Pyramidal saxifrage
Saxifraga granulata	Fair maids of France
	Meadow saxifrage
Saxifraga hieracifolia	Hawkweed saxifrage
Saxifraga hirsuta	Hairy saxifrage
	Kidney saxifrage
Saxifraga hypnoides	Dovedale moss
	Mossy rockfoil
	Mossy saxifrage
Saxifraga moschata	Musky saxifrage
Saxifraga oppositifolia	Purple saxifrage
Saxifraga peltata	Umbrella plant
Saxifraga rosacea	Irish saxifrage
Saxifraga sarmentosa	Mother of thousands
	Strawberry geranium
Saxifraga stellaris	Starry saxifrage
Saxifraga stellata	Starry saxifrage
Saxifraga stolonifera	Mother of thousands
	Strawberry geranium
Saxifraga tridactylotes	Rue-leaved saxifrage
Saxifraga umbrosa	London pride
	Saint Patrick's cabbage
Saxifraga × urbium	London pride
	Nancy pretty
	None-so-pretty
	Saint Patrick's cabbage
Scabiosa atropurpurea	Mournful widow
	Pincushion flower
	Sweet scabious
Scabiosa caucasica	Florist's scabious
	Pincushion flower
Scabiosa columbaria	Small scabious
Scabiosa columbaria ochraleuca	Yellow scabious
Scabiosa lucida	Brilliant scabious
	Shining scabious
Schizocodon soldanelloides	Fringebell
Sedum acre	Biting stonecrop
	Golden moss
	Wall pepper
Sedum adolphi	Golden sedum
Sedum album	White stonecrop
Sedum anglicum	English stonecrop

Sedum dasyphyllum	
	Thick-leaved stonecrop
Sedum forsteranum	
	Rock stonecrop
Sedum maximum	Ice plant
Sedum morganianum	Ass's tail
	Beaver tail
	Burro's tail
	Donkey's tail
	Horse's tail
	Lamb's tail
Sedum pachyphyllum	
	Jelly-bean plant
Sedum reflexum	
	Reflexed stonecrop
	Stone orpine
Sedum rhodiola	Roseroot
Sedum rosea	Roseroot
	Roseroot sedum
Sedum × rubrotinctum	Pork and beans
Sedum spectabile	Ice plant
Sedum telephium	Live forever
	Livelong
	Midsummer men
	Orpine
Sedum villosum	Hairy stonecrop
Sempervivum arachnoideum	
	Cobweb houseleek
	Spider houseleek
Sempervivum montanum	
	Mountain houseleek
Sempervivum soboliferum	
	Hen-and-chickens houseleek
Sempervivum tectorum	
	Bullock's eye
	Common houseleek
	Roof houseleek
Senecio aquaticus	Marsh ragwort
Senecio aureus	Golden groundsel
Senecio bicolor	Silver ragwort
Senecio confusus	
	Mexican flame vine
Senecio erucifolius	Hoary ragwort
Senecio fluviatilis	
	Broad-leaved ragwort
Senecio integrifolius	Field fleawort
Senecio jacobaea	Jacobea
	Ragwort
	Saint James' wort
	Staggerwort
Senecio nemorensis	
	Wood ragwort
Senecio palustris	Marsh fleawort
Senecio petasites	Velvet groundsel
Senecio rowleyanus	String of beads
	String of pearls
Senecio squalidus	Oxford ragwort
Senecio sylvaticus	
	Wood groundsel
Senecio viscosus	Sticky groundsel
Senecio vulgaris	Groundsel
Serratula shawii	Sawwort
Sesbania tripetii	Scarlet wisteria
Shortia soldanelloides	Fringebell
Silaum silaus	Pepper saxifrage
Silene acaulis	Cushion pink
	Moss campion
Silene alba	White campion
Silene alpestris	Alpine campion
Silene armeria	
	Sweet william catchfly
Silene coeli-rosa	Rose of heaven
Silene conica	Sand catchfly
	Striated catchfly
Silene dichotoma	Forked catchfly
Silene dioica	Red campion
Silene gallica	Small catchfly
Silene italica	Italian catchfly
Silene latifolia	White campion
Silene maritima	Sea campion
Silene noctiflora	
	Night-flowering catchfly
Silene nutans	Nottingham catchfly
Silene oculata	Rose of heaven
Silene otites	Spanish catchfly
Silene pendula	Nodding catchfly
Silene quadrifida	Alpine campion
Silene rupestris	Rock campion
Silene vulgaris	Bladder campion
Silene vulgaris maritima	
	Sea campion
Silybum marianum	Blessed thistle
	Holy thistle
	Milk thistle
	Our Lady's milk thistle
	Saint Mary's milk thistle
	Saint Mary's thistle
Sisymbrium altissimum	Tall rocket
Sisymbrium irio	London rocket
Sisymbrium loeselii	
	False London rocket
Sisymbrium officinale	
	Hedge mustard
Sisymbrium orientale	
	Eastern rocket
	Oriental rocket
Sisyrinchium angustifolium	
	Blue-eyed grass
Sisyrinchium bellum	
	California blue-eyed grass
Sisyrinchium californicum	
	Golden-eyed grass
Sisyrinchium douglasii	
	Grass widow
	Purple-eyed grass
	Spring bell
Sisyrinchium stratiatum	
	Satin flower
Solanum capsicastrum	
	Winter cherry

POPULAR GARDEN PLANTS

Tacsonia mollissima

Solanum carolinense
Apple of sodom
Bull nettle
Horse nettle
Poison potato

Solanum crispum
Chilean potato tree

Solanum dulcamara   Bittersweet
Woody nightshade

Solanum jasminoides
Jasmine nightshade
Potato vine

Solanum laciniatum
Kangaroo apple
Poroporo

Solanum nigrum   Black nightshade

Solanum rantonnetii   Potato bush

Solanum sosomeum
Apple of sodom

Solanum wendlandii
Giant potato vine

Soldanella alpina   Alpine snowbell

Soldanella montana
Mountain snowbell
Mountain tassel

Soldanella pusilla   Dwarf snowbell

Solidago canadensis
Canadian golden rod
Golden rod

Solidago odora   Sweet golden rod

Solidago virgaurea
Golden rod
Woundwort

Spathiphyllum wallisii   White sails

Specularia speculum
Venus's looking glass

Specularia speculum-veneris
Venus's looking glass

Spiraea aruncus   Goat's beard

Spiraea lobata
Queen of the prairie

Stachys affinis   Chinese artichoke
Chorogi
Japanese artichoke
Knotroot

Stachys arvensis   Corn woundwort
Field woundwort

Stachys betonica   Betony
Bishop's wort
Wood betony

Stachys byzantina   Lamb's ear
Lamb's lugs
Lamb's tongue
Sow's ear
Woolly betony
Woolly woundwort

Stachys germanica
Downy woundwort

Stachys grandiflora   Woundwort

Stachys lanata   Lamb's ear
Lamb's tongue
Woolly betony

Stachys macrantha   Woundwort

Stachys officinalis   Betony
Bishop's-wort
Wood betony

Stachys olympica   Lamb's ear
Lamb's tongue
Woolly betony

Stachys palustris
Marsh woundwort

Stachys recta   Yellow woundwort

Stachys sylvatica
Hedge woundwort
Wood woundwort

Stapelia gigantea   Giant stapelia
Giant starfish
Giant toad plant
Zulu giant

Stapelia variegata   Starfish plant
Toad cactus
Toad plant

Stapelia hirsuta
Hairy starfish flower
Hairy toad plant
Shaggy starfish

Statice bonduellii   Algerian statice

Statice latifolium
Border sea lavender
Broad-leaved sea lavender
Statice

Statice sinuatum
Notch-leaf statice
Winged statice

Statice suworowii   Rat's-tail statice
Russian statice

Stokesia cyanea   Cornflower aster
Stokes's aster

Stokesia laevis   Cornflower aster
Stokes's aster

Streptocarpus hybridus
Cape primrose

Streptocarpus saxorum
False African violet

Stylomecon heterophylla
Blood drop
Flaming poppy
Wind poppy

Stylophorum diphyllum
Celandine poppy
Wood poppy

Symphytum officinale   Boneset
Bruisewort
Comfrey
Common comfrey
Knitbone

Symphytum orientale
White comfrey

Symphytum peregrinum
Russian comfrey

Symphytum × uplandicum
Russian comfrey

Tacsonia mollissima
Banana passionfruit
Curuba

Tagetes erecta

Tagetes erecta	African marigold
	American marigold
	Aztec marigold
	Big marigold
Tagetes filifolia	Irish lace
Tagetes lucida	Sweet mace
	Sweet-scented marigold
Tagetes minuta	Muster-john-henry
Tagetes patula	French marigold
Tagetes signata	Signet marigold
	Tagetes
Tagetes tenuifolia	Signet marigold
	Tagetes
Tanacetum cinerariifolium	
	Dalmatian pellitory
	Dalmatian pyrethrum
	Pyrethrum
Tanacetum coccineum	Feverfew
	Painted daisy
	Pyrethrum
Tanacetum densum amani	
	Prince of Wales' feathers
Tanacetum parthenium	
	Bachelor's buttons
	Feverfew
Tanacetum vulgare	Buttons
	Tansy
Taraxacum erythrospermum	
	Lesser dandelion
Taraxacum laevicatum	
	Lesser dandelion
Taraxacum officinale	Dandelion
	Fairy clock
	Lion's teeth
	Pee the bed
	Piss the bed
	Swine's snout
Telekia speciosa	
	Large yellow ox-eye
Tetranema mexicana	
	Mexican foxglove
Tetranema roseum	
	Mexican foxglove
	Mexican violet
Thalictrum alpinum	
	Alpine meadow rue
Thalictrum aquilegiifolium	
	Meadow rue
Thalictrum clavatum	Lady rue
Thalictrum coreanum	
	Dwarf meadow rue
Thalictrum dioicum	
	Early meadow rue
	Quicksilver weed
Thalictrum flavum	
	Common meadow rue
	Yellow meadow rue
Thalictrum minus	
	Lesser meadow rue
Thalictrum polygamum	
	King-of-the-meadow
	Muskrat weed
	Tall meadow rue

Thermopsis caroliniana	
	Carolina lupin
Thlaspi alliaceum	
	Garlic pennycress
Thlaspi alpestre	Alpine pennycress
Thlaspi arvense	Fanweed
	Field pennycress
	Frenchweed
	Mithridate mustard
	Pennycress
	Stinkweed
Thlaspi perfoliatum	
	Perfoliate pennycress
Thunbergia alata	
	Black-eyed susan
Thunbergia grandiflora	
	Clock vine
Thymus × citriodorus	
	Lemon-scented thyme
Thymus drucei	
	European wild thyme
	Wild thyme
Thymus herba-barona	
	Caraway thyme
Thymus nitidus	Sicily thyme
Thymus praecox	Hairy thyme
Thymus pulegioides	Large thyme
Thymus richardii nitidus	
	Sicily thyme
Thymus serpyllum	
	Breckland thyme
	Creeping thyme
	European wild thyme
	Wild thyme
Thymus vulgaris	Common thyme
	Garden thyme
Tiarella cordifolia	Coolwort
	Foam flower
Tiarella polyphylla	Foam flower
Tiarella unifoliata	Sugar scoop
Tithonia rotundifolia	
	Mexican sunflower
Tithonia speciosa	
	Mexican sunflower
Torenia fournieri	Blue wings
	Wishbone flower
Trachelium caeruleum	
	Blue throatwort
	Common throatwort
Trachymene caerulea	
	Blue lace flower
Tradescantia albiflora	
	Wandering jew
Tradescantia × andersoniana	
	Spiderworts
	Trinity flowers
Tradescantia fluminensis	
	Wandering jew
Tradescantia virginiana	
	Spiderworts
	Trinity flowers
Tradescantia zebrina	Inch plant

POPULAR GARDEN PLANTS

Verbena tenera

Trollius × *cultorum*
Garden globe flowers
Trollius europaeus Boule d'or
Common globe flower
Globe flower
Trollius × *hybridus*
Garden globeflowers
Trollius laxus
Spreading globe flower
Trollius pumilis
Dwarf globe flower
Tropaeolum canariensis
Canary creeper
Tropaeolum lobbianum
Shield nasturtium
Tropaeolum majus
Garden nasturtium
Indian cress
Tall nasturtium
Tropaeolum minus
Dwarf nasturtium
Tropaeolum peltophorum
Shield nasturtium
Tropaeolum peregrinum
Canary-bird flower
Canary-bird vine
Canary creeper
Tropaeolum polyphyllum
Tropaeolum
Tropaeolum speciosum
Flame creeper
Flame nasturtium
Scotch creeper
Scottish flame flower
Umbilicus rupestris Kidneywort
Navelwort
Pennywort
Wall pennywort
Uvularia grandiflora Bellwort
Cowbells
Haybells
Merry bells
Uvularia perfoliata Strawbells
Vaccaria pyramidata Cockle
Cow herb
Dairy pink
Saponaria
Valeriana dioica Marsh valerian
Valeriana montana
Mountain valerian
Valeriana officinalis Cat's valerian
Common valerian
Garden heliotrope
Phu
Valeriana pyrenaica
Heart-leaved valerian
Pyrenean valerian
Valeriana saxatilis Dwarf valerian
Valeriana walichii Indian valerian
Vancouveria hexandra
American barrenwort
Vancouveria planipetala
Inside-out flower
Redwood ivy

Venidium fastuosum Cape daisy
Monarch of the veldt
Namaqualand daisy
Veratrum album
White false hellebore
White helleborine
Veratrum californicum Corn lily
Veratrum nigrum
Black false hellebore
Black helleborine
Veratrum viride
American white hellebore
Indian poke
Itchweed
White helleborine
Verbascum blattaria Moth mullein
Verbascum lychnitis
White mullein
Verbascum nigrum Dark mullein
Verbascum phlomoides
Orange mullein
Verbascum phoeniceum
Purple mullein
Verbascum pulverulentum
Hoary mullein
Verbascum thapsus Aaron's rod
Adam's flannel
Candlewick
Common mullein
Flannel mullein
Hag's taper
Jacob's staff
Jupiter's staff
Lady's foxglove
Rag paper
Shepherd's club
Velvet mullein
White mullein
Woolly mullein
Verbascum virgatum
Twiggy mullein
Verbena aubletia Creeping vervain
Rose verbena
Verbena bipinnatifida
Dakota vervain
Verbena bracteata
Prostrate vervain
Verbena canadensis Clump vervain
Creeping vervain
Rose verbena
Verbena hastata Blue vervain
Simpler's joy
Verbena × *hortensis*
Garden verbenas
Verbena × *hybrida*
Garden verbenas
Verbena jamaicensis
Jamaica vervain
Verbena lasiostachys Vervain
Verbena officinalis Vervain
Verbena rigida Vervain
Verbena stricta Hoary vervain
Verbena tenera Italian verbena

Verbena tenuisecta

Verbena tenuisecta	Moss verbena
Verbena urticifolia	White vervain
Verbesina enceliodes	Butter daisy
	Golden crown beard
Veronica agrestis	Field speedwell
Veronica alpina	Alpine speedwell
Veronica americana	
	American brooklime
Veronica arvensis	Wall speedwell
Veronica austriaca	
	Austrian speedwell
	Large speedwell
Veronica beccabunga	Brooklime
	Cow cress
	European brooklime
	Water pimpernel
Veronica candida	
	Woolly speedwell
Veronica chamaedrys	Angel's eye
	Bird's eye
	Germander speedwell
Veronica filiformis	
	Round-leaved speedwell
Veronica fruticans	Rock speedwell
Veronica hederifolia	
	Ivy-leaved speedwell
Veronica hulkeana	
	New Zealand lilac
Veronica incana	Woolly speedwell
Veronica montana	
	Mountain speedwell
	Wood speedwell
Veronica officinalis	
	Common speedwell
	Gipsyweed
	Heath speedwell
	Speedwell
Veronica peregrina	
	American speedwell
Veronica perfoliata	
	Digger's speedwell
Veronica persica	
	Buxbaum's speedwell
	Persian speedwell
Veronica prostrata	
	Rockery speedwell
Veronica repens	
	Corsican speedwell
Veronica scutellata	
	Marsh speedwell
Veronica serpyllifolia	
	Thyme-leaved speedwell
Veronica spicata	Spiked speedwell

Veronica verna	Spring speedwell
Veronica virginica	Black root
	Culver's root
Veronicastrum virginicum	
	Black root
	Culver's root
Vinca major	Band plant
	Blue buttons
	Greater periwinkle
	Larger periwinkle
Vinca minor	Common periwinkle
	Lesser periwinkle
	Running myrtle
	Trailing myrtle
Vinca rosea	Madagascar periwinkle
	Rose periwinkle
Viola cornuta	Horned violet
	Tufted pansy
Viola cucullata	Marsh violet
Viola hederacea	Australian violet
	Trailing violet
Viola lutea	Mountain pansy
Viola odorata	Sweet violet
Viola saxatilis	Yellow violet
Viola tricolor	Heartsease pansy
	Wild pansy
Viscaria alpina	Alpine campion
Viscaria elegans	Rose of heaven
Viscaria vulgaris	Catchfly
	German catchfly
Wahlenbergia albomarginata	
	Bellflower
	Tufted harebell
Wahlenbergia hederacea	
	Ivy-leaved bellflower
	Ivy-leaved harebell
Wattakaka chinensis	
	Chinese wax flower
Wattakaka sinensis	
	Chinese wax flower
Xeranthemum anuum	
	Common immortelle
	Immortelle
Zauschneria californica	
	Californian fuchsia
	Humming bird's trumpet
Zinnia angustifolia	
	Narrow-leaved zinnia
Zinnia elegans	Common zinnia
	Youth and old age
Zinnia haageana	Mexican zinnia
	Narrow-leaved zinnia

TREES, BUSHES, AND SHRUBS

It is often said that the difference between trees and shrubs is simple; trees have a single woody stem from which branches grow to form a crown whereas a shrub has several woody stems rising from ground level, forming a crown. But this is over-simplification. The exact shape of a tree can be completely altered by wind action, by a difference in spacing or even by artificial pruning.

A bush is generally defined as a woody plant that is between a shrub and a tree in size.

Silver maple –
Acer saccharinum

Abelmoschus manihot

Abelmoschus manihot	
	Sunset hibiscus
Aberia caffra	Kai apple
	Kau apple
	Kei apple
	Umkokolo
Aberia gardneri	
	Ceylon gooseberry
	Kitembilla
Abies alba	Common silver fir
	European silver fir
	Silver fir
Abies amabilis	Alpine fir
	Beautiful fir
	Cascade fir
	Pacific silver fir
	Red silver fir
	White fir
Abies balsamea	Balm of gilead
	Balsam fir
Abies borisii-regis	Bulgarian fir
	King Boris's fir
Abies bornmuellerana	
	Bornmüller's fir
	Turkish fir
Abies bracteata	Bristle-cone fir
	Santa Lucia fir
Abies cephalonica	Grecian fir
	Greek fir
Abies chensiensis	Shensi fir
Abies cilicica	Cilician fir
Abies concolor	Colorado white fir
	White fir
Abies concolor lowiana	
	Low's white fir
	Pacific white fir
	Sierra fir
Abies delavayi	Delavay's silver fir
Abies delavayi forrestii	Forrest's fir
	Forrest's silver fir
Abies fargesii	Farge's fir
	Sutchuen fir
Abies firma	Japanese fir
	Momi fir
Abies fraseri	Fraser's balsam fir
	She balsam
	Southern balsam fir
Abies grandis	Giant fir
	Grand fir
	Lowland fir
Abies holophylla	Manchurian fir
	Needle fir
Abies homolepis	Nikko fir
Abies koreana	Korean fir
Abies lasiocarpa	Alpine fir
	Rocky mountain fir
	Subalpine fir
Abies lasiocarpa arizonica	
	Arizona cork fir
	Cork fir
Abies macrocana	Moroccan fir

Abies magnifica	Californian red fir
	Red fir
Abies mariesii	Maries' fir
Abies marocana	Maroc fir
Abies nephrolepis	East Siberian fir
	Khingan fir
Abies nordmanniana	Caucasian fir
Abies numidica	Algerian fir
Abies pindrow	West Himalayan fir
Abies pinsapo	Hedgehog fir
	Spanish fir
Abies pinsapo glauca	
	Blue Spanish fir
Abies procera	Bracted fir
	Feather-cone fir
	Noble fir
Abies recurvata	Min fir
Abies religiosa	Sacred fir
Abies sachalinensis	Sakhalin fir
Abies siberica	Siberian fir
Abies spectabilis	East Himalayan fir
	Himalayan fir
Abies squamata	Flaky fir
Abies sutchuenensis	Farges fir
	Sutchuen fir
	Szechwan fir
Abies veitchii	Veitch's silver fir
Abies venusta	Bristle-cone fir
	Santa Lucia fir
Abies webbiana	East Himalayan fir
	Himalayan fir
Abronia umbellata	
	Pink sand verbena
Acacia abyssinica	Ethiopean acacia
Acacia accola	Wallangarra
Acacia acinacea	Gold dust
Acacia acuminata	
	Raspberry-jam tree
Acacia alata	Cedar wattle
	Winged wattle
Acacia albida	
	Ana tree
	Apple-ring acacia
	Winter thorn
Acacia aneura	Mulga
Acacia arabica	Indian gum
Acacia armata	Hedge wattle
	Kangaroo thorn
Acacia baileyana	
	Cootamundra wattle
	Golden mimosa
Acacia berteriana	Bastard logwood
Acacia binervia	Coastal myall
	Sally wattle
Acacia botrycephala	
	Sunshine wattle
Acacia brachybotrya	Grey mulga
Acacia bynoeana	Dwarf nealie
Acacia calamifolia	Broom wattle
	Wallowa

TREES, BUSHES, AND SHRUBS

Acacia woodii

Acacia cardiophylla	
	Wyalong wattle
Acacia catechu	Black cutch tree
	Catechu
	Cutch
	Khair
	Wadalee gum tree
Acacia cornigera	Bull-horn acacia
	Swollen-thorn acacia
Acacia cultriformis	Knife acacia
	Knife-leaf acacia
Acacia cyanophylla	Blue-leaf wattle
	Blue wattle
	Golden willow
	Orange wattle
	Port Jackson willow
Acacia dealbata	Mimosa
	Silver wattle
Acacia decora	Graceful wattle
Acacia decurrens	Green wattle
Acacia dunnii	Elephant's-ear wattle
Acacia elongata	Swamp wattle
Acacia farnesiana	Cassie
	Green wattle
	Huisache
	Opopanax
	Popinac
	Sponge tree
	Sweet acacia
	West Indian blackthorn
Acacia galpinii	Apiesdoring
Acacia giraffae	Camel thorn
Acacia greggii	Catclaw acacia
	Texas mimosa
Acacia gummifera	Barbary gum
	Mogadore gum
	Morocco gum
Acacia homalophylla	
	Fragrant myall
	Gidgee myall
	Myallwood
	Violetwood
	Yarran
Acacia horrida	Cape gum
Acacia howittii	Sticky wattle
Acacia implexa	Lightwood
	Screw-pod wattle
Acacia juniperina	Prickly wattle
Acacia karroo	Karroo thorn
	Sweet thorn
Acacia kettlewelliae	Buffalo wattle
Acacia koa	Koa
Acacia leprosa	Cinnamon wattle
Acacia longifolia	
	Sydney golden wattle
Acacia mearnsii	Black wattle
Acacia melanoxylon	
	Australian blackwood
	Blackwood
	Blackwood acacia

Acacia nilotica	Babul
	Gum acacia
	Gum-arabic tree
	Shittimwood
	Suntwood
Acacia nudicaulis	Bamboo briar
Acacia paniculata	Sunshine wattle
Acacia pendula	Weeping myall
Acacia penninervis	Blackwood
	Mountain hickory
Acacia podalyriifolia	
	Mount Morgan wattle
	Pearl acacia
	Queensland silver wattle
	Queensland wattle
Acacia pravissima	Oven's acacia
	Oven's wattle
Acacia prominens	
	Golden-rain wattle
Acacia pruinosa	Frosty wattle
Acacia pubescens	Hairy wattle
Acacia pycnantha	Golden wattle
Acacia retinodes	Wirilda
Acacia rigens	Needle-bush wattle
Acacia rubida	Red-leaved wattle
	Red-stemmed acacia
Acacia saileyana	
	Cootamundra wattle
Acacia salicina	Cooba
	Willow acacia
Acacia saligna	Golden wreath
	Weeping wattle
Acacia senegal	Cape gum
	Cape jasmine
	Egyptian thorn
	Gum acacia
	Gum-arabic tree
	Senegal gum
	Sudan gum-arabic
Acacia seyal	Gum-arabic tree
	Thirty thorn
	Whistling tree
Acacia spadicigera	
	Bull's horn acacia
Acacia spectabilis	Glory wattle
	Mudgee wattle
Acacia suaveolens	Sweet acacia
Acacia terminalis	Cedar wattle
	Peppermint wattle
Acacia tortilis	Umbrella thorn
Acacia tortuosa	Corkscrew wattle
	Twisted wattle
Acacia verniciflua	Varnish wattle
Acacia verticillata	Prickly moses
	Star acacia
	Star wattle
Acacia vestita	Weeping boree
Acacia victoriae	Bramble acacia
	Bramble wattle
Acacia woodii	Paper bark thorn
	Yellow bark thorn

Acacia xanthophloea

Acacia xanthophloea
　　　　　Fever tree
Acalypha hispida　　Chenille plant
　　　　　Foxtail
　　　　　Phillipine medusa
　　　　　Red-hot cat-tail
Acalypha wilkesiana
　　　　　Beefsteak plant
　　　　　Copper-leaf
　　　　　Fire-dragon
　　　　　Jacob's coat
　　　　　Match-me-if-you-can
Acer argutum　　Deep-veined maple
Acer barbatum　　Florida maple
　　　　　Southern sugar maple
　　　　　Sugar tree
Acer buergeranum　　Trident maple
Acer campestre　　Common maple
　　　　　Field maple
　　　　　Hedge maple
Acer capillipes
　　　Japanese snakebark maple
　　　Red snakebark maple
　　　Snakebark maple
Acer cappadocicum
　　　　　Cappadocian maple
　　　　　Caucasian maple
Acer carpinifolium　　Hornbeam-
　　　　　leaved maple
　　　　　Hornbeam maple
Acer circinatum　　Vine maple
Acer cissifolium　　Vineleaf maple
Acer crataegifolium　　Hawthorn-
　　　　　leaved maple
　　　　　Snakebark maple
Acer davidii
　　　Chinese snakebark maple
　　　Pére David's maple
Acer diabolicum　　Devil's maple
　　　　　Horned maple
Acer distylum　　Lime-leaf maple
Acer forrestii　　Forrest's maple
　　　　　Snakebark maple
Acer ginnala　　Amur maple
Acer glabrum　　Rock maple
　　　　　Rocky mountain maple
Acer glabrum douglasii
　　　　　Douglas maple
Acer griseum　　Paper bark maple
Acer heldreichii　　Balkan maple
　　　　　Heldreich's maple
Acer hersii　　Hers's maple
　　　　　Snakebark maple
Acer hyrcanum　　Balkan maple
Acer japonicum
　　　　　Downy Japanese maple
　　　　　Full-moon maple
　　　　　Japanese maple
　　　　　Smooth Japanese maple
Acer japonicum aconitifolium
　　　　　Coral bark maple
Acer japonicum aureum
　　　　　Golden moon maple

Acer laxiflorum　　Snakebark maple
Acer leucoderme　　Chalk maple
Acer lobelii　　Lobel's maple
Acer macrophyllum
　　　　　Big-leaved maple
　　　　　Canyon maple
　　　　　Oregon maple
Acer martinii　　Martin's maple
Acer maximowiczianum
　　　　　Nikko maple
Acer miyabei　　Miyabe's maple
Acer monspessulanum
　　　　　Montpelier maple
Acer negundo　　Ashleaf maple
　　　　　Box elder
　　　　　Water ash
Acer negundo auratum
　　　　　Golden ashleaf maple
Acer nigrum　　Black maple
Acer nikoense　　Nikko maple
Acer opalus　　Italian maple
Acer palmatum　　Japanese maple
　　　　　Smooth Japanese maple
*Acer palmatum
atropurpureum*
　　　Blood-leaf Japanese maple
Acer palmatum senkaki
　　　　　Coral bark maple
Acer pennsylvanicum　　Moose bark
　　　　　Moosewood
　North American snakebark maple
　　　　　Pennsylvania maple
　　　　　Snakebark maple
　　　　　Striped maple
　　　　　Whistlewood
Acer pictum　　Painted maple
Acer platanoides　　Norway maple
Acer platanoides laciniatum
　　　　　Eagle's claw maple
Acer pseudoilatanus worlei
　　　　　Golden sycamore
Acer pseudoplatanus　　Sycamore
*Acer pseudoplatanus
altropurpureum*
　　　Purple-leaved sycamore
*Acer pseudoplatanus
costorphinense*
　　　　　Costorphine plane
Acer pseudosieboldianum
　　　　　Korean maple
Acer rubrum　　Canadian maple
　　　　　Red maple
　　　　　Scarlet maple
　　　　　Soft maple
　　　　　Swamp maple
Acer rufinerve
　　　Grey-budded snakebark maple
　　　　　Snakebark maple
Acer saccharinum　　Bird's eye maple
　　　　　River maple
　　　　　Silver maple
　　　　　Soft maple
　　　　　White maple

TREES, BUSHES, AND SHRUBS

Albizzia lebbeck

Acer saccharum	Hard maple
	Rock maple
	Sugar maple
Acer sempervirens	Cretan maple
Acer shirasawanum aureum	
	Golden-leaved Japanese maple
Acer spicatum	Mountain maple
Acer tataricum	Tartar maple
	Tatarian maple
Acer tataricum ginnala	
	Amur maple
Acer tetramerum	Birch-leaf maple
Acer trautvetteri	Red-bud maple
	Trautvetter's maple
Acer triflorum	
	Rough-barked maple
Acer truncatum	Shantung maple
Acer velutinum	Persian maple
Acer velutinum vanvolxemii	
	Van Volxem's maple
Acer × zoeschense	Zoeschen maple
Achras sapota	Chicle
	Sapodilla plum
Acmena smithii	Lillypilly
Acoelorrhaphe wrightii	
	Everglades palm
	Saw cabbage palm
	Silver saw palm
Acokanthera oblongifolia	
	Kaffia plum
	Wintersweet
Acrocarpus fraxinifolius	
	Pink cedar
	Red cedar
	Shingle tree
Acrocomia mexicana	Coyoli palm
Actinorhytis calapparia	
	Calappa palm
Adansonia digitata	Baobab
	Dead-rat tree
	Monkey-bread tree
	Upside-down tree
Adansonia gregori	Bottle tree
Adenanthera pavonina	
	Barbados pride
	Coral pea
	Coralwood
	Peacock flower fence
	Red sandalwood tree
	Redwood
	Sandalwood tree
Adenium obesum	Desert rose
Adensonia digitata	Lemonade tree
Adhatoda vasica	Malabar nut
Aegle marmelos	Bael tree
	Ball tree
	Bela tree
	Bengal quince
	Golden apple
	Indian bael
Aesculus californica	
	Californian buckeye chestnut
	Californian horse chestnut

Aesculus × carnea	
	Pink horse chestnut
	Red horse chestnut
Aesculus flava	Sweet buckeye
	Yellow buckeye
Aesculus glabra	Ohio buckeye
Aesculus hippocastanum	
	Common horse chestnut
	Horse chestnut
Aesculus × hybrida	
	Hybrid buckeye
Aesculus indica	
	Indian horse chestnut
Aesculus neglecta	
erythroblastos	
	Sunrise horse chestnut
Aesculus octandra	Sweet buckeye
	Yellow buckeye
Aesculus parviflora	
	Bottlebrush buckeye
	Dwarf buckeye
	Dwarf horse chestnut
	Shrubby pavia
Aesculus pavia	
	American red buckeye
	Red buckeye
Aesculus turbinata	
	Japanese horse chestnut
Agathis alba	Amboina pitch tree
	Mountain agathis
Agathis australis	Kauri pine
Agathis brownii	Queensland kauri
Agathis dammara	Amboina pine
	Dammar
Agathis microstachys	
	Black kauri pine
Agathis robusta	
	South Queensland kauri
Agathis vitiensis	Fijian kauri pine
Agathosma betulina	Round buchu
Agathosma crenulata	Long buchu
Agonis flexuosa	
	Australian willow myrtle
	Peppermint tree
	Willow myrtle
Ailanthus altissima	Copal tree
	Tree of heaven
	Varnish tree
Ailanthus glandulosa	Ailanto
Ailanthus vilmoriniana	
	Downy tree of heaven
Aiphanes caryotifolia	Ruffle pine
	Spiny pine
Albizzia distachya	Plume albizzia
Albizzia julibrissin	Mimosa tree
	Persian albizzia
	Pink siris
	Silk tree
Albizzia lebbeck	
	East Indian walnut
	Lebbek tree
	Siris tree
	Woman's tongue tree

Albizzia lophantha

Albizzia lophantha	Plume albizzia
Albizzia odoratissima	
	Ceylon rosewood
Albizzia rhodesica	Red-paper tree
Albizzia toona	Red siris
Alectryon excelsum	Titoki
Alectryon subcinereus	
	Smooth rambutan
Aleurites cordata	
	Japanese wood oil tree
Aleurites fordii	
	Chinese wood oil tree
	Tung oil tree
	Tung tree
Aleurites moluccana	
	Candleberry tree
	Candlenut tree
	Country walnut
	Indian walnut
	Otaheite walnut
	Varnish tree
Aleurites montana	Mu tree
	Tung
Alhagi camelorum	Camel thorn
Alhagi maurorum	Manna tree
Allamanda cathartica	Allamanda
	Buttercup flower
	Golden trumpet
	Yellow bell
Alnus cordata	Italian alder
Alnus crispa mollis	
	American green alder
	Green alder
	Mountain alder
Alnus glutinosa	Black alder
	Common alder
Alnus glutinosa imperialis	
	Cut-leaved alder
Alnus hirsuta	Japanese hairy alder
	Manchurian alder
Alnus incana	European alder
	Grey alder
	White alder
Alnus incana pendula	
	Weeping alder
Alnus japonica	Japanese alder
Alnus maritima	Seaside alder
Alnus nepalensis	Nepalese alder
Alnus nitida	Himalayan alder
Alnus oregano	Oregon alder
	Red alder
Alnus orientalis	Oriental alder
Alnus rhombifolia	White alder
Alnus rubra	Oregon alder
	Red alder
Alnus rugosa	Hazel alder
	Smooth alder
	Speckled alder
Alnus serrulata	Smooth alder
	Tag alder
Alnus sinuata	Sitka alder
Alnus subcordata	Caucasian alder

Alnus tenuifolia	Mountain alder
Alnus viridis	
	European green alder
	Green alder
Aloysia citriodora	Lemon verbena
Aloysia triphylla	Lemon plant
	Lemon-scented verbena
	Lemon verbena
Alstonia scholaris	
	Australian fever bush
	Bitter bark
	Devil's bit
	Devil tree
	Dita bark
	Pali-mara
Alternant ficoidea	Joseph's coat
	Parrot leaf
Althaea frutex	Althaea
	Bush hollyhock
	Bush mallow
	Hibiscus
	Rose of sharon
	Tree hollyhock
Amelanchier arborea	June berry
Amelanchier laevis	
	Allegheny service berry
	June berry
	Shad berry
	Shad blow
	Shad bush
	Snowy mespil
Amelanchier lamarckii	June berry
	Snowy mespil
Amelanchier ovalis	Snowy mespil
Amhertsia nobilis	Burmese pride
	Pride of Burma
Amorpha canescens	Lead plant
Amorpha fruticosa	Bastard indigo
	False indigo
Anacardium occidentale	Cashew
	Maranon
Andira araroba	Araroba
	Cabbage tree
	Ringworm powder tree
Andromeda polifolia	
	Bog rosemary
	Marsh rosemary
Anemopsis californica	
	Yerba mansa
Annona cherimola	Cherimalla
	Cherimoya
	Custard apple
Annona diversifolia	Anona blanca
	Ilama
Annona glabra	Alligator apple
	Pond apple
Annona montana	
	Mountain soursop
	Wild soursop
Annona muricata	Guanabana
	Prickly custard apple
	Soursop
Annona palustris	Alligator apple

TREES, BUSHES, AND SHRUBS

Arctostaphylos tomentosa

Annona reticulata Bullock's heart
Custard apple

Annona senegalensis
Wild custard apple

Annona squamosa Custard apple
Sugar apple
Sweetsop

Anopterus glandulosus
Tasmanian laurel

Anthockistazam besiaca
Forest fever tree

Anthyllis barba-jovis
Jupiter's beard

Antiaris toxicaria Upas tree

Antidesma bunius Bignay
Chinese laurel

Aphananthe aspera Muku tree

Aralia chinensis
Chinese angelica tree

Aralia cordata Udo

Aralia elata Angelica tree
Japanese angelica tree

Aralia hispida Bristly sarsaparilla

Aralia japonica Castor oil plant
Fig-leaf palm
Formosa rice tree
Japanese fatsia
Paper plant

Aralia nudicaulis
American sarsaparilla
Rabbit root
Sarsaparilla
Spikenard
Wild sarsaparilla

Aralia racemosa
American spikenard

Life-of-man
Petty morel
Spikenard

Aralia sieboldii Fig-leaf palm

Aralia spinosa Angelica tree
Devil's walking-stick
Hercules' club
Prickly ash

Araucaria angustifolium
Brazilian araucaria

Brazilian pine
Candelabra tree
Pirana pine

Araucaria araucana Chile pine
Monkey-puzzle

Araucaria bidwillii Bunya-bunya
Bunya pine

Araucaria columnaris
New caledonia pine

Araucaria cunninghamii
Hoop pine
Moreton Bay pine

Araucaria excelsa
Norfolk Island pine

Araucaria heterophylla
Australian pine
House pine
Norfolk Island pine

Araucaria hunsteinii Klinki pine

Arbutus andrachne
Cyprus strawberry tree
Eastern strawberry tree
Grecian strawberry tree

Arbutus × andrachnoides
Hybrid strawberry tree

Arbutus menziesii Madrona
Pacific madrone

Arbutus unedo Cane apple
Killarney strawberry tree

Archontophoenix alexandrae
Alexandra palm
Northern bungalow palm

Archontophoenix cunninghamiana
Piccabeen bungalow palm
Piccabeen palm

Arctostaphylos andersonii
Heart-leaf manzanita

Arctostaphylos canescens
Hoary manzanita

Arctostaphylos cinerea
Del norte manzanita

Arctostaphylos columbiana
Hairy manzanita

Arctostaphylos crustacea
Brittle-leaf manzanita

Arctostaphylos densiflora
Sonoma manzanita

Arctostaphylos glandulosa
Eastwood manzanita

Arctostaphylos glauca
Big-berry manzanita

Arctostaphylos insularis
Island manzanita

Arctostaphylos manzanita
Bearberry
Manzanita
Parry manzanita

Arctostaphylos mariposa
Mariposa manzanita

Arctostaphylos morroensis
Morro manzanita

Arctostaphylos obispoensis
Serpentine manzanita

Arctostaphylos otayensis
Otay manzanita

Arctostaphylos pajaroensis
Pajaro manzanita

Arctostaphylos patula
Greenleaf manzanita
Green manzanita

Arctostaphylos pungens
Mexican manzanita

Arctostaphylos silvicola
Silver-leaf manzanita

Arctostaphylos tomentosa
Shaggy-bark manzanita

Arctostaphylos uva-ursi
Bear's grape
Common bearberry
Hog cranberry
Kinnikinick
Mealberry
Mountain box
Red bearberry
Sandberry

Arctostaphylos viscida
White-leaf manzanita

Arctous alpinus Black bearberry

Ardisia crenata Coralberry
Spiceberry

Ardisia escallonioides Marlberry

Areca aleracea Betel-nut palm
Cabbage palm

Areca catechu Areca-nut palm
Betel-nut palm
Betel palm
Catechu
Pinang

Areca lutescens Butterfly palm

Arecastrum romanzoffianum
Queen palm

Arenga pinnata Areng palm
Black fibre palm
Gomuti palm
Sugar palm

Argania spinosa Argan tree
Morocco ironwood

Argemone mexicana Devil's fig
Mexican poppy

Aristotelia racemosa
New Zealand wineberry

Aronia arbutifolia Chokeberry
Red chokeberry

Aronia melanocarpa
Black chokeberry

Aronia prunifolia
Purple chokeberry

Artocarpus altilis Breadfruit

Artocarpus heterophyllus Jack fruit

Artocarpus incisus Breadfruit

Artocarpus lakoocha Monkey jack

Asimina triloba Pawpaw

Astragalus gummifer
Gum tragacanth

Astrocaryum aculeatum Tucuma

Atherosperma moschata
Plume nutmeg

Atherosperma moschatum
Black sassafras
Southern sassafras

Athrotaxis cupressoides
Smooth Tasmanian cedar

Athrotaxis laxifolia Summit cedar

Athrotaxis selaginoides
King William pine

Atriplex canescens Grey sage brush

Atriplex halimus Sea orach
Tree purslane

Atriplex portulacoides Sea purslane

Attalea funifira Piassaba

Aucomea klainiana Gaboon

Aucuba japonica Japanese laurel
Spotted laurel

Austrocedrus chilensis
Chilean cedar
Chilean incense cedar

Averrhoa bilimbi Bilimbi

Avicennia nitida Black mangrove

Azalea obtusa Kirishima azalea

Azalea oldhamii Formosan azalea

Azalea procumbens Alpine azalea

Azalea viscosa Swamp honeysuckle

Azara integrifolia Goldspire

Azim tetracantha Needle bush

Backhousia citriodora
Sweet verbena tree

Bactris gasipaes Maraja palm
Peach palm
Pejibaye

Bactris guineensis Prickly pole
Tobago cane

Bactris major Prickly palm

Banksia grandis
Australian honeysuckle
Bull banksia

Banksia integrifolia Coast banksia

Banksia littoralis Swamp banksia

Baphia nitida
Barwood
Camwood

Barringtonia acutangula
Indian oak

Bauhinia purpurea Bull-hoof tree
Butterfly tree
Camel's foot
Orchid tree
Ox-hoof tree

Bauhinia variegata Ebonywood
Purple orchid tree

Beaucarnea recurvata
Elephant's foot
Ponytail

Benzoin aestivale Spice bush

Berberis aristata Nepal barberry

Berberis hypokerina Silver holly

Berberis ilicifolia
Holly-leaved barberry

Berberis morrisonensis
Mount Morrison barberry

Berberis thunbergii
Japanese barberry

*Berberis thunbergii
atropurpurea*
Purple-leaf barberry

Berberis thunbergii aurea
Golden-leaved barberry

Berberis vulgaris Barberry
Common barberry
Pipperidge

TREES, BUSHES, AND SHRUBS

Brahea brandegeei

Berberis vulgaris atropurpurea
Purple-leaf barberry

Berberis wilsoniae
Wilson's barberry
Wilson's berberis

Bertholletia excelsa Brazil nut
Cream nut
Para nut

Betula alba dalecarlica
Swedish birch

Betula albosinensis
Chinese red-barked birch

Betula alleghaniensis Grey birch
Yellow birch

Betula coerulea-grandis
Blue birch

Betula ermanii Erman's birch
Russian rock birch

Betula glandulosa Dwarf birch

Betula grossa
Japanese cherry birch

Betula jacquemontii
Himalayan birch
Jacquemont's birch
White-barked birch
White-barked Himalayan birch

Betula japonica
Japanese white birch

Betula lenta Black birch
Cherry birch
Mahogany birch
Mountain mahogany
Sweet birch

Betula lutea Yellow birch

Betula mandschurica
Japanese white birch

Betula maximowicziana
Japanese large-leaved birch
Japanese red birch
Monarch birch

Betula medwediewii
Transcaucasian birch

Betula nana Dwarf birch
Rock birch

Betula nigra Black birch
Red birch
River birch

Betula occidentalis Water birch

Betula papyrifera Canoe birch
Paper birch
White birch

Betula pendula
Common silver birch
European white birch
Lady-of-the-woods
Silver birch
Warty birch
White birch

Betula pendula dalecarlica Cut-leaf birch
Swedish birch
Weeping Swedish birch

Betula pendula purpurea
Purple birch

Betula pendula tristis
Weeping birch

Betula pendula youngii
Weeping birch
Young's weeping birch

Betula platyphylla Szechuan birch

Betula platyphylla japonica
Japanese white birch

Betula populifolia Fire birch
Grey birch
Oldfield birch
White birch

Betula pubescens
Common white birch
Downy birch
Hairy birch
White birch

Betula utilis Himalayan birch

Bischofia javanica Toog

Bixia orellana Achiote
Annatto
Lipstick tree

Blighia sapida Akee

Bolusanthus speciosus Tree wisteria

Bombax buonopozense
Gold coast bombax

Bombax ceiba
Red silk-cotton tree

Bombax malabaricum Cotton tree
Indian silk-cotton tree
Red silk-cotton tree

Borassus flabellifer Doub palm
Palmyra palm
Tala palm
Toddy palm
Wine palm

Boswellia thurifera Frankincense
Olibanus tree

Bougainvillea glabra Paper flower

Bourreria ovata Strongback

Brachychiton acerifolius
Flame bottle tree
Flame tree
Illawarra flame tree

Brachychiton australis
Broad-leaved bottle tree
Flame tree

Brachychiton discolor Hat tree
Queensland lace bark
Scrub bottle tree

Brachychiton gregorii
Desert kurrajong

Brachychiton populneus Kurrajong

Brachychiton rupestris
Narrow-leaved bottle tree
Queensland bottle tree

Brahea armata Blue fan palm
Blue hesper palm
Grey goddess
Mexican blue palm

Brahea brandegeei
San José hesper palm

Brahea edulis

Brahea edulis	Guadalupe palm
Brahea elegans	Franceschi palm
Brassaia actinophylla	
	Australian ivy palm
	Australian umbrella tree
	Octopus tree
	Queensland umbrella tree
	Queen's umbrella tree
	Starleaf
Breynia disticha	Foliage flower
	Snowbush
Brosimum alicastrum	Breadnut
	Cow tree
Broussonetia papyrifera	
	Paper mulberry
	Tapa-cloth tree
Brownea grandiceps	
	Rose of Venezuela
Brunfelsia americana	
	Lady-of-the-night
Brunfelsia australis	
	Morning-noon-and-night
	Paraguay jasmine
	Yesterday-and-today
Brunfelsia undulata	Rain tree
Brunfelsia uniflora	Manaca
	Vegetable mercury
Bucida buceras	Black olive
Buddleia alternifolia	
	Butterfly bush
	Fountain buddleia
Buddleia davidii	Buddleia
	Butterfly bush
	Orange eye
	Summer lilac
Buddleia globosa	Orange ball tree
Buddleia salviifolia	
	South African sagewood
Bumelia lanuginosa	Black haw
	Chattamwood
	False buckthorn
	Gum elastic
	Shittimwood
Bumelia lycioides	Buckthorn
	Ironwood
	Mock orange
	Shittimwood
	Southern buckthorn
Burchelia bubalina	
	Wild pomegranate
Bursaria spinosa	Box thorn
Bursera microphylla	Elephant tree
	Torote
Bursera simaruba	Gumbo-limbo
	Gum-elemi
	West Indian birch
Butia capitata	Jelly palm
	Pindo palm
Butia eriospatha	Woolly butia palm
Butia yatay	Jelly palm
	Yatay palm
Butyrospermum paradoxum	
	Shea butter tree

Butyrospermum parkii	
	Shea butter tree
Buxus aurea pendula	
	Weeping golden box
Buxus balearica	Balearic box
Buxus macowani	Cape box
Buxus prostrata	Horizontal box
Buxus sempervirens	Boxwood
	Common box
	Dudgeon
Buxus suffruticosa	Edging box
Byrsonima crassifolia	
	Charcoal tree
Caesalpinia braziliensis	
	Brazilwood
Caesalpinia coriaria	Divi-divi
Caesalpinia echinata	Brazilwood
	Peachwood
Caesalpinia ferrea	
	Brazilian ironwood
Caesalpinia gilliesii	
	Bird of paradise
Caesalpinia peltophoroides	
	False brazilwood
	Sibipiruna
Caesalpinia pulcherrima	
	Barbados pride
	Dwarf poinciana
	Flower fence
	Peacock flower
	Pride of Barbados
Caesalpinia sappan	Brazilwood
	Sappanwood
Caesalpinia spinosa	Tara
Caesalpinia vesicaria	Brasiletto
Calamus rotang	Rattan cane
Callicarpa americana	
	Beautyberry
	French mulberry
Callistemon citrinus	Bottlebrush
	Crimson bottlebrush
	Lemon bottlebrush
Callistemon salignus	
	Willow-leaved bottlebrush
Callistemon speciosus	
	Albany bottlebrush
Callistemon viminalis	
	Weeping bottlebrush
Callitris calcarata	
	Black cypress pine
Callitris columellaris	
	Murray river pine
	White cypress pine
Callitris endlicheri	
	Black cypress pine
	Red cypress pine
Callitris glauca	Murray river pine
	White cypress pine
Callitris intratropica	
	Northern cypress pine
Callitris macleayana	
	Port Macquarie pine

Carpenteria californica

Callitris muelleri
Mueller's cypress pine

Callitris oblonga
Tasmanian cypress pine

Callitris preissii
Common cypress pine
Rottnest island pine

Callitris rhomboides
Oyster-bay pine
Port Jackson pine

Callitris robusta
Common cypress pine

Callitris verrucosa   Turpentine pine

Calluna erica   Heather

Calluna vulgaris   Ling
Scotch heather

Calluna vulgaris alba
White heather

Calocedrus bidwillii   Pahautea

Calocedrus chilensis
Chilean incense cedar

Calocedrus decurrens
California incense cedar
Incense cedar

Calocedrus formosana
Formosa incense cedar

Calocedrus macrolepis
Chinese incense cedar

Calocedrus plumosa   Kawaka

Calodendrum capense
Cape chestnut

Calophyllum brasiliense
Calaba tree
Jacareuba
Maria
Saint Mary's wood
Santa Maria

Calophyllum inophyllum
Alexandrian laurel
Indian laurel
Laurelwood

Calothamnus asper
Rough netbush

Calothamnus gilesii   Giles' netbush

Calothamnua validus
Barren's clawflower

Calothamnus villosus
Woolly netbush

Calotropis gigantea
Bowstring hemp
Crown plant
Madar

Calotropis procera   Calotropis
Mudar

Calycanthus fertilis
Carolina allspice

Calycanthus floridus
Carolina allspice
Pineapple shrub
Strawberry shrub

Calycanthus occidentalis
Californian allspice

Calycophyllum candissimum
Degame

Calyptronoma dulcis
Cuban manac

Calyptronoma occidentalis
Long-thatch palm
Jamaican manac

Camellia japonica
Common camellia

Camellia japonica rusticana
Snow camellia

Camellia oleifera   Tea oil plant

Camellia sasanqua
Sasanqua camellia

Camellia sinensis   Tea plant

Camellia sinensis assamensis
Assam tea

Camellia thea   Tea plant

*Camellia × williamsii
c.f.coates*   Fishtail camellia

Cananga odorata   Ilang-ilang

Canarium commune   Kenari

Canella alba   White cinnamon

Canella winterana
White cinnamon
Wild cinnamon

Cantua buxifolia   Magic flower
Sacred flower

Capparis cynophallophora
Jamaica caper tree

Capparis spinosa   Caper bush

Caragana arborescens
Siberian pea tree

Caragana arborescens pendula
Weeping pea tree

Caragana frutex
Russian pea shrub

Caragana jubata
Shag-pine pea shrub

Caragana pygmaea
Dwarf pea tree

Carica cundinamarcensis
Mountain papaw
Mountain pawpaw

Carica papaya   Common papaw
Common pawpaw
Melon tree
Papaw
Papaya
Pawpaw

Carica pubescens   Mountain papaya

Carissa bispinosa   Hedge thorn

Carissa carandas   Karanda

Carissa grandiflora   Amatungulu
Natal plum

Carissa macrocarpa   Amatungulu

Carmichaelia odorata   Lilac broom

Carpenteria californica
Californian anemone bush
Californian mock orange
Tree anemone

Carpinus betulus

Carpinus betulus	
	Common hornbeam
	European hornbeam
	Hornbeam
Carpinus caroliniana	
	American hornbeam
	Blue beech
	Water beech
Carpinus japonica	
	Japanese hornbeam
Carpinus orientalis	
	Eastern hornbeam
	Oriental hornbeam
Carya aquatica	Bitter pecan
	Water hickory
Carya cathayensis	Chinese hickory
Carya cordiformis	Bitter nut
	Pig nut
	Swamp hickory
Carya glabra	Broom hickory
	Hog nut
	Pignut hickory
Carya illinoensis	Pecan
Carya laciniosa	King nut
	Shellbark hickory
Carya myristiciformis	
	Nutmeg hickory
Carya ovalis	Red hickory
Carya ovata	Shagbark hickory
	Shellbark hickory
Carya pallida	Pale hickory
	Sand hickory
Carya pecan	Pecan
Carya tomentosa	Bigbud hickory
	Mockernut
	Squarenut
	White-heart hickory
Caryota mitis	
	Burmese fishtail palm
	Clustered fishtail palm
	Tufted fishtail palm
Caryota urens	Jaggery palm
	Kittul tree
	Sago palm
	Toddy palm
	Wine palm
Casimiroa edulis	Mexican apple
	White sapote
	Zapote blanco
Casimiroa tetrameria	Matasano
Cassia acutifolia	Alexandrian senna
	Senna
Cassia alata	Candlestick senna
	Christmas candle
	Empress candle plant
	Ringworm cassia
	Ringworm senna
Cassia angustifolia	Indian senna
	Tinnevelly senna
Cassia artemisioides	
	Feathery cassia
	Wormwood senna
Cassia auriculata	Avaram
	Tanner's cassia

Cassia chamaecrista	Ground senna
Cassia corymbosa	Buttercup bush
Cassia covesii	Desert cassia
Cassia didymobotrya	Candle bush
	Golden wonder
Cassia elata	Ringworm senna
Cassia eremophila	Desert cassia
Cassia fasciculata	Golden cassia
	Partridge pea
	Prairie senna
Cassia fistula	Golden rain
	Golden shower
	Indian laburnum
	Pudding-pipe tree
	Purging cassia
	Purging fistula
Cassia grandis	Horse cassia
	Pink shower
Cassia hebecarpa	Wild senna
Cassia javanica	
	Appleblossom cassia
	Appleblossom senna
	Pink cassia
Cassia laevigata	Smooth senna
Cassia marilandica	Wild senna
Cassia moschata	Bronze shower
Cassia nictitans	
	Wild sensitive plant
Cassia nodosa	Jointwood
	Pink-and-white shower
Cassia occidentalis	Coffee senna
	Stinking weed
	Styptic weed
Cassia senna	Alexandrian senna
Cassia siamea	Kassod tree
Cassia splendida	Golden wonder
Cassia tora	Sicklepod
	Sickle senna
Cassine glauca	Ceylon tea
Cassine laneana	
	Bermuda olivewood bark
Cassine orientalis	False olive
Cassinia fulvida	Golden bush
	Golden heather
Cassinia vauvilliersii albida	
	Silver heather
Castanea alnifolia	
	Bush chinquapin
	Downy chestnut
Castanea crenata	
	Japanese chestnut
Castanea dentata	
	American chestnut
Castanea mollissima	
	Chinese chestnut
Castanea ozarkensis	
	Ozark chestnut
Castanea pumila	
	Chinquapin chestnut
Castanea sativa	European chestnut
	Spanish chestnut
	Sweet chestnut

Castanea sativa albomarginata
    Variegated sweet chestnut

Castanea sativa aureomarginata
    Variegated sweet chestnut

Castanea sempervirens
    Bush chinquapin

Castanopsis chrysophylla
    Giant chinquapin

Castanopsis cuspidata
    Japanese chinquapin

Castanopsis megacarpa
    Great Malayan chestnut

Castanopsis sempervirens
    Bush chinquapin

Castanospermum australe
    Black bean tree
    Moreton bay chestnut

Castilla elastica
    Castilla rubber tree
    Panama rubber tree

Casuarina cunninghamiana
    Australian river oak
    River oak

Casuarina equisetifolia
    Australian beefwood
    Australian pine
    Beechwood
    Horsetail she-oak
    Horsetail tree
    Mile tree
    Red beefwood
    South sea ironwood
    Swamp she-oak

Casuarina nana   Dwarf she-oak

Casuarina stricta
    Drooping she-oak
    She-oak

Casuarina torulosa
    Australian forest oak

Catalpa bignonioides   Catalpa
    Indian bean tree
    Southern catalpa

Catalpa bignonioides aurea
    Golden-leaved catalpa

Catalpa × erubescens
    Hybrid catalpa

Catalpa fargesii   Farges catalpa

Catalpa × hybrida   Hybrid catalpa

Catalpa ovata   Chinese catalpa
    Yellow catalpa

Catalpa speciosa   Catawba
    Cigar tree
    Indian bean
    Northern catalpa
    Western catalpa

Catesbaea spinosa   Lily thorn

Catha edulis   Abyssinian tea
    Arabian tea
    Cafta
    Chat
    Khat
    Qat
    Somali tea

Cavanillesia plantanifolia   Cuipo

Ceanothus americanus
    Mountain-sweet
    New Jersey tea plant
    Red root
    Wild snowball

Ceanothus arboreus
    Catalina ceanothus
    Catalina mountain lilac
    Felt-leaf ceanothus

Ceanothus caeruleus
    Azure ceanothus

Ceanothus cordulatus   Snowbush

Ceanothus crassifolius
    Hoaryleaf ceanothus

Ceanothus cuneatus   Buckbrush

Ceanothus cyaneus
    San Diego ceanothus

Ceanothus dentatus
    Santa Barbara ceanothus

Ceanothus foliosus
    Wavyleaf ceanothus

Ceanothus gloriosus
    Point Reyes ceanothus

Ceanothus griseus
    Carmel ceanothus

Ceanothus griseus horizontalis
    Carmel creeper
    Yankee point ceanothus

Ceanothus impressus
    Santa Barbara ceanothus

Ceanothus integerrimus
    Deerbrush
    Deerbush

Ceanothus masonii
    Bolinas ridge ceanothus

Ceanothus papillosus
    Wart leaf ceanothus

Ceanothus prostratus
    Mahala-mat
    Squaw carpet

Ceanothus pumilus   Siskiyou-mat

Ceanothus purpureus
    Holly leaf ceanothus

Ceanothus ramulosis
    Coast ceanothus

Ceanothus rigidus
    Monterey ceanothus

Ceanothus sanguineus   Oregon tea
    Wild lilac

Ceanothus sorediatus   Jim brush
    Jim bush

Ceanothus spinosus
    Green-bark ceanothus
    Red-heart

Ceanothus thyrsiflorus
    Blueblossom
    Californian lilac

Ceanothus thyrsiflorus repens
    Creeping blueblossom

Ceanothus tomentosus
    Woolly-leaf ceanothus

Cecropia palmata   Snakewood tree

Cecropia peltata	Trumpet tree
Cedrela odorata	Barbados cedar
	Cigar-box cedar
	Spanish cedar
	West Indian cedar
Cedrela sinensis	Chinese cedar
Cedronella canariensis	
	Balm of gilead
	Canary balm
Cedrus atlantica	Atlas cedar
Cedrus atlantica glauca	
	Blue atlas cedar
	Blue cedar
Cedrus atlantica pendula	
	Weeping atlas cedar
Cedrus brevifolia	Cyprian cedar
	Cyprus cedar
Cedrus deodara	Deodar
	Himalayan cedar
	Indian cedar
Cedrus deodara aurea	
	Golden deodar cedar
	Western Himalayan cedar
Cedrus libani	Cedar of Lebanon
Ceiba pentandra	Kapok tree
	Silk-cotton tree
	White silk-cotton tree
Celtis africanus	White stinkwood
Celtis australis	Honeyberry
	Lote tree
	Mediterranean hackberry
	Nettle tree
	Southern nettle tree
Celtis caucasica	
	Caucasian nettle tree
Celtis iguanaea	Granjeno
Celtis japonica	Chinese hackberry
	Japanese hackberry
Celtis kraussiana	Stinkwood
Celtis laevigata	
	Mississippi hackberry
	Sugarberry
Celtis occidentalis	Hackberry
	Nettle tree
	Sugarberry
Celtis sinensis	Chinese hackberry
	Japanese hackberry
Cephaelis ipecacuanha	
	Ipecacuanha
Cephalanthus occidentalis	
	Button bush
Cephalotaxus fortunii	
	Chinese plum yew
	Cowtail pine
	Plum yew
Cephalotaxus harringtonia	
	Cowtail pine
	Harrington plum yew
	Japanese cowtail pine
	Japanese plum-fruited yew
Cephalotaxus harringtonia drupacea	Drooping cowtail pine
	Japanese plum yew

Cephalotaxus sinensis	
	Chinese cowtail pine
Ceratonia siliqua	Algarroba bean
	Carob
	Locust bean
	Saint John's bread
Ceratopetalum apetalum	
	Lightwood
Ceratopetalum gummiferum	
	Red gum
Cercidiphyllum japonicum	
	Katsura
Cercidium floridum	Palo verde
Cercis canadensis	
	American judas tree
	American redbud
	Redbud
Cercis chinensis	Chinese redbud
Cercis occidentalis	
	Western redbud
Cercis racemosa	Chinese redbud
Cercis siliquastrum	Judas tree
	Love tree
Cercocarpus montanus	
	Mountain mahogany
Ceroxylon alpinum	Wax palm
Cestrum diurnum	Day cestrum
	Day jessamine
Cestrum nocturnum	
	Night jasmine
	Night jessamine
Cestrum parqui	
	Willow-leaved jessamine
Chaenomeles japonica	
	Dwarf quince
	Lesser flowering quince
Chaenomeles speciosa	
	Flowering quince
	Japanese quince
Chaenomeles speciosa nivalis	
	Ornamental quince
Chamaecistus procumbens	
	Alpine azalea
Chamaecyparis formosensis	
	Formosan cedar
	Formosan cypress
Chamaecyparis lawsoniana	
	Lawson's cypress
	Port Orford cedar
Chamaecyparis lawsoniana allumii	Scarab cypress
Chamaecyparis lawsonia nootkatensis	Nootka cypress
Chamaecyparis lawsoniana stewartii	
	Golden lawson's cypress
Chamaecyparis nootkatensis	
	Alaska cedar
	Alaska yellow cedar
	Canoe cedar
	Nootka cypress
	Yellow cypress

Citrullus colocynthis

Chamaecyparis obtusa
Hinoki cypress
Japanese false cypress
Tree of the sun

Chamaecyparis obtusa crippsii
Golden hinoki cypress

Chamaecyparis pisifera
Sawara cypress

Chamaecyparis thyoides
Atlantic white cedar
Southern white cedar
Swamp white cedar
White cedar
White cypress

Chamaedaphne calyculata
Cassandra
Leather leaf

Chamaedorea elegans
Good-luck palm
Parlour palm

Chamaedorea erumpens
Bamboo palm

Chamaedorea seifrizii   Reed palm

Chamaedorea tepejilote   Pacaya

Chamaelaucium uncinatum
Geraldton wax

Chamaerops humilis
Dwarf fan palm
European fan palm

Chilopsis linearis   Desert willow
Flowering willow

Chimonanthus praecox
Wintersweet

Chionanthus retusa
Chinese fringe tree

Chionanthus virginicus
American fringe tree
Fringe tree
Old man's beard
Poison ash
Snowflower
Virginian snowflower

Chiranthodendron pentadactylon
Handflower tree
Mexican hand plant

Chlorophora excelsa   Iroko

Chlorophora tinctoria   Fustic

Chloroxylon swietenia
East Indian satinwood
Satinwood

Choisya ternata
Mexican orange blossom

Chondrodendron tomentosum
Curare
Pareira

Chorisia speciosa   Floss silk tree

Chrysalidocarpus lutescens
Areca palm
Butterfly palm
Golden-feather palm
Madagascar palm
Yellow butterfly palm
Yellow palm

Chrysobalanus icaco   Coco palm
Coco plum
Icaco

Chrysolepis chrysophylla
Golden chestnut

Chrysolepis cuspidata
Japanese chestnut

Chrysolepis megacarpa
Greater Malayan chestnut

Chrysolepis sempervirens
Bush chinquapin

Chrysophyllum cainito   Caimita
Star apple

Chrysophyllum oliviforme
Jamaican damson plum
Satinleaf

Cinchona calisaya   Calisaya
Jesuit's bark
Yellow bark

Cinchona cordifolia
Cartagena bark

Cinchona micrantha   Huanuco

Cinchona officinalis   Quinine

Cinchona succirubra
Peruvian bark

Cinchona succirula   Red cinchona

Cinnamodendron corticosum
False winter's bark
Red canella

Cinnamomum burmanii
Padang cassia

Cinnamomum camphora
Camphor tree

Cinnamomum cassia   Cassia
Cassia bark tree
Chinese cinnamon

Cinnamomum loureirii
Cassia-flower tree
Saigon cinnamon

Cinnamomum zeylanicum
Ceylon cinnamon
Cinnamon
Cinnamon tree

Cistus ladanifer   Gum cistus

Cistus monspeliensis
Montpelier rock rose

Cistus salviifolius
Sage-leaved rock rose

Citharexylum fruticosum
Fiddlewood

Citharexylum spinosum
Fiddlewood
Zitherwood

× *Citrofortunella floridana*
Limequat

× *Citrofortunella mitis*
Calamondin
Panama orange

× *Citrofortunella swinglei*
Limequat

× *Citroncirus webberi*   Citrange

Citrullus colocynthis   Bitter apple

Citrus aurantifolia	Key lime
	Lime
	Mexican lime
	West Indian lime
Citrus aurantifolia × *fortunella*	
	Limequat
Citrus aurantium	Bigarade
	Bitter orange
	Seville orange
	Sour orange
Citrus aurantium bergamia	
	Bergamot
	Bergamot orange
Citrus bergamia	Bergamot orange
Citrus grandis	Pomelo
	Pumelo
Citrus ichangensis	Ichang lemon
Citrus limetta	Sweet lime
Citrus limon	Lemon
Citrus × *limonia*	Lemandarin
	Mandarin lime
	Rangpur lime
Citrus lumia	Sweet lemon
Citrus otaitensis	Otaheite orange
Citrus maxima	Pomelo
	Pompelmous
	Pummelo
	Shaddock
Citrus medica	Citron
Citrus medica cedra	Cedrat lemon
Citrus × *nobilis*	Tangor
Citrus × *nobilis king*	
	King mandarin
	King of Siam
	King orange
Citrus × *nobilis temple*	
	Temple orange
Citrus × *paradisi*	Grapefruit
Citrus paradisi × *citrus reticulata*	
	Ugli fruit
Citrus reticulata	Clementine
	Mandarin orange
	Satsuma orange
	Tangerine
Citrus sinensis	Orange
	Sweet orange
Citrus tachibana	Tachibana orange
Citrus × *tangelo*	Tangelo
Citrus trifoliata	Hardy orange
Cladrastis chinensis	
	Chinese yellow-wood
Cladrastis lutea	
	American yellowwood
	Virgilia
	Yellowwood
Cladrastis platycarpa	
	Japanese yellowwood
Cladrastis sinensis	
	Chinese yellowwood
Cladrastis tinctoria	
	American yellowwood

Cladrastis wilsonii	
	Wilson's yellowwood
Clausena lansium	Wampee
	Wampi
Clerodendrum fargesii	
	Glory bower
	Glory tree
Clerodendrum indicum	
	Tube flower
	Turk's turban
Clerodendrum paniculatum	
	Pagoda flower
Clerodendrum × *speciosum*	
	Java glory bean
	Pagoda flower
Clerodendrum thomsoniae	
	Bag flower
	Bleeding glory-bower
	Bleeding-heart vine
	Glory tree
Clerodendrum trichotomum	
	Glory tree
Clethra acuminata	White alder
Clethra alnifolia	
	Sweet pepperbush
	Summer-sweet
Clethra arborea	
	Folhado
	Lily-of-the-valley tree
Clethra tomentosa	Downy clethra
Clianthus formosus	Glory pea
	Sturt's desert pea
Clianthus puniceus	Lobster claw
	Parrot bill
	Parrot's bill
Cliftonia monophylla	Black titi
	Buckwheat brush
	Buckwheat tree
	Ironwood
	Titi
Clitoria ternatea	Butterfly pea
Cneorum tricoccum	Spurge olive
Coccoloba diversifolia	
	Pigeon plum
	Snailseed
Coccoloba uvifera	Kino
	Platterleaf
	Sea grape
Coccothrinax argentata	
	Florida silver palm
	Silver palm
Coccothrinax argentea	
	Broom palm
	Silver thatch
Coccothrinax crinata	Thatch palm
Cochlospermum frazeri	
	Kapok bush
Cochlospermum religiosum	
	Buttercup tree
	Silk-cotton tree
	Yellow-cotton tree
Cochlospermum vitifolium	
	Buttercup tree
	Rose imperial

TREES, BUSHES, AND SHRUBS

Cornus capitata mas

Cocos nucifera — Coconut palm

Codiaeum variegatum — Croton
Variegated croton

Codonocarpus cotinifolius
Bell-fruit tree

Coffea arabica — Arabian coffee
Coffee

Coffea canephora — Robusta coffee
Wild robusta coffee

Coffea liberica — Liberian coffee

Coffea zanguebariae
Zanzibar coffee

Cola acuminata — Abata cola
Cola tree
Goora nut
Kola
Kola nut

Cola anomala — Bamenda cola

Cola nitida — Cola
Gbanja cola
Kola

Cola verticilliata — Owe cola

Coleonema pulchrum
Confetti bush

Colpothrinax wrightii — Barrel palm
Bottle palm
Cuban belly palm

Combretum erythrophyllum
Bush willow

Combretum microphyllum
Burning bush
Flame creeper

Commifera abyssinica — Myrrh

Commifera molmol — Myrrh

Commifera myrrha — Myrrh

Conium maculatum — Hemlock

Connarus guianensis — Zebra wood

Conocarpus erectus — Buttonwood

Conospermum huegelii
Slender smoke bush
Smoke bush

Conospermum stoechadis
Australian smoke bush

Copernica cerifera — Carnanba palm
Carnauba palm

Copernica macroglossa
Petticoat palm

Coprifera mopane — Turpentine tree

Copernica prunifera
Carnauba wax palm

Coprosma baueri
Looking-glass bush

Coprosma repens
Looking-glass bush
Mirror plant

Corchorus capsularis — Jute
White jute

Corchorus olitorius — Jew's mallow
Jute
Melukhie
Tossa jute

Cordia alliodora — Ecuador laurel
Laurel
Laurel negro

Cordia boissieri — Anacahuita

Cordia dentata — Jackwood

Cordia myxa — Assyrian plum
Selu

Cordia nitida — Red manjack
West Indian cherry

Cordia sebestena — Anaconoa
Cordia
Geiger tree
Geranium tree
Spanish cordia

Cordyline australis — Cabbage tree
Fountain dracaena
Giant dracaena
Grass palm
Palm lily

Cordyline fruticosa
Good-luck plant
Tree of kings

Cordyline indivisa — Blue dracaena

Cordyline terminalis
Good-luck plant
Hawaiian good-luck plant
Tree of kings

Corema album
Portuguese crowberry

Corema conradii
Plymouth crowberry

Coriaria nepalensis — Tanner's tree

Coriaria ruscifolia — Deu

Cornus alba — Red-barked dogwood
Siberian dogwood
Tartar dogwood

Cornus alba aurea
Golden dogwood

Cornus alba elegantissima
Silver dogwood
Silver variegated dogwood

Cornus alba sibirica
Westonbirt dogwood

Cornus alba spaethii
Golden variegated dogwood

Cornus alternifolia — Green osier
Pagoda dogwood

Cornus amomum — Red willow
Silky dogwood

Cornus canadensis — Bunchberry
Crackerberry
Creeping dogwood
Dwarf cornel
Puddingberry

Cornus capitata — Bentham's cornel
Strawberry tree

Cornus capitata florida
Flowering dogwood

Cornus capitata mas
Cordelian cherry

Cornus controversa

Cornus controversa
Giant dogwood
Japanese dogwood
Table dogwood
Wedding-cake tree

Cornus controversa variegata
Variegated Japanese dogwood

Cornus florida
American boxwood
Flowering dogwood
New England boxwood
Virginia dogwood
White dogwood

Cornus florida rainbow
Rainbow dogwood

Cornus florida rubra
Pink dogwood

Cornus glabrata   Brown dogwood

Cornus kousa   Japanese dogwood
Kousa

Cornus kousa chinensis
Chinese dogwood

Cornus mas   Cornel
Cornelian cherry
Sorbet

Cornus nuttallii
Mountain dogwood
Nuttall's dogwood
Pacific dogwood

Cornus officinalis
Japanese cornel
Japanese cornelian cherry

Cornus purpusii   Silky dogwood

Cornus racemosa   Grey dogwood
Panicled dogwood

Cornus rugosa
Round-leaved dogwood

Cornus sanguinea
Blood-twig dogwood
Common dogwood
Dogberry
Dogwood
Pegwood
Shrubby dogwood

Cornus sericea   American dogwood
Red osier dogwood

Cornus stolonifera
Red osier dogwood

Cornus stolonifera flaviramea
Yellow-stemmed dogwood

Cornus stricta   Stiff dogwood

Cornus suecica   Dwarf cornel

Corokia cotoneaster
Wire-netting bush

Coronilla emerus   Scorpion senna

Corpinus betulus
European hornbeam

Correa alba   Botany bay tea tree

Corylopsis paucifolia
Buttercup winter hazel

Corylopsis spicata   Corylopsis
Spike winter hazel

Corylus americana
American filbert
American hazel

Corylus avellana   Cobnut
Common hazel
Filbert

Corylus avellana aurea
Golden hazel

Corylus avellana contorta
Contorted hazel
Corkscrew hazel
Harry Lauder's walking stick

Corylus avellana heterophylla   Cut-leaved hazel

Corylus chinensis   Chinese filbert
Chinese hazel

Corylus colurna   Tree hazel
Turkish filbert
Turkish hazel

Corylus cornuta   Beaked filbert
Beaked hazel

Corylus hamamelis   Witch hazel

Corylus maxima   Filbert
Giant filbert
Kentish cob

Corylus maxima purpurea
Purple-leaved cob
Purple-leaved filbert

Corylus sieboldiana
Japanese hazel

Corylus tibetica   Tibetan hazel

Corynocarpus laevigata
Karaka
New Zealand laurel

Corypha australis
Australian cabbage palm
Australian fan palm
Cabbage palm
Gippsland palm

Corypha elata   Gebang palm

Corypha umbraculifera
Talipot palm

Cotinus americanus
American smoke tree
Chittamwood

Cotinus coggygria   Smoke bush
Smoke tree
Venetian sumach
Wig tree

Cotinus coggygria purpureus
Burning bush
Purple smoke tree

Cotinus obovatus
American smoke tree
Chittamwood

Cotoneaster adpressus praecox   Nan-shan bush

Cotoneaster apiculatus
Cranberry cotoneaster

Cotoneaster frigidus   Cotoneaster
Himalayan tree cotoneaster

Cotoneaster horizontalis
Rock cotoneaster

TREES, BUSHES, AND SHRUBS

Cupressus bakeri mathewsii

*Cotoneaster hybridus
  pendulus*
        Weeping cotoneaster
Cotoneaster microphyllus
        Rose box
Cotoneaster simonsii
        Himalayan cotoneaster
Cotoneaster × watereri
        Weeping cotoneaster
Coula edulis   African walnut
Couroupita guianensis
        Cannonball tree
        Carrion tree
Crataegomespilus dardari
        Bronvaux medlar
Crataegus altaica
        Altai mountain thorn
Crataegus apiifolia
        Parsley-leaved thorn
Crataegus azarolus   Azarole
Crataegus brachyacantha
        Pomette bleue
Crataegus calpodendron
        Blackthorn
        Peacock thorn
        Pear thorn
Crataegus crus-galli
        Cockspur thorn
Crataegus douglasii
        Black hawthorn
Crataegus flava   Summer haw
        Yellow-fruited thorn
        Yellow haw
Crataegus laciniata Oriental thorn
Crataegus laevigata
        Double crimson thorn
        Double pink thorn
        Double white thorn
        English hawthorn
        May
        Midland hawthorn
        Quick-set thorn
        Red hawthorn
        Red may
        Two-styled hawthorn
        White thorn
Crataegus × lavallei
        Hybrid cockspur thorn
Crataegus mollis Downy hawthorn
        Red haw
Crataegus monogyna
        Bread and cheese
        Common hawthorn
        Hawthorn
        May
        Quick
        Quickthorn
        Whitethorn
Crataegus monogyna biflora
        Glastonbury thorn
Crataegus monogyna compacta
        Dwarf hawthorn
Crataegus monogyna pendula
        Weeping hawthorn

Crataegus monogyna praecox
        Glastonbury thorn
Crataegus nigra
        Hungarian hawthorn
        Hungarian thorn
Crataegus orientalis
        Oriental thorn
Crataegus oxyacanthoides
        Midland hawthorn
Crataegus pedicellata Scarlet haw
Crataegus pentagyna
        Five-seeded hawthorn
Crataegus phaenopyrum
        Washington thorn
Crataegus pinnatifida
        Chinese hawthorn
Crataegus praecox
        Glastonbury thorn
Crataegus × prunifolia
        Broadleaf cockspur thorn
        Frosted thorn
Crataegus submollis
        Emmerson's thorn
Crataegus tanacetifolia
        Tansy-leaved thorn
Crataegus tomentosa
        Peacock thorn
        Pear thorn
+ *Crataegomespilus dardarii*
        Bronvaux medlar
Crataera gynandra   Garlic pear
Crateva religiosa
        Sacred garlic pear
Crescentia cujete   Calabash tree
Crinodendron hookeriana
        Lantern tree
Croton cascarilla   Cascarilla
        Wild rosemary
Croton monanthogynus
        Prairie tea
Cryptomeria japonica
        Japanese cedar
        Japanese red cedar
Cudrania tricuspidata
        Chinese silkworm thorn
Cunninghamia konishii Taiwan fir
Cunninghamia lanceolata
        Chinese fir
Cunonia capensis African red alder
        Rooiels
        Spoon tree
× *Cupressocyparis leylandii*
        Leyland cypress
Cupressus abramsiana
        Santa Cruz cypress
Cupressus arizonica
        Arizona cypress
        Smooth-barked Arizona cypress
Cupressus bakeri   Modoc cypress
Cupressus bakeri mathewsii
        Siskiyou cypress

Cupressus cashmeriana
    Kashmir cypress

Cupressus duclouxiana
    Bhutan cypress

Cupressus forbesii   Tecate cypress

Cupressus funebris
    Chinese weeping cypress
    Mourning cypress
    Weeping cypress

Cupressus glabra
    Smooth Arizona cypress
    Smooth-barked Arizona cypress

Cupressus goveniana
    Californian cypress
    Gowen's cypress

Cupressus guadalupensis
    Guadalupe cypress
    Tecate cypress

Cupressus lusitanica   Cedar of Goa
    Mexican cypress
    Portuguese cypress

Cupressus macnabiana
    Macnab's cypress

Cupressus macrocarpa
    Monterey cypress

Cupressus nevadensis
    Piute cypress

Cupressus sargentii
    Sargent's cypress

Cupressus sempervirens
    Italian cypress
    Mediterranean cypress

Cupressus stephensonii
    Californian cypress
    Cuyamaca cypress

Cupressus torulosa   Bhutan cypress
    Himalayan cypress

Cussonia paniculata   Cabbage tree
    Mountain kiepersol

Cussonia spicata   Cabbage tree

Cycas circinalis   Fern palm
    Queen sago
    Sago palm

Cycas media   Nut palm

Cycas revolta   Japanese fern palm
    Japanese sago palm
    Sago palm

Cyclanthera pedata   Achocha

Cydonia oblonga   Common quince
    Quince

Cyphomandra betacea
    Tree tomato

Cyrilla racemiflora   Black titi
    Huckleberry
    Ironwood
    Leatherwood
    Myrtle
    Red titi
    Titi
    White titi

Cyrostachys lakka
    Sealing-wax palm

Cyrostachys renda
    Sealing-wax palm

Cytisus albus   Portuguese broom
    White Portuguese broom

Cytisus battandieri
    Moroccan broom
    Pineapple broom

Cytisus canariensis   Genista

Cytisus decumbens
    Prostrate broom

Cytisus demissus   Dwarf broom

Cytisus genista   Common broom

Cytisus grandiflorus
    Woolly-podded broom

Cytisus monspessulanus
    Montpelier broom

Cytisus multiflorus
    White Spanish broom

Cytisus multiflorus albus
    White Portuguese broom

Cystus palmensis   Tagasaste

Cytisus × praecox
    Warminster broom

Cystus proliferus   Escabon

Cytisus purpureus   Purple broom

Cytisus scoparius   Common broom
    Scotch broom

Daboecia cantabrica
    Connemara heath
    Irish heath
    Saint Daboec's heath

Dacrycarpus dacrydioides
    Kahikatea
    New Zealand white pine

Dacrydium bidwillii
    Mountain pine

Dacrydium colensoi   Westland pine

Dacrydium cupressinum   Imou pine
    Red pine
    Rimu

Dacrydium fonkii
    South American pine

Dacrydium franklinii   Huon pine

Dacrydium intermedium
    Yellow silver pine

Dacrydium laxifolium
    Mountain rimu

Daemonorops draco
    Dragon's blood

Daemonorops grandis   Malay palm

Dais cotonifolia   Posy bush

Dalbergia latifolia   Black wood

Dalbergia sissoo   Sissoo

Damasonium alisma   Star fruit

Danae racemosa
    Alexandrian laurel

Daphne aurantiaca
    Golden-flowered daphne

Daphne cneorum
    Garland flower bush

TREES, BUSHES, AND SHRUBS

Dracaena sanderana

Daphne × houtteana
Purple-leaved daphne

Daphne laureola   Spurge laurel

Daphne mezereum
Cottage mezereon
February daphne
Mezereon

Daphne odora   Winter daphne

Daphne pontica
Twin-flowered daphne

Dasypogon bromeliaefolius
Pineapple bush

Datura suaveolens
Angel's trumpet

Davidia involucrata   Dove tree
Ghost tree
Handkerchief tree
Lady's handkerchief tree
Pocket handkerchief tree

Delonix regia   Fancy annie
Flamboyant tree
Flame of the forest
Flame tree
Gulmohur
Peacock flower
Poinciana
Royal poinciana

Dendromecon rigidum
Tree poppy

Dendrosicyos socotrana
Socotra cucumber tree

Derris elliptica   Derris
Tuba root

Derris scandens   Malay jewel vine

Desmanthus illinoensis
Prairie mimosa
Prickleweed

Detarium senegalense
Dattock tree
Tallow tree

Deutzia gracilis   Deutzia
Japanese snow flower

Dialium guineense
Sierra Leone tamarind
Velvet tamarind

Dictyosperma album
Princess palm

Dictyosperma aureum
Yellow princess palm

Dillenia indica   Elephant apple tree
Hondapara

Dioon edule   Chestnut dioon
Virgin's palm

Diospyros chinensis
Chinese persimmon
Date palm
Japanese persimmon
Kaki
Keg fig

Diospyros digyna   Black sapote

Diospyros ebenaster   Ceylon ebony
East Indian ebony
Ebony
Macassar ebony

Diospyros ebenum   Ceylon ebony
East Indian ebony
Ebony
Macassar ebony

Diospyros kaki
Chinese persimmon
Date plum
Japanese persimmon
Kaki
Keg fig
Persimmon

Diospyros kurzii   Andaman marble
Zebra wood

Diospyros lotus   Date plum
Godsberry

Diospyros melanoxylon
Coromandel ebony

Diospyros mespiliformis
Lagos ebony
West African ebony

Diospyros monbuttensis
Walking-stick ebony
Yoruba ebony

Diospyros nigra   Black persimmon
Black sapote

Diospyros texana
Black persimmon

Diospyros virginiana
American persimmon
Common persimmon
Date plum
Persimmon
Possum apple
Possumwood

Diospyros whyteana   Black bark
Bladder nut

Dipteryx odorata   Tonka bean

Dirca palustris   Leatherwood
Moosewood
Ropebark
Wicopy

Discaria toumatou   Wild irishman

Distylium racemosum   Isu tree

Dombeya × cayeuxii
Pink snowball tree

Dombeya wallichii   Pink ball tree

Dorema ammoniacum
Ammoniacum
Gum ammoniac

Doryalis - see *Dovyalis*

Dovyalis caffra   Kai apple
Kau apple
Kei apple
Umkokolo

Dovyalis hebecarpa
Ceylon gooseberry
Kitembilla

Dracaena arborea   Tree dracaena

Dracaena draco   Dragon tree

Dracaena indivisa   Blue dracaena

Dracaena sanderana
Belgian evergreen
Ribbon plant

Dracaena surculosa

Dracaena surculosa
  Gold-dust dracaena
  Spotted dracaena
Dracaena terminalis
  Good-luck plant
  Hawaiian good-luck plant
  Tree of kings
Drimys lanceolata
  Mountain pepper
  Pepper tree
Drimys winteri   Winter's bark
Dryas octopetala   Mountain avens
Duboisia myoporoides
  Corkwood tree
  Pituri
Duranta ellisia   Brazilian skyflower
  Golden dewdrop
  Pigeon berry
  Skyflower
Duranta repens
  Brazilian skyflower
  Golden dewdrop
  Pigeon berry
  Skyflower
Durio zibethinus   Durian
Duvernoia adhatodioides
  Pistol bush
  Snake bush
Elaeagnus angustifolia   Oleaster
  Russian olive
  Silver berry
  Wild olive
Elaeagnus argentea
  American silverberry
  Silver berry
Elaeagnus commutata
  American silver berry
  Silver berry
Elaeagnus edulis   Cherry elaeagnus
  Gumi
Elaeagnus latifolia   Oleaster
  Wild olive
Elaeagnus multiflora
  Cherry elaeagnus
  Gumi
Elaeagnus pungens
  Thorny elaeagnus
Elaeagnus umbellata
  Autumn olive
Elaeis guineensis   African oil palm
  Macaw-fat
  Oil palm
Elaeis melanococca
  American oil palm
Elaeis oleifera   American oil palm
Elaeocarpus reticulatus
  Blueberry ash
Elaeocarpus serratus   Ceylon olive
Elaeocarpus stauntonii
  Chinese mint bush
Embothrium coccineum
  Chilean fire bush
  Chilean fire tree

Empetrum nigrum   Crowberry
Empetrum rubrum
  South American crowberry
Enallagma latifolia   Black calabash
Encephalartos altensteinii
  Bread tree
  Prickly cyad
Enchylaena tomentosa
  Ruby saltbush
Entada gigas   Nicker bean
  Sword bean
Entandrophragma cylindricum   Sapele
Enterolobium cyclocarpum
  Elephant's ear
Epacris impressa   Australian heath
Eperua falcata   Wallaba
Ephedera distachya
  Shrubby horsetail
Ephedera vulgaris   Desert tea
  Ma huang
Epigaea repens   Mayflower
Eremophila maculata   Emu bush
  Spotted emu bush
Eretia tinifolia   Bastard cherry
Erica arborea   Briar
  Tree heath
Erica australis   Spanish heath
Erica baccans   Berry heath
Erica canaliculata
  Christmas heather
Erica carnea   Snow heather
  Spring heath
  Winter heath
Erica ciliaris   Ciliate heath
  Dorset heath
  Fringed heath
Erica cinerea   Bellflower heather
  Bell heather
  Grey heath
  Purple heather
  Scotch heath
  Twisted heath
Erica doliiformis
  Everblooming French heather
Erica erigena   Irish heath
Erica herbacea   Snow heath
  Spring heath
Erica hyemalis   Cape heath
  French heather
  White winter heather
Erica lusitanica   Portugal heath
  Spanish heath
Erica mackaiana   Mackay's heath
Erica mammosa   Red signal heath
Erica mediterranea
  Irish heath
  Mediterranean heather
Erica scoparia   Besom heath
Erica terminalis   Corsican heath
Erica tetralix   Bog heather
  Cross-leaved heath

TREES, BUSHES, AND SHRUBS

Eucalyptus crucis

Erica vagans Cornish heath
Erinacea anthyllis Blue broom
Branch thorn
Hedgehog broom
Eriobotrya japonica
Japanese loquat
Japanese medlar
Japanese plum
Loquat
Erythea armata Blue fan palm
Blue hesper palm
Blue palm
Erythrina arborea Cardinal spear
Cherokee bean
Coral bean
Erythrina corallodendron
Common coral bean
Coral tree
Erythrina coralloides
Naked coral tree
Erythrina crista-galli Cock's comb
Cockspur coral tree
Coral tree
Cry-baby tree
Erythrina fusca Swamp immortelle
Erythrina glauca
Swamp immortelle
Erythrina herbacea Cardinal spear
Cherokee bean
Coral bean
Erythrina lysistemon
Lucky bean tree
Erythrina monosperma Wilwilli
Erythrina ovalifolia
Swamp immortelle
Erythrina poeppigiana
Mountain immortelle
Erythrina tahitiensis Wilwilli
Erythrina vespertilio
Grey corkwood
Erythrina zeyheri Plough breaker
Prickly cardinal
Erythrophleum guineese
Redwater tree
Sassy bark
Erythrophleum suaveolens
Redwater tree
Erythroxylum coca Coca
Cocaine plant
Spadic
Escallonia macrantha Escallonia
Eucalyptus acacae formis
Wattle-leaved peppermint
Eucalyptus acmenioides
White mahogany
Eucalyptus agglomerata
Blue-leaved stringybark
Eucalyptus aggregata Black gum
Eucalyptus alba Timor white gum
Eucalyptus albens White box
Eucalyptus alpina
Grampian stringybark

Eucalyptus amplifolia
Cabbage gum
Eucalyptus amygdalina
Black peppermint
Eucalyptus andreana
River peppermint
Eucalyptus archeri Alpine gum
Eucalyptus argillacea
Kimberley grey box
Eucalyptus astringens
Brown mallee
Eucalyptus bakeri Malee box
Eucalyptus baxteri
Brown stringybark
Eucalyptus blakelyi
Blakely's red gum
Eucalyptus blaxlandii
Blaxland's stringybark
Eucalyptus bosistoana
Bosisto's box
Eucalyptus botryoides Bungalay
Eucalyptus bridgesiana
Apple box
Eucalyptus burracoppinensis
Burracoppin mallee
Eucalyptus caesia Gungurru
Eucalyptus calophylla Marri
Red gum
Eucalyptus camaldulensis Marri
Murray red gum
Red gum
Red river gum
Eucalyptus camphora
Broad-leaved sally
Eucalyptus capitallata
Brown stringybark
Eucalyptus cephalocarpa
Long-leaved argyle apple
Eucalyptus cinerea Argyle apple
Mealy stringybark
Silver dollar tree
Spiral eucalyptus
Eucalyptus citriodora
Lemon-scented gum
Lemon-scented spotted gum
Eucalyptus cladocalyx Sugar gum
Eucalyptus clavigera Apple gum
Cabbage gum
Eucalyptus coccifera
Mount Wellington peppermint
Tasmanian snow gum
Eucalyptus consideniana Yertchuk
Eucalyptus cordata
Heart-leaved silver gum
Eucalyptus cornuta Yate tree
Eucalyptus corynocalyx
Sugar plum
Eucalyptus cosmophylla Cup gum
Eucalyptus crebra
Narrow-leaved ironbark
Eucalyptus crucis
Southern cross silver mallee

Eucalyptus dalrympleana
Broad-leaved kindling bark
Mountain gum

Eucalyptus dealbata
Tumbledown gum

Eucalyptus deanei   Deane's gum

Eucalyptus delegatensis   Alpine ash
Gum-top stringybark

Eucalyptus desmondensis
Desmond mallee

Eucalyptus diptera   Bastard gimlet
Two-winged gimlet

Eucalyptus diversicolor   Karri
Karri gum

Eucalyptus dives
Broad-leaved peppermint

Eucalyptus dumosa   Congo mallee
Mallee

Eucalyptus elaeophora   Bundy

Eucalyptus elata   River peppermint

Eucalyptus eremophila
Horned mallee

Eucalyptus erythrocorys   Illyarie
Illyarri

Eucalyptus erythronema
Red mallee

Eucalyptus ewartiana
Ewart's mallee
Red-flowered mallee

Eucalyptus eximia
Yellow bloodwood

Eucalyptus fastigiata
Brown barrel
Cut-tail

Eucalyptus fibrosa
Broad-leaved ironbark

Eucalyptus ficifolia
Red-flowering gum

Eucalyptus flocktoniae   Merrit gum

Eucalyptus forrestiana
Forrest's marlock
Fuchsia gum

Eucalyptus fraxinoides   White ash

Eucalyptus gigantea   Alpine ash

Eucalyptus glaucescens
Tingiringi gum

Eucalyptus globoidea
White stringybark

Eucalyptus globulus   Blue gum
Southern blue gum
Stringybark tree
Tasmanian blue gum

Eucalyptus globulus maidenii
Maiden's gum

Eucalyptus gomphocephalus
Tuart gum

Eucalyptus goniocalyx   Bundy
Spotted mountain gum

Eucalyptus grandis   Flooded gum
Rose gum

Eucalyptus grossa
Coarse-leaved mallee

Eucalyptus gummifera
Red bloodwood

Eucalyptus gunnii   Cider gum

Eucalyptus haemastoma
Scribbly gum

Eucalyptus incrassata   Lerp mallee
Ridge-fruited mallee

Eucalyptus johnstonii
Tasmanian brown gum
Yellow gum

Eucalyptus kruseana
Kruse's mallee

Eucalyptus lane-poolei
Salmon white gum

Eucalyptus lansdowneana
Crimson mallee box

Eucalyptus largiflorens   Black box

Eucalyptus lehmannii   Bushy yate
Lehmann's gum

Eucalyptus leucophylla
Kimberley grey box

Eucalyptus leucoxylon
White ironbark

Eucalyptus longicornis   Red morell

Eucalyptus longifolia
River peppermint
Woollybutt

Eucalyptus luehmanniana
Yellow-topped mallee ash

Eucalyptus macarthurii
Camden woollybutt
Paddy river box

Eucalyptus macrandra
Long-flowered marlock

Eucalyptus macrocarpa   Bluebush

Eucalyptus macrorhyncha
Red stringybark

Eucalyptus maculata   Spotted gum

Eucalyptus maidenii   Maiden's gum

Eucalyptus mannifera
Red-spotted gum

Eucalyptus marginata
Jarrah
West Australian mahogany

Eucalyptus megacarpa   Bullick

Eucalyptus megacornata
Warted yate

Eucalyptus melliodora   Yellow box

Eucalyptus micrantha   Snappy gum

Eucalyptus microcorys
Tallow-wood

Eucalyptus microtheca   Coolibah
Flooded box
Jinbul
Moolar

Eucalyptus miniata
Darwin woollybutt

Eucalyptus mitchelliana
Weeping sally

Eucalyptus moluccana   Grey box

Eucalyptus muellerana
Yellow stringybark

Eucalyptus neglecta
Omeo round-leaved gum

Eucalyptus nicholii
Narrow-leaved black peppermint
Nichol's willow-leaved
peppermint

Eucalyptus niphophila   Snow gum

Eucalyptus nitens   Silver top

Eucalyptus nutans   Red moort

Eucalyptus obliqua
Messmate stringybark

Eucalyptus occidentalis
Flat-topped yate

Eucalyptus oldfieldii
Oldfield's mallee

Eucalyptus orbifolia
Round-leaved mallee

Eucalyptus ovata   Swamp gum

Eucalyptus paniculata
Grey ironbark

Eucalyptus papuana   Ghost gum

Eucalyptus parvifolia
Small-leaved gum

Eucalyptus pauciflora
Cabbage gum
White sally

Eucalyptus perriniana
Round-leaved snow gum
Spinning gum

Eucalyptus phoenicia   Scarlet gum

Eucalyptus pilularis   Blackbutt

Eucalyptus piperita
Sydney peppermint

Eucalyptus platyphilla
Timor white gum

Eucalyptus platypus
Round-leaved moort

Eucalyptus polyanthemos
Australian beech
Silver dollar gum

Eucalyptus polybractea
Silver mallee scrub

Eucalyptus populifolia   Bimble box

Eucalyptus populnea   Bimble box

Eucalyptus preissiana
Bell-fruited mallee

Eucalyptus propinqua
Small-fruited grey gum

Eucalyptus ptychocarpa
Swamp bloodwood

Eucalyptus pulchella
White peppermint

Eucalyptus pulverulenta
Money tree
Silver-leaved mountain gum

Eucalyptus punctata   Grey gum

Eucalyptus pyriformis
Pear-fruited mallee

Eucalyptus racemosa   Snappy gum

Eucalyptus radiata
Grey peppermint
Narrow-leaved peppermint
White-top peppermint

Eucalyptus × rariflora   Black box

Eucalyptus regnans
Australian mountain ash
Giant gum
Mountain ash

Eucalyptus resinifer
Red mahogany

Eucalyptus × rhodantha
Rose mallee

Eucalyptus risdonii
Silver peppermint

Eucalyptus robertsonii
Robertson's peppermint

Eucalyptus robusta
Swamp mahogany

Eucalyptus rossii   White gum

Eucalyptus rubida
Candle-bark gum

Eucalyptus rudis   Desert gum
Swamp gum

Eucalyptus salicifolia
Black peppermint

Eucalyptus saligna
Sydney blue gum

Eucalyptus salmonophloia
Salmon gum

Eucalyptus salubris   Gimlet gum

Eucalyptus sepalcralis
Blue weeping gum

Eucalyptus sideroxylon   Mugga
Red ironbark

Eucalyptus sieberi
Black mountain ash

Eucalyptus smithii
Blackbutt peppermint
Gully ash
Gully gum

Eucalyptus spathulata
Swamp mallee

Eucalyptus staigeriana
Lemon-scented ironbark

Eucalyptus steedmanii
Steedman's gum

Eucalyptus stellulata   Black sally

Eucalyptus stowardii
Stoward's mallee

Eucalyptus stricklandii
Strickland's gum

Eucalyptus stricta
Blue mountain mallee

Eucalyptus tenuiramis
Silver peppermint

Eucalyptus tereticornis
Forest red gum
Grey gum

Eucalyptus tetradonta
Darwin stringybark

Eucalyptus tetragona
White-leaved marlock

Eucalyptus tetraptera
  Four-winged mallee
  Square-fruited mallee
Eucalyptus torquata   Coral gum
Eucalyptus umbra
  White mahogany
Eucalyptus urnigera   Urn gum
Eucalyptus vernicosa
  Varnish-leaved gum
Eucalyptus viminalis   Manna gum
  Ribbon gum
Eucalyptus woodwardii
  Woodward's blackbutt
  Yellow-flowered gum
Eucommia ulmoides
  Chinese rubber tree
  Chinese silk thread tree
  Gutta-percha tree
Eucryphia cordifolia   Ulmo
Eucryphia glutinosa   Eucryphia
Eucryphia lucida   Leatherwood
Eucryphia moorai   Plumwood
Eucryphia × *nymansensis*
  nymansay
  Nyman's hybrid eucryphia
Eucryphia pinnatifolia   Eucryphia
Eugenia aggregata
  Rio Grande cherry
Eugenia aquea   Rose apple
Eugenia axillaris   White stopper
Eugenia brasiliensis   Brazil cherry
  Grumichama
  Grumixameira
Eugenia buxifolia   Spanish stopper
Eugenia caryophyllus   Clove tree
Eugenia cheken   Cheken
Eugenia confusa   Ironwood tree
  Red stopper
Eugenia dombeyi   Brazil cherry
  Grumichama
  Grumixameira
Eugenia foetida   Spanish stopper
Eugenia garberi   Ironwood tree
  Red stopper
Eugenia jambolana   Jambolan
  Jambul
  Java plum
Eugenia luschnathiana   Pitomba
Eugenia malaccensis   Malay apple
Eugenia michelii   Barbados cherry
  Brazil cherry
  Cayenne cherry
  Pitanga
Eugenia myrtifolia
  Australian brush cherry
Eugenia myrtoides
  Spanish stopper
Eugenia pitanga   Pitanga
Eugenia smithii   Lilly-pilly
Eugenia ugni   Chilean guava

Eugenia uniflora   Barbados cherry
  Brazil cherry
  Cayenne cherry
  Pitanga
  Surinam cherry
Euodia daniellii   Daniell's euodia
Euodia hupehensis   Euodia
Euonymus alata
  Winged spindle tree
Euonymus americana
  Bursting heart
  Strawberry bush
  Strawberry tree
Euonymus atropurpurea
  Burning bush
  Wahoo elm
Euonymus europaea
  Common spindle tree
  European spindle tree
  Skewerwood
  Spindle tree
Euonymus hamiltonianus
  yedoensis   Japanese spindle tree
Euonymus japonica
  Evergreen spindle tree
  Japanese euonymus
  Japanese spindle tree
Euonymus latifolius
  Broad-leaved spindle
Euonymus occidentalis
  Western burning bush
Eupatorium micranthrum
  Mexican incense bush
Eupatorium weinmannianum
  Mexican incense bush
Euphorbia antisyphilitica
  Candelilla
Euphorbia candelabrum
  Candelabra tree
Euphorbia caput-medusae
  Medusa's head
Euphorbia cereiformis   Milk-barrel
Euphorbia corollata
  Flowering spurge
  Tramp's spurge
  Wild hippo
Euphorbia cyathophora
  Fiddler's spurge
  Fire-on-the-mountain
  Mexican fire plant
  Painted leaf
Euphorbia cyparissias
  Cypress spurge
Euphorbia grandidens
  Big-tooth euphorbia
Euphorbia heptagona   Milk-barrel
Euphorbia heterophylla
  Annual poinsettia
  Fire-on-the-mountain
  Mexican fire plant
  Painted spurge
Euphorbia horrida
  African milk-barrel
Euphorbia ingens   Candelabra tree

Ficus glomerata

Euphorbia ipecacuanhae
  Carolina ipecac
  Carolina spurge
  Ipecac spurge
  Wild ipecac

Euphorbia lactea
  Candelabra cactus
  Dragon bones
  False cactus
  Hat-rack cactus
  Mottled spurge

Euphorbia lathyris   Caper spurge
  Mole plant
  Myrtle spurge

Euphorbia leviana   Milk-barrel

Euphorbia marginata   Ghostweed
  Snow-on-the-mountain

Euphorbia neriifolia
  Hedge euphorbia
  Oleander-leaved euphorbia

Euphorbia pulcherrima
  Christmas star
  Lobster plant
  Mexican flameleaf
  Painted leaf
  Poinsettia

Euphorbia tirucalli   Finger tree
  Indian tree spurge
  Milk bush
  Pencil tree
  Rubber euphorbia

Euphorbia trigona
  Abyssinian euphorbia
  African milk tree

Euptelea polyandra
  Japanese euptelea

Euterpe edulis   Assai palm

Euterpe oleracea   Assai palm

Exocarpus cupressiformis
  Australian cherry

Exochorda grandiflora   Pearl bush

Exochorda racemosa   Pearl bush

Fabiana imbricata   False heath
  Pichi

Fagara flava   Satinwood

Fagus americana   American beech

Fagus crenata   Japanese beech

Fagus englerana   Chinese beech
  Engler's beech

Fagus grandifolia   American beech

Fagus japonica   Japanese beech

Fagus longipetiolata
  Chinese beech

Fagus orientalis   Oriental beech

Fagus sieboldii   Japanese beech
  Siebold's beech

Fagus sylvatica   Common beech

Fagus sylvatica asplenifolia
  Fernleaf beech

Fagus sylvatica cristata
  Cockscomb beech

Fagus sylvatica dawyck
  Dawick beech

Fagus sylvatica fastigiata
  Fastigiate beech

Fagus sylvatica heterophylla
  Cut-leaved beech
  Fern-leaved beech

Fagus sylvatica pendula
  Weeping beech

Fagus sylvatica purpurea
  Copper beech
  Purple beech

Fagus sylvatica rohanii
  Cut-leaf purple beech
  Purple fern-leaved beech

Fagus sylvatica zlatia
  Golden beech

Fallugia paradoxa   Apache plume

Fatsia japonica   Castor oil palm
  False castor oil plant
  Figleaf palm
  Formosa rice tree
  Japanese fatsia
  Japanese figleaf palm
  Paper plant

Feijoa sellowiana   Feijoa
  Pineapple guava

Feronia elephantum
  Elephant apple
  Wood apple

Feronia limonia   Elephant apple
  Wood apple

Ficus altissima   Council tree

Ficus aspera   Clown fig

Ficus aurea   Florida strangler fig
  Golden fig
  Strangler fig

Ficus australis   Little-leaf fig
  Port Jackson fig
  Rusty fig

Ficus belgica   Assam rubber tree
  India rubber tree
  Rubber plant

Ficus benghalensis   Banyan
  East Indian fig
  Indian banyan

Ficus benjamina   Benjamin tree
  Java fig
  Small-leaved rubber plant
  Tropic laurel
  Weeping fig
  Weeping laurel

Ficus capensis   Bush fig
  Cape fig

Ficus carica   Common fig

Ficus deltoides   Mistletoe fig
  Mistletoe rubber plant

Ficus diversifolia   Mistletoe fig
  Mistletoe rubber plant

Ficus dryepondtiana   Congo fig

Ficus elastica   Assam rubber
  India rubber tree
  Rubber plant

Ficus glomerata   Cluster fig

Ficus indica

Ficus indica	Banyan
	East Indian fig tree
	Indian banyan
Ficus infectoria	Spotted fig
Ficus lacor	Spotted fig
Ficus lyrata	Fiddle-leaf fig
Ficus macrocarpa	
	Australian banyan
	Moreton Bay fig
Ficus macrophylla	
	Australian banyan
	Moreton bay fig
Ficus montana	Oak-leaf fig
Ficus mysorensis	Mysore fig
Ficus natalensis	Natal fig
Ficus nekbudu	Kaffir fig
	Zulu fig tree
Ficus pandurata	Fiddle-leaf fig
Ficus perforata	
	West Indian laurel fig
Ficus pretoriae	Wonderboom
Ficus pseudopalma	Dracaena fig
	Philippine fig
Ficus pumila	Climbing fig
	Creeping fig
	Creeping rubber plant
Ficus quercifolia	Oak-leaf fig
Ficus racemosa	Cluster fig
Ficus religiosa	Bo tree
	Peepul
	Sacred fig
Ficus repens	Climbing fig
	Creeping fig
	Creeping rubber plant
Ficus retusa	Chinese banyan
	Glossy-leaf fig
	Indian laurel
	Malay banyan
Ficus rubiginosa	Little-leaf fig
	Port Jackson fig
	Rusty fig
Ficus stipulata	Climbing fig
	Creeping fig
	Creeping rubber plant
Ficus superba	Sea fig
Ficus sycomorus	Bible fig
	Egyptian sycamore
	Mulberry fig
	Pharoah's fig
	Sycamore fig
Ficus utilis	Kaffir fig
	Zulu fig tree
Ficus virens	Spotted fig
Ficus vogelii	
	West African rubber tree
Ficus wightiana	
	Large-leaved banyan
Firmiana simplex	
	Chinese bottle tree
	Chinese parasol tree
	Japanese varnish tree
	Phoenix tree

Fitzroya cupressoides	Alerce
	Patagonian cypress
Flacourtia indica	Batoko plum
	Governor's plum
	Madagascar plum
	Ramontchi
Flacourtia ramontchi	Batoko plum
	Governor's plum
	Madagascar plum
	Ramontchi
Forestiera acuminata	Swamp privet
Forestiera neomexicana	
	Desert olive
Forsythia europaea	
	European golden ball
Forsythia × *intermedia*	Forsythia
Forsythia ovata	Korean forsythia
Forsythia sieboldii	
	Weeping forsythia
Forsythia suspensa	Golden bell
	Weeping forsythia
Forsythia suspensa fortunei	
	Arching forsythia
Fortunella japonica	
	Marumi kumquat
	Round kumquat
Fortunella margarita	Kumquat
	Nagami kumquat
	Oval kumquat
Fothergilla monticola	Witch alder
Fouquieria splendens	Coach-whip
	Jacob's staff
	Ocotillo
	Vine cactus
Frangula alnus	Alder buckthorn
Franklinia alatamaha	
	Franklin tree
Frasera carolinensis	
	American colombo
Fraxinus alba	American white ash
	White ash
Fraxinus americana	American ash
	American white ash
	White ash
Fraxinus angustifolia	
	Narrow-leaved ash
Fraxinus angustifolia veltheimii	Single-leaved ash
Fraxinus anomala	Single-leaved ash
	Utah ash
Fraxinus caroliniana	Carolina ash
	Pop ash
	Water ash
Fraxinus chinensis	Chinese ash
Fraxinus dipetala	Flowering ash
Fraxinus excelsior	Common ash
	European ash
Fraxinus excelsior aurea pendula	
	Golden weeping ash
Fraxinus excelsior diversifolia	
	Single-leaf ash

Fraxinus excelsior jaspidea
Golden ash
Golden-bark ash
Yellow-bark ash

Fraxinus excelsior pendula
Weeping ash

Fraxinus latifolia   Oregon ash

Fraxinus mandshurica
Manchurian ash

Fraxinus mariesii
Chinese flowering ash

Fraxinus nigra   Black ash

Fraxinus oregona   Oregon ash

Fraxinus ornus   Flowering ash
Manna ash

Fraxinus oxycarpa   Caucasian ash

Fraxinus oxycarpa raywood   Claret ash

Fraxinus pennsylvanica   Green ash
Red ash

Fraxinus quadrangulata   Blue ash

Fraxinus spaethiana
Spaeth's flowering ash

Fraxinus syriaca   Syrian ash

Fraxinus texensis   Texas ash

Fraxinus tomentosa   Pumpkin ash

Fraxinus uhdei   Evergreen ash
Shamel ash

Fraxinus velutina   Arizona ash
Velvet ash

Fraxinus xanthoxyloides
Afghan ash
Varnish-leaved gum

Fraxinus xanthoxyloides dimorpha   Algerian ash

Fuchsia arborescens   Tree fuchsia

Fuchsia boliviana
Peruvian fuchsia tree

Fuchsia conica   Hardy fuchsia

Fuchsia excorticata
New Zealand tree fuchsia

Fuchsia magellanica   Hardy fuchsia

Fusanus acuminatus   Quandong

Galpinia transvaalica
Wild pride of India

Garcinia mangostana   Mangosteen

Gardenia florida   Cape jasmine
Common gardenia

Gardenia grandiflora
Cape jasmine
Common gardenia

Gardenia jasminoides
Cape jasmine
Common gardenia

Garrya elliptica   Quinine bush
Silk tassel bush

Garrya fremontii   Fever bush

Garrya macrophylla
Mexican tassel bush

Gaultheria hispida   Snowberry
Tasmanian waxberry
Waxberry

Gaultheria procumbens   Boxberry
Canadian tea
Checkerberry
Creeping winterberry
Partridge berry
Tea berry
Wintergreen

Gaultheria shallon   Lemonleaf
North American salal
Shallon

Gaylussacia baccata
Black huckleberry

Gaylussacia brachycera
Box huckleberry

Gaylussacia dumosa
Dwarf huckleberry

Gaylussacia frondosa   Blue tangle
Dangleberry
Dwarf huckleberry

Gaylussacia resinosa
Black huckleberry

Gaylussacia ursina
Bear huckleberry
Buckberry

Genipa americana   Genipap
Marmalade box

Genista aethnensis
Mount Etna broom

Genista anglica   Needle furze
Petty whin
Pretty whin

Genista canariensis   Florist's genista

Genista dalmatica
Dalmatian broom

Genista germanica
German greenweed

Genista hispanica   Spanish broom
Spanish gorse

Genista januensis   Genoa broom

Genista lydia   Balkan gorse

Genista nyssana   Nish broom

Genista pilosa   Hairy greenweed
Hairy greenwood

Genista raetam   Juniper rush
White broom

Genista sagittalis   Arrow broom

Genista sylvestris pungens
Dalmatian broom

Genista tinctoria   Dyer's broom
Dyer's greenweed
Greenweed
Woadwaxen
Woodwaxen

Genista virgata   Madeira broom

Gevuina avellana   Chilean hazel
Chilean nut

Ginkgo biloba   Duck's foot tree
Maidenhair tree

Gleditsia aquatica   Swamp locust
Water locust

Gleditsia caspica

Gleditsia caspica   Caspian locust

Gleditsia chinensis
                Chinese honey locust

Gleditsia japonica
                Japanese honey locust

Gleditsia sinensis
                Chinese honey locust

Gleditsia triacanthos   Honey locust
                Honeyshuck
                Sweet locust

Gliricidia maculata   Madre
                Nicaraguan cocoa-shade

Gliricidia sepium   Madre
                Nicaraguan cocoa-shade

Glycosmis pentaphylla
                Jamaica mandarin orange

Glyptostrobus lineatus
                Chinese swamp cypress
                Chinese water pine

Gonystylus bancanus   Ramin

Gordonia lasianthus   Black laurel
                Loblolly bay

Gossypiospermum praecox
                West Indian boxwood

Graptophyllum hortense
                Caricature plant

Graptophyllum pictum
                Caricature plant

Grevillea aquifolium
                Holly-leaved grevillea

Grevillea robusta   Silky oak

Grevillea wilsonii   Firewheel

Guaiacum officinale   Lignum vitae
                Tree of life

Guazuma ulmifolia   Bastard cedar

Guizotia abyssinica   Niger seed

Gulielma gassipaes   Peach palm

Gustavia augusta   Stinkwood

Gymnocladus chinensis
                Chinese coffee tree

Gymnocladus dioica   Chicot
                Kentucky coffee tree
                Knicker tree
                Nicker tree

Haematoxylum
  campeachianum   Bloodwood
                Campeachy-wood
                Logwood
                Peachwood

Hagenia abyssinica   Kousso

Hakea bucculenta   Red pokers

Hakea laurina   Sea-urchin tree

Hakea victoriae   Royal hakea

Halesia carolina   Carolina silverbell
                Oppossumwood
                Shittimwood
                Silver bell
                Snowdrop tree
                Wild olive

Halesia monticola
                Mountain silver bell
                Mountain snowdrop tree

Halesia tetraptera   Silver bell
                Snowdrop tree

Halimium lasianthum
  formosum   Sweet cistin

Halimodendron halodendron
                Salt tree

Hamamelis japonica
                Japanese witch hazel

Hamamelis mollis
                Chinese witch hazel

Hamamelis vernalis
                Ozark witch hazel

Hamamelis virginiana
                Common witch hazel

Hamelia patens   Firebush
                Scarlet bush

Hardenbergia monophylla
                Australian lilac

Hardenbergia violacea
                Australian sarsaparilla

Harpephyllum caffrum
                Kaffir plum

Hebe cupressoides   Cypress hebe

Hebe hulkeana   New Zealand lilac

Hedyscepe canterburyana
                Umbrella palm

Helichrysum rosmarinifolium
                Snow in summer

Helichrysum serotinum
                Curry plant

Helichrysum stoechas   Goldilocks

Heliocarpus americanus   Sun fruit

Hemidesmus indica
                Indian sarsaparilla

Heritiera macrophylla
                Looking-glass tree

Herminiera elaphroxylon   Ambash
                Pith tree

Heteromeles arbitifolia
                Christmas berry
                Tollon
                Toyon

Heteropyxis natalensis
                Lavender tree

Hevea brasiliensis   Caoutchouc
                Para rubber tree
                Rubber tree

Hibiscus cannabinus   Bastard jute

                Bimli jute
                Bimlipatum
                Deccan hemp
                Deckaner hemp
                Indian hemp
                Kenaf

Hibiscus diversifolius
                Cape hibiscus
                Wild cotton

Hibiscus elatus   Cuban bast
                Mahoe
                Mountain mahoe
                Tree hibiscus

TREES, BUSHES, AND SHRUBS

Ilex aquifolium

Hibiscus farragei
Desert rose mallow

Hibiscus grandiflorus
Great rose mallow

Hibiscus huegelii　　　Satin hibiscus

Hibiscus militaris
Soldier rose mallow

Hibiscus moscheutos
Common rose mallow
Swamp rose mallow
Wild cotton

Hibiscus mutabilis
Confederate rose
Cotton rose

Hibiscus rosa-sinensis
Chinese hibiscus

Hibiscus sabdariffa　　Indian sorrel
Jamaica sorrel
Roselle

Hibiscus schizopetalus
Coral hibiscus
Fringed hibiscus
Japanese hibiscus
Japanese lantern

Hibiscus sinensis　　　Blacking plant
China rose
Chinese hibiscus
Hawaiian hibiscus

Hibiscus syriacus　　　Althaea
Bush hollyhock
Bush mallow
Hibiscus
Rose of sharon
Tree hollyhock

Hibiscus tiliaceus　　　Mahoe

Hibiscus trionum　　　Bladder ketmia
Flower-of-an-hour

Hippomane mancinella
Manchineel

Hippophae rhamnoides
Sallow thorn
Sea buckthorn

Hoheria lyallii　　　　Lacebark
Ribbonwood

Hoheria populnea　　　Lacebark
Ribbonwood

Hoheria sexstylosa　　Ribbonwood

Holacantha emoryi
Crucifixion thorn

Holodiscus ariifolius　Cream bush
Ocean spray

Holodiscus discolor　　Cream bush
Ocean spray
Spray brush

Homalanthus populifolius
Queensland poplar

Houmiria floribunda
Bastard bullet tree

Hovenia dulcis　Japanese raisin tree

Howea belmoreana
Belmor sentry palm
Curly palm
Curly sentry palm

Howea forsterana
Forster's sentry palm
Kentia palm
Paradise palm
Sentry palm
Thatch-leaf palm

Hura crepitans　　　　Javillo
Monkey-pistol
Monkey's dinner bell
Sandbox tree

Hydrangea arborescens
Seven barks

Hydrangea aspera
Rough-leaved hydrangea

Hydrangea hortensia
Mop-head hydrangea

Hydrangea macrophylla
Lace-cap hydrangea

Hydrangea quercifolia
Oak-leaf hydrangea

Hymenaea courbaril　Anime resin
Locust tree

Hymenosporum flavum
Native frangipani

Hyophorbe lagenicaulis
Bottle palm
Pignut palm

Hyophorbe verschaffeltii
Spindle palm

Hypericum androsaemum
Bible leaf
Tutsan

Hypericum calycinum
Aaron's beard
Rose of sharon

Hypericum hircinum
Stinking tutsan

Hypericum undulatum
Wavy Saint John's wort

Hypericum × inodorum
Tall tutsan

Hyphaene thebaica　　Doom palm
Doum palm
Egyptian
doom palm
Gingerbread palm

Idesia polycarpa　　　Idesia
Igiri tree
Ligiri tree

Idria columnaris　　　Boojum tree

Ilex × altaclarensis camelliifolia
Spineless broadleaved holly

Ilex × altaclarensis golden king　　　　Highclere holly

Ilex × altaclarensis lawsoniana
Lawson's holly

Ilex amelanchier　　　Sarvis holly
Swamp holly

Ilex aquifolium　　　Common holly
English holly
Hulver bush
Oregon holly
Wild holly

Ilex aquifolium argentea
Silver hedgehog holly

Ilex aquifolium aurea
Golden holly

Ilex aquifolium aurea pendula
Golden weeping holly

Ilex aquifolium bacciflava
Yellow-fruited holly

Ilex aquifolium ferox
Hedgehog holly

Ilex aquifolium ferox aurea
Golden-blotched hedgehog holly

Ilex aquifolium flavescens
Moonlight holly

Ilex aquifolium fructuluteo
Yellow-berried holly

Ilex aquifolium pendula
Weeping holly

Ilex × attenuata   Topal holly

Ilex cassine   Cassina
Dahoon holly
Yaupon

Ilex chinensis   Kashi holly

Ilex coriacea   Bay gall bush
Large gallberry
Sweet gallberry

Ilex cornuta   Chinese holly
Horned holly

Ilex crenata   Box-leaved holly
Japanese holly

Ilex decidua   Possumhaw holly

Ilex dipyrena   Himalayan holly

Ilex geniculata   Furin holly

Ilex glabra   Appalachian tea
Bitter gallberry
Gallberry
Inkberry
Winterberry

Ilex integra   Japanese holly
Mochi tree

Ilex laevigata   Smooth winterberry

Ilex latifolia   Lustre-leaf holly
Tarajo

Ilex lucida   Bay gall bush
Large gallberry
Sweet gallberry

Ilex × meservae   Blue holly

Ilex opaca   American holly

Ilex paraguariensis   Maté
Paraguay tea
Yerba-de-maté

Ilex perado   Azorean holly
Canary holly
Madeira holly

Ilex pernyi   Perny's holly

Ilex platyphylla
Canary Islands holly

Ilex purpurea   Kashi holly

Ilex rotunda   Kurogane holly

Ilex serrata   Japanese winterberry

Ilex verticillata   Black alder
Winterberry

Ilex vomitoria   Cassina
Yaupon

Illicium anisatum
Aniseed tree
Chinese anise
Japanese anise
Star anise

Illicium floridanum   Aniseed tree
Poison bay
Purple anise

Illicium religiosum   Star anise

Illicium verum   Chinese star anise
Star anise

Inocarpus edulis
Otaheite chestnut

Ipomoea arborescens
Morning glory tree

Ipomoea fistulosa
Morning glory bush

Ipomoea purga   Jalap

Ipomoea turpethum   Indian jalap
Turpeth

Isopogon anemonifolius
Tall conebush

Itea illicifolia   Hollyleaf sweetspire

Itea japonica   Japanese sweetspire

Itea virginica   Sweetspire
Tassel-white
Virginia sweetspire
Virginia willow

Ixora chinensis   Chinese ixora
Malaya ixora

Ixora coccinea
Flame-of-the-woods
Jungle flame
Jungle geranium

Ixora incarnata
Flame-of-the-woods
Jungle flame
Jungle geranium

Jacaranda acutifolia   Fearn tree
Jacaranda

Jacaranda mimosifolia   Jacaranda

Jacaranda procera   Carob tree

Jacquinia armillaris   Barbasco
Bracelet wood

Jacquinia barbasco   Barbasco

Jasminum gracillimum
Pinwheel jasmine
Star jasmine

Jasminum grandiflorum
Catalonian jasmine
Royal jasmine
Spanish jasmine

Jasminum humile
Himalayan jasmine
Italian jasmine

Jasminum mesnyi
Japanese jasmine
Primrose jasmine
Yellow jasmine

Jasminum multiflorum
Star jasmine

Juniperus jackii

Jasminum nitidum
  Angelwing jasmine
  Confederate jasmine
  Windmill jasmine

Jasminum nudiflorum
  Winter-flowering jasmine

Jasminum officinale
  Common white jasmine
  Poet's jasmine
  White jasmine

Jasminum sambac Arabian jasmine
  Zambak

Jateorhiza calumba Calumba

Jatropha curcus Barbados nut
  French physic nut
  Physic nut
  Purging nut

Jatropha integerrima
  Guatemalan rhubarb
  Peregrina
  Spicy jatropha

Jatropha multifida Coral plant
  Physic nut

Jatropha podagrica
  Australian bottle plant
  Tartoga

Jubaea chilensis Chilean wine palm
  Coquito palm
  Honey palm
  Little cokernut palm
  Wine palm

Jubaea spectabilis
  Chilean wine palm
  Coquito palm
  Honey palm
  Little cokernut palm
  Wine palm

Juglans ailantifolia Heartnut
  Japanese walnut

Juglans californica
  California walnut

Juglans cathayensis
  Chinese butternut
  Chinese walnut

Juglans cinerea Butternut
  White butternut
  White walnut

Juglans hindsii
  Californian black walnut

Juglans jamaicensis
  West Indies walnut

Juglans major Arizona walnut

Juglans mandshurica
  Manchurian walnut

Juglans microcarpa Little walnut
  River walnut
  Texan walnut

Juglans nigra Black walnut

Juglans regia Common walnut
  English walnut
  Madeira walnut
  Nux regia
  Persian walnut

Juglans regia laciniata
  Cut-leaf walnut

Juglans rupestris Little walnut
  River walnut
  Texan walnut

Juniperus ashei Ashe juniper
  Mountain cedar
  Ozark white cedar

Juniperus bermudiana
  Bermuda cedar

Juniperus californica
  Californian juniper

Juniperus cedrus
  Canary Islands juniper

Juniperus chinensis
  Chinese juniper

Juniperus chinensis aurea
  Golden Chinese juniper
  Young's golden juniper

Juniperus chinensis columnaris glauca
  Blue Chinese juniper

Juniperus chinensis kaizuka
  Hollywood juniper

Juniperus communis
  Common juniper

Juniperus communis compressa Noah's ark juniper

Juniperus communis depressa
  Canadian juniper

Juniperus communis hibernica
  Irish juniper

Juniperus communis pyramidalis Swedish juniper

Juniperus communis stricta
  Columnar juniper
  Irish juniper

Juniperus communis suecica
  Swedish juniper

Juniperus conferta Shore juniper

Juniperus deppeana pachyphlaea Alligator juniper
  Chequered juniper

Juniperus depressa
  Canadian dwarf juniper

Juniperus drupacea Habbel
  Plum juniper
  Syrian juniper

Juniperus excelsa Grecian juniper

Juniperus flaccida Mexican juniper

Juniperus foetidissima
  Stinking juniper

Juniperus formosana
  Prickly cypress

Juniperus horizontalis
  Creeping cedar
  Creeping juniper
  Creeping savin juniper

Juniperus horizontalis douglasii Waukegan juniper

Juniperus jackii
  Rocky mountains juniper

Juniperus × media pfitzerana
Pfitzer juniper

Juniperus monosperma
Cherrystone juniper

Juniperus morrisonicola
Mount Morrison juniper

Juniperus occidentalis
California juniper
Sierra juniper
Western juniper

Juniperus osteosperma
Utah juniper

Juniperus oxycedrus   Cade
Prickly juniper
Sharp cedar

Juniperus pachyphloea
Alligator juniper

Juniperus phoenicea
Phoenician juniper

Juniperus pinchotii
Red-berry juniper

Juniperus procera   African juniper
East African juniper

Juniperus procumbens
Creeping juniper

Juniperus recurva
Drooping juniper
Himalayan juniper

Juniperus recurva coxii
Coffin juniper
Cox's juniper

Juniperus rigida   Needle juniper
Stiff-leaved juniper
Temple juniper

Juniperus sabina   Savin

Juniperus sabina tamariscifolia
Spanish savin

Juniperus scopulorum
Colorada red cedar
Rocky Mountains juniper

Juniperus silicicola
Southern red cedar

Juniperus squamata
Scaly-leaved Nepal juniper

Juniperus squamata meyeri
Meyer's blue juniper

Juniperus squamata pygmaea
Pigmy juniper

Juniperus suecica
Scandinavian juniper

Juniperus thurifera   Incense juniper
Spanish juniper

Juniperus utahensis   Desert juniper

Juniperus virginiana
Eastern red cedar
Pencil cedar
Pencil juniper
Red cedar

Juniperus wallichiana
Black juniper

Kallstroemia platyptera
Cork hopbush

Kalmia angustifolia   Dwarf laurel
Lambkill
Pig laurel
Sheep laurel
Wicky

Kalmia cuneata   White wicky

Kalmia glauca
American swamp laurel
Swamp laurel

Kalmia latifolia   American laurel
Calico bush
Ivybush
Mountain laurel
Spoonwood

Kalmia microphylla   Alpine laurel
Western laurel

Kalmia poliifolia   Bog kalmia
Bog laurel
Pale laurel

Kalopanax pictus   Castor aralia
Prickly castor oil tree

Kentia belmoreana
Belmore sentry palm
Curly palm

Kentia forsterana
Forster's sentry palm
Kentia palm
Sentry palm
Thatch-leaf palm

Kerria japonica   Jew's mallow

Kerria japonica plena
Bachelor's buttons

Keteleeria davidiana   Chinese pine

Khaya nyasica   African mahogany
Nyasaland mahogany

Khaya senegalensis
African mahogany
Senegal mahogany

Kigelia africana   Sausage tree
Wild peach

Kigelia pinnata   Sausage tree

Knightia excelsa
New Zealand honeysuckle tree
Rewa rewa

Koelreuteria elegans   Flamegold

Koelreuteria paniculata   China tree
Golden rain tree
Pride of India
Varnish tree

Kokia drynarioides   Kokio

Kokoona zeylanica   Kokoon tree

Kola vera   Kola nut tree

Kolkwitzia amabilis
American beautybush
Beautybush
Wilson's beautybush

+ *Laburnocytisus adamii*
Adam's laburnum

Laburnum alpinum
Alpine laburnum
Scotch laburnum

Laburnum alpinum pendulum
Weeping scotch laburnum

Leopoldinia piassaba

Laburnum anagyroides
    Common laburnum

Laburnum anagyroides aureum
    Golden chain
    Golden-leaved laburnum
    Golden rain

Laburnum vossii   Voss's laburnum

Laburnum vulgare
    Common laburnum
    Golden chain
    Golden rain

Laburnum × watereri
    Golden chain tree
    Voss's laburnum

Lagarostrobus franklinii
    Huon pine

Lagerstroemia indica   Cape myrtle
    Cranesbill myrtle

Lagerstroemia speciosa
    Pride of India
    Queen's cape myrtle

Lagunaria patersonii
    Cow itch tree
    Norfolk Island hibiscus
    Primrose tree
    Queensland pyramidal tree

Larix decidua   Common larch
    European larch

Larix decidua pendula
    Weeping european larch

Larix × eurolepis   Dunkeld larch
    Hybrid larch

Larix gmelinii   Dahurian larch

Larix gmelinii principis-rupprechtii
    Prince Rupprecht larch

Larix griffithiana   Sikkim larch

Larix griffithii   Himalayan larch
    Sikkim larch

Larix kaempferi   Japanese larch
    Money pine

Larix laricina   American larch
    Black larch
    Eastern larch
    Hackmatack
    Tamarack

Larix lyalli   Alpine larch
    Lyall's larch

Larix occidentalis   Western larch

Larix × pendula   Weeping larch

Larix potaninii   Chinese larch

Larix russica   Siberian larch

Larix sibirica   Siberian larch

Larrea divaricata   Creosote bush

Larrea tridentata   Creosote bush

Latania borbonica   Bourbon palm
    Red latan

Latania loddigesii   Blue latan

Latania lontaroides   Red latan

Latania verschaffeltii   Yellow latan

Laurus azorica   Canary Island laurel

Laurus canariensis
    Canary Island laurel

Laurus maderensis
    Canary Island laurel

Laurus nobilis   Bay laurel
    Poet's laurel
    Roman laurel
    Royal bay
    Sweet bay
    True laurel

Laurus nobilis angustifolia
    Willow-leaf bay

Laurus nobilis aureus   Golden bay

Lavandula angustifolia
    Common lavender
    Old English lavender

Lavandula angustifolia vera
    Dutch lavender

Lavandula dentata
    Fringed lavender

Lavandula lanata   Woolly lavender

Lavandula officinalis
    Old English lavender
    Mitcham lavender

Lavandula spica
    Old English lavender
    Mitcham lavender

Lavandula stoechas
    Butterfly lavender
    French lavender

Lavandula vera   Dutch lavender

Lavatera arborea   Tree mallow

Lavatera assurgentiflora
    Californian tree mallow

Lavatera olbia   Bush mallow
    Shrubby mallow
    Tree mallow

Lavatera trimestris   Annual mallow
    Rose mallow

Lawsonia alba   Henna
    Mignonette tree

Lawsonia inermis   Egyptian privet
    Henna
    Mignonette tree

Ledum groenlandicum
    Labrador tea plant

Ledum latifolium
    Labrador tea plant

Ledum palustre
    Marsh ledum
    Wild rosemary

Lecythis zabucayo   Monkey nut
    Paradise nut
    Sapucia nut

Leea coccinea   West Indian holly

Leiophyllum buxifolium
    Box sand myrtle
    Sand myrtle

Leitneria floridana   Corkwood
    Florida corkwood

Leonotis leonorus   Lion's ear

Leopoldinia piassaba   Piassaba

Leptospermum laevigatum
Australian tea tree

Leptospermum lanigerum
Woolly tea tree

Leptospermum scoparium
Manuka
New Zealand tea tree
Tea tree

Leucaena glauca White popinac

Leucodendron argenteum
Silver tree

Leucothoe fontanesiana
Fetterbush

Leycesteria formosa
Flowering nutmeg
Himalayan honeysuckle
Pheasant berry

Leycythis grandiflora Wadadura

Leycythis usitata Monkey pot tree

Libocedrus bidwillii Pahautea tree

Libocedrus decurrens
Incense cedar

Libocedrus plumosa Kawaka tree

Licula grandis Ruffled fan palm

Ligustrum amurense Amur privet

Ligustrum japonicum
Japanese privet
Wax-leaf privet

Ligustrum lucidum Chinese privet
Glossy privet
Nepal privet
Shining privet
Wax-leaf privet
White wax tree

Ligustrum ovalifolium
Californian privet
Japanese privet
Oval-leaved privet

*Ligustrum ovalifolium
argenteum* Silver privet

*Ligustrum ovalifolium
aureum* Golden privet

Ligustrum sinense Chinese privet

Ligustrum vulgare
Common privet
Prim privet

Lindera benzoin Benjamin bush
Benzoin
Fever bush
Spice bush

Lindera melissifolia Jove's fruit

Linnaea borealis Twin flower

Linospadix monostachya
Walking-stick palm

Lippia citriodora Lemon plant
Lemon verbena

Liquidambar formosana
Chinese sweet gum
Formosan gum

Liquidambar orientalis
Oriental sweet gum
Storax

Liquidambar styraciflua
American sweet gum
Bilsted gum
Red gum
Satinwood
Sweet gum

Liriodendron chinense
Chinese tulip tree

Liriodendron tulipifera
Tulip poplar
Tulip tree
Whitewood
Yellow poplar

Litchi chinensis Leechee
Lichi
Litchi
Lychee

Lithocarpus densiflorus
Tanbark oak
Tanoak

Livistona australis
Australian cabbage palm
Australian fan palm
Cabbage palm
Gippsland fountain palm
Gippsland palm

Livistona chinensis
Chinese fan palm
Chinese fountain palm

Lodoicea maldavica Coco-de-mer
Double coconut

Loiseleuria procumbens
Alpine azalea
Mountain azalea

Lonicera ledebourii
North American honeysuckle

Lonicera nitida
Shining honeysuckle
Bush honeysuckle

Lonicera xylosteum
Fly honeysuckle

Luma apiculata
Orange-bark myrtle

Luma chequen Cheken

Lupinus arboreus Tree lupin
Yellow tree lupin

Lycium afrum Kaffir thorn

Lycium balbarum
Matrimony thorn
Matrimony vine

Lycium barbarum
Chinese box thorn
Duke of Argyll's tea tree

Lycium chinense
China tea
Duke of Argyll's tea tree
Chilean tea tree

Lycium gracillianum
Chilean tea tree

Lycium pallidum
Fremont's box thorn

Lyonia ligustrina Male berry
Male blueberry

TREES, BUSHES, AND SHRUBS

Malus ioensis plena

Lyonia lucida — Fetterbush
Tetterbush

Lyonia mariana — Stagger bush

Lyonothamnus floribundus — Catalina ironwood

Lysiloma latisiliqua — Sabicu

Macadamia integrifolia — Australian nut
Macadamia nut
Queensland nut

Macadamia ternifolia — Queensland nut
Small-fruited Queensland nut

Macadamia tetraphylla — Macadamia nut
Rough-shell macadamia nut

Macaranga grandifolia — Coral tree

Maclura aurantiaca — Osage orange

Maclura pomifera — Bow-wood tree
Osage orange

Macropiper excelsum — Kawa-kawa
Pepper tree

Maddenia hypoleuca — Madden cherry

Magnolia acuminata — Blue magnolia
Cucumber tree

Magnolia ashei — Ashe magnolia

Magnolia campbellii — Campbell's magnolia
Pink tulip tree

Magnolia cordata — Yellow cucumber tree

Magnolia delavayi — Chinese evergreen magnolia

Magnolia denudata — Lily tree
Yulan

Magnolia fraseri — Ear-leaved umbrella tree

Magnolia grandiflora — Bull bay
Evergreen magnolia
Laurel magnolia
Loblolly magnolia
Southern magnolia

Magnolia heptapeta — Yulan

Magnolia hypoleuca — Japanese bigleaf magnolia
Japanese cucumber tree

Magnolia kobus — Northern Japanese magnolia

Magnolia lilliflora — Lily-flowered magnolia

Magnolia macrophylla — Large-leaved cucumber tree
Large-leaved magnolia

Magnolia obovata — Japanese magnolia

Magnolia salicifolia — Willow-leaf magnolia

Magnolia sinensis — Chinese magnolia

Magnolia × soulangiana — Hybrid magnolia
Magnolia
Saucer magnolia

Magnolia sprengeri deva — Goddess magnolia

Magnolia stellata — Starry magnolia

Magnolia tripetala — Umbrella tree

Magnolia × veitchii — Veitch's magnolia

Magnolia virginiana — Swamp bay
Sweet bay

Magnolia yulan — Yulan

Mahonia aquifolium — Blue barberry
Holly barberry
Holly mahonia
Mountain grape
Oregon grape

Mahonia japonica — Japanese mahonia

Mahonia lomariifolia — Yunnan mahonia

Mahonia nervosa — Oregon grape
Water holly

Mahonia repens — Creeping barberry

Malaviscus arboreus — Pepper hibiscus
Sleepy mallow
Turk's cap

Mallotus philippinensis — Kamala tree
Kamila tree

Malpighia coccigera — Dwarf holly
Miniature holly
Singapore holly

Malpighia glabra — Barbados cherry

Malpighia urens — Cow-itch cherry

Malus angustifolia — American crab apple
Southern wild crab apple
Wild crab apple

Malus baccata — Siberian crab apple

Malus coronaria — Garland crab apple
Sweet crab apple
Sweet-scented crab
Wild sweet crab

Malus domestica — Cultivated apple
Orchard apple

Malus florentina — Hawthorn-leaf crab apple

Malus floribunda — Japanese crab apple
Purple chokeberry
Showy crab apple

Malus fusca — Oregon crab apple

Malus halliana — Hall's crab apple

Malus hupehensis — Chinese crab apple
Hupeh crab apple

Malus ioensis — Prairie crab apple

Malus ioensis plena — Bechtel crab apple

Malus × magdeburgensis
Magdeburg apple

Malus prunifolia
Plum-leaved apple

Malus pumila Commercial apple
Paradise apple

Malus × purpurea
Purple crab apple

Malus × robusta
Siberian crab apple

Malus sieboldii Toringo crab

Malus × soulardii Soulard crab

Malus spectabilis
Chinese crab apple
Chinese flowering apple
Hai-tung crab apple

Malus sylvestris
Common crab apple
John Downie crab apple
Lichfield crab apple
Wild crab apple

Malus tschonoskii Pillar apple

Malvaviscus arboreus
Sleeping hibiscus
Sleepy mallow
Wax mallow

Mammea americana Mamey
Mammee
Mammee apple
South American apricot

Mandevilla suaveolens
Chilean jasmine

Mangifera indica Mango

Manihot dulcis Sweet cassava

Manihot esculenta Bitter cassava
Cassava
Mandioca
Manioc
Sweet-potato tree
Tapioca
Yuca

Manihot glaziovii Ceara rubber

Manihot utilissima Bitter cassava

Manilkara bidentata Balata

Manilkara zapota Chicozapote
Marmalade plum
Nazeberry
Nispero
Sapodilla
Sapodilla plum

Margyricarpus pinnatus Pear fruit
Pearl berry
Pearl fruit

Margyricarpus setosus Pear fruit
Pearl berry
Pearl fruit

Mauritia flexuosa Ita palm
Tree-of-life

Mauritia setigera Ita palm
Tree-of-life

Maximiliana caribaea
Cucurite palm
Inaja palm

Maximiliana maripa
Cucurite palm
Inaja palm

Maximiliana regia Cucurite palm
Inaja palm

Maytenus boaria Mayten

Medicago arborea Moon trefoil

Melaleuca armillaris
Bracelet honey myrtle

Melaleuca elliptica
Granite bottlebrush

Melaleuca ericifolia
Swamp paperbark

Melaleuca huegelii Honey myrtle

Melaleuca lanceolata Moonah

Melaleuca lateritia
Robin redbreast bush

Melaleuca leucadendron Cajaput
River tea tree
Weeping tea tree
White tea tree

Melaleuca nematophylla
Wiry honey myrtle

Melaleuca nesophylla
Western tea myrtle

Melaleuca pubescens Moonah

Melaleuca quinquenervia
Paperbark tree
Punk tree
Swamp tea tree
Tea tree

Melaleuca rhaphiophylla
Swamp paperbark

Melaleuca squarrosa
Scented paperbark

Melastoma malabathricum
Indian rhododendron

Melia azedarach Azediracta
Bead tree
Chinaberry
China tree
Indian lilac
Japanese bead tree
Paradise tree
Persian lilac
Pride of China
Pride of India
Syrian bead tree

Melia dubium Ceylon mahogany
White cedar

Melia indica Indian neem tree

Melianthus major Honey bush

Melicoccus bijugatus Genipe
Honeyberry
Mamoncillo
Spanish lime

Melicytus ramiflorus Mahoe
Whiteywood

Meryta sinclairii Puka

Mespilus germanica Medlar

Mesurea ferrea Ironwood

TREES, BUSHES, AND SHRUBS

Myrtus chequen

Metasequoia glyptostroboides
Dawn redwood
Shui-hsa
Water fir

Metrosideros excelsa
New Zealand christmas tree
Pohutukawa

Metrosideros robusta   Iron tree
New Zealand christmas tree
Rata

Metrosideros tomentosa
New Zealand christmas tree

Metroxylum sagu   Sago palm

Michelia champaca   Champaca
Fragrant champaca
Orange champaca

Michelia figo   Banana shrub

Microcitrus australasica
Australian finger lime

Microcitrus australis
Australian round lime

Microcoelum weddellianum
Weddel palm

Microcycas calocoma   Palma corcho

Mimulus aurantiacus
Shrubby musk

Mimusops balata   Beefwood

Mimusops elengi   Medlar
Spanish cherry

Mitchella repens   Checkerberry
Partridge berry
Squaw berry
Squaw vine

Monodora myristica
African nutmeg
Calabash nutmeg
Jamaica nutmeg

Moquila utilis   Pottery tree

Morinda citrifolia   Awl tree
Indian mulberry

Morinda royoc   Royoc

Moringa oleifera   Ben oil tree
Horseradish tree

Moringa pterygosperma
Horseradish tree

Morus alba   Silkworm tree
White mulberry

Morus alba pendula
Weeping mulberry

Morus australis   Aino mulberry

Morus microphylla
Mexican mulberry

Morus nigra   Black mulberry
Common mulberry

Morus rubra   American mulberry
Red mulberry

Muntingia calabura   Calabur

Murraya exotica   Orange jessamine

Murraya koenigii   Curry-leaf tree
Karapincha

Murraya paniculata   Chinese box
Cosmetic bark tree
Curry-leaf tree
Orange jasmine
Satinwood

Musa acuminata   Banana
Edible banana
Plantain

Musa fehi   Fehi banana

Musa nana   Dwarf banana

Musa ornata   Flowering banana

Musa × paradisiaca   Edible banana
Plantain

Musa textilis   Abaca
Manila hemp

Musa troglodytarum   Fehi banana

Musanga cecropioides
Umbrella tree

Mussaenda eythrophylla
Red flag bush

Myoporum insulare   Boobyalla

Myoporum laetum   Ngaio

Myoporum sandwicense
Bastard sandalwood
Naio

Myoporum tenuifolium
Waterbush

Myoporum tetrandrum
Tasmanian waterbush

Myrciaria cauliflora   Jaboticaba

Myrica californica
California bayberry
California wax myrtle

Myrica carolinensis
Candleberry myrtle
Wax myrtle

Myrica cerifera
Candleberry myrtle
Tallow shrub
Waxberry
Wax myrtle

Myrica faya   Candleberry myrtle

Myrica gale   Bog myrtle
Gale
Meadow fern
Sweet gale

Myrica pennsylvanica   Bayberry
Candleberry
Northern bayberry
Swamp candleberry

Myristica fragrans   Nutmeg

Myroxylon balsamum   Tolu tree

Myroxylon pereirae
Balsam of Peru
Peruvian balsam
Tolu balsam

Myrsine africana
African boxwood
Cape myrtle

Myrtus bullata   Ramarama

Myrtus chequen   Chilean myrtle

Myrtus communis Common myrtle
Greek myrtle
Myrtle
Swedish myrtle
Myrtus communis tarentina
Tarentum myrtle
Myrtus luma Orange-bark myrtle
Myrtus ugni Chilean guava
Mysporum parvifolium
Creeping boobialla
Nandina domestica
Chinese sacred bamboo
Heavenly bamboo
Nanteen
Sacred bamboo
Nannorrhops ritchiana
Mazari palm
Napoleona heudottii
Napolean's button
Nauclea latifolia African peach
Nectandra rodiaei Greenheart
Nemopanthus mucronatus
Catberry
Mountain holly
Neopanax arboreus
Five fingers tree
Nephelium lappaceum Pulasan
Rambutan
Nephelium litchi Lychee
Nephelium malaiense
Mata kuching
Nephelium mutabile Pulasan
Nerium oleander
Common oleander
Oleander
Rose bay oleander
Neviusia alabamensis
Alabama snow wreath
Nicotiana glauca Shrub tobacco
Noltea africana Soap bush
Normanbya normanbyi
Black palm
Nothofagus antarctica
Antarctic beech
Guindo
Nirre
Nothofagus betuloides
Coigue de magellanes
Oval-leaved southern beech
Nothofagus cliffortioides
Mountain beech
Nothofagus cunninghamii
Myrtle beech
Tasmanian beech
Nothofagus dombeyi Coigue
Dombey's southern beech
Nothofagus fusca
New Zealand red beech
Red beech
Nothofagus glauca Hualo
Roblé de maule
Nothofagus gunnii
Tanglefoot beech

Nothofagus menziesii Silver beech
Nothofagus moorei
Australian beech
Nothofagus obliqua Roblé beech
Roblé pellin
Nothofagus procera Raoul beech
Rauli beech
Nothofagus pumilo Lenga
Roblé blanco
Nothofagus solandri Black beech
*Nothofagus solandri
cliffortioides* Mountain beech
Nothofagus truncata Hard beech
Notospartium carmicheliae
Pink broom
Nuttallia cerasiformis Oso-berry
Nuytsia floribunda Christmas tree
Fire tree
Nyctanthes arbor-tristis
Indian night jasmine
Night jasmine
Tree-of-sadness
Nypa fruticans Mangrove palm
Nipa palm
Nypa palm
Nyssa aquatica Cotton gum
Large tupelo
Tupelo gum
Water tupelo
Wild olive
Nyssa candicans Ogechee lime
Nyssa sylvatica Black gum
Black tupelo
Pepperidge
Sour gum
Tupelo
Upland tupelo
Ochna japonica Bird's eye bush
Mickey-mouse plant
Ochna serrulata Bird's eye bush
Mickey-mouse plant
Ochroma pyramidale Balsa wood
Corkwood
Ocotea bullata Black stinkwood
Greenheart
Ocotea radiaei Greenheart
Oemleria cerasiformis Indian plum
Oso-berry
Oenocarpus batava Patana palm
Olea africana Wild olive
Olea europaea Common olive
Olive
Olea ferruninea Indian olive
Olea laurifolia Black ironwood
Olearia argophylla Muskwood
Olearia illicifolia Maori holly
Mountain holly
Olearia macrodonta
New Zealand holly
Olearia nummularifolia
Daisy bush

Passiflora suberosa

Olearia phlogopappa
Daisy bush
Tasmanian daisy bush

Olinia emarginata
Transvaal hard pear

Olmediella betschlerana
Costa Rican holly
Manzanote
Puerto Rican holly

Olneya tesota   Desert ironwood

Oncoba spinosa   Snuffbox tree

Oncosperma tigillarium
Nibung palm

Oplopanax horridus   Devil's club

Orbignya barbosiana   Babassu

Orbignya cohune   Cohune palm

Orbignya speciosa   Babassu palm

Osmanthus americanus
American olive
Devil wood
Wild olive

Osmanthus fragrans   Fragrant olive
Sweet olive
Tea olive

Osmanthus heterophyllus
Chinese holly
False holly
Holly-leaved olive

Osmaronia cerasiformis
Oso-berry

Osmunda regalis   Royal fern

Ostrya carpinifolia
European hop-hornbeam
Hop hornbeam

Ostrya japonica
Japanese hop-hornbeam

Ostrya virginiana
American hop-hornbeam
Eastern hop-hornbeam
Ironwood
Leverwood

Oxandra lanceolata   Lancewood

Oxycoccus macrocarpus
American cranberry

Oxycoccus oxycoccus
European wild cranberry

Oxycoccus palustris
European wild cranberry

Oxydendrum arboreum
Sorrel tree
Sourwood tree
Titi

Ozothamnus rosmarinifolius
Snow in summer

Ozothamnus thyrsoidens
Snow in summer

Pachira aquatica   Guiana chestnut
Provision tree
Water chestnut
Wild cocoa tree

Pachira insignis   Wild chestnut

Pachysandra procumbens
Alleghany spurge

Pachysandra terminalis
Mountain spurge

Paeonia delavayi   Tree peony

Paeonia suffruticosa   Moutan
Moutan paeony
Tree peony

Palaquium gutta   Gutta percha

Paliurus aculeatus   Christ's thorn
Jerusalem thorn

Paliurus australis   Christ's thorn

Paliurus spina-christi
Christ's thorn
Crown of thorns
Jerusalem thorn

Paliurus virgatus   Christ's thorn

Palmae hyphaene   Doum palm
Gingerbread palm

Palmetto causiarum
Puerto Rican hat palm

Pandanus leram
Nicobar breadfruit
Screwpine

Pandanus odoratissimus
Breadfruit
Pandang

Pandanus tectorius   Pandanus palm
Thatch screw palm

Pandanus utilis
Common screw palm

Pandanus veitchii
Veitch's screw pine

Pandorea jasminiodes
Australian bower plant
Bower plant

Parinari curatellifolia
Mobala plum

Parinari excelsa   Guinea plum

Parinari macrophylla
Gingerbread plum
Gingerbread tree

Parkia biglobosa   African locust

Parkia filicoidea
African locust bean

Parkia speciosa   Petai

Parkinsonia aculeata
Jerusalem thorn
Mexican palo verde

Parmentiera cereifera
Candle tree
Panama candle tree

Parmentiera edulis   Cuachilote
Guajilote

Parrotia persica   Iron tree
Persian ironwood

Passiflora caerulea
Blue passion flower
Brazilian passion flower
Passion flower

Passiflora suberosa   Meloncillo

Paulownia tomentosa
Empress tree
Foxglove tree
Karri tree
Paulownia
Princess tree

Peltophorum pterocarpum
Copper-pod tree
Yellow flame tree

Persea americana　　Aguacate
Alligator pear
Avocado pear
Palta

Persea borbonia　　Florida mahogany
Laurel tree
Red bay
Swamp red bay
Sweet bay
Tisswood

Persea indica　　Indian laurel

Peumus boldus　　Boldo
Chilean boldo tree

Phellodendron amurense
Amur cork tree

Phellodendron chinensis
Chinese cork tree

Phellodendron japonicum
Japanese cork tree

Philadelphus coronarius
Mock orange
Syringa

Phillyrea decora　　Jasmine box

Phillyrea latifolia　　Phillyrea

Phillyrea vilminiana
Jasmine box

Phoenix abyssinica
Ethiopian date palm

Phoenix canariensis　　Canary palm
Canary date palm
Canary Island date

Phoenix dactylifera　　Date palm

Phoenix paludosa
Mangrove date palm

Phoenix reclinata
Senegal date palm

Phoenix roebelenii
Miniature date palm
Pygmy date palm
Roebelin palm

Phoenix rupicola　　Cliff date palm
East Indian wine palm
India date palm
Wild date palm

Phoenix sylvestris　　India date palm
Wild date palm

Phoenix zeylanica
Ceylon date palm

Phorium tenax　　New Zealand flax

Photinia glabra　　Japanese photinia

Photinia serrulata
Chinese hawthorn

Phygelius capensis　　Cape figwort

Phyllanthus acidus
Gooseberry tree
Indian gooseberry
Otaheite gooseberry

Phyllanthus emblica　　Emblic
Myrobalan

Phyllirea vilmoriniana
Jasmine box

Phyllocladus alpinus
Alpine celery-topped pine
Celery pine

Phyllocladus aspleniifolius
Adventure bay pine
Celery pine
Celery-topped pine

Phyllocladus glaucus　　Toatoa tree

Phyllocladus rhomboidalis
Celery-top pine

Phyllocladus trichomanoides
Celery pine
Tanekaha tree

Physocarpus opulifolius　　Ninebark

Phytelephas macrocarpa
Ivory-nut palm
Ivory palm
Tagua

Phytolacca dioica　　Phytolacca

Picea abies　　Christmas tree
Common spruce
Norway spruce

Picea abies carpathica
Carpathian spruce

Picea abies maxwellii
Maxwell spruce

Picea abies nidiformis　　Nest spruce

Picea alcoquiana　　Alcock's spruce

Picea asperata　　Chinese spruce
Dragon spruce

Picea bicolor　　Alcock's spruce

Picea brachytyla　　Sargent's spruce

Picea brewerana
Brewer's weeping spruce
Siskiyou spruce
Weeping spruce

Picea cembroides　　Nut pine

Picea engelmannii
Engelmann's spruce

Picea glauca　　Cat spruce
White spruce

Picea glauca albertiana
Alberta white spruce

Picea glauca albertiana nana
Dwarf alberta spruce

Picea glehnii　　Saghalin spruce
Sakhalin spruce

Picea jezoensis　　Yeddo spruce
Yezo spruce

Picea jezoensis hondoensis
Hondo spruce

Picea koyamai　　Koyama spruce

Picea likiangensis　　Likiang spruce

Pinus jeffreyi

Picea likiangensis purpurea
Purple spruce

Picea mariana Black spruce
Bog spruce
Double spruce

Picea maximowiczii
Japanese spruce

Picea montigena
Candelabra spruce

Picea monophylla Nut pine
Single-leaf pinyon pine
Stone pine

Picea morrisonicola
Mount Morrison spruce
Taiwan spruce

Picea nigra Black spruce
Bog spruce
Double spruce

Picea obovata Siberian spruce
Picea omorika Serbian spruce
Picea orientalis Oriental spruce
Picea polita Japanese spruce
Tiger-tail spruce

Picea pungens Blue spruce
Colorado spruce

Picea pungens glauca Blue spruce
Colorado blue spruce

Picea rubens American red spruce
Red spruce

Picea rubra Red spruce
Picea schrenkiana Schrenk's spruce
Picea sitchensis Silver spruce
Sitka spruce

Picea smithiana Himalayan spruce
Indian spruce
Morinda spruce
West Himalayan spruce

Picea spinulosa
East Himalayan spruce
Sikkim spruce

Picea torana Tiger-tail spruce
Picraena excelsa Bitter ash
Jamaica quassia

Picramnia antidesma Cascara
Picramnia pentandra Bitterbush
Picrasma quassioides Picrasma
Pilocarpus jaborandi Jaborand
Jaborandi

Pilostegia viburnoides
Climbing hydrangea

Pimenta dioica Allspice
Pimento

Pimenta officinalis Allspice
Jamaica pepper
Pimento

Pimenta racemosa Bay tree
Bay-
rum tree

Pinckneya pubens Bitter bark
Fever tree
Georgia bark tree

Pinus albicaulis White-bark pine

Pinus aristata Bristle-cone pine
Hickory pine

Pinus armandii Armand's pine
Chinese white pine
David's pine

Pinus attenuata Knobcone pine
Pinus ayacahuite
Mexican white pine

Pinus balfouriana Foxtail pine
Pinus banksiana Grey pine
Jack pine
Scrub pine

Pinus brutia Calabrian pine
Pinus bungeana Lacebark pine
Pinus canariensis
Canary Island pine

Pinus caribaea Caribbean pine
Cuban pine

Pinus cembra Arolla pine
Russian cedar
Swiss stone pine

Pinus cembroides Mexican nut pine
Mexican stone pine
Nut pine
Pinyon pine

Pinus cembroides monophylla
One-leaved nut pine

Pinus cembroides edulis
Two-leaved nut pine

Pinus clausa Sand pine
Pinus contorta Beach pine
Lodgepole pine
Shore pine

Pinus contorta latifolia
Inland lodgepole pine
Lodgepole pine

Pinus coulteri Big-cone pine
Pinus densiflora Japanese red pine
*Pinus densiflora oculus-
draconis* Dragon-eye pine

Pinus echinata Short-leaf pine
Yellow pine

Pinus edulis Nut pine
Pinyon pine
Two-leaved nut pine

Pinus elliottii Elliott's pine
Slash pine

Pinus engelmannii Apache pine
Pinus excelsa Bhutan pine
Pinus flexilis Limber pine
Pinus gerardiana Chilghoza pine
Gerard's pine
Nepal nut pine

Pinus glabra Cedar pine
Spruce pine

Pinus greggii Gregg's pine
Pinus halepensis Aleppo pine
Jerusalem pine

Pinus × holfordiana Holford pine
Pinus insularis Benguet pine
Khasya pine

Pinus jeffreyi Jeffrey's pine

Pinus kerkusii

Pinus kerkusii	Tenasserim pine
Pinus khasya	Khasya pine
Pinus koraiensis	Korean pine
Pinus lambertiana	Giant pine
	Sugar pine
Pinus leucodermis	Bosnian pine
	Okinawa pine
Pinus luchuensis	Luchu pine
Pinus merkusii	Tenasserim pine
Pinus monophylla	Nut pine
	Single-leaf pine
	Stone pine
Pinus montezumae	
	Montezuma pine
	Rough-barked Mexican pine
Pinus monticola	
	Californian mountain pine
	Mountain white pine
	Western white pine
Pinus mugo	Dwarf mountain pine
	Mountain pine
	Swiss mountain pine
Pinus mugo pumilo	
	European scrub pine
Pinus muricata	Bishop pine
Pinus nigra	Austrian pine
Pinus nigra caramanica	
	Austrian pine
	Crimean pine
Pinus nigra cebennensis	
	Cevennes pine
	Pyrenean pine
Pinus nigra maritima	
	Corsican pine
Pinus nigra nigra	Austrian pine
Pinus occidentalis	Cuban pine
Pinus palustris	Georgia pine
	Long-leaf pine
	Pitch pine
	Southern pine
	Southern pitch pine
	Southern yellow pine
Pinus parviflora	
	Japanese white pine
Pinus patula	Jelecote pine
	Mexican pine
	Mexican weeping pine
	Spread-leaved pine
Pinus peuce	Macedonian pine
Pinus pinaster	Bournemouth pine
	Cluster pine
	Maritime pine
Pinus pinea	Italian stone pine
	Stone pine
	Umbrella pine
Pinus ponderosa	Ponderosa pine
	Western yellow pine
Pinus ponderosa arizonica	
	Arizona pine
Pinus pumila	Dwarf Siberian pine
	Dwarf stone pine
	Japanese pine

Pinus pungens	Hickory pine
	Prickly pine
	Table mountain pine
Pinus quadrifolia	Parry pine
Pinus radiata	Monterey pine
	Remarkable cone pine
Pinus resinosa	American red pine
	Canadian red pine
	Norway pine
	Red pine
Pinus rigida	Northern pitch pine
	Pitch pine
Pinus roxburghii	Indian cher pine
	Long-leaved Indian pine
Pinus sabiniana	Digger pine
Pinus strobus	Deal pine
	Eastern white pine
	Weymouth pine
	White pine
Pinus sylvestris	Scotch fir
	Scot's pine
Pinus rubra	Highland pine
Pinus tabuliformis	Chinese pine
Pinus taeda	Frankincense pine
	Loblolly pine
	Oldfield pine
Pinus teocote	Twisted-leaf pine
Pinus thunbergii	Black pine
	Japanese black pine
Pinus toeda	Frankincense pine
Pinus torreyana	Soledad pine
	Torrey pine
Pinus uncinata	Mountain pine
Pinus virginiana	Jersey pine
	Poverty pine
	Scrub pine
	Spruce pine
Pinus wallichiana	Bhutan pine
	Blue pine
	Himalayan pine
	Himalayan white pine
Piper angustifolium	Matico
Piper betel	Betel
	Betel pepper
Piper betle	Betel
	Betel pepper
Piper methysticum	Kava
Piptanthus laburnifolius	
	Evergreen laburnum
	Nepal laburnum
Piptanthus nepalensis	
	Evergreen laburnum
Piscidia erythrina	Fish-poison tree
	Jamaica dogwood
Piscidia piscipula	
	Jamaican dogwood
	West Indian dogwood
Pisonia alba	Lettuce tree
Pisonia grandis	
	Brown cabbage tree
Pisonia umbellifera	
	Bird-catcher tree
	Para-para

TREES, BUSHES, AND SHRUBS

Podocarpus spicatus

Pistacia atlantica
Mount Atlas mastic tree
Pistacia chinensis Chinese pistachio
Pistacia lentiscus Chios mastic tree
Mastic tree
Pistacia terebinthus
Chian turpentine tree
Cyprus turpentine tree
Terebinth
Turpentine tree
Pistacia texana American pistachio
Lentisco
Pistacia vera Green almond
Pistachio
Pistacia nut
Pithecellobium dulce Huamuchii
Manila tamarind
Opiuma
Pithecellobium flexicaule
Texas ebony
Pithecellobium guadalupense
Blackbead
Pithecellobium unguis-cati
Blackbead
Cat's claw
Pittosporum crassifolium Karo
Pittosporum eugenioides
Lemonwood
Tarata
Pittosporum phillyraeoides
Narrow-leaved pittosporum
Willow pittosporum
Pittosporum rhombifolium
Queensland pittosporum
Pittosporum tenuifolium Kohuhu
New Zealand pittosporum
Pittosporum
Tawhiwhi
Pittosporum tobira
Australian laurel
Japanese pittosporum
Mock orange
Tobira
Pittosporum undulatum
Mock orange
Victorian box
Pittosporum viridiflorum
Cape pittosporum
Plagianthus regius
Ribbonwood tree
Planera aquatica Water elm
Planera ulmifolia Water elm
Platanus × acerifolia
London plane
Platanus × hispanicus
London plane
Platanus hybrida London plane
Platanus occidentalis
American plane
American sycamore
Buttonball
Buttonwood
Eastern sycamore
Western plane

Platanus orientalis Chennar tree
Oriental plane
West Asian plane
Platanus orientalis insularis
Cyprian plane
Platycladus orientalis
Oriental arborvitae
Pleiogynium cerasiferum
Burdekin plum
Queensland hog plum
Plueria rubra Pagoda tree
Plumeria acuminata Frangipani
Plumeria rubra Frangipani tree
Nosegay tree
Red jasmine
Temple tree
West Indian jasmine
Podalyria calyptrata
Sweet pea bush
Water blossom pea
Podalyria sericea Satin bush
Podocarpus amarus Black pine
Podocarpus andinus Chilean yew
Plum fir
Plum-fruited yew
Podocarpus chilinus
Willow podocarp
Podocarpus dacrydioides Kahika
Kahikatea
Red pine
White pine
Podocarpus elatus Brown pine
She pine
White pine
Podocarpus elongatus
African yellowwood
Fern podocarpus
Weeping podocarpus
Podocarpus falcatus
Common yellowwood
Oteniqua yellowwood
Podocarpus ferrugineus Miro
Rusty podocarp
Podocarpus gracilior
African fern pine
Podocarpus latifolius
Real yellowwood
Podocarpus macrophyllus
Buddhist pine
Japanese yew
Kusamaki tree
Large-leaved podocarp
Southern yew
Podocarpus nagi
Broadleaf podocarpus
Nagi
Podocarpus nivalis Alpine totara
Podocarpus nubigenus
Chilean totara
Manio
Podocarpus salignus
Willow podocarp
Podocarpus spicatus Matai
New Zealand black pine

Podocarpus totara

Podocarpus totara	Mahogany pine
	Totara
Poinciana gilliesii	Bird of paradise
Polygala cowellii	Tortuguero
	Violet tree
Polygonum baldschuanicum	
	Russian vine
Polyscias balfouriana	
	Balfour's aralia
Polyscias filicifolia	Fern-leaf aralia
Polyscias fruticosa	Ming aralia
Polyscias guilfoylei	Coffee tree
	Geranium-leaf aralia
	Wild coffee
Poncirus trifoliata	Hardy orange
	Japanese bitter orange
	Trifoliate orange
Pongamia pinnata	Karum tree
	Poonga oil tree
Populus adenopoda	Chinese aspen
Populus alba	Abele
	Silver-leaved poplar
	White poplar
Populus angulata	Carolina poplar
Populus angustifolia	
	Willow-leaved poplar
Populus balsamifera	
	Balsam poplar
	Hackmatack
	Tacamahak
Populus × *berolinensis*	
	Berlin poplar
Populus × *canadensis*	
	Carolina poplar
	Hybrid black poplar
Populus candicans	Balm of gilead
	Ontario poplar
Populus canescens	Grey poplar
Populus deltoides	Cottonwood
	Eastern cottonwood
	Necklace poplar
Populus eugenei	Carolina poplar
Populus fremontii	Fremont poplar
Populus gileadensis	Balm of gilead
Populus grandidentata	
	Big-toothed aspen
Populus heterophylla	
	Black cottonwood
	Downy poplar
	Swamp cottonwood
Populus lasiocarpa	
	Chinese necklace poplar
	Chinese poplar
	Necklace poplar
Populus laurifolia	
	Siberian balsam poplar
Populus maximowiczii	
	Japanese poplar
Populus nigra	Black poplar
	Lombardy poplar
Populus nigra betulifolia	
	Downy black poplar
	Manchester poplar

Populus nigra italica	Italian poplar
	Lombardy poplar
Populus nigra italica foemina	
	Female lombardy poplar
Populus nigra plantierensis	
	Western lombardy poplar
Populus regenerata	Railway poplar
Populus robusta	
	False lombardy poplar
Populus sargentii	
	Great plains cottonwood
Populus serotina	
	Black italian poplar
Populus serotina aurea	
	Golden poplar
Populus sieboldii	Japanese aspen
Populus tacamahaca	
	Balsam poplar
Populus tremula	Aspen
	European aspen
	Trembling aspen
Populus tremula pendula	
	Weeping aspen
Populus tremuloides	
	American aspen
	Quaking aspen
	Quiverleaf
	Trembling aspen
Populus tremuloides pendula	
	Parasol de Saint. Julien
Populus trichocarpa	
	Black cottonwood
	Western balsam poplar
Portulacaria afra	Elephant bush
Posoqueria latifolia	
	Needleflower tree
Pouteria campechiana	Canistel
	Eggfruit
	Sapote amarillo
	Sapote borracho
	Ti-es
Pouteria sapota	Mamey colorado
	Mamey sapote
	Mammee sapote
	Marmalade plum
	Sapote
Pritchardia pacifica	Fiji fan palm
Prosopsis chilensis	Algarrobo
Prosopsis glandulosa	Mesquite
Prosopsis juliflora	Algarrobo
	Mesquite
Prosopsis pubescens	Screw bean
	Tornillo
Protea aurea	Waterlily protea
Protea barbigera	
	Giant woolly protea
Protea cynaroides	Giant protea
	King protea
Protea grandiceps	Peach protea
Protea mellifera	Honey flower
	Honey protea
	Sugarbush
Protea repens	Sugarbush

TREES, BUSHES, AND SHRUBS

Prunus ivensii

Prumnopitys ferruginea	Miro
Prumnopitys taxifolia	Matai
New Zealand black pine	
Prunus alleghaniensis	
	Alleghany plum
	American sloe
	Sloe
Prunus amanagawa	
	Erect Japanese cherry
	Lombardy poplar cherry
Prunus americana	
	American red plum
	American wild plum
	August plum
	Goose plum
	Hog plum
Prunus × amygdalopersica	
pollardii	Pollard's almond
Prunus amygdalus	
	Common almond
Prunus andersonii	Desert peach
Prunus angustifolia	
	Chickasaw plum
	Sand plum
Prunus armeniaca	Apricot
	Common apricot
Prunus avium	Bird cherry
	Gean
	Mazzard
	Sweet cherry
	Wild cherry
Prunus avium plena	Double gean
	Double white cherry
Prunus avium sylvestris	Gean
Prunus besseyi	
	Rocky mountain cherry
	Sand cherry
	Western sand cherry
Prunus (blireana	
	Double cherry-plum
Prunus brigantina	
	Briançon apricot
Prunus campanulata	
	Bell-flowered cherry
	Formosan cherry
	Taiwan cherry
Prunus canescens	Greyleaf cherry
Prunus capuli	Mexican cherry
Prunus caroliniana	
	American mock orange
	Cherry laurel
	Mock orange
	Wild orange
Prunus cerasifera	Cherry plum
	Flowering plum
	Greenglow plum
	Myrobalan plum
Prunus cerasifera	
atropurpurea	Purple plum
Prunus cerasifera nigra	
	Black-leaved plum
	Blaze

Prunus cerasifera pissardii	
	Purple flash
	Purple-leaved plum
Prunus cerasoides rubea	
	Kingdon Ward's carmine cherry
Prunus cerasus	Pie cherry
	Sour cherry
Prunus cerasus semperflorens	
	All saints cherry
Prunus × cistena	
	Crimson dwarf cherry
	Purple-leaf sand cherry
Prunus communis	
	Common almond
Prunus conradinae	Chinese cherry
Prunus cornuta	
	Himalayan bird cherry
Prunus × dasycarpa	Black apricot
	Purple apricot
Prunus davidiana	Chinese peach
Prunus depressa	Sand cherry
Prunus domestica	Damson
	Plum
	Prune
	Wild plum
Prunus domestica institia	Bullace
	Damson
Prunus domestica italica	
	Greengage
Prunus dulcis	Almond
	Common almond
	Sweet almond
Prunus dulcis amara	
	Bitter almond
Prunus dulcis roseoplena	
	Double almond
Prunus × effusus	Duke cherry
Prunus emarginata	Bitter cherry
	Oregon cherry
Prunus fasciculata	Desert almond
	Wild almond
	Wild peach
Prunus fremontii	Desert apricot
Prunus fruticosa	Dwarf cherry
	Ground cherry
Prunus glandulosa	
	Chinese bush cherry
	Flowering almond
Prunus gracilis	Oklahoma plum
	Prairie cherry
Prunus × gondouinii	Duke cherry
Prunus ilicifolia	Evergreen cherry
	Holly-leaved cherry
	Islay plum
	Mountain holly
	Wild cherry
Prunus incana	Willow cherry
Prunus incisa	Fuji cherry
Prunus insititia	Bullace
	Damson
Prunus ivensii	Weeping cherry

Prunus jacquemontii

Prunus jacquemontii	
	Afghan cherry
	Flowering almond
Prunus japonica	Flowering almond
	Japanese bush cherry
	Japanese plum
Prunus laurocerasus	Cherry laurel
	Common laurel
	English laurel
	Laurel
Prunus laurocerasus serbica	
	Serbian laurel
Prunus leveilleana	
	Korean hill cherry
Prunus litigiosa	Tassel cherry
Prunus lusitanica	Portugal laurel
Prunus lyonii	Catalina cherry
Prunus maackii	Manchurian cherry
Prunus mahaleb	Mahaleb
	Perfumed cherry
	Saint lucie cherry
Prunus mandshurica	
	Manchurian apricot
Prunus maritima	Beach plum
	Shore plum
Prunus mugus	Tibetan cherry
Prunus mume	Japanese apricot
Prunus munsoniana	
	Wild-goose plum
Prunus mutabilis stricta	
	Chinese hill cherry
Prunus nigra	Canada plum
Prunus nipponica	
	Japanese alpine cherry
Prunus padus	Bird cherry
	European bird cherry
	Hagberry
Prunus padus colorata	
	Purple-leaved bird cherry
Prunus pennsylvanica	Bird cherry
	Fire cherry
	Pin cherry
	Wild red cherry
Prunus persica	Common peach
	Peach
Prunus persica nectarina	
	Nectarine
Prunus persica nucipersica	
	Nectarine
Prunus prostrata	Mountain cherry
	Rock cherry
Prunus pumila	
	Dwarf American cherry
	Dwarf cherry
	Sand cherry
Prunus reverchonii	Hog plum
Prunus rufa	Himalayan cherry
Prunus salicifolia	Capulin cherry
Prunus salicina	Japanese plum
Prunus sargentii	
	Japanese hill cherry
	Sargent's cherry

Prunus serotina	
	American black cherry
	Black cherry
	Rum cherry
Prunus serrula	Birchbark cherry
	Paperbark cherry
	Tibetan cherry
Prunus serrulata	Hill cherry
	Japanese cherry
	Oriental cherry
Prunus serrulata hupehensis	
	Chinese hill cherry
	Hupeh cherry
Prunus serrulata kanzan	
	Japanese double pink cherry
Prunus serrulata pubescens	
	Korean hill cherry
Prunus serrulata spontanea	
	Hill cherry
Prunus sibirica	Siberian apricot
Prunus simonii	Apricot plum
	Simon's plum
Prunus speciosa	Oshima cherry
Prunus spinosa	Blackthorn
	Sloe
Prunus subcordata	Oregon plum
	Pacific plum
	Sierra plum
Prunus subhirtella	Higan cherry
	Rosebud cherry
	Spring cherry
Prunus subhirtella ascendens	
	Rosebud cherry
Prunus subhirtella autumnalis	
	Autumn cherry
	Higan cherry
	Winter cherry
	Winter-flowering cherry
Prunus subhirtella pendula	
	Weeping cherry
	Weeping rosebud cherry
Prunus subhirtella pendula rosea	Weeping spring cherry
Prunus × sultana	Wickson plum
Prunus tai-haku	
	Great white cherry
Prunus tenella	
	Dwarf Russian almond
	Russian almond
Prunus tomentosa	
	Chinese bush cherry
	Downy cherry
	Hansen's cherry
	Nanking cherry
Prunus triflora	Japanese plum
Prunus triloba	Flowering almond
Prunus virginiana	Choke cherry
	Virginian bird cherry
Prunus virginiana demissa	
	Western choke cherry

TREES, BUSHES, AND SHRUBS

Quercus acuta

Prunus yedoensis
Japanese flowering cherry
Potomac cherry
Yoshino cherry

Pseudobombax ellipticum
Shaving-brush tree

Pseudocydonia sinensis
Chinese quince

Pseudolarix amabilis   Golden larch

Pseudolarix kaempferi
Golden larch

Pseudopanax crassifolius
Lancewood

Pseudotsuga japonica
Japanese douglas fir

Pseudotsuga macrocarpa
Big-cone spruce
Large-coned douglas fir

Pseudotsuga menziesii   Douglas fir
Oregon douglas fir

Pseudotsuga menziesii caesia
Frazer river douglas fir
Grey douglas fir

Pseudotsuga menziesii glauca
Blue douglas fir

Pseudotsuga sinensis
Chinese douglas fir

Pseudotsuga wilsoniana
Wilson's douglas fir

Psidium cattleianum
Strawberry guava

Psidium friedrichsthalianum
Costa Rican guava

Psidium guajava   Apple guava
Common guava
Guava
Yellow guava

Psidium guineense   Guava

Psidium montanum
Mountain guava
Spice guava

Psychotria emetica   False ipecac

Psychotria nervosa   Wild coffee

Psychotria sulzneri   Wild coffee

Ptelea trifoliata   Hop tree
Shrubby trefoil
Stinking ash
Swamp dogwood
Water ash
Wingseed

Pterocarpus angolensis
Transvaal teak
West African barwood

Pterocarpus erinaceus   Barwood
Senegal rosewood
West African kino

Pterocarpus indicus
Burmese rosewood
Padauk

Pterocarpus marsupium
Bastard teak
Kinos

Pterocarpus santalinus
Red sandalwood
Red saunders
Sanderswood

Pterocarya fraxinifolia
Caucasian wing nut

Pterocarya × rehderana
Hybrid wing nut

Pterocarya rhoifolia
Japanese wing nut

Pterocarya stenoptera
Chinese wing nut

Pterostyrax hispida   Asgara
Epaulette tree

Ptychosperma elegans
Alexander palm
Solitary palm

Ptychosperma macarthurii
Hurricane palm
Macarthur palm

Punica granatum   Pomegranate

Punica granatum nana
Dwarf pomegranate

Pyracantha atalantoides
Gibb's firethorn

Pyracantha coccinea
Buisson ardent
Firethorn

Pyracantha crenulata
Nepalese white thorn

Pyrus amygdaliformis
Almond-leaved pear

Pyrus austriaca   Austrian pear

Pyrus caucasica   Caucasian pear

Pyrus communis   Common pear
Wild pear

Pyrus cordata   Plymouth pear

Pyrus kawakamii   Evergreen pear

Pyrus nivalis   Snow pear

Pyrus pashia   Kumaon pear

Pyrus pyraster   Wild pear

Pyrus pyrifolia   Asian pear
Chinese sand pear
Japanese pear
Kumoi
Nashi
Oriental pear
Sand pear

Pyrus salicifolia   Silver pear
Willow-leaved pear

Pyrus salicifolia pendula
Weeping pear
Weeping willow-leaved pear

Pyrus salvifolia   Sage-leaved pear

Pyrus ussuriensis   Chinese pear
Sand pear

Quassia amara   Bitterwood
Surinam quassia

Quercus acuta
Japanese evergreen oak
Japanese red oak

Quercus acutissima

Quercus acutissima	Chestnut oak
	Japanese chestnut oak
	Sawtooth oak
Quercus agrifolia	
	Californian field oak
	Californian live oak
	Encina
Quercus alba	White oak
Quercus aliena	Oriental white oak
Quercus alnifolia	
	Cyprus golden oak
	Golden oak
Quercus bicolor	
	American white oak
	Swamp white oak
Quercus borealis	Red oak
Quercus calliprinos	Palestine oak
Quercus canariensis	Algerian oak
	Mirbeck's oak
Quercus castaneifolia	Chestnut-leaved oak
Quercus cerris	Bitter oak
	Mossy cup oak
	Turkey oak
Quercus chrysolepis	
	Californian live oak
	Canyon live oak
	Golden-cup oak
	Maul oak
Quercus coccifera	Kermes oak
Quercus coccinea	Scarlet oak
Quercus conferta	Hungarian oak
Quercus conocarpa	Singapore oak
Quercus dentata	Daimyo oak
Quercus douglasii	Blue oak
Quercus dumosa	
	Californian scrub oak
	Scrub oak
Quercus durata	Leather oak
Quercus ellipsoidalis	Jack oak
	Northern pin oak
Quercus emoryi	Emory oak
Quercus engelmannii	
	Engelmann's oak
Quercus faginea	Portuguese oak
Quercus falcata	Spanish oak
	Spanish red oak
Quercus frainetto	Hungarian oak
	Italian oak
	Macedonian oak
Quercus gambelii	Gambel's oak
	Shin oak
Quercus × ganderi	Gander's oak
Quercus garryana	Oregon oak
	Oregon white oak
	Western oak
Quercus glandulifera	Konara oak
Quercus glauca	Ring-cupped oak
Quercus havardii	Havard oak
	Shinnery oak
Quercus × heterophylla	
	Bartram oak
Quercus × hispanica	
	Luccombe oak
Quercus × hispanica lucombeana	Exeter oak
Quercus ilex	Evergreen oak
	Holly oak
	Holm oak
Quercus ilicifolia	Bear oak
	Scrub oak
Quercus imbricaria	Laurel oak
	Shingle oak
Quercus incana	Bluejack oak
	High-ground willow oak
	Sand jack
	Turkey oak
Quercus ithaburensis	Valonia oak
Quercus kelloggii	
	Californian black oak
	Kellogg oak
Quercus laevis	Catesby oak
	Turkey oak
Quercus laurifolia	Darlington oak
	Laurel oak
Quercus libani	Lebanon oak
Quercus lobata	
	Californian white oak
	Valley oak
Quercus lusitanica	
	Portuguese oak
Quercus × ludoviciana	
	Ludwig's oak
Quercus lyrata	Overcup oak
	Swamp post oak
Quercus macdonaldii	
	Macdonald oak
Quercus macedonica	
	Macedonian oak
Quercus macranthera	
	Caucasian oak
Quercus macrocarpa	Burr oak
	Mossy cup oak
Quercus macrolepis	Valonia oak
Quercus maingayi	Maingay's oak
Quercus marilandica	
	Blackjack oak
	Blackthorn oak
	Jack oak
Quercus michauxii	
	Swamp chestnut oak
Quercus mirbeckii	Algerian oak
Quercus mongolica	Mongul oak
Quercus muehlenbergii	
	Chestnut oak
	Yellow chestnut oak
	Yellow oak
Quercus myrsinaefolia	
	Bamboo-leaved oak
Quercus nigra	Possum oak
	Water oak
Quercus palustris	Pin oak
	Spanish oak
Quercus petraea	Durmast oak
	Sessile oak

Rhododendron catawbiense

Quercus phellos	Willow oak
Quercus phillyraeoides	Ubame oak
Quercus pontica	Armenian oak
	Pontine oak
Quercus prinoides	Chinquapin oak
	Dwarf chestnut oak
Quercus prinus	Bamboo-leaved oak
	Basket oak
	Chestnut oak
	Rock chestnut oak
	Swamp chestnut oak
Quercus pubescens	Downy oak
	Green oak
Quercus pyrenaica	Pyrenean oak
Quercus rober	Common oak
	English oak
	Pendunculate oak
	Truffle oak
Quercus rober concordia	Golden oak
Quercus rober pendula	Weeping oak
Quercus rober purpurescens	Purple English oak
Quercus robur asplenifolia	Fern-leaved oak
Quercus robur fastigiata	Cypress oak
Quercus robur filicifolia	Cut-leaf oak
Quercus rubra	American red oak
	Northern red oak
	Red oak
Quercus sadlerana	Deer oak
Quercus shumardii	Shumard's oak
	Shumard's red oak
Quercus stellata	Post oak
Quercus suber	Cork oak
Quercus texana	Texas red oak
Quercus tomentella	Island oak
Quercus trojana	Macedonian oak
Quercus (turneri	Turner's oak
Quercus undulata	Rocky mountain scrub oak
Quercus vaccinifolia	Huckleberry oak
Quercus variabilis	Chinese cork oak
	Oriental cork oak
Quercus velutina	Black oak
	Quercitron
	Yellow-bark oak
Quercus velutina rubrifolia	Champion's oak
Quercus virginiana	Live oak
	Southern live oak
Quercus warburgii	Cambridge oak
Quercus wislizenii	Interior live oak

Quillaia saponaria	Soap bark tree
	Soap tree
Raphi farinifera	Raffia palm
Raphia ruffia	Raffia palm
Raphiolepis indica	Indian hawthorn
Raphiolepis japonica	Yeddo hawthorn
Raphiolepis umbellata	Indian hawthorn
	Yeddo hawthorn
Ravenala madagascariensis	Traveller's palm
	Traveller's tree
Ravensara aromatica	Madagascar nutmeg
Rhamnus alaternus	Italian buckthorn
Rhamnus alpina	Alpine buckthorn
Rhamnus californica	Coffee berry
Rhamnus caroliniana	Carolina buckthorn
	Indian cherry
Rhamnus catharticus	Common buckthorn
	Purging blackthorn
	Ramsthorn
Rhamnus crocea	Redberry buckthorn
Rhamnus davurica	Dahurian buckthorn
Rhamnus frangula	Alder buckthorn
	Black dogwood
Rhamnus infectoria	Avignon berry
Rhamnus pumila	Dwarf buckthorn
Rhamnus purshiana	Bearberry
	Californian buckthorn
	Cascara sagrada
Rhamnus saxitilis	Rock buckthorn
Rhapidophyllum hystrix	Blue palmetto
	Needle palm
	Porcupine palm
Rhapis excelsa	Bamboo palm
	China cane
	Fern rhapis
	Ground rattan cane
	Lady palm
	Little lady palm
	Miniature fan palm
	Partridge cane
	Slender lady palm
Rhapis humilis	Reed rhapis
	Slender lady palm
Rhizophora mangle	American mangrove
	Red mangrove
Rhododendron arboreum	Rhododendron
	Tree rhododendron
Rhododendron catawbiense	Mountain rose bay

Rhododendron ferrugineum
Alpen rose

Rhododendron hirsutum
Hairy alpen rose

Rhododendron luteum
Common yellow azalea
Yellow azalea

Rhododendron maximum
Great laurel
Rose bay

Rhododendron obtusum
Kirishima azalea

Rhododendron oldhamii
Formosan azalea

Rhododendron ponticum
Great laurel
Rhododendron
Rose bay

Rhododendron poukhanense
Korean azalea

Rhododendron simsii
Indian azalea

Rhododendron viscosum
Clammy azalea
Swamp azalea
Swamp honeysuckle
White swamp azalea

Rhodomyrtus tomentosa
Downy myrtle
Hill gooseberry
Hill guava

Rhodosphaera rhodanthema
Queensland yellowwood
Yellowwood

Rhopalostylis sapida
Feather-duster palm
Nikau palm

Rhus aromatica
Fragrant sumac
Lemon sumac
Polecat bush
Sweet-scented sumac
Sweet sumac

Rhus chinensis
Nutgall tree

Rhus copallina
Dwarf sumac
Mountain sumac
Shining sumach
Wing-rib sumac

Rhus coriaria
Elm-leaved sumac
Sicilian sumac
Tanner's sumac

Rhus cotinus
Smoke tree
Venetian sumac

Rhus glabra
Scarlet sumac
Smooth sumac
Upland sumac
Vinegar tree

Rhus hirta
Stag's horn sumac

Rhus integrifolia
Lemonade berry
Lemonade sumac
Sourberry

Rhus laurina
Laurel sumac

Rhus microphylla
Correosa
Desert sumac
Scrub sumac
Small-leaved sumac

Rhus ovata
Sugarbush
Sugar sumac

Rhus potaninii
Chinese varnish tree

Rhus radicans
Cow-itch
Markry
Mercury
Poison ivy
Poison oak

Rhus succedanea
Wax tree

Rhus trilobata
Skunk bush

Rhus typhina
Stag's horn sumac
Velvet sumac
Virginian sumac

Rhus verniciflua
Japanese lacquer tree
Japanese varnish tree
Lacquer tree
Varnish tree

Rhus vernix
Poison dogwood
Poison elder
Poison sumac
Swamp sumac

Rhus virens
Evergreen sumac
Lentisco
Tabacco sumac

Rhyticocos amara
Overtop palm

Ribes alpinum
Alpine currant
Mountain currant

Ribes americanum
American blackcurrant

Ribes aureum
Buffalo currant
Golden currant

Ribes bracteosum
Californian blackcurrant

Ribes lacustre
Swamp currant

Ribes nigrum
Blackcurrant

Ribes odoratum
Buffalo currant
Golden currant

Ribes sanguineum
Flowering currant

Ribes speciosum
Flowered gooseberry

Ribes uva-crispa
Common gooseberry
Gooseberry
Goosegog

Ribes viburnumifolium
Evergreen currant

Ricinus communis
Castor bean
Castor oil plant
Palma christi
Wonder tree

Robinia hispida
Bristly locust
Mossy locust
Rose acacia

Robinia pseudoacacia	Black locust tree
	Common acacia
	False acacia
	Locust tree
	Robinia
	Yellow locust tree
Robinia pseudoacacia appalachia	Shipmast acacia
Robinia pseudoacacia frisia	Golden acacia
	Golden locust
Robinia pseudoacacia inermis	Mop-head acacia
Robinia pseudoacacia umbraculifera	Mop-head acacia
Robinia viscosa	Clammy locust
Robus fructicosus	Blackberry
Rosa acicularis	Needle rose
Rosa agrestis	Field briar
Rosa × alba	Jacobite rose
	White rose of York
Rosa alpina	Alpine rose
Rosa arkansana	Arkansas rose
Rosa arvensis	Ayrshire rose
	Field rose
	Trailing rose
Rosa banksiae	Banksian rose
	Lady Banks's rose
Rosa banksiae lutea	Yellow banksian rose
Rosa blanda	Meadow rose
	Smooth rose
Rosa × borboniana	Bourbon rose
Rosa bracteata	Macartney rose
Rosa brunonii	Himalayan musk rose
Rosa canina	Briar rose
	Common briar
	Dog briar
	Dog rose
Rosa centifolia	Cabbage rose
	Provence rose
Rosa centifolia cristata	Crested moss rose
Rosa centifolia mucosa	Moss rose
Rosa chinensis	China rose
	Monthly rose
Rosa chinensis minima	Fairy rose
Rosa chinensis viridiflora	Green rose
Rosa damascena	Damask rose
Rosa damascena trigintipetala	Kazanlik rose
Rosa damascena versicolor	York and Lancaster rose
Rosa eglanteria	Eglantine rose
	Sweet briar
Rosa foetida	Austrian briar
	Austrian yellow rose
Rosa foetida bicolor	Austrian copper rose

Rosa foetida persiana	Persian yellow rose
Rosa gallica	French rose
	Provence rose
	Red rose
Rosa gallica officinalis	Apothecary's rose
	Lancaster red rose
	Lancaster rose
Rosa gallica versicolor	Rosa mundi
Rosa gymnocarpa	Baldhip rose
	Redwood rose
	Wood rose
Rosa hemisphaerica	Sulfer rose
	Sulphur rose
Rosa laevigata	Cherokee rose
Rosa lutea hoggii	Hog's double yellow rose
Rosa majalis	Cinnamon rose
	May rose
Rosa moschata	Musk rose
Rosa multiflora grevillei	Seven sisters rose
Rosa × noisettiana	Champney rose
	Noisette rose
Rosa × odorata	Tea rose
Rosa officinalis	Apothecary's rose
	Old red damask rose
Rosa omeiensis	Mount Omei rose
Rosa palustris	Swamp rose
Rosa pimpinellifolia	Burnet rose
	Scotch briar
	Scotch rose
Rosa primula	Incense rose
Rosa roxburghii	Burr rose
	Chestnut rose
	Chinquapin rose
Rosa rubiginosa	Eglantine rose
	Sweet briar
Rosa rugosa	Hedgehog rose
	Japanese rose
	Ramanas rose
	Turkestan rose
Rosa sempervirens	Evergreen rose
Rosa sericea	Mount Omei rose
Rosa setigera	Prairie rose
Rosa sherardii	Northern downy rose
Rosa stellata mirifica	Sacramento rose
Rosa sulphurea	Sulfer rose
	Sulphur rose
Rosa tomentosa	Downy rose
Rosa villosa	Apple rose
	Soft-leaved rose
Rosa villosa duplex	Wolley-dod's rose
Rosa wichuraiana	Memorial rose
Rosmarinus officinalis	Rosemary
Roystonea borinquena	Puerto Rican royal palm

Roystonea elata

Roystonea elata   Florida royal palm

Roystonea hispaniolana
     Spanish royal palm

Roystonea oleracea   Cabbage palm
     Caribbean royal palm
     South American royal palm

Roystonea regia   Cuban royal palm
     Royal palm

Rubus cockburnianus
     Chinese bramble

Rubus deliciocus
     Rocky Mountains bramble

Rubus fructicosus   Bramble

Rubus illecebrosus
     Strawberry-raspberry

Rubus laciniatus
     Fern-leaved bramble

Rubus parviflorus   Salmon berry

Rubus phoenicolasius   Wineberry

Ruprechtia coriacea   Biscochito

Ruscus aculeantus
     Butcher's broom

Russelia equisetiformis
     Coral plant
     Fountain bush

Russelia juncea   Coral plant
     Firecracker
     Fountain bush

Ruta graveolens   Rue

Sabal adansonii   Bush palmetto
     Dwarf palmetto
     Scrub palmetto

Sabal bermudana
     Bermuda palmetto

Sabal causiarum
     Puerto Rican hat palm

Sabal etonia   Scrub palmetto

Sabal jamaicensis   Bull thatch
     Jamaica palmetto

Sabal mauritiiformis   Trinidad palm

Sabal mexicana   Texas palmetto

Sabal minor   Bush palmetto
     Dwarf palmetto
     Scrub palmetto

Sabal palmetto   Blue palmetto
     Cabbage palmetto
     Cabbage tree
     Palmetto

Sabal repens   Scrub palmetto

Sabal uresana   Sonoran palmetto

Sabinea carinalis   Carib wood

Salacea edulis   Salac

Salix acutifolia   Caspian willow
     Purple-twig willow

Salix adenophylla   Furry willow

Salix aegyptiaca
     Calif of Persia willow
     Musk willow

Salix alba   White willow

Salix alba argentea   Silver willow

Salix alba britzensis
     Coralbark willow

Salix alba caerulea
     Blue willow
     Cricket-bat willow

Salix alba chermesina
     Coral-bark willow
     Orange-twig willow
     Scarlet willow

Salix alba coerulea   Bat willow

Salix alba sericea   Silver willow

Salix alba vitellina   Egg-yolk willow
     Golden willow

Salix × ambigua   Puzzle willow

Salix amygdaloides
     Almond willow
     Peach-leaved willow

Salix andersoniana
     Green mountain sallow

Salix aquatica   Water sallow

Salix arbuscula   Little tree willow
     Mountain willow

Salix arctica   Arctic willow

Salix arenaria   Sand willow

Salix aurita   Eared willow
     Round-eared willow
     Trailing sallow

Salix babylonica   Weeping willow

Salix babylonica pekinensis
     Peking willow

*Salix babylonica pekinensis
pendula*   Weeping willow

*Salix babylonica pekinensis
tortuosa*   Contorted willow
     Corkscrew willow
     Dragon's claw willow
     Twisted willow

*Salix babylonica pekinensis
tortuosa aurea*   Golden curls tree

Salix basfordiana   Basford willow

Salix bebbiana   Beak willow

Salix × blanda   Niobe willow
     Wisconsin weeping willow

Salix caesia   Blue willow

Salix candida   Sage willow

Salix caprea   Florist's willow
     Goat willow
     Great sallow
     Palm willow
     Pussy willow
     Sallow

Salix caprea chermesina
     Coralbark willow

Salix caprea pendula
     Kilmarnock willow
     Weeping sally
     Weeping willow

Salix × chrysocoma
     Weeping willow

Salix cinerea   Common sallow
     Grey sallow
     Grey willow

Sambucus pubens

Salix cinerea atrocinerea	
	Grey sallow
Salix cinerea oleifolia	
	Common sallow
Salix cordata	Furry willow
	Heart-leaved willow
Salix crassifolia	
	Thick-leaved sallow
Salix daphnoides	Violet willow
Salix discolor	Large pussy willow
	Pussy willow
Salix elaeagnos	Hoary willow
Salix elegantissima	
	Thurlow weeping willow
Salix exigua	Coyote willow
Salix fragilis	Brittle willow
	Crack willow
Salix fragilis decipiens	
	Cardinal willow
Salix gracilistyla	
	Japanese pussy willow
Salix hastata	
	Halberd-leaved willow
Salix helvetica	Swiss willow
Salix herbacea	Dwarf willow
	Least willow
Salix humilis	Grey willow
	Prairie willow
	Small pussy willow
Salix interior	Sandbar willow
Salix laevigata	Polished willow
	Red willow
Salix lanata	Woolly willow
Salix lapponum	Downy willow
	Lapland willow
Salix lasiandra	Pacific willow
Salix lasiolepis	Arroyo willow
Salix lucida	Shining willow
Salix magnifica	
	Magnolia-leaved willow
Salix matsudana	Peking willow
Salix matsudana pendula	
	Weeping Peking willow
Salix matsudana tortuosa	
	Contorted willow
	Corkscrew willow
	Dragon's claw willow
Salix melanostachys	
	Black pussy willow
Salix mutabilis	Puzzle willow
Salix myrsinifolia	
	Dark-leaved willow
Salix myrsinites	
	Whortle-leaved willow
	Wortle willow
Salix nigra	Black willow
Salix nigricans	Dark-leaved willow
Salix nivalis	Tufted willow
Salix pellita	Satiny willow
Salix pentandra	Bay willow
	Laurel willow

Salix petraea	Rock sallow
Salix phylicifolia	Tea-leaf willow
Salix purpurea	Basket willow
	Purple osier
Salix purpurea pendula	
	Weeping purple osier
	Weeping purple willow
Salix pyrifolia	Balsam willow
Salix repens	Creeping willow
Salix reticulata	Netted willow
Salix × rubens basfordiana	
	Basford willow
Salix salviaefolia	
	Sage-leaved willow
Salix scoulerana	Scouler willow
Salix sepulcralis chrysocoma	
	Golden weeping willow
	Weeping willow
Salix sericea	Silky willow
Salix triandra	Almond willow
	French willow
Salix urbaniana	Japanese willow
Salix uva-ursi	Bearberry willow
Salix viminalis	Basket willow
	Common osier
	Osier
Salix xerythroflexuosa	
	Golden curls willow
Salvadora persica	Toothbrush tree
Salvia fulgens	Mexican red sage
Salvia officinalis	Common sage
Salvia rutilans	Pineapple sage
Samanea saman	Monkeypod tree
	Rain tree
	Saman tree
	Zamang tree
Sambucus caerulea	Blue elder
Sambucus canadensis	
	American elder
	Sweet elder
Sambucus canadensis aurea	
	Golden American elder
Sambucus ebulus	Dane's elder
	Danewort
	Dwarf elder
	Wallwort
Sambucus nigra	Black elder
	Bourtree
	Common elder
	European elder
	Pipe tree
Sambucus nigra aurea	
	Golden elder
Sambucus nigra laciniata	
	Cut-leaved elder
	Fern-leaved elder
	Parsley-leaved elder
Sambucus pubens	
	American red elder
	Red-berried elder
	Stinking elder

Sambucus racemosa	Alpine elder
	European red elder
	Red-berried elder
Sandoricum indicum	Sandal tree
Sandoricum koetjapa	Sentol
Santalum album	Sandalwood
	White sandalwood
Santalum rubrum	
	Red sandalwood
Santolina chamaecyparisus	
	Lavender cotton
Santolina incana	Lavender cotton
Santolina rosmarinifolia	Holy flax
Sapindus drummondii	
	Soapberry tree
	Wild China tea
Sapindus marginatus	
	Wild China soapberry
Sapindus mukorossi	
	Chinese soapberry
Sapindus saponaria	
	False dogwood
	Jaboncillo
	Soapberry
Sapium hippomane	Milk tree
Sapium salicifolium	Tallow tree
Sapium sebiferum	
	Chinese tallow tree
	Vegetable tallow tree
Saraca indica	Asoka tree
	Sorrowless tree
Sarcococco buxacaea	
	Christmas box
Sassafras albidum	Ague tree
	Sassafras
Satureia montana	Winter savory
Saxegothaea conspicua	
	Prince Albert's yew
Schefflera arboricola	
	Umbrella tree
Schinus molle	
	American mastic tree
	Australian pepper tree
	Californian pepper tree
	Mastic tree
	Molle
	Pepper tree
	Peruvian mastic tree
	Peruvian pepper tree
	Pirul
Schinus terebinthifolia	
	Brazilian pepper tree
	Christmas berry tree
Schizophragma viburnoides	
	Climbing hydrangea
Schleichera oleosa	Ceylon oak
	Gum-lac
	Lac tree
Schotia afra	Hottentot's bean
	Kaffir bean
Schotia brachypetala	Tree fuchsia
Schotia latifolia	
	Elephant hedge bean tree

Sciadophyllum brownii	
	Galapee tree
Sciadopitys verticillata	
	Japanese umbrella pine
	Parasol pine
	Umbrella pine
Semecarpus anacardium	
	Dhobi's nut
	Marking-nut tree
	Varnish tree
Sequoia sempervirens	
	Californian coast redwood
	Californian redwood
	Coast redwood
	Redwood
Sequoiadendron giganteum	
	Californian big tree
	Giant sequoia
	Mammoth tree
	Sierra redwood
	Wellingtonia
Serenoa repens	Sabal
	Saw palmetto
	Scrub palmetto
Seriocarpus conyzoides	
	Silk fruit tree
Sesbania formosa	
	White dragon tree
Sesbania tripetii	Wisteria tree
Severinia buxifolia	
	Chinese box orange
Shepherdia argentea	Buffalo berry
	Silver buffalo berry
Shepherdia canadensis	
	Buffalo berry
	Soapberry
Shorea robusta	Sal
Sicana odorifera	Cassa-banana
Siliphium terebinthaceum	
	Turpentine tree
Simaba cedron	Cedron
Simarouba amara	Bitter damson
	Dysentery bark
	Simarouba
Simarouba glauca	Aceituno
	Bitterwood
	Paradise tree
Smilax glauca	Saw briar
Smilax hispida	Hag briar
Smilax ornata	Jamaica sarsaparilla
Smilax rotundifolia	Green briar
	Horse briar
Socratea exorhiza	Zanona palm
Solanum aviculare	Kangaroo apple
Solanum crispum	Potato tree
Solanum jasminoides	Potato vine
Solanum laciniatum	
	Kangaroo apple
Solanum topiro	Cocona
Soleanea berteriana	Motillo

TREES, BUSHES, AND SHRUBS

Staphylea pinnata

Sophora japonica
  Chinese scholar tree
  Japanese pagoda tree
  Pagoda tree
  Scholar's tree

Sophora japonica + pendula
  Contorted pagoda tree

Sophora secundiflora   Mescal bean
  Texas mountain laurel

Sophora tetraptera   Kowhai
  New Zealand laburnum
  New Zealand sophora

Sophora tomentosa   Silverbush

Sorbus alnifolia
  Alder-leaved rowan

Sorbus americana
  American mountain ash
  Dogberry
  Missey-mooney
  Roundwood

Sorbus anglica Cheddar whitebeam

Sorbus aria   Chess apple
  Whitebeam

Sorbus aria wilfred fox
  European whitebeam

Sorbus aucuparia
  Common mountain ash
  Common rowan
  European mountain ash
  Mountain ash
  Quickbeam
  Quicken tree
  Rantry
  Rowan

Sorbus aucuparia asplenifolia
  Cut-leaved mountain ash

Sorbus aucuparia pendula
  Weeping mountain ash

Sorbus aucuparia rossica-major
  Russian mountain ash

Sorbus aucuparia xanthocarpa
  Yellow-berried mountain ash

Sorbus austriacus
  Austrian whitebeam

Sorbus cashmiriana
  Kashmir rowan

Sorbus chamaemespilus
  Alpine whitebeam
  False medlar

Sorbus commixta Japanese rowan

Sorbus commixta embley
  Chinese scarlet rowan

Sorbus cuspidata
  Himalayan whitebeam

Sorbus devoniensis   French hales

Sorbus discolor
  Chinese scarlet rowan

Sorbus domestica   Service tree
  True service tree

Sorbus fennica Finnish whitebeam

Sorbus folgneri Chinese whitebeam
  Folgner's whitebeam

Sorbus graeca   Greek whitebeam

Sorbus hupehensis
  Chinese mountain ash
  Chinese rowan
  Hupeh rowan

Sorbus × hybrida
  Finnish whitebeam

Sorbus intermedia
  Swedish whitebeam

Sorbus lancastriensis
  Lancashire whitebeam

Sorbus latifolia
  Broad-leaved whitebeam

Sorbus × latifolia
  Fontainebleau service tree

Sorbus minima   Least whitebeam

Sorbus mougeotii
  Pyrenean whitebeam

Sorbus reducta
  Creeping mountain ash
  Pygmy rowan

Sorbus rupicola   Cliff whitebeam
  Rock whitebeam

Sorbus sargentiana Sargent's rowan

Sorbus suecica Swedish whitebeam

Sorbus × thuringiaca
  Bastard service tree
  Hybrid rowan

Sorbus torminalis   Chequer tree
  Wild service tree

Sorbus vilmorinii Vilmorin's rowan
  Vilmorin's sorbus

Sparmannia africana
  African hemp
  House lime

Spartium junceum   Spanish broom
  Weaver's broom

Spathodea campanulata
  African tulip tree
  Flame of the forest
  Fountain tree
  Tulipan
  Tulip tree

Spiraea salicifolia   Bridgewort

Spondias cytherea   Ambarella
  Golden apple
  Otaheite apple
  Wi tree

Spondias dulcis   Otaheite apple

Spondias lutea   Golden apple
  Jamaica plum
  Yellow mombin

Spondias mombin   Hog plum
  Jobo tree
  Yellow mombin

Spondias purpurea   Jocote
  Purple mombin
  Red mombin
  Spanish plum

Stahlia monosperma   Cobana
  Polisandro

Staphylea holocarpa   Bladdernut

Staphylea pinnata   Anthony nut
  European bladdernut

Staphylea trifolia
American bladdernut

Stenocarpus salignus Reefwood

Stenocarpus sinuatus
Firewheel tree
Queensland firewheel tree
Wheel of fire

Sterculia acerfolia Flame tree

Sterculia foetida Indian almond

Sterculia rupestris
Queensland bottle tree

Stewartia malacodendron
Silky camellia

Stewartia ovata Mountain camellia

Stewartia pseudocamellia
Deciduous camellia
Japanese stewartia

Stewartia sinensis
Chinese stewartia
Chinese stuartia

Strychnos ignatii Ignatius bean

Strychnos nux-vomica
Nux-vomica tree
Strychnine

Strychnos potatorum Clearing nut
Water-filter nut

Strychnos spinosa Natal orange

Strychnos toxifera Curare

Styrax americanus
American storax
Mock orange

Styrax benzoin Benzoin

Styrax hemsleyana
Hemsley's storax

Styrax japonica Japanese snowbell
Snowbell tree

Styrax obassia Big-leaved storax
Fragrant snowbell

Styrax officinalis
Mediterranean storax
Storax

Swietenia candollea
Venezuelan mahogany

Swietenia macrophylla
Broad-leaved mahogany
Honduras mahogany
Mahogany

Swietenia mahogoni
Madeira redwood
Spanish mahogany
West Indian mahogany

Syagrus coronata Licuri palm
Ouricuri palm

Symphoricarpos albus Snowberry

Symphoricarpos occidentalis
Wolfberry

Symphoricarpos orbiculatus
Coralberry
Indian currant

Symphoricarpos rivularis
Snowberry

Symphoricarpos rubra vulgaris Coralberry
Indian currant

Symphoricarpos vulgaris
Coralberry
Indian currant

Symplocos paniculata
Asiatic sweetleaf
Sapphire berry

Symplocos tinctoria Horse sugar
Sweetleaf

Synadenium grantii
African milkbush

Syncarpia glomulifera
Turpentine tree

Synsepalum dulcificum
Miraculous berry
Miraculous fruit

Syringa amurensis Amur lilac

Syringa × chinensis Chinese lilac
Rouen lilac

Syringa emodi Himalayan lilac

Syringa josikaea Hungarian lilac

Syringa laciniata Cut-leaf lilac

Syringa meyeri palibin
Korean lilac

Syringa microphylla superba
Daphne lilac

Syringa oblata Lilac

Syringa × persica Persian lilac

Syringa reticulata
Japanese tree lilac

Syringa swegiflexa Pink pearl lilac

Syringa velutina Korean lilac

Syringa villosa Late lilac

Syringa vulgaris Common lilac

Syringa yunnanensis Yunnan lilac

Syzygium aqueum
Water rose apple tree

Syzygium aromaticum Clove tree

Syzygium coolminianum
Blue lilly-pilly

Syzygium cumini Black plum
Jambolan plum
Jambool
Jambu
Java plum

Syzygium grande Sea apple

Syzygium jambos Malabar plum
Rose apple

Syzygium malaccense Malay apple
Pomerac jambos
Rose apple

Syzygium paniculatum
Australian brush cherry
Brush cherry

Syzygium pyenanthum
Wild rose-apple

Syzygium samarangense Jambool
Jambosa
Java apple
Wax apple

TREES, BUSHES, AND SHRUBS

Thuya koraiensis

Tabebuia argentea
Paraguayan trumpet tree
Silver trumpet tree
Tree of gold

Tabebuia chrysantha
Golden trumpet tree

Tabebuia flavescens   Green ebony

Tabebuia pallida
Cuban pink trumpet tree
White cedar

Tabebuia pentaphylla   Pink poui
Pink tecoma

Tabebuia riparia
Jamaican trumpet tree
Whitewood

Tabebuia rosea   Pink poui
Pink trumpet tree
Rosy trumpet tree

Tabebuia serratifolia   Apamata
Yellow poui

Tabernaemontana coronaria
Adam's apple
Cape jasmine
Nero's crown

Tabernaemontana divaricata
Adam's apple
Cape gardenia
Cape jasmine
Flower-of-love

Taiwania flousiana   Coffin tree

Tamarindus indica   Tamarind
Tamarindo

Tamarix anglica   Tamarisk

Tamarix aphylla   Athel

Tamarix gallica   French tamarisk
Manna bush
Salt cedar

Taxodium ascendens   Pond cypress
Upland cypress

Taxodium ascendens nutans
Nodding pond cypress

Taxodium distichum   Bald cypress
Deciduous cypress
Swamp cypress

Taxodium distichum pendens
Weeping swamp cypress

Taxodium mucronatum
Ahuehuete
Mexican cypress
Mexican swamp cypress

Taxus baccata   Common yew
English yew

Taxus baccata aurea   Golden yew

Taxus baccata aureovariegata
Golden Irish yew

Taxus baccata dovastoniana
Westfelton yew

Taxus baccata fastigiata   Irish yew

Taxus baccata fastigiata aurea
Golden Irish yew

Taxus baccata fructo-luteo
Yellow-berried yew

Taxus brevifolia   American yew
Californian yew
Pacific yew
Western yew

Taxus canadensis   American yew
Canadian yew

Taxus celebica   Chinese yew

Taxus chinensis   Chinese yew

Taxus cuspidata   Japanese yew

Taxus floridana   Florida yew

Taxus media   Anglo-Japanese yew

Taxus media hicksii   Hicks' yew

Taxus wallichiana   Himalayan yew

Tecoma stans   Yellowbells
Yellow bignonia
Yellow elder
Yellow trumpet tree

Tectona grandis   Teak

Telopea oreades
Gippsland waratah

Telopea truncata
Tasmanian waratah

Terminalia alata   Indian laurel

Terminalia bellirica   Myrobalan

Terminalia catappa   Indian almond
Kamani
Myrobalan
Olive bark tree
Tropical almond

Terminalia ivorensis   Indigbo

Terminalia superba   Afara

Tetracentron sinense   Spur-leaf tree

Tetraclinis articulata   Alerce
Arar tree

Tetrapanax papyriferus
Rice-paper tree

Teucrium chamaedrys
Wall germander

Teucrium fructicans
Shrubby germander

Teucrium marum   Cat thyme

Theobroma cacao   Cacao
Chocolate tree
Cocoa

Thespesia populnea   Bhendi tree
Mahoe
Portia tree

Thevetia peruviana   Be-still tree
Lucky nut
Yellow oleander

Thrinax argentea   Broom palm
Silver broom palm

Thrinax morrisii   Key palm

Thrinax parviflora
Florida thatch palm
Palmetto thatch

Thujopsis dolabrata   Hiba
Hiba arbor-vitae

Thuya koraiensis   Korean thuya

Thuya occidentalis

Thuya occidentalis	
	American arbor-vitae
	Arbor-vitae
	Northern white cedar
	White cedar
	Yellow cedar
Thuya orientalis	
	Chinese arbor-vitae
	Chinese thuya
	Chinese white cedar
Thuya plicata	Giant cedar
	Western arbor-vitae
	Western red cedar
Thuya standishii	
	Japanese arbor-vitae
	Japanese thuya
Thuyopsis dolabrata	Hiba
	Hiba arbor-vitae
Tilia americana	
	American basswood
	American lime
	Basswood
	Whitewood
Tilia amurensis	Amur lime
Tilia cordata	Linden
	Small-leaved lime
Tilia × euchlora	Caucasian lime
	Crimean lime
Tilia × europaea	Common lime
	Linden
Tilia × europaea pallida	
	Royal lime
Tilia heterophylla	
	White basswood
Tilia insularis	Korean lime
Tilia japonica	Japanese lime
	Japanese linden
Tilia mandshurica	
	Manchurian lime
Tilia × moltkei	Von Moltke's lime
Tilia mongolica	Mongolian lime
Tilia oliveri	Chinese lime
	Oliver's lime
Tilia petiolaris	Pendant silver lime
	Weeping silver lime
Tilia platyphyllos	
	Broad-leaved lime
	Large-leaved lime
Tilia platyphyllos rubra	
	Red broad-leaved lime
Tilia tomentosa	Silver lime
Tilia × vulgaris	Common lime
Tipuana tipu	Pride of Bolivia
	Rosewood
	Tipu tree
Toona odorata	Spanish cedar
	West Indian cedar
Toona sinensis	Chinese cedar
	Toon
Torreya californica	
	Californian nutmeg
	Stinking yew

Torreya nucifera	Japanese nutmeg
	Japanese torreya
	Kaya
Torreya taxifolia	Foetid yew
	Stinking cedar
	Yew-leaved torreya
Trachycarpus fortunei	
	Chusan palm
	Hemp palm
	Windmill palm
Treculia africana	
	African bread tree
Trevesia palmata	Snowflake aralia
	Snowflake tree
	Tropical snowflake
Triphasia trifolia	Limeberry
Triplaris americana	Ant tree
	Long john
	Palo santo
Triplaris surinamensis	
	Guayabo hormiguero
Triplochiton scleroxylon	Obeche
Tristania conferta	Brisbane box
	Brush box
Tristania laurina	Kanooka
Trochodendron aralioides	
	Wheel tree
Tsuga canadensis	
	Canadian hemlock
	Canada pitch
	Eastern hemlock
Tsuga caroliniana	
	Carolina hemlock
Tsuga chinensis	Chinese hemlock
Tsuga diversifolia	
	Northern Japanese hemlock
Tsuga dumosa	Himalayan hemlock
Tsuga formosana	
	Formosan hemlock
Tsuga heterophylla	
	Western hemlock
Tsuga × jeffreyi	Jeffrey's hemlock
Tsuga mertensiana	
	Mountain hemlock
Tsuga sieboldii	Japanese hemlock
	Siebold's hemlock
	Southern Japanese hemlock
Turpinia occidentalis	
	Cassada wood
	Cassava wood
Ugni molinae	Chilean guava
	Murtillo
	Uni
Ulex europaeus	Common gorse
	Furze
	Gorse
	Prickly broom
	Whin
Ulex europaeus plenus	
	Double flowered gorse
Ulex europaeus strictus	Irish gorse
Ulex gallii	Western gorse

TREES, BUSHES, AND SHRUBS

Vaccinium parvifolium

Ulex minor	Dwarf furze
	Dwarf gorse
Ulmus alata	Small-leaved elm
	Wahoo elm
	Winged elm
Ulmus americana	American elm
	Water elm
	White elm
Ulmus angustifolia	Cornish elm
	Goodyer elm
Ulmus angustifolia cornubiensis	Cornish elm
Ulmus carpinifolia	
	European field elm
	Smooth-leaved elm
Ulmus carpinifolia cornubiensis	Cornish elm
Ulmus carpinifolia sarniensis	
	Guernsey elm
	Smooth-leaved elm
	Wheatley elm
Ulmus crassifolia	Cedar elm
Ulmus davidiana japonica	
	Japanese elm
Ulmus fulva	Sweet elm
Ulmus glabra	Scotch elm
	Wych elm
Ulmus glabra camperdown	
	Camperdown elm
	Weeping elm
Ulmus glabra horizontalis	
	Weeping wych elm
Ulmus glabra pendula	
	Tabletop elm
	Weeping elm
Ulmus × hollandica	Dutch elm
Ulmus × hollandica smithii	
	Downton elm
Ulmus × hollandica vegata	
	Huntingdon elm
Ulmus japonica	Japanese elm
Ulmus laevis	European white elm
Ulmus major	Dutch elm
Ulmus mexicana	Mexican elm
Ulmus minor	Smooth-leaved elm
Ulmus minor stricta cornubiensis	Cornish elm
Ulmus minor stricta sarniensis	Guernsey elm
	Jersey elm
	Wheatley elm
Ulmus parvifolia	Chinese elm
Ulmus procera	Common elm
	English elm
Ulmus pumila	Dwarf elm
	Siberian elm
Ulmus rubra	Indian elm
	Moose elm
	Red elm
	Slippery elm
Ulmus × sarniensis	Jersey elm
	Wheatley elm

Ulmus serotina	Red elm
	September elm
Ulmus stricta	Cornish elm
Ulmus stricta sarniensis	Jersey elm
Ulmus thomasii	Cork elm
	Rock elm
Ulmus × vegata	Chichester elm
	Huntingdon elm
Umbellularia californica	
	Californian bay
	Californian laurel
	Californian olive
	Californian sassafras
	Myrtle
	Oregon myrtle
	Pepperwood
Ungnadia speciosa	False buckeye
	Mexican buckeye
	Spanish buckeye
	Texan buckeye
Uragoga ipecacuanha	
	Ipecacuanha
Vaccinium angustifolium	
	Lowbush blueberry
Vaccinium arboreum	Farkleberry
Vaccinium arctostaphylos	
	Caucasian whortleberry
Vaccinium atrococcum	
	Black highbush blueberry
Vaccinium caespitosum	
	Dwarf bilberry
Vaccinium canadense	Sour top
	Velvet leaf
Vaccinium corymbosum	
	Blueberry
	Highbush blueberry
	Swamp blueberry
	Whortleberry
Vaccinium elliottii	
	Elliott's blueberry
Vaccinium hirsutum	
	Hairy huckleberry
Vaccinium macrocarpon	
	American cranberry
	Cranberry
Vaccinium myrsinites	
	Evergreen blueberry
Vaccinium myrtillus	Bilberry
	Blaeberry
	Huckleberry
	Whinberry
	Whortleberry
Vaccinium ovatum	
	Box blueberry
	California huckleberry
	Shot huckleberry
Vaccinium oxycoccus	Cranberry
	European wild cranberry
Vaccinium padifolium	
	Madeira whortleberry
Vaccinium parvifolium	
	Red bilberry
	Red huckleberry

BOTANICAL NAMES

Vaccinium stamineum   Deerberry
    Squaw huckleberry

Vaccinium uliginosum
    Bog bilberry
    Bog whortleberry
    Moorberry

Vaccinium virgatum
    Rabbit-eye blueberry

Vaccinium vitis-idaea   Cowberry
    Cranberry
    Foxberry
    Red bilberry

Vangueria edulis
    Madagascar tamarind
    Tamarind-of-the-Indies

Vangueria esculenta
    Forest wild medlar

Veitchia merrillii   Christmas palm
    Manila palm

Verbena triphylla   Lemon verbena

Viburnum acerifolium
    Arrowwood
    Dockmackie
    Maple-leaved viburnum
    Possum haw

Viburnum alnifolium
    Devil's shoestrings
    Dogberry
    Dog hobble
    Hobblebush
    Hobble marsh
    Mooseberry
    Moosewood
    Tanglefoot
    Trip-toe
    Wayfaring tree
    White mountain dogwood
    Witch hobble

Viburnum cassinoides   Swamp haw
    Teaberry
    Wild raisin
    Withe-rod

Viburnum dentatum   Arrowwood
    Southern arrowwood

Vibernum dilatatum
    Linden vibernum

Viburnum japonicum
    Japanese viburnum

Viburnum lantana   Hoarwithy
    Meal tree
    Twistwood
    Wayfaring tree

Viburnum lantanoides
    American wayfaring tree
    Hobble bush
    Witch hobble

Viburnum lentago
    Black haw
    Cowberry
    Nannyberry
    Nanny plum
    Sheepberry
    Sweetberry
    Tea plant
    Wild raisin

Viburnum macrocephalum
    Chinese snowball

Viburnum molle   Black alder
    Poison haw

Viburnum nudum
    Smooth withe-rod
    Smooth withy-rod

Viburnum odoratissimum
    Sweet viburnum

Viburnum opulus   Crampback
    Cranberry bush
    European cranberry bush
    Guelder rose
    Red elder
    Swamp elder
    Water elder

Viburnum opulus sterile
    Snowball tree

Viburnum pauciflorum
    Mooseberry

Viburnum plicatum
    Japanese snowball

Viburnum plicatum tomentosum
    Lace-cap viburnum
    Lace-cup viburnum

Viburnum prunifolium   Black haw
    Sheepberry
    Stag bush
    Sweet haw
    Sweet viburnum

Viburnum rufidulum   Blue haw
    Rusty nannyberry
    Southern black haw

Viburnum setigerum
    Tea viburnum

Viburnum tinus   Laurustinus

Viburnum trilobum   Crampbark
    Cranberry bush
    Cranberry tree
    Grouseberry
    Highbush cranberry
    Pimbina
    Squawbush
    Summerberry

Viburnum wrightii   Leatherleaf

Vitex agnus-castus   Chaste tree
    Hemp tree
    Indian spice tree
    Monk's pepper tree
    Sage tree
    Wild pepper tree

Vitex lucens   Pururi tree

Vonitra fibrosa   Piassava palm

Warszewiczia coccinea   Chaconia
    Waterwell tree
    Wild poinsettia

Washingtonia filifera
    Californian fan palm
    Desert fan palm
    Petticoat palm
    Washington palm

TREES, BUSHES, AND SHRUBS

Zombia antillarum

Washingtonia robusta
Mexican fan palm
Mexican Washington palm
Thread palm

Weigela florida — Weigela

Weigela florida purpurea
Purple weigela

Weinmannia racemosa
Kamahi tree
Towai tree

Westringia fruticosa
Australian rosemary

Westringia rosmariniformis
Victoria rosemary

Widdringtonia cupressoides
African cypress
Berg cypress
Mountain cypress
Sapree wood

Widdringtonia juniperoides
Clanwilliam cedar

Widdringtonia schwarzii
Willimore cedar
Willowmore cedar

Widdringtonia whytei
Mlanje cedar

Wisteria sinensis
Chinese kidney bean

Xanthorrhiza simplicissima
Yellow root

Xanthorrhoea arborea
Botany Bay gum

Xanthorrhoea australis
Grass tree

Xanthorrhoea preissii — Black boy

Xanthoxylum americanum
Prickly ash
Toothache tree

Xanthoxylum clava-herculis
Southern prickly ash

Xanthoxylum fagara — Wild lime

Ximenia americana — Hog plum
Tallowwood

Xylopia aethiopica Guinea pepper

Yucca aloifolia — Dagger plant
Spanish bayonet
Spanish dagger

Yucca baccata — Banana yucca
Blue yucca
Datil
Spanish bayonet
Wild date

Yucca brevifolia — Joshua tree

Yucca carnerosana Spanish dagger

Yucca elata — Palmella
Soap tree
Soapweed

Yucca elephantipes Spineless yucca

Yucca filamentosa — Adam's needle
Needle palm
Silk grass
Spoonleaf yucca

Yucca glauca — Soapweed
Soapwell

Yucca gloriosa — Adam's needle
Lord's candlestick
Palm lily
Roman candle
Yucca

Yucca recurvifolia — Century plant

Yucca treculeana — Palma pita

Yucca whipplei — Our lord's candle

Zanthoxylum americanum
Northern prickly ash
Prickly ash
Toothache tree
Yellowwood

Zanthoxylum clava-herculis
Hercules' club
Pepperwood
Sea ash
Southern prickly ash

Zanthoxylum fagara — Wild lime

Zanthoxylum flavum
West Indian silkwood

Zanthoxylum piperitum
Japan pepper

Zelkova abelicea — Abelitzia
Cretan zelkova

Zelkova carpinifolia
Caucasian elm
Russian elm

Zelkova crenata — Caucasian elm

Zelkova serrata Japanese zelkova
Keaki
Sawleaf zelkova

Zelkova sinica — Chinese zelkova

Zelkova sinica verschaffeltii
Cutleaf zelcova

Zieria smithii — Sandfly bush

Ziziphus jujuba — Chinese date
Chinese jujube tree
Common jujube tree

Ziziphus lotus — Lotus tree

Ziziphus mucronata — Buffalo thorn

Ziziphus mauritania
Cottony jujube
Indian jujube

Zombia antillarum — Zombi palm

WILD FLOWERS

It is not possible in a volume such as this to include every species of wild flower; even one limited to European or North American floras would run to a substantial number of books, therefore preference has been given to the more common and widely distributed plants.

It is now illegal to dig up wild plants in many parts of the world, and the gathering of wild flowers by collectors and amateur 'flower lovers' coupled with the widespread use of modern herbicides has resulted in the near-extinction of many species. Wild flowers should be left undisturbed to be enjoyed by all.

Common poppy –
Papaver rhoeas

Aceras anthropophorum

Aceras anthropophorum	
	Man orchid
Achillea millefolium	
	Milfoil
	Yarrow
Achillea ptarmica	Sneezewort
Acinos arvensis	Basil thyme
Aconitum napellus	Monkshood
Acorus calamus	Sweet flag
Actaea spicata	Baneberry
Adoxa moschatellina	Moschatel
Aegopodium podagraria	
	Goutweed
	Ground elder
Aethusa cynapium	Fool's parsley
Agrimonia eupatoria	Agrimony
	Common agrimony
Agrimonia procera	
	Fragrant agrimony
	Scented agrimony
Ajuga reptans	Bugle
Alchemilla alpina	
	Alpine lady's mantle
Alisma lanceolatum	
	Lanceolate water plantain
Alisma natans	
	Floating water plantain
Alisma plantago-aquatica	
	Water plantain
Alisma ranunculoides	
	Lesser water plantain
Alliaria petiolata	Garlic mustard
Allium ampeloprasum	Wild leek
Allium oleraceum	Field garlic
Allium schoenoprasum	Chives
Allium scorodoprasm	Sand leek
Allium sphaerocephalon	
	Round-headed leek
Allium triquetrum	
	Triangular-stemmed garlic
Allium vineale	Crow garlic
	Wild onion
Althaea hirsuta	
	Hispid marsh mallow
Althaea officinalis	Marsh mallow
Anacamptis pyramidalis	
	Pyramidal orchid
Anagallis arvensis	
	Scarlet pimpernel
Anagallis foemina	Blue pimpernel
Anagallis minima	Chaff weed
Anagallis tenella	Bog pimpernel
Anaphalis margaritacea	
	Everlasting
	Pearly everlasting
Anemone apennina	Blue anemone
Anemone nemorosa	
	Wood anemone
Angelica sylvestris	Wild angelica
Antennaria dioica	
	Mountain everlasting
Anthemis arvensis	
	Corn chamomile
Anthemis cotula	
	Stinking chamomile
Anthemis tinctoria	
	Yellow chamomile
Anthriscus caucalis	Bur chervil
Anthriscus cerefolium	
	Garden chervil
Anthriscus sylvestris	Cow parsley
Anthyllis cicer	Wild lentil
Anthyllis frigidus	
	Yellow alpine milk vetch
Anthyllis vulneraria	Kidney vetch
Antirrhinum majus	Snapdragon
Aphanes arvensis	Parsley piert
Apium inundatum	
	Lesser marshwort
Apium nodiflorum	
	Fool's watercress
Aquilegia vulgaris	Columbine
Arabis hirsuta	Hairy rock cress
Arctium lappa	Great burdock
Arctium minus	Lesser burdock
Arctium nemorosum	
	Dark burdock
Arctium pubens	Burdock
	Common burdock
Aristolochia clematitis	Birthwort
Armeria maritima	Sea pink
	Thrift
Artemisia absinthium	Wormwood
Artemisia campestris	
	Breckland wormwood
Artemisia maritima	
	Sea wormwood
Artemisia norvegica	
	Norwegian mugwort
Artemisia stellerana	Dusty miller
Artemisia verlotorum	
	Chinese mugwort
Artemisia vulgaris	Mugwort
Arthrocnemum	
	Perennial glasswort
Arum italicum	Italian cuckoo-pint
	Italian lords-and-ladies
Arum maculatum	Cuckoo-pint
	Lords-and-ladies
Asarina procumbens	Asarina
Asarum europaeum	Asarabacca
Aster tripolium	Sea aster
Astragalus danicus	
	Purple milk vetch
Astragalus glycyphyllos	
	Wild liquorice
Astrantia major	Astrantia
Atriplex laciniata	Frosted orach
	Sandy orach
Atriplex littoralis	Beach orach
	Grassy orach
	Shoreline orach

Centaurea solstitialis

Atriplex patula	Common orach
	Orach
Atriplex prostrata	
	Arrowhead orach
	Spear-leaved orach
Atropa bella-donna	
	Deadly nightshade
Attium ursinum	Ramsons
	Wood garlic
Baldellia ranunculoides	
	Lesser water plantain
Ballota nigra	Black horehound
	Horehound
Bellis perennis	Daisy
Berula erecta	Lesser water parsnip
Beta vulgaris maritima	Sea beet
Betonica officinalis	Betony
Bidens cernua	
	Nodding bur marigold
Bidens frondosa	Beggarticks
Bidens tripartita	Bur marigold
	Trifid bur marigold
Blackstonia perfoliata	Yellowwort
Brassica napus	Rape
	Swede
Brassica nigra	Black mustard
Brassica oleracea	Wild cabbage
Brassica rapa	Turnip
Bryonia dioica	White bryony
Bunium bulbocastanum	
	Great pignut
Bupleurum falcatum	
	Sickle-leaved hare's ear
Bupleurum rotundifolium	
	Thorow-wax
Bupleurum subovatum	
	False thorow-wax
Bupleurum tenuissimum	
	Slender hare's ear
Butomus umbellatus	
	Flowering rush
Cakile maritima	Sea rocket
Calamintha sylvatica	Calamint
	Common calamint
Calla palustris	Bog arum
Callitriche stagnalis	
	Bog water starwort
	Common water starwort
	Mud water starwort
Calluna vulgaris	Common heather
	Ling
Caltha palustris	Kingcup
	Marsh marigold
	Mollyblobs
Calystegia pulchra	
	Hairy bindweed
Calystegia sepium	Bindweed
	Common bindweed
	Hedge bindweed
Calystegia silvatica	
	Great bindweed
	Large bindweed

Calystegia soldanella	
	Sea bindweed
Campanula glomerata	
	Clustered bellflower
Campanula hederacea	
	Ivy-leaved bellflower
Campanula lactiflora	
	Giant bellflower
Campanula latifolia	
	Giant bellflower
	Great bellflower
Campanula medium	
	Canterbury bell
Campanula patula	
	Spreading bellflower
Campanula rapunculoides	
	Creeping bellflower
Campanula rotundifolia	Harebell
	Scottish bluebell
Campanula trachelium	
	Nettle-leaved bellflower
Canvallaria majalis	
	Lily-of-the-valley
Capsella bursa-pastoris	
	Shepherd's purse
Cardamine amara	
	Large bitter cress
Cardamine flexuosa	
	Wavy bitter cress
Cardamine hirsuta	
	Hairy bitter cress
Cardamine pratensis	
	Cuckoo flower
	Lady's smock
Cardaria draba	Hoary cress
Carduus acanthoides	
	Welted thistle
Carduus dissectum	Meadow thistle
Carduus helenioides	
	Melancholy thistle
Carduus nutans	Musk thistle
	Nodding thistle
Carduus palustris	Marsh thistle
Carduus pratensis	Meadow thistle
Carduus pycnocephalus	
	Plymouth thistle
Carduus tenuiflorus	
	Slender-flowered thistle
Carlina vulgaris	Carline thistle
Centaurea calcitrapa	Star thistle
Centaurea cyanus	Cornflower
Centaurea nemoralis	
	Brown knapweed
Centaurea nigra	
	Common knapweed
	Hardheads
	Knapweed
Centaurea scabiosa	
	Great knapweed
Centaurea solstitialis	
	Yellow star thistle

Centaurium capitatum

Centaurium capitatum
Tufted centaury
Centaurium erythraea
Common centaury
Centaurium littorale   Sea centaury
Centaurium pulchellum
Lesser centaury
Centranthus ruber   Red valerian
Cephalanthera damasonium
White helleborine
Cephalanthera longifolia
Narrow-leaved helleborine
Ceratophyllum demersum
Rigid hornwort
Chaenorhinum minus
Small toadflax
Chaerophyllum temulentum
Rough chervil
Chamaemelum nobile   Chamomile
Sweet chamomile
Chelidonium majus
Greater celandine
Chenopodium album   Fat hen
Chenopodium bonus-
henricus   Good King Henry
Chrysanthemum
leucanthemum   Ox-eye daisy
Chrysanthemum parthenium
Bachelor's buttons
Feverfew
Chrysanthemum segetum
Corn marigold
Chrysanthemum vulgare   Tansy
Chrysosplenium alternifolium
Alternate-leaved golden saxifrage
Chrysosplenium oppositifolium
Golden saxifrage
Cicerbita alpina
Alpine blue sowthistle
Alpine lettuce
Cicerbita macrophylla
Blue sowthistle
Cichorium intybus   Chicory
Circaea alpina
Alpine enchanter's nightshade
Alpine nightshade
Circaea lutetiana
Enchanter's nightshade
Cirsium acaule   Dwarf thistle
Cirsium acaute   Ground thistle
Cirsium acautonauct
Ground thistle
Cirsium arvense   Creeping thistle
Cirsium dissectum   Meadow thistle
Cirsium eriophorum   Woolly thistle
Cirsium heterophyllum
Melancholy thistle
Cirsium palustre   Marsh thistle
Cirsium tuberosum
Tuberous meadow thistle
Cirsium vulgare   Spear thistle

Clematis alpina   Alpine clematis
Clematis vitalba   Old man's beard
Traveller's joy
Clinopodium vulgare   Wild basil
Coeloglossum viride   Frog orchid
Colchicum autumnale
Meadow saffron
Conium maculatum   Hemlock
Conopodium majus   Pignut
Consolida ambigua   Larkspur
Consolida regale   Forked larkspur
Convolvulus arvensis
Field bindweed
Lesser bindweed
Conyza canadensis
Canadian fleabane
Corallorhiza trifida
Coralroot orchid
Coronilla varia   Crown vetch
Corydalis bulbosa
Purple corydalis
Corydalis claviculta
Climbing corydalis
Corydalis lutea   Yellow corydalis
Cramble maritima   Sea kale
Crepis biennis   Rough hawk's beard
Crepis capillaris
Smooth hawk's beard
Crepis foetida
Stinking hawk's beard
Crepis mollis   Soft hawk's beard
Crepis nicaeensis
French hawk's beard
Crepis paludosa
Marsh hawk's beard
Crepis vesicaria
Beaked hawk's beard
Crithmum maritimum
Rock samphire
Samphire
Crocus nudiflorus
Naked autumn crocus
Crocus purpureus   Purple crocus
Spring crocus
Crocus vernus   Purple crocus
Spring crocus
Cuscuta epithymum   Dodder
Lesser dodder
Cuscuta europaea   Greater dodder
Cyclamen hederifolium   Cyclamen
Cymbalaria muralis
Ivy-leaved toadflax
Cypripedium calceolus
Lady's slipper
Dactylorhiza fuchsii
Common spotted orchid
Dactylorhiza maculata
Heath spotted orchid
Datura stramonium   Thorn apple
Daucus carota   Wild carrot

Euphorbia lathyrus

Daucus carota gummifer	Sea carrot
Dianthus armeria	Deptford pink
Dianthus barbatus	Sweet william
Dianthus caryophyllus	Wild carnation
Dianthus deltoides	Maiden pink
Dianthus gratianopolitanus	Cheddar pink
Dianthus plumarius	Wild pink
Digitalis grandiflora	Large yellow foxglove
Digitalis lutea	Small yellow foxglove
Digitalis purpurea	Foxglove
Diplotaxis muralis	Wall rocket
Diplotaxis tenuifolia	Wall rocket
Dipsacus fullonum	Common teasel / Teasel
Dipsacus pilosus	Small teasel
Dipsacus sylvestris	Common teasel / Teasel
Doronicum pardalianches	Leopard's bane
Draba muralis	Wall whitlowgrass
Drosera anglica	Great sundew
Drosera intermedia	Long-leaved sundew / Oblong-leaved sundew
Drosera rotundifolia	Round-leaved sundew / Sundew
Dryas octopetala	Mountain avens
Elodea canadensis	Canadian waterweed
Elodea nuttallii	Nuttall's waterweed
Empetrum nigrum	Crowberry
Endymion non-scriptus	Bluebell
Epilobium alsinifolium	Chickweed willow herb
Epilobium anagallidifolium	Alpine willow herb
Epilobium angustifolium	Rosebay willow herb
Epilobium hirsutum	Great willow herb
Epilobium lanceolatum	Spear-leaved willow herb
Epilobium montanum	Broad-leaved willow herb
Epilobium obscurum	Short-fruited willow herb / Thin-runner willow herb
Epilobium palustre	Bog willow herb / Marsh willow herb
Epilobium parviflorum	Hairy willow herb

Epilobium roseum	Pale willow herb / Pedicelled willow herb
Epilobium tetragonum	Square-stemmed willow herb
Epipactis atropurpurea	Dark red helleborine
Epipactis atrorubens	Dark red helleborine
Epipactis helleborine	Broad-leaved helleborine
Epipactis latifolia	Broad-leaved helleborine
Epipactis leptochila	Narrow-lipped helleborine
Epipactis palustris	Marsh helleborine
Epipactis purpurata	Violet helleborine
Epipactis sessilifolia	Violet helleborine
Epipogium aphyllum	Ghost orchid
Eranthis hyemalis	Winter aconite
Erica ciliaris	Ciliate heather
Erica cinera	Bell heather / Fine-leaved heath / Purple heather
Erica hibernica	Irish heath
Erica mackaiana	Mackay's heath
Erica tetralix	Cross-leaved heath
Erica vagans	Cornish heath
Erigeron acer	Blue fleabane
Erigeron boreatis	Alpine fleabane
Erigeron canadensis	Canadian fleabane
Erigeron mucronatus	Mexican fleabane
Erodium cicutarium	Common storksbill
Erodium dunense	Dune storksbill
Erodium maritimus	Sea storksbill
Erodium moschatum	Musky storksbill
Erophila verna	Common whitlowgrass / Whitlowgrass
Eryngium campestre	Coastal holly / Field eryngo
Eryngium maritimum	Sea holly
Eupatorium cannabinum	Hemp agrimony
Euphorbia amygdaloides	Wood spurge
Euphorbia cyparissias	Cypress spurge
Euphorbia dulcis	Sweet spurge
Euphorbia exigua	Dwarf spurge
Euphorbia helioscopia	Sun spurge
Euphorbia hyberna	Irish spurge
Euphorbia lathyrus	Caper spurge

Euphorbia paralias

Euphorbia paralias	Sea spurge
Euphorbia peplis	Purple spurge
Euphorbia peplus	Petty spurge
Euphorbia pilosa	Hairy spurge
Euphorbia platyphyllos	Broad-leaved spurge
Euphorbia portlandica	Portland spurge
Euphorbia stricta	Upright spurge
Euphorbia uralensis	Russian spurge
Euphrasia anglica	English sticky eyebright
Euphrasia arctica	Greater eyebright
Euphrasia brevipila	Short-haired eyebright
Euphrasia cambrica	Dwarf Welsh eyebright
Euphrasia confusa	Little kneeling eyebright
Euphrasia curta	Hairy-leaved eyebright
Euphrasia hirtella	Small-flowered sticky eyebright
Euphrasia micrantha	Common slender eyebright
Euphrasia montana	Mountain sticky eyebright
Euphrasia nemorosa	Common eyebright
Euphrasia occidentalis	Broad-leaved eyebright
Euphrasia pseudokerneri	Chalk hill eyebright
Euphrasia rivularis	Snowdon eyebright
Euphrasia rostkoviana	Large-flowered sticky eyebright
Euphrasia salisburgensis	Irish eyebright Narrow-leaved eyebright
Euphrasia scottica	Slender Scottish eyebright
Euphrasia tetraquetra	Broad-leaved eyebright
Fallopia convolvulus	Black bindweed
Filipendula ulmaria	Meadowsweet
Filipendula vulgaris	Dropwort
Foeniculum vulgare	Fennel
Fragaria moschata	Hautbois strawberry
Fragaria vesca	Wild strawberry
Fritillaria meleagris	Fritillary Snake's head fritillary
Fumaria capreolata	White ramping fumitory
Fumaria officinale	Common fumitory Fumitory
Gagea lutea	Yellow star of Bethlehem
Galanthus nivalis	Snowdrop
Galega officinalis	Goat's rue
Galeopsis angustifolia	Red hemp-nettle
Galeopsis speciosa	Large-flowered hemp-nettle
Galeopsis tetrahit	Common hemp-nettle Hemp-nettle
Galium album	Hedge bedstraw
Galium aparine	Cleavers Goose grass
Galium boreale	Northern bedstraw
Galium cruciata	Crosswort
Galium debile	Pond bedstraw
Galium odoratum	Sweet woodruff
Galium palustre	Marsh bedstraw
Galium parisiense	Wall bedstraw
Galium pumilum	Slender bedstraw
Galium saxatile	Heath bedstraw
Galium spurium	False cleavers
Galium tricornutum	Corn bedstraw
Galium uliginosum	Fen bedstraw
Galium verum	Lady's bedstraw
Gentiana amarella	Autumn felwort Autumn gentian
Gentiana campestris	Field felwort Field gentian
Gentiana germanica	Scarce autumn felwort
Gentiana nivalis	Small alpine gentian
Gentiana pneumonanthe	Marsh gentian
Gentiana verna	Spring gentian
Gentianella campestris	Field felwort Field gentian
Gentianella germanica	Scarce autumn felwort
Geranium columbinum	Dove's foot cranesbill Long-stemmed cranesbill
Geranium dissectum	Cut-leaved cranesbill
Geranium lucidum	Shining cranesbill
Geranium molle	Dove's foot cranesbill Soft cranesbill
Geranium phaeum	Dusky cranesbill
Geranium pratense	Meadow cranesbill

Inula conyza

Geranium purpureum
Lesser herb robert

Geranium pusillum
Small-flowered cranesbill

Geranium pyrenaicum
Hedgerow cranesbill
Pyrenean cranesbill

Geranium robertianum
Herb robert
Stinking bob

Geranium rotundifolium
Round-leaved cranesbill

Geranium sanguiuneum
Blood-red geranium
Bloody geranium

Geranium sylvaticum
Wood cranesbill

Geranium versicolor
Streaky cranesbill

Geum rivale   Nodding water avens
Water avens

Geum urbanum   Herb bennet
Wood avens

Glaucium flavum
Yellow horned poppy

Glaux maritima   Saltwort
Sea milkwort

Glechoma hederacea   Ground ivy

Goodyera repens
Creeping lady's tresses

Gymnadenia conopsea
Fragrant orchid

Habenaria albida
Small white orchid

Habenaria bifolia
Lesser butterfly orchid

Habenaria chlorantha
Butterfly orchid

Halimione portulacoides
Sea purslane

Hedera helix   Ivy

Helianthemum apenninum
White rockrose

Helianthemum canum
Hoary rockrose

Helianthemum nummularium
Common rockrose

Helichrysum arenarium
Everlasting

Helleborus foetidus
Stinking hellebore

Helleborus viridis
Green hellebore

Hemintia echioides
Bristly oxtongue

Heracleum mantegazzianum
Giant hogweed

Heracleum sphondylium
Cow parsnip
Hogweed

Herminium monorchis
Musk orchid

Hesperis matronalis   Dame's violet

Hieracium aurantiacum
Fox-and-cubs
Orange hawkweed

Hieracium pilosella
Mouse-ear hawkweed

Hieracium umbellatum
Hawkweed

Hieracium vulgatum
Common hawkweed

Himantoglossum hircinum
Lizard orchid

Hippocrepis comosa
Horseshoe vetch

Hippuris vulgaris   Mare's tail

Hottonia palustris   Water violet

Humulus lupulus   Hop

Hyacinthoides non-scriptus
Bluebell

Hydrocharis morsus-ranae
Frogbit

Hyoscyamus niger   Henbane

Hypericum androsaemum   Tutsan

Hypericum calycinum
Rose of sharon

Hypericum elodes
Bog Saint John's wort
Marsh Saint John's wort

Hypericum hircinum
Stinking tutsan

Hypericum hirsutum
Hairy Saint John's wort

Hypericum humifusum
Creeping Saint John's wort

Hypericum linarifolium
Narrow-leaved Saint John's wort

Hypericum montanum
Mountain Saint John's wort

Hypericum perforatum
Perforate Saint John's wort

Hypericum pulchrum
Beautiful Saint John's wort

Hypericum tetrapterum
Square-stemmed Saint John's wort

Hypericum undulatum
Wavy Saint John's wort

Hypochoeris glabra
Smooth cat's ear

Hypochoeris maculata
Spotted cat's ear

Hypochoeris radicata   Cat's ear
Common cat's ear

Impatiens capensis   Orange balsam

Impatiens glandulifera
Indian balsam

Impatiens noli-tangere
Touch-me-not
Wild balsam

Impatiens parviflora   Small balsam
Small yellow balsam

Inula conyza
Ploughman's spikenard

Inula crithmoides

Inula crithmoides
Golden samphire
Inula helenium — Elecampane
Iris foetidissima — Stinking iris
Iris pseudacorus — Yellow flag
Yellow iris
Iris spuria — Blue iris
Violet iris
Jasione montana — Sheep's bit
Kickxia elatine — Fluellen
Sharp-leaved fluellen
Kickxia spuria
Round-leaved fluellen
Knautia arvensis — Field scabious
Lactuca macrophylla
Blue sowthistle
Lactuca perennis — Blue lettuce
Lactuca saligna — Least lettuce
Lactuca scariola — Prickly lettuce
Lactuca serriola — Prickly lettuce
Lactuca virosa — Acrid lettuce
Great lettuce
Lamiastrum galeobdolon
Yellow archangel
Lamium album — White dead-nettle
Lamium amplexicaule — Henbit
Lamium hybridum
Cut-leaved dead-nettle
Lamium maculatum
Spotted dead-nettle
Lamium moluccellifolium
Intermediate dead-nettle
Lamium purpureum
Purple dead-nettle
Red dead-nettle
Lapsana communis — Nipplewort
Lathraea squamaria — Toothwort
Lathyrus aphaca — Yellow vetchling
Lathyrus hirsutus — Hairy vetchling
Lathyrus japonicus — Sea pea
Seaside pea
Lathyrus latifolius
Broad-leaved everlasting pea
Everlasting pea
Lathyrus maritimus — Sea pea
Seaside pea
Lathyrus montanus — Bitter vetch
Lathyrus niger — Black bitter vetch
Black vetch
Lathyrus nissolia — Grass vetchling
Lathyrus palustris — Marsh pea
Lathyrus pratensis
Yellow meadow vetchling
Lathyrus sylvestris — Wild pea
Lathyrus tuberosus — Tuberous pea
Tuberous vetchling
Legousia hybrida
Venus's looking glass
Lemna gibba — Fat duckweed
Gibbous duckweed

Lemna minor — Common duckweed
Lesser duckweed
Lemna trisulca — Ivy duckweed
Ivy-leaved duckweed
Leontodon autumnalis
Autumn hawkbit
Smooth hawkbit
Leontodon hispidus
Rough hawkbit
Leontodon leysseri — Hawkbit
Leontodon taraxacoides — Hawkbit
Lepidium campestre
Field pepperwort
Lepidium heterophyllum
Downy pepperwort
Smith's pepperwort
Lepidium latifolium — Dittander
Lepidium ruderale
Narrow-leaved pepperwort
Lepidium sativum — Garden cress
Leucanthemum vulgare
Ox-eye daisy
Leucojum aestivum
Summer snowflake
Leucojum vernum
Spring snowflake
Leucorchis albida
Small white orchid
Levisticum officinale — Lovage
Leycesteria formosa
Flowering nutmeg
Ligusticum scoticum — Scots lovage
Lilium martagon — Turk's cap lily
Lilium pyrenaicum — Pyrenean lily
Limonium vulgare
Comon sea lavender
Sea lavender
Limosella aquatica — Mudwort
Linaria arenaria — Sand toadflax
Linaria arvensis — Field toadflax
Linaria pelisseriana
Jersey toadflax
Linaria purpurea — Purple toadflax
Linaria repens — Pale toadflax
Linaria supina — Prostrate toadflax
Linaria vulgaris — Common toadflax
Linum bienne — Pale flax
Linum catharticum — Fairy flax
White flax
Linum perenne — Blue flax
Perennial flax
Liparis loeselii — Fen orchid
Listera cordata — Lesser twayblade
Listera ovata — Common twayblade
Twayblade
Lithospermum arvense
Corn gromwell
Field gromwell
Lithospermum officinale
Common gromwell
Gromwell

Myosotis ramosissima

Lithospermum purpurocaeruleum	Blue gromwell
Littorella lacustris	Shore weed
Littorella uniflora	Shore weed
Lobelia dortmanna	Water lobelia
Lobelia urens	Blue lobelia
Lonicera caprifolium	Perfoliate honeysuckle
Lonicera periclymenum	Honeysuckle
Lonicera xylosteum	Fly honeysuckle
Lotus angustissimus	Long-fruited bird's foot trefoil
Lotus corniculatus	Bird's foot trefoil
Lotus hispidus	Hairy bird's foot trefoil
Lotus uliginosus	Greater bird's foot trefoil / Marsh bird's foot trefoil
Lunaria annua	Honesty
Lunaria rediviva	Perennial honesty
Lupinus arboreus	Tree lupin
Lupinus luteus	Yellow lupin
Lupinus nootkatensis	Lupin
Luronium natans	Floating water plantain
Lychnis dioica	Red campion
Lychnis flos-cuculi	Ragged robin
Lychnis viscaria	Red catchfly / Sticky catchfly
Lycopus europaeus	Gipsywort
Lysimachia nemorum	Yellow pimpernel
Lysimachia nummularia	Creeping jenny
Lysimachia thyrsiflora	Tufted loosestrife
Lysimachia vulgaris	Yellow loosestrife
Lythrum hyssopifolia	Grass-poly / Hyssop-leaved loosestrife
Lythrum portula	Water purslane
Lythrum salicaria	Purple loosestrife
Maianthemum bifolium	May lily
Malva moschata	Musk mallow
Malva neglecta	Dwarf mallow
Malva pusilla	Small mallow
Malva sylvestris	Common mallow
Marrubium vulgare	White horehound
Matricaria matricarioides	Pineapple weed
Matricaria recutita	Scented mayweed / Wild chamomile
Matthiola sinuata	Sea stock
Meconopsis cambrica	Welsh poppy
Medicago arabica	Spotted medick
Medicago falcata	Sickle medick / Yellow medick
Medicago lupulina	Black medick
Medicago minima	Bur medick / Small medick
Medicago polymorpha	Fimbriate medick / Toothed medick
Medicago sativa	Lucerne
Melampyrum pratense	Common cow wheat
Melilotus alba	White melilot
Melilotus altissima	Tall melilot / Yellow melilot
Melilotus indica	Small melilot
Melilotus officinalis	Ribbed melilot
Mentha aquatica	Water mint
Mentha arvensis	Corn mint
Mentha longifolia	Horse mint
Mentha × piperita	Pepper mint
Mentha pulegium	Penny royal
Mentha spicata	Spear mint
Mentha spicata × suaveolens	French mint
Mentha suaveolens	Round-leaved mint
Menyanthes trifoliata	Bogbean
Mercurialis annua	Annual mercury
Mercurialis perennis	Dog's mercury
Mimulus guttatus	Monkey flower
Mimulus luteus	Blood-drop emlets / Blotched monkey flower
Mimulus moschatus	Musk
Misopates orontium	Lesser snapdragon
Moneses uniflora	One-flowered wintergreen / Single-flowered wintergreen
Monotropa hypopitys	Yellow bird's nest
Muscari atlanticum	Grape hyacinth
Muscari neglectum	Grape hyacinth
Muscari racemosum	Grape hyacinth
Myosotis alpestris	Alpine forget-me-not
Myosotis arvensis	Common forget-me-not / Field forget-me-not / Forget-me-not
Myosotis caespitosa	Lesser water forget-me-not
Myosotis discolor	Changing forget-me-not / Yellow forget-me-not
Myosotis ramosissima	Early forget-me-not

Myosotis scorpioides

Myosotis scorpioides
Water forget-me-not

Myosotis secunda
Creeping forget-me-not
Marsh forget-me-not

Myosotis sicula
Jersey forget-me-not

Myosotis stolonifera
Short-leaved forget-me-not

Myosotis sylvatica
Wood forget-me-not

Myosoton aquaticum
Pond chickweed
Water chickweed

Myriophyllum alterniflorum
Alternate-flowered water milfoil

Myriophyllum spicatum
Spiked water milfoil

Myriophyllum verticillatum
Whorled water milfoil

Narcissus × bifloris
Primrose peerless

Narcissus hispanicus
Spanish daffodil

Narcissus majalis   Pheasant's eye

Narcissus obvallaris   Tenby daffodil

Narcissus pseudonarcissus
Wild daffodil

Narthecium ossifragum
Bog asphodel

Nasturtium officinale   Watercress

Neottia nidus-avis
Bird's nest orchid

Nepeta cataria   Catmint

Nuphar lutea   Yellow waterlily

Nuphar pumila   Least waterlily
Lesser waterlily

Nymphaea alba   White waterlily

Nymphoides peltata
Fringed waterlily

Odontites verna   Red bartsia

Oenanthe aquatica
Fine-leaved water dropwort

Oenanthe crocata
Hemlock water dropwort

Oenanthe fistulosa
Common water dropwort
Tubular water dropwort
Water dropwort

Oenanthe fluviatilis
River water dropwort

Oenanthe lachenalii
Parsley water dropwort

Oenanthe pimpinelloides
Corky-fruited water dropwort

Oenanthe silaifolia
Meadow water dropwort

Oenothera biennis
Common evening primrose
Evening primrose

Oenothera erythrosepala
Large-flowered evening primrose

Oenothera stricta
Erect evening primrose
Fragrant evening primrose

Onobrychis viciifolia   Sainfoin

Ononis natrix
Large yellow restharrow

Ononis reclinata   Small restharrow

Ononis repens
Common restharrow
Restharrow

Ononis spinosa   Prickly restharrow

Onopordon acanthium
Scottish thistle

Ophrys apifera   Bee orchid

Ophrys arachnites
Late spider orchid

Ophrys aranifera
Early spider orchid

Ophrys fuciflora
Late spider orchid

Ophrys insectifera   Fly orchid

Ophrys muscifera   Fly orchid

Ophrys sphegodes
Early spider orchid

Orchis hircina   Lizard orchid

Orchis laxiflora   Jersey orchid

Orchis mascula   Early purple orchid

Orchis militaris   Soldier orchid

Orchis morio
Green-winged orchid

Orchis purpurea   Lady orchid

Orchis pyramidalis
Pyramidal orchid

Orchis simia   Monkey orchid

Orchis ustulata   Burnt orchid
Dwarf orchid

Origanum vulgare   Marjoram

Ornithogalum pyrenaicum
Bath asparagus
Spiked star of Bethlehem

Ornithogalum umbellatum
Star of Bethlehem

Ornithopus perpusillus   Birdsfoot
Least birdsfoot

Orobanche alba   Red broomrape

Orobanche amethystea
Carrot broomrape
Coastal broomrape

Orobanche apiculata
Lesser broomrape

Orobanche caryophyllacea
Clove-scented broomrape

Orobanche elatior   Tall broomrape

Orobanche hederae   Ivy broomrape

Orobanche major   Tall broomrape

Orobanche maritima
Carrot broomrape
Coastal broomrape

Orobanche minor
Lesser broomrape

Potamogeton gramineus

Orobanche picridis
Picris broomrape

Orobanche purpurea
Purple broomrape

Orobanche ramosa
Branched broomrape

Orobanche rapum-genistae
Greater broomrape

Orobanche reticulata
Thistle broomrape

Orthilia secunda
Serrated wintergreen

Oxalis acetosella   Wood sorrel

Oxalis corniculata
Procumbent yellow sorrel
Yellow sorrel

Oxalis europaea
Upright yellow sorrel

Oxalis pes-caprae
Bermuda buttercup

Oxycoccus microcarpus
Small cranberry

Oxycoccus palustris   Cranberry

Oxyria digyna   Mountain sorrel

Papaver agremone
Long rough-headed poppy
Prickly poppy

Papaver dubium
Long smooth-headed poppy

Papaver rhoeas   Common poppy
Common red poppy
Field poppy

Papaver somniferum
Opium poppy

Parentucellia viscosa
Yellow bartsia

Parietaria judaica
Mortar pellitory
Pellitory-of-the-wall

Paris quadrifolia   Herb paris

Parnassia palustris
Grass of parnassus

Pastinaca sativa   Wild parsnip

Pedicularis foliosa   Leafy lousewort

Pedicularis palustris
Marsh lousewort
Red rattle

Pedicularis sylvatica   Lousewort

Petasites fragrans
Winter heliotrope

Petasites hybridus   Butterbur

Phyteuma orbiculare
Round-headed rampion

Phyteuma spicatum
Spiked rampion

Phyteuma tenerum
Round-headed rampion

Picris echioides   Bristly oxtongue

Picris hieracioides
Hawkweed oxtongue

Pilosella aurantiaca
Orange hawkweed

Pilosella officinarum
Mouse-ear hawkweed

Pimpinella major
Greater burnet saxifrage

Pimpinella saxifraga
Burnet saxifrage

Pinguicula grandiflora
Greater butterwort
Large-flowered butterwort

Pinguicula lusitanica
Pale butterwort
Pink butterwort

Pinguicula vulgaris
Common butterwort

Plantago arenaria
Branched plantain

Plantago coronopus
Buck's horn plantain

Plantago lanceolata
Ribwort plantain

Plantago major   Great plantain

Plantago maritima   Sea plantain

Plantago media   Hoary plantain
Lamb's tongue

Platanthera bifolia
Lesser butterfly orchid

Platanthera chlorantha
Butterfly orchid
Greater butterfly orchid

Polygala calcarea   Chalk milkwort

Polygala serpyllifolia
Heath milkwort

Polygala vulgaris
Common milkwort

Polygonatum multiflorum
Solomon's seal

Polygonatum odoratum
Angular solomon's seal
Lesser solomon's seal

Polygonatum verticillatum
Whorled solomon's seal

Polygonum amphibium
Amphibious bistort

Polygonum aviculare   Knotgrass

Polygonum bistorta
Common bistort

Polygonum hydropiper
Water pepper

Polygonum lapathifolium
Pale persicaria

Polygonum persicaria   Redshank

Potamogeton alpinus
Reddish pondweed
Red pondweed

Potamogeton coloratus
Fen pondweed

Potamogeton crispus
Curled pondweed

Potamogeton epihydros
American pondweed

Potamogeton gramineus
Various-leaved pondweed

Potamogeton heterophyllus
Various-leaved pondweed

Potamogeton lucens
Shining pondweed

Potamogeton natans
Broad-leaved pondweed
Floating pondweed

Potamogeton nodosus
Loddon pondweed

Potamogeton oblongus
Bog pondweed

Potamogeton pectinatus
Fennel pondweed

Potamogeton perfoliatus
Perfoliate pondweed

Potamogeton polygonifolius
Bog pondweed

Potamogeton praelongus
Long-stalked pondweed

Potamogeton rufescens
Reddish pondweed
Red pondweed

Potentilla anglica
Procumbent cinquefoil

Potentilla anserina   Silverweed

Potentilla argentea
Hoary cinquefoil

Potentilla crantzii   Alpine cinquefoil

Potentilla erecta
Common tormentil
Tormentil
Upright cinquefoil

Potentilla fruticosa
Shrubby cinquefoil

Potentilla palustris
Marsh cinquefoil

Potentilla reptans
Creeping cinquefoil

Potentilla rupestris
Rock cinquefoil

Potentilla sterilis
Barren strawberry

Potentilla tabernaemontani
Spring cinquefoil

Primula elatior   Oxlip

Primula farinosa
Bird's eye primrose

Primula veris   Cowslip

Primula vulgaris   Primrose

Prunella laciniata
Cut-leaved self heal

Prunella vulgaris   Self heal

Pseudorchis albida
Small white orchid

Pulicaria dysenterica
Common fleabane
Fleabane

Pulicaria vulgaris   Lesser fleabane
Small fleabane

Pulsatilla vernalis   Pasque flower

Pulsatilla vulgaris   Pasque flower

Pyrola media
Intermediate wintergreen
Medium wintergreen

Pyrola minor
Common wintergreen
Lesser wintergreen

Pyrola rotundifolia
Large wintergreen
Round-leaved wintergreen

Pyrola uniflora
One-flowered wintergreen
Single-flowered wintergreen

Radiola linoides   Allseed

Ranunculus acris
Common meadow buttercup
Meadow buttercup

Ranunculus aquatilis
Common water crowfoot
Water crowfoot

Ranunculus arvensis
Corn buttercup

Ranunculus auricomus
Goldilocks
Goldilocks buttercup

Ranunculus baudotii
Seaside crowfoot

Ranunculus bulbosus
Bulbous buttercup

Ranunculus circinatus   Circul
leaved crowfoot

Ranunculus ficaria
Lesser celandine

Ranunculus fluitans
River crowfoot
River water crowfoot

Ranunculus hederaceus
Ivy-leaved crowfoot

Ranunculus omiophyllus
Moorland crowfoot

Ranunculus paladosus
Jersey buttercup

Ranunculus parviflorus
Small-flowered buttercup

Ranunculus repens
Creeping buttercup

Ranunculus sardous
Pale hairy buttercup

Ranunculus sceleratus
Celery-leaved crowfoot

Ranunculus trichophyllus
Dark-hair crowfoot
Thread-leaved water crowfoot

Ranunculus tripartitus
Little three-lobed crowfoot

Raphanus maritimus   Sea radish

Raphanus raphanistrum
Wild radish

Raphanus sativus   Radish

Reseda lutea   Wild mignonette

Reseda luteola   Dyer's rocket
Weld

Reseda phyteuma
Corn mignonette

WILD FLOWERS

Sedum forsteranum

Reynoutria japonica
Japanese knotweed
Round knotweed
Rhinanthus angustifolius
Greater yellow rattle
Rhinanthus minor Hayrattle
Yellow rattle
Rhinanthus serotinus
Greater hayrattle
Rhodiola rosea Roseroot
Roemeria hybrida
Violet horned poppy
Romulea columnae Sand crocus
Warren crocus
Romulea parviflora Sand crocus
Warren crocus
Rosa agrestis
Narrow-leaved sweet briar
Rosa arvensis Field rose
Trailing rose
Rosa canina Dog rose
Rosa dumalis
Short-pedicelled rose
Rosa micrantha Lesser sweet briar
Rosa pimpinellifolia Burnet rose
Rosa rubiginosa Sweet briar
Rosa sherardii
Northern downy rose
Rosa stylosa Long-styled rose
Rosa tomentosa Downy rose
Harsh downy rose
Rosa villosa Soft-leaved rose
Rubia peregrina Wild madder
Rubus caesius Dewberry
Rubus chamaemorus Cloudberry
Rubus idaeus Raspberry
Rubus saxatilis Stone bramble
Rumex acetosa Common sorrel
Sorrel
Rumex acetosella Sheep's sorrel
Rumex conglomeratus
Clustered dock
Rumex crispus Curly dock
Rumex hydrolapathum
Water dock
Wetland dock
Rumex obtusifolius
Broad-leaved dock
Nettle-cure dock
Ruscus aculeatus Butcher's broom
Sagina nodosa Knotted pearlwort
Sagina procumbens
Procumbent pearlwort
Prostrate pearlwort
Sagittaria sagittifolia Arrowhead
Salicornia europaea Glasswort
Salsola kali Prickly saltwort
Spiny saltwort
Salvia horminoides Clary
Wild sage
Salvia pratensis Meadow clary

Salvia verbenaca Wild clary
Samolus valerandi Brookweed
Sanguisorba minor Lesser burnet
Salad burnet
Sanguisorba officinalis
Great burnet
Saponaria officinalis Soapwort
Saussurea alpina Alpine sawwort
Saxifraga aizoides
Yellow mountain saxifrage
Saxifraga cespitosa Tufted saxifrage
Saxifraga granulata
Bulbous saxifrage
Meadow saxifrage
Saxifraga hirculus
Yellow bog saxifrage
Saxifraga hypnoides
Mossy saxifrage
Saxifraga nivalis
Clustered alpine saxifrage
Saxifraga oppositifolia
Purple saxifrage
Saxifraga rivularis
Alpine rivulet saxifrage
Saxifraga rosacea Irish saxifrage
Saxifraga spathularis
Saint Patrick's saxifrage
Saxifraga stellaris Starry saxifrage
Saxifraga tridactylites
Rue-leaved saxifrage
Scabiosa arvensis Field scabious
Scabiosa columbaria
Small scabious
Scabiosa succisa
Devil's bit scabious
Scilla autumnalis Autumn squill
Scilla verna Spring squill
Scleranthus annuus Annual knawel
Scorzonera humilis Viper's grass
Scrophularia auriculata
Water figwort
Scrophularia ehrhartii
Scarce water figwort
Scrophularia nodosa
Common figwort
Figwort
Scrophularia scorodonia
Balm-leaved figwort
Scrophularia umbrosa
Green figwort
Scarce water figwort
Scrophularia vernalis
Yellow figwort
Scutellaria galericulata Skullcap
Scutellaria minor Lesser skullcap
Sedum acre Biting stonecrop
Wall pepper
Sedum album White stonecrop
Sedum anglicum English stonecrop
Sedum forsteranum
Rock stonecrop

Sedum reflexum	Reflexed stonecrop
Sedum rosea	Roseroot
Sedum telephium	Livelong
	Orpine
Sedum villosum	Hairy stonecrop
Senecio aquaticus	Marsh ragwort
Senecio erucifolius	Hoary ragwort
Senecio fluviatilis	
	Broad-leaved ragwort
Senecio intergrifolius	
	Field fleawort
Senecio jacobaea	
	Common ragwort
	Ragwort
Senecio paludosus	
	Great fen ragwort
Senecio palustris	Marsh fleawort
Senecio spathulifolius	
	Spathulate fleawort
Senecio squalidus	Oxford ragwort
Senecio sylvaticus	
	Heath groundsel
	Wood groundsel
Senecio viscosus	Sticky groundsel
Senecio vulgaris	Groundsel
Serratula tinctoria	Sawwort
Sibbaldia procumbens	Sibbaldia
Silaum silaus	Pepper saxifrage
Silene acaulis	Moss campion
	Mountain campion
Silene alba	White campion
Silene conica	Striated catchfly
Silene dioica	Red campion
Silene gallica	Small catchfly
Silene italica	Italian catchfly
Silene noctiflora	
	Evening catchfly
	Night-flowering catchfly
Silene nutans	Nodding catchfly
	Nottingham catchfly
Silene otites	Spanish catchfly
Silene vulgaris	Bladder campion
Silene vulgaris maritima	
	Sea campion
Silybum marianum	Milk thistle
Sinapsis alba	White mustard
Sinapsis arvensis	Charlock
Smyrnium olustratum	Alexanders
Smyrnium perfoliatum	
	Perforate alexanders
Solanum dulcamara	Bittersweet
	Woody nightshade
Solanum luteum	Hairy nightshade
Solanum nigrum	Black nightshade
Solanum sarrachoides	
	Green nightshade
Solidago altissima	Tall golden rod
Solidago canadensis	
	Canadian golden rod
	Tall golden rod

Solidaga virgaurea	Golden rod
Sonchus arvensis	Corn sowthistle
	Perennial sowthistle
Sonchus asper	Prickly sowthistle
Sonchus oleraceus	
	Common sowthistle
	Smooth sowthistle
	Sowthistle
Sonchus palustris	Fen sowthistle
Specularia hybrida	
	Venus's looking glass
Spergula arvensis	Corn spurrey
Spergularia marina	
	Lesser sea spurrey
Spergularia media	
	Greater sea spurrey
Spergularia rubra	Sand spurrey
Spergularia rupicola	
	Rock sea spurrey
Spiranthes aestivalis	
	Summer lady's tresses
Spiranthes autumnalis	
	Autumn lady's tresses
	Lady's tresses
Spiranthes romanzoffiana	
	Threefold lady's tresses
Spiranthes spiralis	
	Autumn lady's tresses
	Lady's tresses
Spirodela polyrhiza	
	Great duckweed
Stachys alpina	Alpine woundwort
Stachys arvensis	Field woundwort
Stachys betonica	Betony
Stachys germanica	
	Downy woundwort
Stachys palustris	
	Marsh woundwort
Stachys sylvatica	
	Hedge woundwort
	Wood woundwort
Statice armeria	Sea pink
	Thrift
Stellaria alsine	Bog stitchwort
Stellaria graminea	
	Lesser stitchwort
Stellaria holostea	
	Greater stitchwort
Stellaria media	Chickweed
	Common chickweed
Stellaria neglecta	
	Greater chickweed
Stellaria nemorum	
	Wood stitchwort
Stellaria palustris	
	Marsh stitchwort
Stratiotes aloides	Water soldier
Suaeda maritima	Annual sea-blite
	Sea-blite
Subularia aquatica	Awlwort
Succisa pratensis	Devil's bit scabious

Urtica urens

Symphytum asperum
Rough comfrey

Symphytum officinale   Comfrey
Common comfrey

Symphytum orientale
White comfrey

Symphytum peregrinum
Blue comfrey

Symphytum tuberosum
Tuberous comfrey

Symphytum × uplandicum
Russian comfrey

Tamus communis   Black bryony

Tanacetum parthenium
Bachelor's buttons
Feverfew

Tanacetum vulgare   Tansy

Taraxacum erythrospermum
Lesser dandelion

Taraxacum laevigatum
Lesser dandelion

Taraxacum officinale   Dandelion

Taraxacum palustre
Little marsh dandelion
Narrow-leaved marsh dandelion

Taraxacum spectabile
Bog dandelion

Teesdalia nudicaulis
Shepherd's cress

Tetragonolobus maritimus
Dragon's teeth

Teucrium botrys
Cut-leaved germander

Teucrium chamaedrys
Wall germander

Teucrium scordium
Water germander

Teucrium scorodonia   Wood sage

Thalictrum alpinum
Alpine meadow rue

Thalictrum flavum
Common meadow rue
Meadow rue

Thalictrum minus
Lesser meadow rue

Thlaspi alpestre   Alpine penny cress

Thlaspi arvense   Field penny cress

Thlaspi perfoliatum
Perfoliate penny cress

Thymus drucei
Common wild thyme

Thymus pulegioides
Larger wild thyme
Large thyme

Thymus serphyllum   Wild thyme

Tofieldia pusilla   Scottish asphodel

Torilis arvensis
Spreading hedge parsley

Torilis japonica
Erect hedge parsley
Hedge parsley
Upright hedge parsley

Torilis nodosa
Knotted hedge parsley

Tragopogon minor
Lesser goat's beard

Tragopogon porrifolius   Salsify

Tragopogon pratensis   Goat's beard

Trientalis europaea
Chickweed wintergreen

Trifolium arvense
Hare's foot clover

Trifolium aureum   Large hop trefoil

Trifolium campestre   Hop trefoil

Trifolium dubium
Lesser yellow trefoil

Trifolium fragiferum
Strawberry-headed clover

Trifolium glomeratum
Flat-head clover

Trifolium hybridum   Alsike clover

Trifolium incarnatum
Crimson clover

Trifolium medium
Zigzag clover

Trifolium micranthum
Least yellow trefoil
Slender trefoil

Trifolium molinerii
Large lizard clover

Trifolium ochroleucon
Creamy clover

Trifolium ornithopodioides
Fenugreek

Trifolium pratense   Red clover

Trifolium squamosa
Teasel-headed clover

Trifolium repens   Dutch clover
White clover

Trifolium resupinatum
Reversed clover

Trifolium scabrum   Rough clover

Trifolium stellata
Star-headed clover

Trifolium striatum   Knotted clover
Soft clover

Trifolium strictum   Upright clover

Trifolium subterraneum
Subterranean clover

Trifolium suffocatum
Suffocated clover

Tripleurospermum indorum
Scentless mayweed

Tripleurospermum maritimum
Scentless chamomile
Sea mayweed

Trollius europaeus   Globeflower

Tuberaria guttata   Spotted rockrose

Tulipa sylvestris   Wild tulip

Tussilago farfara   Coltsfoot

Urtica dioica   Common nettle
Stinging nettle

Urtica urens   Sandy nettle
Small nettle

Utricularia intermedia

Utricularia intermedia
Intermediate bladderwort
Utricularia major
Western bladderwort
Utricularia minor
Lesser bladderwort
Utricularia neglecta
Western bladderwort
Utricularia vulgaris
Common bladderwort
Greater bladderwort
Vaccinium microcarpum
Small cranberry
Vaccinium myrtillus   Bilberry
Whortleberry
Vaccinium oxycoccus   Cranberry
Vaccinium uliginosum
Bog bilberry
Bog whortleberry
Vaccinium vitis-idaea   Cowberry
Valeriana dioica   Lesser valerian
Marsh valerian
Valeriana officinalis
Common valerian
Valerian
Valeriana pyrenaica
Pyrenean valerian
Valerianella carinata
Keel-fruited cornsalad
Valerianella dentata
Narrow-fruited cornsalad
Valerianella locusta
Common cornsalad
Lamb's lettuce
Valerianella rimosa
Broad-fruited cornsalad
Verbascum blattaria
Moth mullein
Verbascum lychnitis
White mullein
Verbascum nigrum   Black mullein
Dark mullein
Verbascum pulverulentum
Hoary mullein
Verbascum thapsus
Common mullein
Great mullein
Verbascum virgatum
Slender mullein
Verbena officinalis   Vervain
Veronica agrestis   Field speedwell
Green field speedwell
Veronica alpina   Alpine speedwell
Veronica anagallis-aquatica
Blue water speedwell
Veronica arvensis   Wall speedwell
Veronica beccabunga   Brooklime
Veronica catenata
Pink water speedwell
Veronica chamaedrys
Germander speedwell
Veronica filiformis
Round-leaved speedwell

Veronica fruticans
Shrubby speedwell
Veronica hederifolia
Ivy-leaved speedwell
Veronica montana
Wood speedwell
Veronica officinalis
Heath speedwell
Veronica peregrina
American speedwell
Veronica persica
Common field speedwell
Persian speedwell
Veronica polita   Grey speedwell
Veronica repens   Corsican speedwell
Veronica scutellata   Marsh speedwell
Veronica serpyllifolia
Thyme-leaved speedwell
Veronica spicata   Spiked speedwell
Veronica triphyllos
Fingered speedwell
Veronica verna   Spring speedwell
Vicia angustifolia
Narrow-leaved vetch
Vicia bithynica   Bithynian vetch
Vicia cracca   Tufted vetch
Vicia hirsuta   Hairy vetch
Vicia lathyroides   Spring vetch
Vicia lutea   Yellow vetch
Vicia orobus   Wood bitter vetch
Vicia sativa   Common vetch
Vicia sepium   Bush vetch
Vicia sylvatica   Wood vetch
Vicia tenuissima   Slender tare
Vicia tetrasperma   Smooth tare
Vinca major   Greater periwinkle
Vinca minor   Lesser periwinkle
Viola arvensis   Field pansy
Viola canina   Dog violet
Viola curtisii   Seaside pansy
Viola hirta   Hairy violet
Viola kitaibeliana   Dwarf pansy
Viola lactea   Pale heath violet
Viola lutea   Mountain pansy
Viola odorata   Sweet violet
Viola palustris   Bog violet
Viola reichenbachiana
Woodland violet
Viola riviniana
Common dog violet
Common violet
Viola rupestris   Teesdale violet
Viola stagnina   Fen violet
Viola tricolor   Tricolor pansy
Viscum album   Mistletoe
Wahlenbergia hederacea
Ivy-leaved bellflower
Wolffia arrhiza   Rootless duckweed